About the Author

MICHAEL KORDA is the *New York Times* bestselling author of *Journey to a Revolution, Horse People, Country Matters, Ulysses S. Grant,* and *Charmed Lives.* He lives with his wife, Margaret, in Duchess County, New York.

IKE

BY MICHAEL KORDA

Journey to a Revolution

Ulysses S. Grant

Marking Time

Horse People

Making the List

Country Matters

Another Life

Man to Man

The Immortals

Curtain

The Fortune

Queenie

Worldly Goods

Charmed Lives

Success

Power

Male Chauvinism

BY MARGARET AND MICHAEL KORDA

Horse Housekeeping

Cat People

IKE

An American Hero

Michael Korda

HARPER PERENNIAL

NEW YORK • LONDON • TORONTO • SYDNEY • NEW DELHI • AUCKLAND

HARPER PERENNIAL

Grateful acknowledgment is made for permission to print the following previously published material:

Quotations from President Dwight Eisenhower's *At Ease, Waging Peace,* and *Mandate for Change,* used by the courtesy and kind permission of John S. D. Eisenhower.
Quotations from *Mrs. Ike,* used by the courtesy and kind permission of Ms. Susan Eisenhower.
Quotations from *Eisenhower, Vol. 1: Soldier, General of the Army, President-Elect, 1890–1952* by Stephen E. Ambrose. Copyright © by Stephen E. Ambrose. Reprinted by permission of Simon & Schuster Adult Publishing Group.
Quotations from *Eisenhower: Soldier and President* by Stephen E. Ambrose. Copyright © 1990 by Stephen E. Ambrose. Reprinted by permission of Simon & Schuster Adult Publishing Group.
Quotations from the war speeches "Into Battle" and "Never Give In" of Prime Minister Winston Churchill. Reproduced with permission of Curtis Brown Ltd, London, on behalf of The Estate of Winston Churchill. Copyright Winston S. Churchill.
Quotations from *The War Memoirs of Charles de Gaulle: Unity* by Charles de Gaulle. English translation by Richard Howard. Copyright © 1959 by Charles de Gaulle. Originally appeared in French as *L'Unité.* Copyright © 1959 by Libraire Plon.
Extracts from *Past Forgetting* by Kay Summersby Morgan are reprinted with the permission of Harold Ober Associates Incorporated. Copyright © 1976 by the Estate of Kay Summersby Morgan.
Lyrics from "They Like Ike" by Irving Berlin. Copyright © 1950 by Irving Berlin. Copyright renewed. International copyright secured. All rights reserved. Reprinted by permission.
Excerpts from *Eisenhower: A Soldier's Life* by Carlo D'Este. Copyright © 2002 by Carlo D'Este. Reprinted by permission of Henry Holt and Company, LLC.
Excerpts from "The General," from *Collected Poems of Siegfried Sassoon* by Siegfried Sassoon, copyright 1918, 1920 by E. P. Dutton. Copyright 1936, 1946, 1947, 1948 by Siegfried Sassoon. Used by permission of Viking Penguin, a division of Penguin Group (USA) Inc.

An extension of this copyright appears on page 780.

A hardcover edition of this book was published in 2007 by HarperCollins Publishers.

FIRST HARPER PERENNIAL EDITION PUBLISHED 2008.

Designed by Amy Hill
Photography consultant: Kevin Kwan

Library of Congress Cataloging-in-Publication Data is available upon request.

ISBN 978-0-06-075666-6 (pbk.)

08 09 10 11 12 ID/RRD 10 9 8 7 6 5 4 3 2

For Margaret, with all my love

And for Lynn Nesbit,
for more reasons that I can list

War is an option of difficulties.

—Major-General James Wolfe
1727–1759

There never was a good war, or a bad peace.

—Benjamin Franklin

The past is a foreign country; they do things differently there.

—L. P. Hartley

If you can talk with crowds and keep your virtue,
Or walk with Kings—nor lose the common touch . . .
Yours is the Earth and everything that's in it,
And—which is more—you'll be a Man, my son!

—Rudyard Kipling

In war men are nothing; one man is everything. . . . Only the commander understands the importance of certain things, and he alone conquers and surmounts all difficulties. An army is nothing without the head.

—Napoleon

Contents

Illustrations

Maps

Author's Note

It is not easy to get military ranks correct in dealing with several nations over a long period of time. To avoid confusion, therefore, I have attempted throughout to give the people mentioned in the book their correct rank *at the time about which I am writing.*

In the case of British figures, I have also tried to deal with the inevitable changes in name that were caused along the way by such honors as knighthoods and peerages. Thus Lieutenant-General Bernard L. Montgomery becomes General Sir Bernard Montgomery and eventually Field Marshal the Viscount Montgomery, and so on.

Where British military ranks are concerned I have included the hyphen with which some of the Army and Royal Navy ranks are correctly written, as in "major-general," or "regimental sergeant-major." "Field marshal" and "admiral of the fleet" are exceptions to the rule, and correctly written without a hyphen. Also, the Royal Air Force does not use hyphens, except in "air vice-marshal." I have not used a hyphen in U.S. military titles, since they are correctly written without one.

Regarding German ranks, I have used the German mostly when no clear-cut equivalent exists in English (as in S.S. ranks like SS Standartenführer), but sometimes also when the German simply makes more sense and prevents confusing a German general with a British or an American one. As a rule, I have aimed for clarity rather than consistency.

I use "Nazi" only when referring to people or institutions that were part of the NSDAP, or Nazi Party. I have avoided the use of "Nazi" instead of "German," as in "Nazi planes" or "Nazi tanks" or "Nazi army." Our enemy in World War II, as in World War I, was Germany, and the Germans. Some of the Germans were Nazis; most of them were not.

With military organizations and units I have tried to make matters as clear as I can. Most armies are roughly similar, in the sense of being divided, in ascending order of size, into companies, battalions, regiments, brigades, divisions, corps, armies, and army groups. Where there are significant differences I will point them out. In the interest of simplicity, I have not followed the military pattern of designating corps with Roman numerals, etc. Instead, numbers through ninety-nine are spelled; 100 and above are in Arabic numerals.

Michael Korda

PART I

The Making of a Hero

The Eisenhower family, 1926 *(left to right)*: Roy, Arthur, Earl, Edgar, David, Ike, Milton, Ida.

CHAPTER 1

"Ike"

Ours is neither a nation nor a culture much given to extended hero worship.

Ralph Waldo Emerson understood his countrymen only too well when he wrote, "Every hero becomes a bore at last." After all, within his own lifetime Emerson would see both Andrew Jackson and Ulysses S. Grant cut down to size. There is no place in American life for the enduring national cult of a hero, no equivalent of France's national passion for Napoleon (a cult strangely enough by no means limited to the French), England's sentimental hero worship of Nelson (and, increasingly, of Winston Churchill), Russia's glorification of Peter the Great.

Perhaps it is the price of being a democracy, and of a deep, inherent distrust of the very idea of an elite—we are all egalitarians at heart, or at least feel a need to pay homage to the idea of equality. We have a natural tendency to nibble away at the great figures of the past; to dig through their lives for flaws, mistakes, and weaknesses; to judge them severely by the standards and beliefs of the present, rather than those that prevailed when they were alive. Thus Washington has been marginalized as a dead white male, and as a slave owner, and remembered more for his ill-fitting false teeth than his generalship. Thus Jefferson has been downgraded from his lofty position as the author of the Declaration of Independence to being treated not merely as a slave owner, but as a spendthrift and hypocrite who slept with his own female slaves. Thus Grant's tomb stood for many decades forlorn and almost unvisited on Riverside Drive and 122nd Street in New York City, despite the fact that it was, until the end of the nineteenth century, a bigger tourist attraction than the Statue of Liberty.

It is a simple fact of American life, this urge to splash graffiti on the pantheon of our heroes. In other countries—or cultures—the building up of national heroes is a full-time job, respected and well rewarded, in France with membership in the Académie Française and the Légion d'Honneur, in the United Kingdom with knighthoods and a cozy place in the cultural establishment; but in ours, whole profitable segments of the media and publishing industries prosper by tearing them down. *Sic transit gloria mundi* might as well be our national motto.

In his own lifetime, for example, Dwight D. Eisenhower underwent a rapid transition from world-class, five-star hero to being ridiculed as an old fuddy-duddy in the White House, out of touch with what was happening in the country, more interested in his golf score than in politics, deaf to the pleas of civil rights leaders (or at least hard of hearing), and, toward the end of his eight years as president, overshadowed by the youth and glamour of the young John F. Kennedy.

It was not just journalists, or editorial cartoonists like Herblock in the *Washington Post*, or intellectuals of the New Frontier, who made fun of Ike—even historians of World War II began to turn their heavy guns on him, particularly admirers of General George S. Patton. Patton's advocates formed a stubborn and robust revisionist cult that would reach its peak when Patton became the hero of Richard Nixon's favorite movie; they held Ike to blame for failing to turn his fractious subordinate loose to seize Berlin before the Russians did, and by that mistake ensuring a divided Germany and the cold war—even though Patton was too far south to have done this.

Like those revisionists who insist that Blücher, instead of Wellington, won the Battle of Waterloo by arriving on the battlefield with his Prussians at the end of the day, or those who believe that Lee would have won the Battle of Gettysburg if only he had listened to Longstreet's advice, Patton's admirers—sixty years after the fact—are still smarting over their hero's complaints. As to the validity of their views, one cannot do better than to quote the duke of Wellington himself, who, when a stranger came up to him on Piccadilly and said, "Mr. Jones, I presume?" is said to have replied, "If you presume that, sir, you'd presume any damned thing."

The guns had hardly fallen silent in Europe before Eisenhower's rivals and subordinates sat down to write their memoirs or edit their diaries for publication. Most of them were sharply critical of Eisenhower. On the British side of the Grand Alliance the newly ennobled field marshals Lord Alanbrooke and Viscount Montgomery, and General Lord Ismay, if they agreed on nothing else, were united in their view that Eisenhower was no strategist. Indeed General Sir Alan Brooke, as he was then, the acerbic Chief of the Imperial General Staff (CIGS), would remark acidly in his dairy in 1943 on the subject of his plan for the invasion, "Eisenhower has got absolutely no strategical outlook."[1]

On the losing side, Hitler's generals, when they came to write their memoirs, were almost all critical of Eisenhower's caution, slowness, conventional tactics, and failure to develop the single thrust that might have brought the war to an end by the winter of 1944—strong stuff from those whom he defeated.

As for the senior American and British airmen—particularly the "bomber barons," among whom the most important and outspoken was Air Chief Marshal Sir Arthur ("Bomber") Harris, Air Officer Commanding, RAF Bomber Command—they expressed in their war memoirs the conviction that had they been given a free hand and unlimited resources the war could have been won in 1944 without an invasion at all; that Eisenhower, in short, had merely wasted time, manpower, money, and fuel, all of which would have been more usefully employed destroying German cities (the RAF strategy), or the German rail network and oil industry (the strategy of the U.S. Army Air Force).

In the United States, enthusiasts for a Pacific-first strategy—centered on the figure of General Douglas MacArthur and the U.S. Navy admirals, most of whom had never wholeheartedly accepted Franklin Roosevelt's agreement with Churchill that the defeat of Nazi Germany, not Japan, must be the Allies' first priority—accused Eisenhower of spinelessly allowing himself to be charmed, bullied, or hoodwinked by the wily British; it was not only the French who distrusted *l'Albion perfide*. Perhaps fortunately, the Soviet marshals alone did not offer much in the way of criticism of Ike, presumably afraid to express any opinions on the matter that might contradict those of Stalin.

Of course to some degree all wars end in a war of words—World War I produced innumerable ill-tempered memoirs from the generals on both sides of the conflict, justifying their own decisions and lamenting the blunders of their colleagues, not to speak of politicians blaming the generals for what had gone wrong on the battlefield, and generals blaming the politicians for interfering in military decisions about which no civilian was qualified to hold an opinion and for botching the peace.

In the United Kingdom, certain events of World War I, most particularly the First Battle of the Somme and the failure of the landings in the Dardanelles, caused a veritable heavy artillery barrage of books which still rumbles on today, and which aroused bitter feelings that were not submerged even by the outbreak of the next world war in 1939.

In France, there were numerous equally sore points on the subject of World War I among the rival generals and politicians. Many of these disputes were exacerbated by a universal national feeling that the British had not pulled their weight in the fighting and had then conspired to cheat France out of its just rewards at the peace conference—*l'Albion perfide* again.

In Germany, of course, the burning question was who was responsible for losing the war; and in all the countries involved, even the infant Soviet Union, whole forests were felled to print books seeking to pin elsewhere the blame for starting it in the first place.

Thus the attempt to take some of the luster off Eisenhower's five stars was nothing out of the ordinary in history, though it must be said that he bore the attacks and the second-guessing of his British colleagues with a degree of patience that might almost have qualified him for sainthood, however much he steamed about them in private. His own war memoirs are notably fair-minded, even on the subject of people he had every reason to resent, and he did not attempt to refute criticism of himself, or encourage others to do so on his behalf.

Still, in Eisenhower's case the natural tendency of the historical pendulum to swing in the opposite direction after a certain amount of time coincided with the growing feeling during his second term in office that he was out of his depth as president, and that like another victorious American general, Ulysses S. Grant, he regarded the presidency more as a reward for his victory in the field than as the summit of his ambition or a full-time job. His soothing grandfatherly style, his unconcealed distaste for the rough-and-tumble of politics, and his garbled syntax played well enough in the heartland but did not appeal to right-wing Republicans who regarded Ike as a newcomer to the GOP (and distrusted his bias toward the United Kingdom and Europe), or to the young and the big-city intellectual elites, who had supported Adlai Stevenson and were now beginning to feel the gravitational pull of the young, vigorous, and articulate junior senator from Massachusetts.

To put matters in perspective, the last years of Eisenhower's presidency coincided with the Beats, Mort Sahl, Marilyn Monroe, and Elvis—things were happening out there beyond the White House and Camp David that would change America more radically than anything since the Civil War, and it cannot be said that Eisenhower was aware of them, or had he been aware, that he would have approved.

The 1950s still conjure up today a world that seems far removed from our own, even to those of us who came of age in it: men wearing hats, and suits with narrow lapels; huge tail-finned cars gleaming with chrome; the dizzyingly rapid growth of the suburbs; the fiction of women as happy, contented housewives; McCarthyism; the height of the cold war; an age when cigarettes weren't considered harmful, the pill had yet to be invented, skirts came down below the knee, and conformity—at least on the surface—ruled.

Eisenhower presided over this America, benign, avuncular, occasionally exasperated. His idea of a good evening was said to be watching old westerns on television, he and Mrs. Eisenhower seated side by side in front of the set with a television tray in front of each of them; his idea of a good meal, a Scotch on the rocks and a steak. Those who disliked the Eisenhowers spread rumors that Mrs. Eisenhower drank more than was good for her and was sometimes unsteady on her feet, while those who admired them believed that she merely suffered from an inner ear imbalance and a certain shyness, or caution, in the presence of politicians, natural in a woman who had spent her married life as an Army wife, and was therefore used to a world in which officers junior to her husband—and their wives—knew their place and took good care to stay in it.

Although Eisenhower, with his big grin, looked like a gregarious soul, this was in part a facade, or a protective mechanism, like a lot of things about him. Decades after serving two terms as his vice president, Richard Nixon still complained that in eight years the Eisenhowers had seldom invited the Nixons to a private dinner at the White House, just the four of them together—a subject about which Mrs. Nixon was said to express, to the very end of her life, what was, for her, a very rare degree of personal resentment. Of course in many respects Eisenhower remained a general, even as president, and generals don't usually invite junior officers to dine with them. Nixon may well have appeared to him to be the equivalent of a junior officer, and in any case political small talk, which was Nixon's specialty, was the last thing Eisenhower wanted to hear in the evening, or even during the day—more likely, he simply preferred to spend his evenings in front of the television set when he could, like so many of his fellow Americans who had elected him president, and perhaps it simply never occurred to him that the Nixons' feelings might be hurt, or even that their feelings mattered one way or the other.

A lifetime in the Army and several years of supreme command inevitably make a certain imprint on a man, and lead to things that mere civilians find it hard to understand. Robert Kennedy used to tell a story about his brother's shock, when he reached the White House, at how much it cost to run, and how many people worked there without anybody seeming to know what, if anything, they did. President Kennedy had, to a serious degree, a rich man's horror at the possibility that he might be accused of extravagance or waste—a subject about which those who have inherited wealth are often particularly sensitive. He told his brother Robert to go into the question of White House staffing with a fine-tooth comb, and to find ways to economize.

During the course of this investigation—for Robert Kennedy plunged into it with his usual zeal and thirst for details—he discovered that nobody could explain the presence of an Army master sergeant who had an office in the basement of the White House. He decided to look into this himself, and eventually discovered in the basement a grizzled old Army NCO in a large office, surrounded by canvases and optical equipment. Eisenhower, it turned out, had been persuaded by Winston Churchill to take up painting as a hobby to calm his nerves. He enjoyed painting, once he took it up, and was good with colors, but apparently he never developed a skill for drawing (a difficulty from which Churchill also suffered). On Churchill's advice, he began by working from postcards, and had one of his orderly sergeants draw the outline of the picture to scale on the canvas.

When he was elected president, the sergeant was posted to the White House and installed in the basement, just in case Ike wanted to paint. For a five-star general, this was nothing unusual or out of the ordinary—there were, after all, thousands of sergeants in the U.S. Army fully occupied doing things even less useful than waiting until Eisenhower wanted a canvas prepared for his oil paints, and nobody in the Army then or now is likely to question an order from a five-star general.

The Kennedy brothers, however, were civilians at heart, despite their wartime service in the U.S. Navy, and they pretended to be shocked by Ike's wastefulness; so the sergeant was swiftly removed from his cozy lair in the White House and sent packing. But the story aptly illustrates another side of Ike's personality—he had, despite simple tastes and a frugality natural to anyone who had lived for many years on a junior officer's salary, the sense of entitlement of any victorious general. Since the beginning of warfare, the military leaders of successful coalitions have always been substantially rewarded by their grateful countries. In the late seventeenth century John Churchill was made duke of Marlborough, and Blenheim Palace was built and furnished for him at public expense; in the nineteenth century Wellington received not only a dukedom, the Garter, a grand London town house on Hyde Park Corner, and a great country estate, but a sizable fortune to support it all. After World War I, Sir Douglas Haig, whose disastrous military strategy in Flanders was widely believed to have cost the United Kingdom several hundred thousand unnecessary casualties for no discernible gain in ground, received an earldom and was made a wealthy man by his grateful nation. Even in the Soviet Union, those of Stalin's marshals lucky enough to survive his purges and win him victories in World War II were rewarded with lavish dachas, limousines, orders and decorations mounted with diamonds and precious gemstones, the lifetime support of uniformed aides and servants, and well-paid sinecures for themselves and their families in the Communist Party and in industry. Only the United States—as the Grants complained in the nineteenth century—had no tradition of rewarding successful generals, except by electing them president.

Neither Eisenhower nor Mrs. Eisenhower would have wanted a dukedom for him, of course, even had it been possible to offer him one, and for some time Eisenhower was not sure whether he wanted the presidency; but even so, by 1945 Ike was no longer a simple hayseed from the Midwest, if he had ever been that, and Mamie understandably expected some compensation for having spent the war alone in a small apartment while he was chauffeured around (alas, all too publicly!) by a glamorous former fashion model. Eisenhower had been a mere lieutenant colonel in 1940 (and a very newly minted one at that), but by 1943 he was on close terms with Roosevelt, Churchill, De Gaulle, and King George VI. He had his own four-engine aircraft and crew at his disposal, not to speak of the beautiful Kay Summersby and her four-door Packard; and although he made a point of living simply, he was nevertheless surrounded by aides and uniformed servants for whom his smallest wishes were commands, and who competed to find for him comfortable, elegant, and restful surroundings. His headquarters in Casablanca, Algiers, London's posh Grosvenor Square, Bushey Park, and Fontainebleau were in no way Spartan, thanks in part to Kay Summersby, who had considerable experience in the beau monde, knew how to entertain, could hold her own in conversations with Roosevelt or Churchill, and switched from the role of driver to Eisenhower's hostess as required, in much the same effortless manner that she eventually moved from a British uniform to a faultlessly tailored American WAC officer's uniform, always slim, cool, sophisticated, and completely devoted to the Supreme Commander of Allied Forces in Europe.

There is, let it be said at once, nothing wrong about this—no historical proof exists that generals benefit from being uncomfortable, or produce better strategy by eating bad food, or by being deprived of attractive female company. On the contrary, the commander of a great army—still more the commander of a great coalition, with its infinitely more vexing problems—surely needs an atmosphere in which he can think clearly, and concentrate on victory, as opposed to worrying about his dinner, or wondering whether his bed is dry and clean. It has always been so, and no doubt always will be.

Generals, even those with fewer than five stars, are not like lesser mortals or ordinary people, and although Eisenhower certainly managed, in Kipling's words, to "Walk with kings—nor lose the common touch," he was neither as simple nor as easygoing a character as the press (and Republican political strategists) liked to present him. A man who has successfully commanded millions of men in battle, who has made perhaps the most difficult and far-reaching military decision of all time, and who accepted the formal surrender of Nazi Germany, must have a core of steel; a streak of ruthlessness; the ability to make cold, hard, objective decisions; and an imperial sense of command, however well disguised they may be by a big grin and a firm handshake.

This, after all, was a man who had been acclaimed all over the world as the leader of a victorious coalition; who had received deferential congratulations from every world leader, including Stalin, De Gaulle, and George VI; who had been honored at the lord mayor's banquet in London, where huge crowds gathered in the streets outside to cheer "good old Ike," then perhaps the most popular American, at home and abroad, of all time. He retired from the Army to become president of Columbia University—a precedent for generals' becoming university presidents had been set by none other than General Robert E. Lee, who accepted the presidency of Washington College (now Washington and Lee University) in Lexington, Virginia, after the Civil War—secure in the knowledge that the White House was his almost for the asking, as it had been for Grant, and, for the first time, made financially secure by the sale of his memoirs to Doubleday and *Life* magazine for unprecedented amounts of money, which his brother Milton and his newfound friends among the rich Republicans invested wisely on his behalf. The man who left America in 1942 almost unknown except among his fellow West Pointers and the small group of officers around General George C. Marshall had, by 1945, reached a level of fame that few men have ever equaled, and went on to maintain a good deal of his popularity through two terms as his country's president. The election buttons bearing the motto "I like Ike," which sprouted on the lapels of Americans in 1951 (and which echoed Irving Berlin's popular song in his hit musical *Call Me Madam*), spoke nothing but the truth: the whole world liked Ike—even the Germans by that time, not to speak of those Democrats who voted for him instead of Adlai Stevenson.

And yet, although Eisenhower has been portrayed respectfully, if not with any great degree of accuracy, on television and in films, and although his name is widely recognized and respected—and although there exists a good-size permanent publishing campaign, led by his son Colonel John Eisenhower, devoted to protecting Ike's military reputation from the attacks of historians who think he should have let Patton take Berlin, or resisted the division of his forces between Montgomery in the north and Patton in the south after the liberation of Paris in 1944, let alone British war historians who tend to side with Alanbrooke's criticisms of him—Eisenhower's role in history, both as the commander of the allied forces against Nazi Germany and as the president who ended the Korean War and guided America through the most dangerous years of the cold war, no longer seems deeply imprinted on the American national consciousness, and certainly not in what passes for history teaching in American education.

Of course there is a natural ebb and flow to historical reputations, however exalted. Sometimes a reputation can be revived by a single great book, as David McCullough did for Harry S. Truman and John Adams; sometimes not, as remains the case with Franklin Roosevelt—but beyond that America is very different from the country it was in 1945, or even 1960. Americans may put magnetic signs on their cars in the shape of a yellow ribbon bearing the legend "We support our troops," but they no longer wish to serve among these troops, are no longer obligated to do so, and—at any rate in the middle and upper classes—don't want to see their own children in uniform. The military has become a world apart, something you see on television or in the movies, not a shared experience as it was in World War II and for some years after it.*

What is more, both popular culture and the academic world, strongly shaped by those who opposed the war in Vietnam, have come to portray war not in terms of generals and their strategy, but in terms of "ordinary" people, soldiers on both sides, civilians, the victims of war, etc. The spectacular success of Tom Brokaw's book *The Greatest Generation* and of Steven Spielberg's film *Saving Private Ryan* is in part because they show the war through the eyes of "the little people" and assume that the war was won by these people's sacrifices. This, of course, is true enough, except that without strategy, direction, planning, discipline, and the willpower of a commander, no battle or war is ever won.

Even so-called "people's wars," like that of the Chinese Communists against the Nationalists, or of the North Vietnamese against South Vietnam and its patrons and allies—the French, then the Americans—required a great leader and first-rate generals, such as North Vietnam produced in Ho Chi Minh and General Giap. In the end, somebody always has to plan, prepare, decide, exhort, and command. History, like warfare, can never be completely democratized.

This transformation in the way Americans think about their own his-

* The author did two years of compulsory military service in the Royal Air Force ("national service," as it was called) followed by five years in the Reserve, joining the RAF in 1951 at the age of seventeen and a half, after leaving school. Some form of compulsory military service existed in every European country, including neutral, pacific Switzerland, and this was true for most young men in the United States as well.

tory (and are taught it, to the extent that they are still taught it at all) has been accompanied by a perceived need to ignore or denigrate "dead white males," as well as by a deliberate turning away from the "great man" school of history, so beloved of Carlyle; and since Eisenhower is dead, was white, and was indubitably a great man his roles as a general, supreme commander, and president now seem less important to historians than the social issues of that period.

Then too, since 1945 Americans have become used to small wars, in which not much is at stake. It is hard for those who did not live through World War II to realize how close we came on several occasions to losing it, and how utterly different the world would have been had Hitler—and the Japanese—won it instead. "The long night of barbarism, unbroken by even a star of hope,"[2] was how Churchill, in 1940, described the prospects for humanity if Hitler won, and, as usual on the subject of Hitler, he was absolutely right. The death camps, the Holocaust, the brutal mass murder of civilians and prisoners, the barbaric medical experiments, and the wholesale killing of the mentally and physically handicapped all had yet to be discovered and documented, but from the very beginning of Hitler's rise to power Churchill grasped what was at stake—the survival of civilization, of humanity, of common decency.

In May 1940, as France collapsed under the weight and speed of the German attack and the British army retreated from Belgium and northern France to the beaches at Dunkirk, Great Britain came within a hair's breadth of losing the war, and in the long summer of 1940 that followed fewer than 2,000 young Royal Air Force fighter pilots stood between Hitler and the victory that was almost within his grasp. They were alone, and, in the end, they were enough, at any rate to ensure that Britain would not be invaded by the Germans in 1940; but at the same time there was no way Britain could win the war without bringing America in. When Randolph Churchill barged into his father's bedroom to congratulate him on becoming prime minister, he asked how his father intended to win the war. Calmly, while shaving, the new prime minister replied that he proposed to hold out somehow until America entered the war, and of course that was exactly what happened.

But in 1940, nobody could have predicted how long this would take, or even if it would, in the end, actually happen—certainly nobody could have guessed that Britain would have to hold out until December 7, 1941, when the Japanese attack on Pearl Harbor (and Hitler's ill-advised declaration of war against the United States) brought the Americans in, more than two years after the war had begun.

With the arrival of the Americans (and the unexpected survival of the Soviet Union in the winter of 1941, when the German army was almost in sight of the Kremlin) it was at last possible to hope that the war might be won, though history—and the astonishing fighting efficiency and professionalism of the German army—would still provide the Allies with numerous opportunities for losing it. The British, having hung on to survival by their fingernails for two years, and lived through the bombing of their cities (nearly 50,000 British civilians would be killed by German bombs during the course of the war), had a fairly natural fear that they would be relegated to second place in the new alliance, and as American troops, most of them still untrained and untried in battle, flooded into their small island, bringing with them such novelties as canned and frozen food, nylon stockings, and chewing gum, many people complained, in a popular phrase of the time, that the Americans were "overpaid, oversexed, and over here."

Grateful as the British were to have, at last, a powerful ally (and one that spoke the same language), their sensitivities were aroused by a whole range of disagreements, including, but by no means limited to, America's instinctive anticolonial attitude toward the British Empire, its dominant wealth and industrial power, and, of course, the burning question of when, where, how, and at whose command Hitler was to be fought.

Perhaps the one thing that could have guaranteed a victory for Germany after 1942 would have been the appointment of the wrong commander for the allied forces in Europe. The British recognized, from the beginning, albeit reluctantly, that since American troops, however inexperienced, would soon outnumber their own, supreme command would have to go to an American. The British generals could not be described as enthusiastic about this prospect, but they were resigned to it. Had the choice fallen on, say, General Douglas MacArthur, who was arrogant, vain, and instinctively anti-British—or on somebody like Patton, who was all these things and worse—the alliance might never have prospered, or even had a chance of succeeding.

That the choice fell on Dwight D. Eisenhower was perhaps the Allies' most singular piece of good fortune in World War II. From the very first moment, he charmed the British, and put them at their ease. They might bewail his lack of strategic brilliance—they complained about it during the war and after the war, and are still complaining about it—but nobody then or now underrated his unique ability to command a coalition. Few expected "Ike," as the British, despite their reserve, quickly got used to calling him, to be another Rommel or Rundstedt, but even his critics praised his fairness, his energy, his patience, his common sense, his authority, and above all his matchless ability to deal with even the most difficult of prima donnas—Montgomery, say; or the rival French generals, De Gaulle and Giraud; or Winston Churchill. To use a favorite phrase of his, Eisenhower also knew how to "keep his eye on the ball." The aim was victory—national feelings, friendships, and interservice rivalries would have to be ruthlessly suppressed on its behalf. Despite wounded feelings on the part of his American generals, he would eventually accept a British air marshal, A. W. Tedder, as his deputy commander; a British admiral as his naval commander; another British air marshal as his air force commander; and a British general as his commander of ground forces. When he heard that an American general had sworn at a British general, he called the British general in and asked what had happened. "He only called me a son-of-a-bitch, sir," the British general said, taken aback by Eisenhower's fury and anxious not to make a fuss, but Eisenhower was better informed and implacable. "I heard he called you 'A *British* son-of-a-bitch,' " he said angrily. "I will make him god-damn well *swim* home!"[3] and the American general was promptly reduced to his substantive rank of lieutenant colonel and sent back to the United States in disgrace.

He not only practiced what he preached but enforced it ruthlessly at every level, from four-star generals to privates, and it worked. Of course, there were unavoidable complaints, "incidents," and problems; but the alliance held together, with the armed forces of all three countries (for the

French were soon adding their own complaints, demands, and historic national concerns to those of the British) united in their respect for their supreme commander.

It need not have been so. For most of World War I, each army retained its own supreme commander—Field Marshal Sir Douglas Haig did not willingly share his plans with Marshal Joffre, or with Joffre's successors, and when the Americans entered the war General John Pershing too kept his cards hidden from his allies (and an ace or two up his sleeve, some of them complained). Eventually, however, Lloyd George and Clemenceau between them managed to impose Marshal Ferdinand Foch on the allied armies as supreme commander just in time to defeat the last great German offensive on the western front in 1918 and win the war; but this had not been easy to accomplish in the teeth of national pride, and it was, in the duke of Wellington's famous phrase about Waterloo, "a damn close run thing." Churchill and Roosevelt, both of whom had played significant roles in World War I, were determined that at least some of the more obvious mistakes of that war should be avoided in the next, and unity of command was one thing they insisted on. Because of Eisenhower, it paid off.

His sensitivity toward the British—not always repaid in kind—derived in part from the unusual twists and turns of his career. It was a source of great regret to him that he had not made it to France in 1917–1918, and had therefore never directly experienced war. Like any professional army officer, commanding troops in battle was what he had been trained for—every other aspect of military life paled by comparison—yet he never got the opportunity to do so. Eisenhower was, from the beginning, a staff man, a problem solver, a military bureaucrat. He spent the war years training troops to go overseas, and he was good at the job—so good that despite the numerous requests he made to go to France and get into action himself, the War Department absolutely refused to let him go, and finally ordered him sharply to stop complaining and serve where he was ordered to. Thus he was deprived of the chance to prove himself in combat, as well as the possibility of winning the glory, honors, medals, and decorations that count so much in a soldier's life; and he felt it keenly. He did not, in fact, get to Europe until 1928, when he was appointed to the relatively humdrum task of writing the guidebook for the American Battle Monuments Commission, which he performed with his usual efficiency.

On the other hand, his visits to the allies' war cemeteries were to have unexpected results when the next war broke out and he reached high command. From the moment the United States entered the war the American chiefs of staff—even General George C. Marshall himself—were impatient with what they took to be British hesitation and doubt about the invasion of Europe. Marshall was determined to invade France and engage the German army in combat as soon as possible—indeed, he hoped to do so in 1942, and was only with great difficulty persuaded to postpone the invasion until 1943.

In part, British doubts on the subject were practical, and based on hard experience of battle against the Wehrmacht in Norway, France, North Africa, Greece, and Crete—General Alan Brooke and the British chiefs of staff knew only too well how formidable the German army was, and believed, correctly, that landing three divisions in France in the face of up to forty experienced German divisions would be suicidal. American troops were inexperienced and only partly trained; American equipment, while lavish, was by no means equal at that stage of the war to that of the German army, particularly in tanks and artillery; and the landing craft and naval forces necessary for an invasion did not yet exist.

The British had been planning for the invasion of France since early in 1941, spurred on by Churchill, and had even carried out a small-scale dress rehearsal in 1942 when they attempted to seize the port of Dieppe. The results of that debacle demonstrated once and for all that the chances of capturing and holding a port in German-occupied France were nil, that such tanks as were then available could not land safely and operate on sandy (as opposed to shingle) beaches or survive German antitank guns, and that the German army's response to any such attempt was swift, sure, and brutal. Many of the Canadians who had been chosen to carry out the raid at Dieppe were slaughtered on the beaches or captured. This outcome not only embittered relations between the British and the Canadian governments for some time but also provided the first chink in the hitherto bulletproof reputation of Admiral Lord Louis Mountbatten, Chief of Combined Operations.

The committee charged with planning the invasion drew from the raid on Dieppe the realistic conclusion that it would have to land on shingle beaches, rather than rely on seizing a port, and that an almost unimaginable amount of work, training, preparation, and innovation would have to take place before it was feasible.

This was not a message that the Americans wanted to hear, and they tended to ascribe it to British timidity, rather than to reason and experience. In the end the decision to land in North Africa late in 1942, which was made *faute de mieux*, at Churchill's urging, provided a very clear picture of just how hard the Germans could fight even when they were outnumbered, a long way from home, and at the end of a dangerously long and fragile supply line, and gave everybody on the Allied side a chance to consider what the results might have been of a landing in France in 1942 against a much larger German army with short interior supply lines. Still, the American chiefs of staff did not forgive the British for deflecting them to North Africa, and from there to Sicily and Italy, and remained doubtful of Britain's commitment to the cross-Channel invasion.

Eisenhower did not share these doubts about British resolve. He had not led troops over the top and into combat in World War I, and he still very much lamented the fact, but he had examined the war cemeteries, and there, before the interminable rows of headstones, he had come to understand and sympathize with the British determination not to repeat the useless slaughter of that war. The United Kingdom's war dead in 1914–1918 reached nearly 750,000, the number of its gravely wounded exceeded 1.6 million. On the first *day* of the First Battle of the Somme in 1916, the British army's casualties were over 60,000—the blackest day in the history of the British army—as wave after wave of troops went over the top and advanced steadily across no-man's-land toward the German trenches, bayonets fixed and gleaming, through barbed wire, mud, mines, and the heaviest artillery barrage of all time into massed machine-gun fire that cut them down like so much wheat before the scythe.

The British were committed to victory in World War II, and indeed they had been fighting the Germans for two years before America came in, but nobody in Britain, from King George VI down, was prepared to repeat the experience of World War I. A protracted stalemate and a mass slaughter, like that of Flanders and the western front, were simply not a possibility that any British government could contemplate or that the British public would tolerate.

The British generals knew that, and planned accordingly. They had fought in the trenches as junior officers, as had politicians as different from one another as the suave foreign secretary Anthony Eden and the colorless Lord Privy Seal and leader of the Labour Party, Clement Atlee—both of whom had been captains in the trenches during World War I and won the Military Cross for bravery. Winston Churchill himself, after resigning from the government because of the failure of the Dardanelles campaign, had commanded an infantry battalion in the trenches, with Archibald Sinclair, later to become leader of the Liberal Party, as his adjutant. There was hardly a home or a family in Britain that had not experienced what was then still known as the Great War, or had not suffered the loss of a loved one, and soldiers, statesmen, and civilians alike were determined to avoid a repetition of that tragedy. On monuments, and in colleges, schools, clubs, factories, and businesses all over the country, plaques listed members or workers who had been killed from 1914 to 1918, a veritable avalanche of names in every town and village of the British Isles to remind people of how widespread—and pointless—the sacrifice had been. A whole generation had been swept away by that war; its absence was an inescapable fact of British life, a reality which nobody could forget, and of which one was (and still is) reminded countless times a day by the words "In remembrance" and a heartbreakingly long list of carved or painted names.

This ever-present reality had a profound effect on British public life and policy between the wars. It drove successive British governments to attempt to "appease" rather than oppose Germany, while, in terms of defense policy, it fueled the exaggerated claims of the air marshals that the next war would be won (or lost) in the air—or, a still more doubtful claim, that the mere threat of bombing would prevent war. Both the public and the postwar governments wanted to believe that if a future war could not be avoided, it would be fought in the air, not in the mud—that a few well-trained aviators would win it quickly and neatly, with a minimum of casualties, and above all that there would be no repetition of Flanders.

That spirit was captured perfectly by the poet John Betjeman, writing in 1940, in the voice of a well-to-do London lady at prayer:

> Gracious Lord, oh bomb the Germans.
> Spare their women for Thy Sake,
> And if that is not too easy
> We will pardon Thy Mistake.
> But, gracious Lord, whate'er shall be,
> Don't let anyone bomb me. . . .*

The founder of the Royal Air Force, Lord Trenchard—one of those singular military visionaries who appear from time to time in English life (one thinks of Kitchener, Fisher, Baden-Powell, T. E. Lawrence, and Wingate)—in fact had based his idea of a separate air force independent of the Army or the Navy on the claim that even the threat of bombing would make a future war unlikely, an early version of the deterrence theory that would become the cornerstone of nuclear policy in the cold war.

That "Boom" Trenchard's few aircraft were slow and cumbersome, awkward creations of varnished wood, stretched wire, and doped fabric that could hardly have reached the German coastline on a clear day, let alone carry bombs large enough to damage anything larger than a toolshed, did nothing to diminish his enthusiasm or the appeal of his ideas. He was, in any case, in the spirit of most visionaries, far better at what we now call PR than at flying, and he had the capacity to summon up in other people's imaginations vast fleets of bombers—"aerial battleships"—darkening the sky and terrifying the Germans.

A skilled bureaucratic infighter, Trenchard could and did annihilate his

* John Betjeman, "In Westminster Abbey," *Collected Poems*, 1979. I owe a debt to Professor John Lukacs for bringing this poem (and much else) to my attention in his masterly book *Five Days in London: May 1940*.

critics by proving that a bomber force would be far cheaper to build and maintain than a large army. He thus preempted both the press and the Treasury, and committed a succession of peace-seeking British governments to spending huge sums on a bomber force that, in the end, would not become a serious weapon against Germany until 1942.

Trenchard's vision had other consequences. In its straitened circumstances after World War I, the United Kingdom cut its defense budget to the bone, and then some. The Royal Navy was of course almost sacrosanct, though the admirals complained loudly and bitterly (and correctly, as it turned out) that they needed more modern ships; so the money that was spent on developing a bomber force (and later on modern fighter aircraft and radar, although over the violent protests of the Air Ministry, since air force orthodoxy at the time was that possession of a bomber force would render fighters unnecessary) was chiefly at the expense of modernizing the Army. Great Britain, which had invented the tank, spent little or nothing to develop new tanks—or, just as important, to develop a new kind of army that understood how to use them. "The cavalry," as one critic of military policy remarked, "has been mechanized in the sense that their horses have been taken away from them."

Thus the British Expeditionary Force (BEF) went to France in 1939 with much the same weapons—and training, organization, and outlook—with which it had left France in 1918. Faced with Hitler's panzer divisions in May 1940, the BEF was lucky to be able to retreat to the Dunkirk beaches, and it owed its survival to the traditional discipline and steadiness under fire of Britain's regular Army, to the Royal Navy and the fighter pilots of the Royal Air Force, and perhaps most important of all to the formidable combination of ancient regimental pride and the spit-and-polish professionalism of the regular Army's noncommissioned officers.

But Dunkirk, however heartwarming to the British, was a disaster, not the way to win a war. As Churchill warned the House of Commons, "We must be very careful not to assign to this deliverance the attributes of a victory. Wars are not won by evacuations."[4] George VI, always receptive to the national mood, might breathe a sigh of relief in a letter to his mother, Queen Mary, that Britain stood alone at last—"Personally, I feel happier now that we have no allies to be polite to and pamper"[5]—but Churchill's view of the situation was darker and more realistic.

Although the RAF had proved itself superior to the Luftwaffe, and the Royal Navy was powerful enough to render every German attempt to use the Kriegsmarine's capital ships on the high seas sacrificial, the war could not be won by the Navy or the Air Force. In order to defeat Germany, its army would have to be beaten in the field, and this the British army was in no position to do. Throughout 1941 the Army fought back and forth across the Western Desert in North Africa to defend Egypt and the Suez Canal; but its initial victories against the Italians were nullified by the arrival of General Erwin Rommel and the German Afrika Korps, and by the need to divert forces first to Greece and then to Crete, where they met with utter disaster. It was not until 1942 that the combination of American Lend-Lease tanks and the meticulous preparation imposed on the British Eighth Army by Lieutenant-General Bernard Montgomery led to a decisive victory over Rommel at the Battle of El Alamein. Egypt and the canal were secure at last, a British army had finally beaten the Germans, a British general (the first one successful enough to become known affectionately to the British public by his nickname, in this case "Monty") had outfought Rommel—Churchill, for the first time in the war, was able to order church bells rung all over Britain to celebrate a victory. But the Western Desert was not Europe. As Churchill himself would say in a major speech on North Africa in the House of Commons, "This is not the end. It is not even the beginning of the end. But it is, perhaps, the end of the beginning."[6]

It was indeed "the end of the beginning," for it was quickly followed in November by the landing of a combined American and British army in French North Africa—the largest and most complicated amphibious landing to date in the history of warfare. This was intended to trap Rommel's forces between Mongomery's advancing army in the east and the Anglo-American forces in the west, under the command of an American general whose name was as yet very far from a household word even at home, and not at all easy to pronounce—Lieutenant General Dwight D. Eisenhower, who would very shortly be known to everyone in the world as "Ike."

CHAPTER 2

"I Hope to God I Know What I'm Doing"

Greenham Common, June 5, 1944: Ike talks to troopers of the 101st Airborne on the eve of D-day.

History has to be written chronologically, but in life it is sometimes the big moments that count. In the spirit of Winston Churchill, who, echoing Ralph Waldo Emerson, remarked, "A foolish consistency is the hobgoblin of little minds," I propose to depart from strict chronology at this point. In any case, as what Eisenhower did in World War II and the importance of his role begin to recede except for those of us old enough to remember these things, it makes sense to step out of chronological order for a few pages and describe what was surely the most critical and difficult decision of the war—and perhaps the most important moment in his life.

D-day was the climactic moment of the war. The huge and bloody battles on the eastern front between the Germans and the Russians dwarfed anything that had taken place, or that would take place, between the Germans and the Anglo-American armies, but D-day was Hitler's last chance to win the war. The war on the eastern front was a war of ideologies; a war of vast numbers; a war in which mass murder, brutality, and suffering on a scale never reached before or since were commonplace. From June 1941 to May 1945 the total number of military and civilian casualties in the Soviet Union is estimated to have been between 23 million and 28 million.[1] And by the summer of 1944, after the Germans' failure to take Moscow in the winter of 1941, and the defeat of the German army at Stalingrad in 1942 and in the Battle of the Kursk Salient in 1943—two of the largest and most costly land battles of all time—it was clear that Hitler's fatal gamble had failed. Whatever else Hitler might achieve, the Soviet Union did not collapse like a house of cards once he had "kicked the door down," as he had predicted, as czarist Russia had collapsed before the German army in 1917.

It may or may not be true that "the past is a foreign country," but what is certain is that in looking backward in time our view of past events must always be significantly distorted by the fact that we know how they turned out. We know that Germany lost the war in 1945, and was in the process of losing it in 1944, and perhaps even in 1943, but nobody then could know for sure that it was going to happen that way.

Of no people was this more true than the Germans themselves. Quite apart from the fearful punishment that could be inflicted on them for "defeatism," most of them did not assume that the war was being lost, or would be lost. Propaganda, patriotism, discipline, fear of the Gestapo, the ubiquity of the Nazi Party, and admiration for—even worship of—the Führer prevented many, perhaps most, Germans from reaching that conclusion, despite the mounting evidence of monumental casualties, the wholesale destruction of Germany's cities, and the retreat of the German army in the east.

Hitler's grip on his own people remained as firm and mysterious as was his hold over the members of his "court"—for his regime was always far more of a despotic court than a government, one in which opposition, even doubt or lack of enthusiasm, was instantly punished. If a full field marshal of the German army could complain that whenever he sought to change Hitler's mind about something one look from those glaucous, hypnotic blue eyes was enough to make him go "weak in the knees," we should not be surprised at the hold Hitler still managed to retain over the German people. If even so clearheaded, powerful, and logical a man as Albert Speer, Hitler's beloved wunderkind Reich Minister of Armaments and former architect, was unable until almost the very end, when it was too late, to admit to the Führer that he believed the war was lost, how can we expect more of ordinary people, whose fate was by then in any case fatally linked with Hitler's own?

The Führer had always led the German people where they wanted to go. That was the secret of his appeal to the masses. They had wanted to throw off the "shackles of Versailles"; to punish the Jews, since somebody other than the Germans had to be held to blame for losing World War I and for the social dislocation and financial collapse which followed that defeat; to reverse and avenge the decision of 1918; to humble France and England; and to destroy bolshevism—and Hitler took them at their word. Now, in 1944, he was leading them to their own ruin, following his own slogan, which they had applauded over and over again for years, *Weltmacht oder Niedergang*—"World power or destruction"—and they were in no position to stop following him now that destruction seemed the more likely of the two outcomes. Even to *think* about the future objectively, as by expressing doubt in Germany's victory, was a kind of crime in Nazi Germany. Hitler had never demanded objectivity, and was still less enthusiastic about it now—what he wanted was adulation, obedience, enthusiasm, and the people's "unshakeable," "fanatic" belief in himself, and of course in the victory he had promised them.[2]

For the German people, lulled by constant propaganda; by what the historian Hugh Trevor-Roper called, with acid precision, Nazism's whole system of "bestial Nordic nonsense"; and by the nationwide fear of expressing even the slightest pessimism, there was no option but to believe in victory. Roosevelt's insistence on Germany's "unconditional" surrender, announced at the Casablanca conference in 1943, very much against Churchill's wishes, made the Germans all the more determined to fight on to the end. They knew what to expect from the Russians; after the Casablanca declaration they expected nothing better from the western allies; and in any case they were in no doubt about what would happen to them if they opposed Hitler. They had no choice but to believe that the Führer's genius would enable him to snatch victory from the jaws of defeat, as it had so many times before, perhaps by means of new secret weapons, perhaps by a brilliant diplomatic coup de main at the very last moment, perhaps by a decisive military victory—if not against the Russians, then against the Americans and the British when they landed in France, as everyone knew they would. The Führer had produced miracles before, and the whole German nation waited, expectantly, for him to do so again.

No major country is defeated until it believes itself to be so, and the Germans were, in the summer of 1944, very far from believing that. Besides, a glance at the map makes it very clear how far Germany was in fact from defeat. It still occupied—with a ruthless iron hand—most of Europe, from Norway and Denmark in the north to the Balkans and Greece in the south, from beyond the borders of prewar Poland in the east to the Pyrenees in the west, as well as all of Italy north of Rome. Hungary was an ally, in the process of becoming a Nazi puppet state, while Croatia and Slovakia were already puppet states, each with its own dictator, with Romania and Bulgaria as allies, however reluctant. Despite the round-the-clock bombing, German war production reached its peak in 1944, under the inspired guidance of Speer. The latest versions of the German Tiger and Panther tanks—designed in part by the future maker of luxury sports cars, Dr. Porsche—were superior to anything the Allies had, and were beginning to appear in quantity; Germany was already mass-producing jet fighters and even rocket-propelled fighters that rendered British and American fighters obsolete overnight, as well as the V-1 flying bomb and V-2 ballistic rocket with which to renew the bombardment of England; new and more powerful long-range U-boats were being produced; everywhere German industry, efficiency, and ingenuity were at work, buttressed by a vast slave labor force provided by the SS and kept hard at work by the most savage brutality. If Germany was defeated, it would not be for lack of weapons, or manpower.

In the meantime, many things in the Third Reich proceeded as usual, among them the endless stream of trains with their sealed cattle cars, carrying the Jews of Europe night and day—from Greece, France, Holland, Belgium, the Balkans, and northern Italy—to their death.

The lieutenants of SS Oberstumbannführer Adolf Eichmann worked overtime to fill the cars—indeed, as his colleague SS Oberstumbannführer Rudolf Hoess, the Kommandant of Auschwitz, complained to anybody who would listen, the death camp had never been so busy as it was in 1944. It was the peak year of the killings. The staff was overworked and overwhelmed by the number of Jews that flooded in. So much Zyklon-B gas was needed that its manufacturers had trouble keeping up with the demand. The state-of-the-art gas-fired crematory ovens that Hoess had insisted on could no longer deal with the number of corpses he was producing daily, and he was forced, much against his will, to add old-fashioned burning pits, the primitive technique he had rejected when he inspected the first death camps before the construction of Auschwitz. The bodies that could not be disposed of in the crematoriums were carried on carts to ditches dug by backhoes, to be laid on steel rails in an exact pattern calculated to promote maximum burning efficiency, the fat, as it drained off, being carefully ladled back onto the bodies by the inmates charged with the process so as to produce a hotter fire, though not before the inmates who worked in "Kanada,"* the anteroom to the crematory, had removed the gold teeth and dental fittings from the mouth of each victim and collected them for transferal to the Reichsbank. All the inmates involved in this process were of course "selected" at regular intervals, and sent to the gas chambers themselves. This kind of efficiency cost Hoess dear; his head ached from the many problems he faced—he was, in any case, a martyr to migraine headaches—and at Nuremberg after the war, to the great discomfort of many of the more important and senior Nazis being tried there, he remembered every detail of the extermination process at Auschwitz; indeed it was hard to shut him up on the subject. Like the head of any

* So named after the Yukon gold rush, when the gold-gathering process was moved there because it was easier and quicker to knock the gold dental fittings out of the mouths of dead inmates than live ones.

major industrial concern, Hoess could rattle off the numbers by heart, except that instead of money, his end product was death—he was the CEO of industrialized mass murder.

Had Hoess but known it, the worst was still to come, for in the autumn and winter of 1944 the Germans, having deposed Admiral Horthy, the regent of Hungary, and installed a government of Hungarian Nazis in his place, moved with desperate urgency to murder the Jews of Hungary before the Russians could get there and save them. Eichmann himself, from his hotel suite in Budapest, initiated the operation, which resulted in nearly 500,000 men, women, and children being packed into cattle cars and sent to Auschwitz for Hoess to deal with. Every night the chimneys of Birkenau—the heart of the death camp—belched lurid flames and greasy black smoke, as Hoess struggled to keep up with the tide of humanity sent to him from Hungary. It was, he thought, his finest moment, a Wagnerian finale to operations at Auschwitz, which he would soon be ordered to destroy.

For by the end of 1944 the orgy of murder changed to a desperate struggle to destroy the evidence. Hoess would not be alone in his task. All over eastern Europe, as the Germans retreated, the SS and police units that had formed the Einzatskommando—mobile killing groups, which had roamed all over the vast area behind the German army shooting Jews, over 1 million of them, and burying them in huge mass graves, often before large crowds of curious or enthusiastic onlookers—were now obliged to dig the bodies up and burn them as the army retreated. Funeral pyres of immense size marked the German retreat, adding another dramatic Wagnerian touch.

Given all this, it was inevitable that word had long since seeped out about the murder of the Jews. Railroad men, who saw the trains and knew where they were going, also knew what was being done there, since the trains came back empty; soldiers on the eastern front were witnesses to the mass shootings, and not only described them in letters home but often sent home horrific snapshots of the killing process; German bureaucrats at every level dealt with matters like canceling the life insurance policies of murdered Jews so the big insurance companies wouldn't have to pay out, transferring their apartments to Aryan families who had been bombed out of their homes, and disposing of Jewish-owned businesses to Aryans—not to forget the number of people involved in transporting the monthly hoard of gold teeth and dental fittings back to Germany and handing them over to the appropriate bank officials to be weighed.

The extermination of the Jews was by no means a state secret like the building of the atomic bomb in America. First of all, it was public policy; Hitler had always said he would do it, and his promises to do so had been greeted by many with enthusiasm. Then too, the number of people involved in carrying out the "final solution," as it came to be called, was too large for there not to be talk about it. If it was not "common knowledge," that was only because a very large number of Germans did not want to know it.

But by the summer of 1944 knowledge enough had seeped into the German consciousness to make people nervous. Even those who were innocent; even those who had mental reservations on the subject (the fashionable phrase of the time was "going into inner exile"); even those who didn't know, or doubted, or refused to face the truth, nevertheless were made uneasy by what they heard or guessed—it was simply "in the air." The Allies, it was feared, would exact a terrible revenge if they won the war, and few people were unaware of the fact that there were many things in Germany (and in occupied Europe) that Allied troops would almost certainly find and that would be hard, if not impossible, to explain—not just the "final solution," of course, but also the horrors meted out in concentration camps to opponents of the regime since 1933, the systematic murder of vast numbers of Russian prisoners of war, the barbarous medical experiments, and the widespread killing of the institutionalized mentally and physically handicapped Germans in the interests of "racial health and hygiene."

Like the decision to demand an unconditional surrender of the Germans, the residual guilt of twelve years of Nazi government inevitably made most Germans all the more determined to fight to the end, and, if possible, win. Too many terrible things had been done by them, or in their name, to make a victory by the Allies welcome—and by 1944 the knowledge was too widespread to ignore, even if one did not know all the details, or even believe them. Ernst von Salomon, a right-wing writer who was fiercely anti-Nazi, caught the mood when he described a woman official of the Nazi Party telling a group of stolid middle-class Germans that they had nothing to fear if the Russians came, since their beloved Führer had prepared for each of them in that event a swift and painless death by gas. Far from being surprised or indignant, the audience members nodded their heads sagely, as if this was perfectly sensible news. Salomon imagined them saying to each other as they walked out of the meeting hall, "Think of all that good gas that was wasted on the Jews!"[3]

Thus, although the present view is that Germany was already beaten in mid-1944, that was not by any means the case. The Germans did not believe they were beaten (or refused to acknowledge the possibility), Germany was still a formidable military power, and Hitler's control over Germans of every rank was still complete and almost unchallenged.

Hitler viewed the coming invasion with mixed feelings. Like any German, he dreaded the idea of a war on two fronts—exactly the mistake that had defeated Germany in World War I, and that he had promised to avoid—but he was unlike his generals in that his mind, always difficult to predict or fathom, could imagine certain possibilities opening up in the event of an invasion. Hitler often described himself as a "sleepwalker," meaning that he moved somnambulistically toward great decisions, but the truth is that he preferred to wait and let great decisions proceed at their own pace toward him. In 1933 he had not demanded the chancellorship; he had waited for President von Hindenburg and the right-wing camarilla around him to offer it, and so they did. In 1938 he had made his demands known on the subject of the Sudetenland, then sat back and waited until Neville Chamberlain flew to Germany to offer him what he wanted, and more. Now in the early summer of 1944, he had no way of preventing the Anglo-American invasion, but he waited, hoping to take advantage of it when it happened.*

* Younger readers may detect here a certain resemblance to events and characters in J. R. R. Tolkien's *The Lord of the Rings*. It is likely that Tolkien, a veteran of the Somme in World War I, was aware of this when he wrote it. His three books are not a simple political allegory, of course, but it is not difficult to see in Sauron, the dread Lord of Mordor, many of the eerier attributes of Hitler, or to suppose that the Orcs represent a kind of allegorized version of the Waffen SS. As for the Hobbits, they can easily be seen as the plucky, robust, cheerful, unmilitary English, summoning up their courage at the last moment to resist evil.

From the point of view of that devious, dangerous, and manipulative mind, with its extraordinary depths; cold, brutal realism; and cynical understanding of how to exploit human weakness, the prospect of the invasion may well have seemed less frightening than it did to his generals. To begin with, if the invasion succeeded, he foresaw increasing tension between America and the Soviet Union—had he lived to see it, the cold war would not have surprised him. There were plenty of people in the United States, he knew, who didn't want to see the Soviet Union gobble up all of Eastern and much of Central Europe, and who thought that Roosevelt was foolish or misled to risk it. Could that tension be sufficiently exacerbated to give Nazi Germany a balancing role? It would require superhuman feats of diplomacy to accomplish, and it was at best a risky gamble, but had the Führer not always gambled for big stakes? When he made the decision to attack Poland in 1939, he had said to Göring, enraptured by his own daring, that they were gambling *va banque* that the British and French would not come in—or, as we would say in English, "all or nothing"—and Göring, who was then still Hitler's "iron man," not yet in disgrace because of his sybaritic lifestyle and the failure of the Luftwaffe, had replied, "But have we not always played *va banque, mein Führer*?"[4] It was true. If Hitler relied on anything, it was his own willpower, nerve, and luck, and these had never failed him.

Still, much the best thing would be the total destruction of the Allied invasion forces on the beach, and this did not seem an impossible task. For over three years a vast labor force, much of it slave labor, had worked to build the "West Wall," a string of fortifications that protected the northern coast of occupied France. For nearly a year, ever since Generalfeldmarschall Erwin Rommel had been placed in command of Army Group B, he had roamed the beaches restlessly trying to plug up weak points in the defenses, inventing and installing millions of undersea devices to destroy tanks and landing craft as they approached the beaches, placing millions of miles of barbed wire and millions of mines of every conceivable size and type.

It had been Hitler's intention to use the entire force of the new jet aircraft to bomb the invasion beaches, flying at speeds that would make them invulnerable to Allied fighters and ground fire; but as usual the Luftwaffe had prevented this—Göring again!—by producing too many of the planes as fighters, rather than as the fast light bombers Hitler wanted. Still, the capacity of the Wehrmacht to destroy the Allied invasion forces was not in doubt on either side of the English Channel. There was certainly enough German infantry in France, more than twenty-five divisions, in addition to which the aged but still formidable Generalfeldmarschall Gerd von Rundstedt, Commander in Chief West, disposed of three panzer (armored) divisions south of the Loire, and three north of it, in Army Group B, plus a further four panzer divisions—the strongest, most experienced, and best-equipped armored divisions in the Wehrmacht—held close to the Pas de Calais in reserve as Panzer Group West, which were not to be released except on the direct order of the Führer himself.

A seaborne landing would inevitably mean that the Allied forces on the beaches would have only a limited quantity of armor and antitank guns at first, and any airborne troops dropped would be, in effect, light infantry—the four armored divisions of Panzer Group West would have no difficulty in annihilating them, provided of course that they could reach the invasion beach in the first twenty-four to forty-eight hours. That was the difficulty, given the extent of Allied air superiority, the likelihood that the French resistance movement would disrupt the railway system, and the magnitude of the preinvasion bombardment; and it was—apart from his habitual mistrust of his own generals—the main reason why the Führer had reserved for himself the decision about committing Panzer Group West, overriding the vehement objections of both field marshals.

Then too, where would the invasion land? That was the big question, after all, possibly the most important question of the war, and it remained unanswered. Hitler and most of the German generals (including Rundstedt himself) thought that the Pas de Calais was the most likely place, partly because it was only eighteen miles from England at its nearest point, and therefore offered the shortest Channel crossing; and partly because the Allies' intelligence had devoted, over the years, unimaginable effort and guile to convincing the Germans that this was so, and that main landing would be preceded by a smaller one elsewhere, intended to provoke them to attack in the wrong place.

Rommel had developed a certain *Fingerspitzengefühl*, a nervous tingling in the fingers, on the subject of Normandy, and had devoted a good deal of attention to fortifying the defenses on its beaches, but the prevailing opinion of the Oberkommando der Wehrmacht (OKW), the German High Command, was that the Allies would land somewhere close to the Pas de Calais, preceded by a diversionary landing, perhaps in Normandy. It was Hitler's determination not to waste Panzer Group West on a dummy invasion that persuaded him to keep his hands on it despite the arguments of two of his most respected generals. Everything—the very fate of Germany—hinged on the ability of Panzer Group West to attack the main invasion force while it was still spread out on the beaches.

Politically, the opportunities that might arise if the invasion was annihilated on the beaches were, to use a favorite word of Hitler's, *kolossal*. Quite apart from stilling any dissension in Germany and causing the inhabitants of those countries occupied by the Germans to lose all hope of liberation, who could guess what the broader consequences might be?*

In terms of fighting manpower it was no secret that the British were already drawn very thin indeed—there was very little chance that they could put together another invasion army to replace the one they might lose, certainly not for a year or more. Was it not possible to imagine then that in these circumstances Churchill himself might lose a vote of confidence and fall? A bloody failure of the long-awaited invasion might also result in Roosevelt's losing his bid for an unprecedented fourth term, and who could say what the results of that might be? A Republican president might feel it was time to make the Pacific America's first priority. Finally, there was Stalin to be considered. He had demanded an invasion, a "sec-

* All these considerations ignore the atomic bomb, of course, but it must be remembered that in the late spring of 1944 the first test was still over a year away. There was no guarantee that it would work; nor do we have any way of knowing whether the United States would have used the atomic bomb against the Germans. As for the Germans themselves, they were so far behind the Allies in the development of a nuclear weapon that the German scientists in captivity at the end of the war refused to believe the news about Hiroshima when they heard it. It is fortunate that Hitler did not pursue an atomic bomb with the same enthusiasm he had for rockets. In *Mein Kampf* he had written, with chilling prescience, "Should the Jew . . . triumph over the people of this world, his crown will be the funeral dance of mankind, and this planet will once again follow its orbit through the ether, without any human life on its surface, as it once did millions of years ago." No doubt if he had had the weapon this is how he would have used it.

ond front," for 1942; when it was postponed to 1943 he was enraged, still more so when it was postponed again to the summer of 1944. Russians were fighting and dying by the millions, while the British and the Americans continued to fight minor campaigns on the periphery of Europe. Now, if the second front the Russians had demanded and waited for during almost four bloody years was smashed on the beaches, and if it might have to be postponed until 1945 or even 1946, might not Stalin, despite his losses, be tempted to make a separate peace? Generous terms could be offered to him—terms so generous that he could hardly say no. After all, had not Stalin, against everybody's expectations but the Führer's, agreed to sign a pact with Germany and divide up Poland in 1939? Perhaps for all of Poland, a solid chunk of the Balkans, a piece of Hungary, peace in the east could be had. . . .

But only if the invasion was crushed.

Things did not look so very different from 10 Downing Street than in the more rarefied air of Hitler's "Eagle's Nest," high above the Obersalzburg; or from his gloomy headquarters on the eastern front, the "Wolf's Lair," buried deep in the grim, dark Teutonic pine forest of Rastenburg. The possibility of the failure of Overlord, the latest code name of the invasion, weighed heavily on the mind of Winston Churchill, as it had from the beginning.

As long ago as the dark days of 1941, when Britain still stood alone, Churchill had set up a committee to plan for the invasion of France, remote as that possibility then seemed. Perhaps as a relief from dealing with the grim realities of Nazi-dominated Europe, his fertile mind turned toward the new weapons that would be required. He ordered the design of large seagoing ships to carry infantry and tanks directly to the beaches—the landing craft infantry (LCI), and the landing ship, tank (LST), known to its crews as "large slow target," that would soon be produced in huge quantities in America. Churchill rescued the eccentric and difficult general Percy Hobart from the Home Guard, where he had been serving as a corporal, restored him to his general's rank, and set him to developing specialized tanks for the invasion—one with a huge flail mounted in front to cut a path through minefields, one with a flamethrower instead of a gun mounted in the turret, another with an inflatable canvas girdle and two small propellers that in theory enabled a heavy tank to swim to shore. These and many even more improbable inventions became known as "Hobart's funnies" but would prove invaluable on D-day.[5]

The prime minister's enthusiasm for the invasion—and for the dirty tricks and secret weapons he planned to use against the Germans—was unmistakable and infectious, but with the passage of time a certain note of caution began to make itself heard. Churchill had been alarmed when the Americans wanted to invade France in 1942, and had worked hard and successfully to persuade Roosevelt that North Africa was a safer place to begin. In 1943, their positions were reversed—it was Roosevelt who had to admit that the need for manpower (and landing craft) in the Pacific made it impossible to send enough American troops to the United Kingdom for a cross-Channel invasion that year. This decision did not dismay Churchill, but it infuriated Stalin, who had been promised a second front in 1943, and responded to the postponement with undisguised rage, contempt, and insults at the unwillingness of the western Allies to come to grips with Germans on European soil. As early as 1942, when Churchill and Averell Harriman (who was representing President Roosevelt) had flown to Moscow to give Stalin the bad news that the Allies would be landing in North Africa, not France, Stalin's reaction had been vitriolic, shaking even so hardened a soul as the prime minister's.

Now, as the invasion season of 1944 approached and the huge preparations for D-day reached their climax, Churchill was uncharacteristically filled with forebodings. The invasion would involve 7,000 ships, more than 10,000 aircraft, and 170,000 men—nothing like it had ever been attempted before in history. Nobody knew better than Churchill that if the invasion failed it would be difficult, perhaps impossible, to raise a new British army for a second try—it would mean the end of Great Britain as a major power. Once before in his life, at Gallipoli in 1915, he had staked his political future on a seaborne landing, and its failure had dogged him until 1939. In his darker moments, it continued to haunt him now. Even the position of the British invasion troops, on the left, meant that as the Allies pivoted inland (provided they succeeded in getting onshore), the British would eventually be advancing across the same ground where they had fought in 1916 and 1917, full of familiar place-names that still stood in the collective British mind for the death of hundreds of thousands of young men.

Above all, in Churchill's view, the Americans' obsession with Overlord, as the invasion was now known, made them blind to other opportunities that were there for the grasping. As late as 1943 the prime minister was still urging the merits of a landing in Norway. Even in 1944, as the date of the invasion approached, he kept coming back again and again to the Italian campaign, which was bogged down for want of manpower, although a determined thrust to the north might, he argued, enable the western Allies to reach Vienna and Central Europe and bring down Hitler before the Russians did. An attack through the Balkans might bring the allied armies into Central Europe, threatening Berlin and preventing the Soviet Union from overrunning Eastern Europe. Any of these strategies would open up a myriad of political opportunities, and none of them would involve anything as supremely risky as a cross-Channel invasion, which was, the prime minister thought, like going into a casino and staking your entire fortune on one number, all or nothing.

Churchill was committed to the invasion, but with an increasing reluctance, and despite nagging doubts, and it had been one of Eisenhower's most urgent tasks since his appointment as supreme commander to keep the prime minister's eye on the ball and to prevent him from seeking easier opportunities by attacking what he liked to describe as the "soft underbelly" of Nazi Germany, rather than its hard carapace. "I was not convinced that this was the only way of winning the war," Churchill would write later, modestly underplaying the strength and persistence of his arguments at the time. But the American position on Overlord (and on Anvil, the subsequent landing in the south of France about which nobody was particularly enthusiastic) was absolute, an article of faith, and no adventures elsewhere, however tempting, were to be allowed to interfere with it. "Thou shalt have no other gods before me" might as well have been the motto of the American chiefs of staff, led by General Marshall.

By May 15, after a final, detailed review of the invasion plans, Churchill felt able to announce, to Eisenhower's dismay, "I am hardening on the enterprise."[6] He meant—despite attempts on his part, after the war, to alter the meaning of his words—that he was at last coming around to the American view of it, and putting aside such false idols in his mind as Norway or the Balkans. As part of the price for agreeing that Eisenhower should be the supreme commander, the British had insisted on having Montgomery appointed as the overall commander of all ground forces—British, Canadian, and American—until such time as they broke out of the beachhead, a decision which pleased neither the American generals nor Montgomery's British rivals. The all-star presentation took place at St. Paul's School, in London, the headquarters of General Montgomery, before a select audience that included the prime minister, "looking puffy and dejected at first,"[7] and King George VI in the front row. Eisenhower spoke with calm earnestness, conveying his enthusiasm and his confidence of success. Montgomery crisply summed up the plan for breaking out of the narrow confines of the beachhead; he was followed by Admiral Sir Bertram Ramsay, the naval commander, who had been responsible for organizing the evacuation of the BEF from Dunkirk in 1940, and finally by Air Chief Marshal Sir Trafford Leigh-Mallory, everybody's least favorite air marshal, to present the air plan. The effect was to erase all doubts from the minds of those who were in the audience, including, most important of all, the prime minister.

The Americans present could not help noticing, with a certain lingering resentment, that not only Eisehower's deputy, Air Chief Marshal Tedder, but also his ground, naval, and air force commanders were all British; but the Britons present were equally conscious of the fact that two out of three soldiers who landed on D-day would be either British or Canadian. By D+1 or D+2 (as the days following D-day were coming to be known) the balance would begin to shift toward the Americans, since huge numbers of American troops were stationed in southern England, almost 2 million of them by May 1944, and if the invasion succeeded as planned, American forces would swiftly outnumber the British and Canadians on the beachhead.

It was hard not to feel confident as each commander listed his resources.

The naval vessels would range from battleships like HMS *Warspite* and USS *Arkansas*, whose heavy guns would be used to pulverize the German fortifications overlooking the beaches at key points, down to the thousands of small landing craft (most of them the brainchildren of a gifted boat designer in New Orleans named Higgins and manned by U.S. Coast Guardsmen) that would be lowered over the sides of the troopships to take the men to the beach. The RAF and the U.S. Army Air Force would bomb railway yards all across northern France to slow down the movement of German divisions toward the beachhead. (The bomber barons, British and American, had fiercely resisted this interruption of bombing targets in Germany to the bitter end, until Eisenhower was finally obliged to call on Churchill and tell him that if the "heavies"—the four-engine bombers—were not placed under his command before D-day he would resign as supreme commander and go home.) Three full airborne divisions, two American on the right, and one British on the left, would be dropped in Normandy the night before the landing to seize key points, along with a whole variety of "dirty tricks," including inflatable dummy paratroopers with firecrackers to simulate rifle fire that were to be dropped far and wide behind the beaches to confuse the Germans. Vast and peculiar constructions made their way toward the south of England by sea without being discovered by the Germans: floating concrete caissons and obsolete ships which would be sunk to create artificial harbors and breakwaters off the beaches, known as Mulberries; and an undersea pipeline for supplying fuel to the beach, known as Pluto (for "pipeline under the ocean"), to name only two. There was intense attention to detail. For example, when Lieutenant General Omar Bradley of the U.S. Army expressed some lingering doubt that his beaches would support the weight of tanks, he found in his office twenty-four hours later a young British naval officer who, having spent the night on one beach in question, presented him with a dripping core sample which proved that the beach was firm shingle down to bedrock. British midget submarines had been weaving their way in and out of the mined obstacles that Rommel had placed in the water, putting men on shore in the darkness to test every assumption of the D-day planners.

All over France, the French resistance groups, many of them with a

British officer from the SOE or an American from the Office of Strategic Services (OSS) attached to them to arrange for air drops of arms and ammunition, began the dangerous task of sabotaging the railways and roads to slow down or block the advance of the German armored divisions, whose prompt arrival on the beachhead would be calamitous. This was a whole other war, one fought by civilians in which capture by the Germans meant prolonged torture by the Gestapo, followed by execution; a war in which men and women were routinely provided with a cyanide capsule to bite down on in case they were taken prisoner or wounded, and were grateful for it; a war in which it was usual for the Germans to take hostages at random from the French civilian population and shoot them in reprisal for acts of sabotage.

Of course there would be problems and mistakes—given the size and the sheer daring of the invasion, everybody recognized that. At one point German E-boats (fast motor torpedo boats, the equivalent of the American PT boat and the British Fairmile MTB) had gotten in among the landing craft during a nighttime training exercise in British waters and wreaked havoc among the troops—nearly 700 American soldiers were killed. At another point, two British commandos who had been examining the French beaches north of the Seine at night were captured by a German patrol. One of them, Lieutenant Lane, was invited to tea on his way to a prisoner-of-war camp by none other than Field Marshal Rommel himself. Lane, very fortunately, had no idea where the invasion would land, but when Rommel asked him his opinion, said that if he were planning it he would take the shortest route, to the Pas de Calais. Rommel replied somewhat more doubtfully that he thought so too.*[8] A bombing raid by the RAF on the railroad marshaling yards of Paris resulted in the death of nearly 700 French civilians, a foretaste of much worse to come—one of Churchill's major concerns was the number of French civilians who would

* This unusual gesture on the part of Rommel reflects both his lively curiosity and his old-fashioned view of the social side of warfare. As other German field marshals complained, he had fought only against the British and the Americans, never on the eastern front against the Russians. It may also have been his way of showing his disapproval of Hitler's order that captured commandos should be shot, and protecting the two British commandos from the firing squads of the Gestapo and the SS.

be killed. Soon the most difficult problem of D-day would surface, with the arrival in the United Kingdom from Algiers of an irate General De Gaulle, who had not been told about the date of the invasion and was indignant at the speech Eisenhower planned to broadcast to the French people, as well as at the plan to put liberated areas of France under Allied military government and issue occupation scrip to replace French francs. This particular idea—to treat France as if it were an enemy country instead of a liberated ally—was a favorite in Washington, where everybody from Roosevelt down disliked De Gaulle and hoped to replace him with some more malleable and "democratic" French figure. But Eisenhower, using his great power as supreme commander (and the very considerable power of his charm), elected to go against the wishes of the president and the State Department, and set out to mollify De Gaulle. The speech would be modified; the occupation scrip vanished; De Gaulle's status as the head of state of the French government in exile would be respected; and De Gaulle himself would arrive in France a week after D-day aboard a French warship.

All over England the vast task of loading the invasion fleet was going on. Rural roads were lined with ammunition dumps, and huge numbers of vehicles, from heavy tanks to jeeps, motorcycles, and bicycles, were being assembled—for the invasion was like an intricate jigsaw puzzle: everything needed had to be packed and loaded so that it all would come off in the right order on the beaches without a hitch. Tanks had to be laboriously backed into the LSTs, so they could come down the ramp onto the beach with their heavier frontal armor facing the enemy and their guns ready to fire; ammunition and medical supplies had to be placed so they would arrive on the beach at the same time as the first troops; hundreds of thousands of two-way radios had to have their frequencies set so they could communicate with each other on D-day. The tasks were endless. All over the United Kingdom, RAF and U.S. Army Air Force ground personnel were busy painting broad black and white stripes on the wings and around the rear fuselage of all the aircraft that would fly on D-day—more than 10,000 of them—since this was the air recognition code for the invasion. Any aircraft appearing over the invasion fleet or the beaches without these stripes would be shot down.

Nothing, it seemed, had been left to chance. The only thing that could go wrong—the only thing over which the supreme commander had no control—was the weather.

All the same, late on the evening of June 5, when the great fleet bearing almost 170,000 men was already at sea, and the paratroopers were already on their way through the dark night sky to their drop zones in Normandy, and the midget submarines were rising to the surface to mark the boundaries of the invasion beaches, Winston Churchill, getting ready to go to bed, said to his wife, Clementine, tears running down his cheeks, "Do you realize that by the time you wake up in the morning, twenty thousand men may have been killed?"[9]

On no pair of shoulders did these concerns rest more heavily than those of the supreme commander himself. Eisenhower commanded more than 3 million men—nearly 1.7 million of them American; 1 million British and Canadian; the rest Free French, Polish, Norwegian, Czech, Belgian, and Dutch. This was the largest international alliance ever assembled in history. Everything that could be done had been done—he had even managed to prevent a last-minute attempt on Churchill's part to watch the landings from the bridge of the cruiser HMS *Belfast,* one of the British warships. Rather than tackle the prime minister personally, he delegated to his chief of staff, the flinty Major General W. Bedell Smith, the task of telephoning the king and asking his majesty to deal with the problem. The king promptly sent his prime minister a brief note to say that if Churchill went, his majesty would feel it was his duty to go too; and Churchill, grumbling and disappointed, backed down.

Only two things could stop the invasion. The first was the possibility that the Germans might have discovered where and when the landings would take place, but from intercepted German radio traffic and the clandestine broadcasts from French resistance groups there appeared to be no evidence of this. The enormous fleet put to sea without any sign at all that the Germans had even noticed it—another consequence of the Allies' control of the air, and another example of the failure of Göring's Luftwaffe. Had some of the new German jet fighters been modified as high-altitude, high-speed photoreconnaissance aircraft like the RAF's PR Mosquitoes and Spitfires, the fate of the invasion might have been very different; but on Hitler's orders the jet fighters were being converted into light bombers instead.

The second thing that could stop the invasion was, of course, the weather, and this was Eisenhower's chief concern. Oddly enough, to those familiar with the weather in and around England, the invasion had always been planned on the assumption that weather conditions would be favorable. May, June, and July were chosen as the ideal months, since they offered the best chance for good weather, as well as the most hours of daylight; but as anybody who remembers making a Channel crossing in the summer knows, even a warm, sunny day is no guarantee against appallingly rough seas.

It was not just a question of weather, either, in a country that was as world-famous for its bad weather as for its bad food. The invasion could take place only when the tides were at their lowest ebb, in order to expose the mined obstacles that Rommel had placed in the water in such staggering quantities, even though the price for this timing was to increase the distance the troops would have to cross over the open beaches in the face of enemy machine-gun fire once they were onshore. Also, there had to be sufficient moonlight the night before the invasion to enable the transport and glider pilots to find the drop zones for the paratroopers, and for the paratroopers and glider pilots to see where they were landing. Low visibility might be an advantage for the troops landing on the beaches, but it would hamper the airmen flying the bombers and fighter-bombers, and perhaps make the parachute drops impossible. High winds would raise surf on the beaches, and again might make it necessary to cancel the parachute drops; but on the other hand, a totally windless morning would shroud the beaches in smoke and make it difficult for the warships to reach enemy targets accurately with their big guns. In short, to have any chance of succeeding, the invasion required a remarkably complex combination of factors, and these would be available only on certain days of each month. June offered only three possible dates: June 5, 6, and 7.

The invasion had originally been scheduled for May, then moved to the first week of June, and eventually fixed for dawn on June 5. As the ships were loaded and the troops moved south, Eisenhower himself gradually moved from his own headquarters outside London to Southwick, outside Portsmouth, where Montgomery and Ramsay had their headquarters, closer to the invasion fleet. By June 1, the whole intricate operation was in motion: the great battleships at were sea, the men were on the move according to rigid timetables, the vehicles were loaded, and the systematic bombing of key points of the French railroad system was already in progress. The only difficulty was the weather: low clouds, high winds, intermittent heavy rain, and rough seas, with more of the same forecast for the foreseeable future. June 2 was worse, and June 3 worse still.

Late on June 3, Eisenhower reluctantly made the decision to postpone the invasion by twenty-four hours because of the weather, from June 5 to June 6, even though much of the invasion force was already at sea, and the troops were packed in like sardines aboard the ships with all their equipment, adding seasickness to their woes. The smaller ships turned around and steamed back to their home ports; the larger ones stayed at sea, to the acute misery of the troops. Eisenhower was aware of their misery—he wanted the troops to reach the beaches in good fighting condition—but he was even more conscious of the danger that every hour the fleet was at sea increased the chance that it might be discovered by a German E-boat or submarine.

Early on the morning of June 4 the weather still offered no visible improvement. Rain poured down; wind rattled the windowpanes of the headquarters at Southwick; low clouds scudded past, driven by high winds—it was one of those days that explain why England is famous all over the world for the quality of its raincoats, umbrellas, and waterproof boots. The sea was a dark gunmetal gray, with heavy surf breaking on the beaches and a heavy swell farther out, where the ships of the invasion fleet tossed, rolled, and pitched, to the intense discomfort of the men aboard. Uncomfortable at the best of times, the troopships were now a nightmare, with heads backed up; men vomiting anyplace they could; and the sour odor of dense cigarette smoke, vomit, and unwashed bodies permeating the low, crowded, dimly lit compartments in which the troops and their bulky equipment and weapons were densely packed. Many of the men actually looked forward eagerly to the moment when they could clamber down the rope nets into the Higgins boats below and head for shore. They were wrong, however—as they would shortly discover. Being packed into a tiny, open Higgins boat in a rough sea, with waves breaking over the sides, was no picnic; nor was moving down the ramp onto the beach while carrying almost 100 pounds of gear and ammunition straight into steady, well-aimed German machine-gun fire—that is, if you were lucky enough to make it to the beach, rather than being dumped in water above your head by a coxswain too eager to get his boat away to wait until it touched ground, or worse yet being blown into the sea when the boat touched a mine or was hit by German artillery fire.

All these things were on Eisenhower's mind on the morning of June 4. Although Southwick House, the Operations Center at naval headquarters, was a "stately home," until recently an elegant and luxurious mansion, the supreme commander had elected to set up his battle headquarters—consisting of his own trailer, a cluster of tents for conferences, and the trailers for his personal staff—in a park a few miles away. The comparative solitude was soothing and gave Eisenhower a chance to take long walks and think. But he was by no means isolated. The elegant white-and-gold walls of Southwick House were now covered with big maps, on which Wrens (members of the Women's Royal Navy Service) plotted the position of each unit of the invasion fleet, to the constant sound of ringing telephones and clattering teletypes. A red telephone in the spare, cramped bunk bedroom of his trailer linked Eisenhower to a secure scrambled line to Washington; a green telephone linked him to 10 Downing Street; a black one linked him to his chief of staff, General Smith.

Despite all this, Eisenhower's authority as supreme commander was absolute—he was under no obligation to consult with anyone. His orders read simply, "You will enter the continent of Europe and, in conjunction with the other United Nations, undertake operations aimed at the heart of Germany and the destruction of her armed forces. . . ." The decision to invade, to postpone the invasion, even to call off the invasion if need be, was his and his alone.

Perhaps the only sign of Eisenhower's nervous tension was the increasingly rapid rate of his chain-smoking. Every ashtray in the trailer was "full to overflowing"[10] with smoldering ashes, one of his staff noticed, as the general restlessly lit one cigarette from another without stubbing out the first one, pausing only to take another gulp of black coffee. From time to time, he stood up, put his hands on his hips, and stared out the window of the trailer at the rain, or stood at the door of the trailer and looked up at the sky.

He knew that by the end of the day he would have to make the decision about whether to go or not to go on June 6. Through most of the day Eisenhower stayed in his trailer, occasionally leaving to take a walk, then climbing slowly up the metal steps back into the trailer, waiting restlessly for the weather to change. It was still pouring that evening, when he finally left for Southwick House, with the big trees around his trailer groaning and creaking in the wind.

Waiting for him in the library at Southwick House were his senior commanders and their chiefs of staff—twelve men in all. All of them were in uniform, of course, and only one was dressed informally: General Montgomery wore his trademark baggy corduroy trousers and roll-neck pullover. The big library had once been an elegant room, but the bookshelves were empty now, and most of the furniture had been removed. A big table covered with green baize cloth was placed in the center of the room. Heavy blackout curtains covered the French windows to the garden but did nothing to keep out the noise of the wind and rain.

Promptly at nine-thirty Eisenhower entered the room and asked the others to sit down. There was no small talk. They all knew why they were there, and shared the same concerns. There was a haze of tobacco smoke in the room—it was an age when almost everybody smoked, except Montgomery, who normally never allowed anybody to smoke in his presence except the king, Churchill, and Eisenhower. The door opened and the three senior meteorologists came in, led by their chief, RAF Group Captain J. N. Stagg.

Stagg began his briefing on a note of cautious optimism—a common British approach to weather, even on the part of people less skilled at forecasting it than he was. Conditions, he admitted, looked terrible and had been disappointing, but he had good news: weather aircraft far out over the Atlantic beyond Iceland had detected a change in the making. A high-pressure front had been observed, which was moving rapidly east. Tomorrow, June 5, would see periods of gradual clearing over southern England and the invasion beaches, with dropping winds. These conditions would improve considerably through the night of June 5, enabling the bombers to operate and the airborne operations to take place, and would continue through the morning of June 6; but later in the day conditions would begin to deteriorate again. In short, there would be a "window" of about twenty-four hours or less.

Surely no weather report in history was ever so vigorously discussed and challenged as Stagg's. All the senior officers present wanted to know more details, demanded more precision, or asked for something more definite to go on, but Stagg was politely, regretfully firm. Things would look better late in the day tomorrow; weather conditions would be marginally acceptable for the paratroopers on the night of June 5, and for the landings on the invasion beaches on the morning of June 6; and after that the weather would begin to close down again. This was all he could promise.

Once Stagg and his staff had been dismissed, Eisenhower asked his commanders to give him their individual opinions. Predictably, Eisenhower's air commander Leigh-Mallory was pessimistic about the airborne operations, having already expressed his fear that they would result in a disastrous bloodbath.

Leigh-Mallory had been a pessimist about the airborne operations from the beginning, even if the weather conditions were perfect—to the great annoyance of the British and American airborne commanders, who were impatient for an opportunity to demonstrate that large-scale parachute and glider operations at night were feasible. In the British and the American armies there was deep suspicion about airborne troops, who were widely suspected of draining the best and bravest junior officers and men away from regular infantry units to create expensive, glamorized elite formations that had so far served very little useful purpose in the war.* Failure to play a major—and successful—role in the coming invasion would very likely lead to the airborne units' being broken up to provide replacements for regular infantry units, so not just honor but survival was at stake. The airborne commanders not only thought that Leigh-Mallory was an alarmist but tended to regard his role as being like that of the doorman of a London hotel asked to call a fleet of taxis for a big party—his job, as they saw it, was to provide the aircraft needed to get them where they were going, not to worry about whether or not conditions were right for the airborne landings or tell them what they already knew: that nighttime parachute landings were risky.

Montgomery, Eisenhower's ground commander, was as usual succinct. He had already given his opinion when the invasion had been postponed for twenty-four hours on June 3, and had not changed his mind. "I would say go," he said simply.

Tedder, Eisenhower's deputy, for once agreed with his fellow air marshal Leigh-Mallory. This must have dismayed Eisenhower, who placed a good deal of confidence in Tedder's judgment. Leigh-Mallory's opinion would not have surprised him—indeed, Eisenhower had already had a long and painful conversation with his air commander, during which he had warned Leigh-Mallory that while he was of course entitled to his opinion about the airborne operations, he must keep it to himself and not let it affect the troops' morale—but Tedder's doubts about letting the invasion go forward on June 6 despite the iffy weather were something Eisenhower would have to take more seriously.

If the decision had been put to a vote—which was not Eisenhower's way of making military decisions—that would be one for going and two for postponing again. But in the military an opinion is not a vote; in the end, the commander makes the decision himself. Besides, Eisenhower knew his own subordinates well, and he understood that poor visibility

* The Germans used their parachute troops—which were part of the Luftwaffe, not the army—to take Crete in 1941. This was a daring and triumphantly successful operation, but it resulted in such crippling losses that they were never used as airborne troops again, and spent the rest of the war fighting as an elite light infantry division instead.

and bad weather are taken more seriously by airmen than by army generals and admirals, and that, whatever Tedder may have thought of Leigh-Mallory, air marshals also had a natural tendency to stick together.

It was left to Admiral "Bert" Ramsay, the naval commander, to give the group the naval facts of life. If the invasion was postponed again, he pointed out, it would be necessary to bring many of the ships back to refuel. A naval operation of that size, spread across ports all over southern England, might easily be picked up by the Germans. In any case, it would take far more than twenty-four hours to refuel the ships and get them back in their proper places again. If the invasion did not take place on June 6, he saw no practical likelihood that it could take place on June 7—it would have to be postponed until late in June or sometime in July.

This went down like the proverbial lead balloon. Postponing the invasion by another twenty-four hours was bad enough, but postponing it until the end of June or the first week of July would be catastrophic. Politically, it would reawaken Stalin's not altogether dormant suspicion that the invasion would never take place at all; and both Roosevelt and Churchill were anxious to avoid such a reaction. Militarily, the risks were worse. Could 170,000 men be brought back to England and returned to their camps and billets without being noticed by a German spy, or without starting a rumor? And what would the effect be on the men's morale? Most of them had already been briefed about the invasion beaches, which meant they would have to be held virtually incommunicado on their airfields, in their barracks, or on their ships. Given human nature, it was unlikely that nobody would manage to break out. Postponement until July, Ike thought, was simply "too bitter to contemplate," and probably impossible.

There was a long silence, broken only by the spatter of rain against the windows, the noise of the wind, and the ticking of a clock. Eisenhower sat at the baize-covered table, his hands clasped in front of him, while the others waited for him to speak. There was nothing more to be discussed; there was no new information on the weather to be had—it was Eisenhower's decision now, and even Montgomery, usually keen to press his own case, sat silently, looking at him.

Nearly five minutes passed, and then Eisenhower said, in a low voice,

"I am quite positive we must give the order. . . . I don't like it, but there it is. . . . I don't see how we can do anything else."[11]

Then he stood up and walked to the door. The order had been given. The invasion would take place on June 6. Nothing could stop it now.

When asked what kind of generals he liked, Napoleon is said to have replied, "Lucky ones."

Nobody in the room could have known it, but luck was about to strike Eisenhower.

Across the Channel, the Germans were suffering, once again, from the difficulties of the Luftwaffe. Unlike the British and the Americans, the Germans had no aircraft that could safely patrol the Atlantic far out beyond Iceland to report regularly on the weather. As a result, the high-pressure front that Group Captain Stagg had reported to Eisenhower was unknown to them. Their meteorological forecast was for the bad weather to continue uninterrupted right through June 5 and 6 and beyond—poor visibility, low cloud cover, rain, high winds, a heavy swell at sea, and strong surf on the French beaches.

The weather forecast was terrible for the foreseeable future, and Field Marshal von Rundstedt concluded that nobody in his right mind would order an invasion as long as it remained so. That being the case, he set off with his immediate staff for a four-day inspection tour, and allowed Field Marshal Rommel to take a few days of leave, as Rommel had requested. Rommel wished to spend his wife's birthday, June 6, at home with her in Germany, and he left early in the morning on June 4 in his Horch touring car, sitting in front, with his chauffeur, Daniel, at the wheel and two of his staff officers seated behind him. His birthday present for his beloved Lucie was beside him on the seat: a pair of elegant suede shoes he had bought in Paris—for even in the fourth year of the German occupation Paris remained the center of women's fashion and haute couture, the place where everybody who could afford to do so still came to buy shoes, gloves, perfume, hats, and lingerie. Even the keen commander of the Twenty-First Panzer Division, General Feuchtinger, a fanatic Nazi and a particular favorite of Hitler's (and therefore, of the German generals in France, one of those most disliked by his colleagues) decided that in view of the weather it was safe to spend the night of June 5 and 6 in Paris, with his mistress.

As for the Führer, he too was taking a vacation, at his home in Berchtesgaden, high above the Obersalzburg, accompanied by his staff and his mistress, Eva Braun.

The bad weather that was giving Eisenhower so much anxiety had lulled his opponents into a fatal false sense of security. As the decisive moment of World War II approached, nobody on the German side was minding the store.

Eisenhower's tiny bedroom in his trailer was littered with ashtrays and with paperback western novels—virtually his only form of relaxation. Having made his decision, he left his commanders to get on with implementing it and went back to the trailer for a few hours of sleep. He was not the kind of man to waste time second-guessing once a decision was made. He knew that at Southwick House the tension through the night would be electric, as orders were passed out setting the whole huge operation in motion again. He had no need to be there, and no wish to be part of the noise and drama.

All over Britain that night, the decision that Eisenhower had made at nine-forty-five, according to the clock in the library at Southwick House, had triggered a frenzy of activity. RAF bombers were already revving up their engines to bomb vital targets in France by night, while all across southern England each one of the thousands of transport aircraft had to be given last-minute checks, as well as the gliders aligned in long rows, just so, with their long towlines neatly arranged in a zigzag pattern in front of them; everywhere ground crews worked through the night, loading bombs, rockets, and long belts of ammunition into combat aircraft for the morning of June 6—the belts of .50-caliber ammunition for the heavy machine guns of the American bombers were twenty-seven feet long, whence the expression "the whole nine yards." Troops moved south toward the embarkation ports, by rail and truck, following a complicated schedule that would take them and their equipment in phases to the beaches once these had been secured on D-day. The British midget submarines that had been in place off the beaches, resting on the bottom since June 4, when the invasion had been postponed for twenty-four hours—their crews now suffering from cramps, exhaustion, and lack of air inside the tiny, crowded hulls—were given, at last, the order to surface before dawn on June 6 and raise a telescopic mast with a signal light to mark the boundaries of each of the five invasion beaches for the ships to come.

Out at sea, an intricate seaborne ballet would soon be taking place, for by the night of June 5 every one of the 7,000 ships had to be in its exact position in the five separate fleets as they steamed toward the beaches. First, well ahead of the rest, would come the minesweepers, to clear channels to the beaches; then the ships carrying the infantry; then the bigger LSTs bearing the tanks and the vehicles; and behind them the big transports, the hospital ships, the communications ships, and the antiaircraft ships. On either side of the five separate invasion fleets swift destroyers would patrol against any possible attack by E-boats or submarines. Farther out still would come the big gray warships, battleships, cruisers, and monitors, to provide the sustained bombardment of the German gun positions commanding the beaches. In France itself, resistance groups would go about sabotaging key points of the transportation system and get ready to put down markers for the parachute and glider landings to the east and west of the invasion beaches. Eisenhower had set all this in motion, but there was nothing further he could do about it except hope that Group Captain Stagg's weather forecast was correct.

Eisenhower woke early in the morning, to the familiar sound of rain and wind. He had slept soundly, but apparently the thought that the invasion might fail was on his mind, for the moment he was dressed, he sat down at his little desk in the trailer and wrote out a kind of brief aide-mémoire of what he would tell the news correspondents if it did. This is one of the most extraordinary documents of the war, dignified, modest, and truthful. Eisenhower takes on himself full responsibility for the failure—it is impossible to imagine Montgomery, Patton, or MacArthur writing a similar note:

> Our landings in the Cherbourg-Havre area have failed to gain a satisfactory foothold and I have withdrawn the troops. My decision

> to attack at this time and place was based on the best information available. The troops, the air, and the Navy did all that bravery and devotion to duty could do. If any blame or fault attaches to the attempt it is mine alone.[12]

He folded the piece of paper carefully, put it in one of the breast pockets of his short uniform jacket (already known as an Eisenhower jacket), and buttoned the flap. It was there if it turned out to be needed—he would not be at a loss for what to say if the worst happened. Then he went off to a final, last-minute briefing, at which Group Captain Stagg, as is so often the case before any open-air event in England, whether it is Wimbledon, the Trooping of the Colors, Ascot, or an invasion, gave yet another cautiously optimistic weather forecast while rain beat on the windows and dark clouds scudded by in the sky. It was pro forma—the decision had been made. Eisenhower listened politely—he was never impatient with people doing their job—but without any sign of interest.

He stayed most of the day at his trailer, receiving a steady stream of what were beginning to be called VIP visitors—among them Churchill, De Gaulle, and Field Marshal Jan Smuts of South Africa, who had played the role of wise man to the British War Cabinet during two world wars—and an even steadier stream of urgent messages. More important than the visitors, even more important than the messages, since at this point there was nothing more he could do, was the very gradual improvement in the weather. Just as Stagg had predicted, the rain slowly diminished during the day, the cloud cover began to lighten, and the wind dropped. Nobody could say that it was good weather, even for England, but it was certainly better weather.

In the late afternoon on June 5, Eisenhower decided that he would visit the airborne troops, who would soon be loading up. He could not visit them all, but he would see as many as he could. It was about them that Eisenhower had the sharpest concern. Nobody had ever attempted a night drop on this scale before, and Leigh-Mallory had predicted that casualties might be as high as 80 percent. From the beginning, Eisenhower had insisted on the airborne drops, despite the possibility of catastrophic losses, since these drops were the only way to secure both flanks of the invasion during the first few crucial hours, when the troops on the beach would be at their most vulnerable, trying to fight their way through the German defenses inland. Eisenhower had faced the problem squarely: if the three airborne divisions—one British, two American—had to be sacrificed to make the invasion possible, so be it. He hated the idea; he hoped that Leigh-Mallory was wrong; but the airborne drops were indispensable to success. On the other hand, he felt the need to go and see the men whose lives he might be sacrificing, and look them in the eyes; and he felt, too, perhaps, that the sight of him would reassure them.

Shortly before six o'clock, he stepped into his olive-drab Packard, with Kay Summersby, who by now had become his driver, hostess, social secretary, and confidante, at the wheel. He told her not to fly the flag from the hood of the car and to cover up the plates bearing his four stars—he wanted no pomp or ceremony. As they set off, the weather began to clear rapidly, the sky turning a glorious red as the sun began to set. "Red sky at night, sailor's delight,"[13] Kay Summersby recited to herself as she watched it, and felt a stirring of optimism.

It was nearly fifty miles from Portsmouth to Newbury, where the U.S. 101st Airborne was spread out at several airfields, some of the men getting ready to board the waiting C-47s, others to board the gliders that would be towed by the transport aircraft. Greenham Common was the first airfield Eisenhower visited, and thirty-one years later Kay Summersby would remember the troops cheering, whistling, and shouting, "Good old Ike!" as the supreme commander appeared.

"Ike," she remembered, "got out and just started walking among the men. When they realized who it was, the word went from group to group like the wind blowing across a meadow, and then everyone went crazy—the roar was unbelievable. . . . There they were, these young paratroopers in their bulky combat kits with their faces blackened so that they would be invisible in the dark of the French midnight. Anything that could not be carried in their pockets was strapped on their backs or to their arms and legs. Many of them had packages of cigarettes strapped to their thighs. They looked so young and brave. I stood by the car and watched as the

General walked among them. . . . He went from group to group and shook hands with as many men as he could. He spoke a few words to every man as he shook his hand, and he looked the man in the eye as he wished him success. 'It's very hard really to look a soldier in the eye,' he told me later, 'when you fear that you are sending him to his death.' "[14]

From Greenham Common, Kay Summersby drove Eisenhower to three more of the 101st Airborne's airfields, but then there was no time left to continue, so she drove him back to Greenham Common to watch the aircraft there take off. It was growing dark now, but an American correspondent saw tears running down Eisenhower's cheeks as he watched, one after another, the C-47s roll down the runway and vanish into the night.

On the way back to Southwick in the car, he said to Kay Summersby, "I hope to God I know what I'm doing."[15]

CHAPTER 3

"What a Man . . . Did as a Boy"

It was Ulysses S. Grant who commented, on the subject of biography, "What I want to know is what a man did as a boy." This is a pretty universal desire—even when we don't know much about the childhood of a major historical figure, a whole cottage industry usually springs up for the purpose of inventing edifying or revealing anecdotes to explain the actions of the grown-up person. This is particularly true of military heroes, even in Egyptian, Greek, and biblical times. Many are the tales still told of the young Napoleon Bonaparte organizing his childhood friends into platoons to take a play fort or fight a mock battle, but they were mostly the inventions of paid propagandists once he became emperor. It has always been so.

Thus, in the absence of touching and morally instructive little stories about George Washington's childhood once he was a famous man, the indefatigable Parson Weems simply invented them, like the one about young George chopping down the cherry tree ("Father, I cannot tell a lie."), and did it so well that they became an indelible part of Washington's life for generations of schoolchildren.

More serious modern biographers are not exempt from this belief that the child must be father to the man. English-language biographers of Hitler have sought diligently for anecdotes about his childhood that would either explain his behavior as an adult politician (the child as a victim of abuse seeking his revenge) or serve as warning signals of what was to come when he grew up (the child as a demonic presence). Much has been made of the fact that Hitler's beloved mother died painfully of breast cancer while under the care of a Jewish doctor, or that the young Hitler was sternly disciplined and physically punished by his domineering father. Yet given the state of nineteenth-century medicine, many mothers died in agony under medical care that could offer them little in the way of hope or relief; and overbearing, authoritarian father figures who imposed corporal punishment on their sons were so common in the Austro-Hungarian Empire that they were more the rule than the exception. Countless Austrian children had a childhood similar to Hitler's without being transformed thereby into monsters, while as for anti-Semitism, it was so widespread among Austrians that it was generally held to be a perfectly respectable opinion, on the left as well as the right—in fact, it cut across party, social, and class lines and was the one thing, apart from hatred of the Slavs, that most Austrians of German descent had in common.

In Eisenhower's case, like that of George Washington, there wasn't a lot to go on, but once he became a presidential candidate a need was felt to provide him with a sunny childhood full of stories that would suggest his future rise to five-star general and national hero. Republican public relations men, the mid-twentieth-century equivalent of Parson Weems with his drum going from village green to village green to tell stories about Washington as a boy, worked overtime to produce a portrait of an idyllic childhood in the rural American Midwest, with the young Ike as the "barefoot boy with cheek of tan," a kind of Kansas Tom Sawyer, and the *Reader's Digest*, *Colliers*, the *Saturday Evening Post*, *Look*, and *Life* sold it enthusiastically to the American public.

Eisenhower did not collaborate in this effort, or perhaps even notice it, but the fact remains that he was the last American presidential candidate whose boyhood story was part of his appeal to voters. When his turn came, Nixon's boyhood seemed to be too full of painful disappointments and embarrassments to make for good election material; and Kennedy's was all too obviously that of a wealthy and overprivileged young man with an ambitious, unscrupulous nouveau riche father, and a number of family skeletons in the closet that were better left locked up there—nobody in the Kennedy family or among those who advised them on such subjects supposed that Jack Kennedy's childhood would make edifying reading for schoolchildren. Eisenhower's childhood, by contrast, might have been written for the purpose.

Left to one side, perhaps because it was simply too hard to explain, was the fact that Eisenhower came from solid Mennonite stock. Kay Summersby was in the habit of attributing Eisenhower's "stubborn streak"—an unmistakable feature of his character—to his "Swiss-German ancestors," and she was not entirely wrong. During World War II, many other people were inclined to attribute Eisenhower's success as a general to his German ancestry—including, understandably enough, the senior German generals themselves, who found it consoling to imagine that they had been beaten by a fellow German—but in fact there was nothing in Eisenhower's ancestry that would have encouraged or in principle even permitted him to follow a military career. The Mennonites were stubborn, all right—they had to be stubborn to survive, in both the Old World and the New—but the most important part of their belief was their uncompromising pacifism. Followers of Menno Simons, a charismatic Swiss preacher, Mennonites were taught to take the Bible literally and as the ultimate authority on everything, and above all to abjure war and killing.

In the Holy Roman Empire (as Austria and Germany were then called) of the sixteenth century these beliefs led to their persecution by both church and state, and many of them took refuge in Switzerland. Mennonites were stiff-necked, sturdy, hardworking, thrifty, and for the most part good farmers, part of the burgeoning world of extreme Protestant sects that flourished in the German-speaking states of Europe after the Reformation, to the horror of both Lutherans and Catholics, and that included the Amish, the Anabaptists, the Dunkers, Seventh-Day Adventists, and many more.

Cruelly persecuted during and after the Thirty Years' War, these sects survived by sheer stubbornness and blind faith, until the colonization of North America, which had already attracted the English Puritans, gave them a place to flee to. When William Penn managed to secure royal assent to a Quaker colony in America—the Quakers were yet another sect which church and state in Great Britain were only too happy to see the back of—Penn set off on a tour of Europe to recruit settlers from other religious sects. The Mennonites, like the Amish, were among the first to respond to his summons to the "new Jerusalem" across the Atlantic, and to the promise of cheap land in good farming country. Penn was eager to attract farmers, however unorthodox their beliefs and whatever language they spoke, since the Quakers were mostly merchants and craftsmen, and the Pennsylvania colony seemed likely to starve to death without somebody to produce its food.

To this day, anyone who travels through the Pennsylvania countryside can hardly fail to notice the incredible proliferation of religious sects, most of them going back to the early eighteenth century. Of these, the Amish are now the most famous, in their new role as a tourist attraction. In Penn's day, the Mennonites were of greater numbers and importance, but over the years they were absorbed more easily into the community and became correspondingly less visible, since their interpretation of the Bible, strict as it was, did not preclude the use of modern inventions. Like the Amish, however, they clung to their formal clothing—black coats and wide-brimmed black hats for the men, high-necked black dresses and cloaks, worn with an apron and a starched white bonnet (known as a "prayer cap"), for the women. In the New World, they remained for the most part German-speaking farmers—very prosperous ones—and, like the Amish and most of the other Pennsylvanian religious sects, always deeply committed pacifists.

The first Eisenhauer (as the name would remain spelled for some time) arrived in Pennsylvania in 1741, and settled near Harrisburg. The Eisenhauers prospered modestly from generation to generation, eventually joining the

River Brethren, a more extreme wing of the Mennonite sect, which had broken away from the main group over a number of thorny theocratic disputes. Jacob Eisenhauer, Eisenhower's grandfather, who, when obliged to speak English, still had a pronounced Pennsylvania Dutch accent, was strongly opposed to slavery and an admirer of President Lincoln, but as a pacifist did not fight in the Civil War.

Although Jacob was a prosperous farmer—his farm in Pennsylvania was "appraised at $13,000, and his personal property, including savings, at $6,000," large amounts at the time—he was persuaded, like many members of the River Brethren, to move west to Kansas after the Civil War, partly because of the opportunity offered to settlers of large tracts of cheap land; partly because religious sects were becoming a minority in an increasingly secularized Pennsylvania; and partly, no doubt, because many of his neighbors resented the fact that most Mennonites had not fought in the Union Army. A few of the River Brethren had already scouted out promising farmland in Kansas in the 1870s, and had begun to settle near the town of Abilene.

This was a curious but fortunate choice. Abilene had briefly been a wide-open cattle town, when it was first founded as a railhead. Cattle were driven up from Texas for shipment to Chicago and the East; the cowboys who herded them were paid off; and a slew of hastily built hotels, saloons, gambling houses, and brothels lined its unpaved, muddy streets to service the cowboys' needs. Murders and hangings were frequent; famous lawmen and notorious outlaws settled there in its heyday;* then, as was typical, the railway was extended farther south and west, and Abilene ceased to be a boomtown and a den of vice and settled into a respectable existence as a prairie farm town with more churches than saloons. A population of pious, serious, hardworking, God-fearing, German-speaking farmers, bearded like Old Testament prophets, was just what Abilene needed in the late 1870s, when the cowboys, whores, gamblers, and gunmen (and the fast money that went with them) had moved on. The River Brethren soon began

* Among them Wyatt Earp, Billy the Kid, and Wild Bill Hickok, as well as such western figures as Calamity Jane. It would be some years before Abilene could fairly be described, as it was in the old cowboy song, as "Abilene, Abilene, purtiest town I've ever seen."

arriving by train from Pennsylvania to stake out their claims to the land, and among them were Dwight D. Eisenhower's paternal grandfather Jacob Eisenhauer, who arrived with his family in April 1878, and lived for some time in the communal Emigrant House until he had purchased land and put up a house of his own.

As was always the case with religious subsects, a certain number of Mennonites formed splinter groups as a result of doctrinal differences; and after a time these broke off and moved farther west, eventually forming farming communities in West Texas that still survive and prosper. But enough Mennonites stayed near Abilene to give the town and the surrounding countryside a whole new character, and to bring widespread prosperity to the region.

Like his biblical namesake, Jacob prospered too, just as he had in Pennsylvania. If Kansas was not quite the promised land, it came close. Jacob was an elder of his church and preached regularly (in German); he farmed, bought more land and real estate, helped found a successful creamery, and eventually founded a bank; and, again like Jacob in the Old Testament, had a wife who bore him many sons—seven, of whom six lived. Tall, bearded (among the River Brethren men wore full beards, but shaved their upper lip), muscular and hardworking, Jacob was at once a patriarch and a pioneer, an uncompromising leader of his religious sect who was at the same time a shrewd, worldly businessman and dealt easily (and with mutual respect) with his non-Mennonite neighbors. Like all Mennonites he neither smoked nor drank, but there was nothing of the zealot about him, and he seems to have had a great capacity for enjoying life, as well as making the best of it.

He was wealthy enough to give each of his children $2,000 and 160 acres of farmland on the child's marriage—a generous and substantial gift in the late nineteenth century—and conservative enough to expect that the children (or the daughters' husbands) would farm the land they had been given. In this regard, his eldest son, David, would disappoint him.

Opinions evidently differ on the subject of David, but it seems clear enough that while he shared his father's stubbornness—it was a family trait—he had ambitions that went beyond the relatively small world of

Mennonite families in Kansas. He was determined to get a college education, and more interested in engineering than in farming. Unusually, for the day, he taught himself Greek, in order to be able to read the Bible in Greek, no small accomplishment, and somehow persuaded his father, Jacob, to let him attend Lane University, a small college founded by the United Brethren in Christ, an evangelical group not in any way associated with the Mennonites. Religious groups and sects, most of them of Baptist or Methodist origin, were proliferating so rapidly in Kansas in the decades after the Civil War that its religious life already seemed more complicated than Pennsylvania's; in any case, the different sects had arrived in Pennsylvania from Europe fully formed, whereas the fertile soil of Kansas seemed to be producing a whole new range of indigenous sects.

The fact that David attended a college run by an evangelical group is perhaps a first indication of his determination to break out of the strict Mennonite mold. Lane was not—nor did it pretend to be—Yale or Harvard, but it offered an opportunity for what we would now call a liberal education, as well as vocational training[1] in mechanics and engineering; and it was, unlike the major colleges and universities of the East, coeducational from its inception, something of an innovation in the nineteenth century.

There, he met a pretty, bright, determined young woman named Ida Elizabeth Stover, who had entered Lane to study music and the humanities—something of an accomplishment, since she too came from a fundamentalist religious background in which a college education for a young woman was not encouraged. The Stovers, though not Mennonites, had made the journey from Europe to escape religious persecution and warfare, first to Virginia and then, eventually, to Kansas, where Ida's brother was a preacher. Ida, whose firm will and independent spirit were apparent to everyone who met her, taught school, saved her money, and with her savings and a small inheritance was able to purchase a piano, which would remain the most prized possession of her life, and attend Lane University—a daring step at the time for a young woman from a deeply religious background. Ida was not only a religious pacifist; because she had grown up in Virginia after the Civil War, her opposition to war and killing, was, if anything, even stronger than David's. The notion that one of her sons would become a world-famous general would have surprised and dismayed Ida, and indeed she had mixed feelings when she lived to see it happen.

Despite obvious differences in temperament—David was stubborn, quick to take offense, and, so far as one can judge, humorless, while Ida was attractive, outgoing, and popular—they were determined to get married, and neither religious nor personality differences were allowed to stand in their way. They were married by a River Brethren preacher; if nothing else, this is a sign of how much Ida was prepared to give up for the marriage, for she had to adopt the garb for married women of the River Brethren, a black floor-length dress buttoned severely up to the chin, a black cloak, an apron, and a white bonnet that covered the hair and all but concealed the face. Ida would soon become the first River Brethren woman to give up the bonnet, but her clothes remained modest, old-fashioned, and severe to the end of her long life.[2]

How intense their religious belief was is hard to determine. David eventually drifted away from the Mennonite Church and the River Brethren; and Ida joined the Bible Student Movement, which later came to be known as Jehovah's Witnesses. Religious faith was a strong part of her life, but was less strong for him, one guesses. He was happy enough to leave the Mennonite Church—and Mennonite ways—behind, while she had probably never wanted to become a part of this church in the first place, but sensibly understood that there was no other way of securing Jacob's blessing on the marriage.

A bigger stumbling block was the fact that David had made his mind up not to farm. This decision must have caused Jacob considerable pain and misgivings, since the Mennonites had a quasi-religious belief in farming and Jacob had been a farmer all his life; but in the end he agreed to mortgage the farmland he had intended for David and give David the proceeds in order to open a store—what was referred to at the time as "grocery and mercantile" and what we would refer to as a general store—in the nearby (and, as it turned out, infelicitously named) town of Hope, where Jacob owned several properties.

David went into business with a partner, Milton Good (the store was called Good and Eisenhower—the German spelling of the family name had long since been dropped). Good was an experienced salesman and a jovial fellow; this was probably just as well, since David had no experience at keeping a shop, and was not by any stretch of the imagination an outgoing personality with a gift for selling. Whether the partnership was David's idea or a condition of his father's is unknown, but in any case it was not a success. Although David would pass on to his sons the legend that Good had absconded with the cash from the till and put the store into bankruptcy, the truth seems to have been that the two partners were simply incompatible, that Good got fed up and walked out on the partnership, and that David was unable to run the store alone, or even with the help of one of his brothers, a self-taught part-time veterinarian and preacher of the gospel. No doubt it was easier for David to live with the idea that he had been cheated than that he had simply failed. David's sons would always believe that their father had worked hard for years to repay his debts, but this also seems a case of gilding the lily—the only significant debt was to David's father, Jacob, who eventually forgave the prodigal son and wrote it off.

Hope was still a small town, so David's failure as a merchant did not go unnoticed. He apparently decided not to stay there and accept his humiliation, and set off alone for Texas. Reading between the lines it is possible to guess that David was obliged to descend from what we would now call the white-collar class, as a local merchant and the son of a prominent and well-to-do farmer and banker, into the ranks of blue-collar workers, though these labels did not exist at the end of the nineteenth century; and that he did not want anybody in Hope to see the transformation. Though people liked to boast that Kansas at the turn of the twentieth century was basically classless, unlike the older towns and cities of the East, what they meant was that there was no distinctive upper class living on inherited wealth. In fact, there was still a wide gulf between those who performed manual labor at a day rate or an hourly rate and those who owned their own business. It was indeed more easily possible to rise quickly on the social scale in the West than in the East, and many did; but falling on the social scale was as painful in Hope as anywhere else, and must have seemed—worse than that—shameful to David Eisenhower.

He went south to Denison, Texas, then a railroad town, and found work there as an "engine wiper," leaving his young wife and their first child in Hope with one of his brothers. Even given David's interest in engineering and things mechanical, being an engine wiper was about as menial a job as you could get and still be working around machinery—it was not just a blue-collar job but an oil-and-dirt-under-the-fingernails job, oiling bearings, filling grease nipples, and wiping down moving parts with a greasy rag; the working day was long, wages were low, and working conditions were miserable. David lived in a rooming house, in itself a difficult comedown for a man who had lived in his own apartment above the store in Hope. It would be more than a year before Ida could join him, and over four years before he and his family would return to Kansas from this impoverished and dreary exile to which David had condemned himself.

Not only had he left Ida behind with a baby, Arthur, she was pregnant with their second son, Edgar. David managed to rent a small, shabby, soot-stained house just across from the railroad yards where he worked—it has been described as "on the wrong side of the tracks," but in Denison, where the railroad was the whole reason for the town, there was in fact no "right" side of the tracks. There she gave birth to Edgar, and, in October 1890, to David Dwight Eisenhower,* the future general and president.

A house that was little better than a shack right next to the railroad tracks in Denison, Texas, shaken and dusted with soot every time a train rumbled and squealed past, was no place to bring up a family. Also, the Eisenhowers were dirt-poor. They had a boarder, but even so, it must have been hard, with two small children, to make ends meet on David's wages; and although Ida's religion urged pious resignation, it seems unlikely that

* Carlo D'Este, whose giant, masterful biography *Eisenhower*: *A Soldier's Life* is a monument of scholarship, like his equally thorough *Patton*: *A Genius for War*, points out that Ida not only reversed her third son's first and middle names, but did the same to her own. Born Elizabeth Ida Stover, she renamed herself Ida Elizabeth. For reasons best known to herself, she disliked the idea of her son being called Dave; or perhaps, angry at her husband's business failure and at being dragged to Denison, Texas, to live in poverty, she simply did not want a son named after him. She insisted on calling the infant Dwight David Eisenhower (Dwight was the first name of a much-admired preacher of the day), and gradually the whole family fell in with her wishes. Nineteenth-century mothers had a will of their own in these matters. The author's uncle, the film producer Sir Alexander Korda, was named by his father Leslie Alexander Korda (in Hungarian Korda Lajos Sandor); but his mother, once she became aware of it, insisted on calling him Sandor Lajos instead, and made it stick. For the rest of his life he would be known as Alex.

she suffered her fate gladly, or entirely without complaint. Her husband's stubbornness held him in place, however; he had chosen Denison, and here he would stay, no matter how unpromising things looked. Eventually, though, his father, Jacob, with the perfect timing and the unexpected, omnipotent gesture of the Lord God in the Old Testament—one almost imagines him descending in a blaze of fire from a grubby coach of the Kansas, Missouri, and Texas Railway—journeyed to Denison to visit his oldest son, took pity on him and his family, and called them back to "the land of milk and honey" in Abilene, Kansas. There he found David a job working in the River Brethren's creamery that he had helped to establish when he first arrived from Pennsylvania.

Old Testament stories spring to one's mind automatically when it comes to Jacob Eisenhower, who had not only the appearance but the manner, the gravity, and the taste for grand gesture of the patriarchs, as well as the boundless interest in family and the determination that his seed would multiply and be fruitful, especially in sons. It is hard not to think of an earlier Jacob, and his son Joseph—he of the "coat of many colors"—for instance, even though their circumstances were different:

> And Joseph made ready his chariot, and went up to meet [Jacob] his father to Goshen, and presented himself unto him; and he fell on his neck, and wept on his neck a good while. And [Jacob] said unto Joseph, Now let me die, since I have seen thy face, because thou art yet alive.

David Eisenhower seems to have settled back into life in Kansas—and the Eisenhower family—quickly, and with relief. He was not well paid at Belle Springs Creamery, and the best he could do for Ida and the children was to rent a small house, but it was a big improvement over Denison, even if it too was on the wrong side of the tracks, on the south side of Abilene, instead of the north where the big Victorian mansions were. For two adults and what would soon be six small children, it was crowded, but family life did not necessarily include privacy in those days. More important to David, one suspects, was the fact that although he was merely a mechanic or "engineer" at the creamery, he had, in effect, rejoined the middle class as a salaried man with a responsible job.

Photographs of him after his return from Texas show a fiercely mustachioed figure, with a snowy white shirt and stiff collar, a dark suit, and a watch chain across his vest, and even in old age he usually retained the dark, formal dress of an office worker or "manager," even when he was on his porch at home, as if he wanted to efface the memory of the manual laborer he had been during his years in Denison.

The dream of having his own business, of being his own boss, of life as a merchant, if it did not die, must have gradually faded, as David came to recognize that it was not to be; but his growing family and his reputation for honesty and responsibility may have been a substantial compensation for what other people regarded as his failure in the world of commerce. He must have felt that whatever else he made of his life, he had at least avoided becoming a farmer, in his father Jacob's image.

As for his David's son Dwight, he would have no childhood memories of Denison, and would always think of Abilene as his hometown. In later years Texans would often claim him as their own, particularly the citizens of Denison, and also of nearby Tyler, to which he had no connection at all, and he always acknowledged them with the broad grin that was his trademark, but he had no real feeling for the place of his father's humiliation.[3]

That is not to say that his father's experience there did not teach him a lesson, however. Dwight Eisenhower would grow up to have his father's stubborn streak and temper, as well as a degree of determination and courage all his own, but on the other hand he would never be one to rock the boat. Dwight would learn early on the value of team play; he would always work hard to fit in; he developed what was from the very beginning a remarkable natural ability to attract the attention of prominent older men—he was never tempted to do as his father had done and walk away from everything he was and had been brought up to be. Not only did he stick to things, but he was at his most comfortable within organizations and groups—at school, on sports teams, at West Point, in the Army. Getting along with people would always be his most easily recognizable talent. There would never be within him the temptation or the need to go off "into the wilderness," away from family, friends, and his structured career, in search of something else. Whatever else he may have learned from his father, Dwight would never be a loner as David had been.

To Dwight Eisenhower, Abilene would come to represent an idyllic place for the rest of his life. David and Ida Eisenhower were poor, but respectably poor, in a place and during an age when money didn't count for everything, and in a small town where a reputation for honesty, reliability, and decency still counted for a lot more. Moreover, the Eisenhower family name stood for something in Abilene; Jacob had made sure of that. There seem to have been no black sheep in the Eisenhower family: the men did not drink or run up debts, and their word was as good as their bond; the women were good mothers, and deeply religious, without being killjoys. David and Ida were stern but loving parents, who gave time and attention unstintingly to their sons, and were determined to provide the best possible education and every chance to rise in life.

The Abilene of Dwight Eisenhower's childhood was still a town of dirt streets; of horses (the automobile would soon bring huge changes, but inventors like Henry Ford and Gottlieb Daimler were still tinkering away at the idea in bicycle sheds on both sides of the Atlantic); of gaslight and kerosene lanterns, rather than electricity; of sidewalks made from lumber, which washed away during severe storms. Indoor plumbing was unknown—a boy's first serious chore was to bring bucketsful of water into the house from the well. The dirt roads led quickly out of town to farmland and beyond to the vast, gently rolling prairie, still unchanged from the days, not so long ago, when nothing had moved over it but the great herds of buffalo and the Indian tribes that lived off them.

There were still men in Abilene who remembered that earlier time, and on rare occasions Abilene's previous spirit as a cattle boomtown reemerged—years later, Dwight would remember seeing a pistol fight in the street outside his house when he was in his teens; with the keen eye for military detail of the future general, he observed that one of the participants fired a nickel-plated revolver.[4] But it was by and large a peaceable town that got along with a single policeman, as well as a dry town in a dry state.

Despite the Mennonite presence, there were plenty of men in their fifties or sixties in Abilene who had fought in the Civil War, and Dwight seems to have listened to their stories—in a more leisurely age, men had more time to sit on porches or in stores and talk—with particular attention, even when he was still very young. "The war," as it was always referred to, as if there had been no other, was not yet history; it was still as real and as familiar as if it had ended only yesterday, and nobody needed to be taught or reminded of its great events. Dwight's mother had been an infant in her cradle in 1862 when Stonewall Jackson's army marched past her parents' house to fight at Cross Keys and Port Republic, and nearly four when General Sheridan descended on the Shenandoah Valley destroying everything in his path, from crops and barns to houses and bridges—a scorched-earth policy in reverse that crippled the Confederacy in the last months of the Civil War. Much as Ida Eisenhower hated war, her stories of what she had heard about from grown-ups and seen with her own eyes in those years were at least as exciting to a small child as those of the men who had participated in it, and may therefore have had an effect on the young Dwight that was the opposite of what she intended.

David and Ida Eisenhower would eventually have seven sons, of whom one, Paul, died in infancy—his death was the event that would influence Ida's conversion to the Bible Study Movement. David sustained his growing family on what was, even for the time and place, a meager salary. He worked a twelve-hour day attending to the machinery of the creamery, but still managed to provide for all his boys a family life that they remembered with gratitude and nostalgia. Dwight Eisenhower would describe his father as "breadwinner, Supreme Court, and Lord High Executioner," while his mother was "tutor and manager of our household." He never recalled hearing "a cross word pass between them," and he felt that they "maintained a genuine partnership in raising their six sons,"[5] which does in fact seem to have been the case. Ida was a believer in self-discipline, reason, and persuasion, though not above the occasional "slap on the hand with a ruler or anything handy and lightweight," whereas David, who dealt on his return from work with any more serious infractions, "was never one for spoiling any child by sparing the rod." In a simpler age, corporal punishment was the norm, and none of the Eisenhower boys seems to have been the worse for it, or to have resented it. The boys recognized early on that their father had a quick temper (like his son Dwight, whose outbursts of temper in later life often took the unwary, lulled by the big grin, by surprise), but also that his anger was reserved for deliberate disobedience. Pacifist though David might be, he did not believe in turning the other cheek, at least so far as his sons were concerned. It was an age when boys were expected to get into trouble now and then, and to defend themselves in the occasional schoolyard fistfight—once, when young Dwight came running home from school pursued by a more "belligerent" boy, his father, taking in the situation from the porch, despite his well-established edict against fighting, simply said, "Chase that boy out of here," which Dwight, somewhat to his own surprise, promptly did.[6] Thus Dwight learned "that attack is the best form of defense," and that "a pounding from an opponent is not to be dreaded as much as constantly living in fear of another," or so he claimed, at any rate, when it came time to write his memoirs.

Of the brothers, Arthur, the eldest, was "studious and ambitious," and gave his parents little trouble. Edgar was a "natural athlete," who often roughhoused with Dwight, getting them both into trouble. Roy was a bit of loner. Earl and Milton, the youngest, were less robust (and troublesome) than Edgar and Dwight. Together, however, they must have been a handful in a tiny house.

Dwight's most often quoted family memory was of a journey to Topeka, where some of Ida's relatives lived, when he was five. Left to his own devices in a house full of strangers, Dwight set out to explore the farm, and found his way to the barnyard blocked by a large, aggressive goose. Anybody who has had any dealings with geese can vouch for their ill temper and willingness to charge—intimidating behavior to an adult, and surely terrifying to a five-year-old. Dwight kept on trying to enter the barnyard, tears in his eyes, and the goose kept blocking his way, wings flapping and beak pecking furiously at the small boy, until Dwight's uncle, Luther Stover, came to his rescue by presenting him with an old broom. Young Dwight wiped away his tears, advanced on the enemy, and smacked the goose as hard as he could with the broom handle, clearing the way into the barnyard, and learning, in the process, as he remembered the incident a lifetime later, "never to negotiate with an adversary except from a position of strength."[7]

In fact, however, the most important incident of his childhood was not his struggle with an angry goose, but his brother Earl's loss of an eye. Dwight had been whittling with a knife in the toolshed, and put the knife down on a windowsill, out of Earl's reach. While Dwight's attention was diverted, Earl, who was then three, managed to climb up on a chair and try to grasp the knife, which slipped from his hand and fell, piercing the child's eye. In this case, there was no draconian punishment—both parents seem to have felt that "accidents will happen" and that this was lesson enough. But Dwight could not help feeling that he had been left in charge of a younger brother and could have—ought to have—prevented the accident. As late as 1966, Dwight Eisenhower was still blaming himself for what had happened, and wrestling with "my feeling of regret . . . heightened by a sense of guilt."[8] In old age he was still alarmed when his grandchildren played with anything that might cause an eye injury. The accident in the toolshed surely helped to form one of his most salient characteristics, far more obvious (and productive) than his inherited quick temper—his extreme sense of responsibility. In later life, Eisenhower might be impulsive sometimes, and always quick to blow off steam, but he never shirked responsibility, or failed to take responsibilities seriously, however burdensome and unwelcome they might be. In his own unsparing view of the matter, he had failed to act responsibly once, with serious and irreparable consequences for Earl, and he was determined that it would never happen again.

It is impossible not to sense the closeness of the Eisenhower family, their strong sense of "it's us against them." The boys might fight together, but they stuck up for one another, and even more important, stuck together as much as the difference in their ages would permit. On the rare occasion when Ida gave them a piece of pie or cake between meals, one boy got to cut the treat up and the others got to choose which piece they wanted before him, thus enforcing a kind of fairness that seems to have been a family characteristic. It is quite clear that Ida's strong sense of fairness, her instinctive and deeply felt moral values, her determination to do the right thing, were passed on to her boys, more by example than by a slap with a ruler or a hairbrush. They were, even by the standards of Abilene at the turn of the twentieth century, poor, and they were taught early on the importance of thrift. "The Indian on our penny would have screamed if we had held it tighter," Eisenhower commented, looking back to the days when eggs were five cents a dozen, bread was three cents a loaf, and a penny would buy a boy a fistful of candy. They may have been cramped for space in the tiny rented house that was the best David could do on his salary, with a backyard hardly big enough "to swing a cat in, if it were a small one," but they were a family others might envy for their closeness.

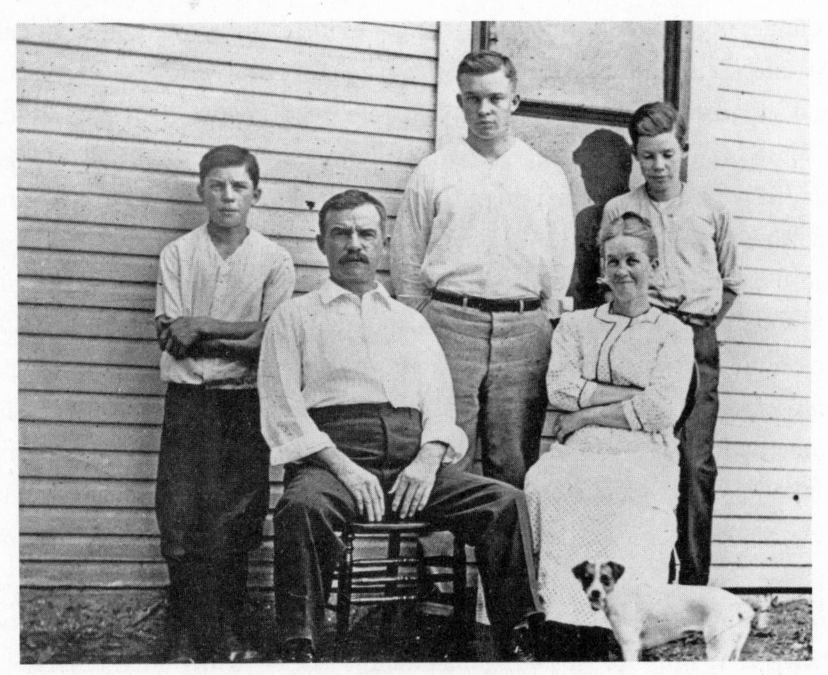

David and Ida, Abilene.

Not everyone in the Eisenhower family was poor, of course. Only two streets away, clearly visible from David and Ida's own house, was the much larger home of David's brother Abraham, on a large plot of land with a good-size barn. Abraham Lincoln Eisenhower (his name was another mark of Jacob's great admiration for President Lincoln) was a successful veterinarian, who needed space for the animals under his care. Abe, as he was known by everybody, had preferred horses to farming even as a boy, and eventually gave up farming altogether to become a full-time vet, although so far as anybody can tell without bothering to take any formal training in the profession, let alone a degree. Still, people back in those days were less concerned with degrees than we are now, and probably didn't care much whether or not Abe had the right to put the initials DVS after his name, provided he cured their livestock for them. Doctors, veterinarians, and dentists were scarce in plains towns like Abilene, and a man who put up his shingle with the right initials after his name, and wore a vest with a watch chain across it, was likely to make a decent living in the days before people expected miracles from medicine, whether human or animal.

In any event, Abe Eisenhower, unlike his brother David, had boundless self-confidence and a talent for energetic self-promotion worthy of the hero of *The Music Man*. "All the cases that have come under his care have been fully restored to perfect soundness," he claimed for himself in a local newspaper, and while that sounds a little on the optimistic side, he must have had either a pretty good record of success or the kind of bravado which makes people forget about the animals that died or continued to go lame under his care. For a time he hung his shingle on his brother David's store, so long as it was in business; but like most vets of that day, he was peripatetic—he went from farm to farm in his gig, with a leather satchel full of remedies, most of which were familiar in Roman times. Abe seems to have been the maverick of the Eisenhower family, who combined charm, chutzpah, and the Midas touch with a sizable dose of snake oil.

About the time his nephew Dwight was eight years old, Abe decided to give up veterinary medicine and move west as an itinerant preacher of the gospel. He may have decided that there was more money to be made from saving souls than from saving colicky horses—he seems to have lived more in the spirit of P. T. Barnum than of his father, Jacob Eisenhower—but he brought to his new profession the same zeal and restless energy that he had brought to veterinary medicine. He and his wife made their way to

Cherokee Indian country in what was then still Oklahoma Territory, where his specially constructed, horse-drawn "gospel wagon" and his habit of standing on top of it and shouting "This way to heaven!" soon won him a solid reputation as a saver of souls. After a period of living in a sod-covered dugout, Abe and his wife, who were childless, would eventually found, in Thomas, Oklahoma, eighteen miles from the nearest railroad, an orphanage which grew and flourished.[9]

Before his departure for Oklahoma Territory, Abe sold his house to his brother David. For a family with six growing boys, it was still pretty cramped quarters, but at least it was a two-story house, with enough space upstairs to allow Ida to have a live-in "hired girl." There were two upstairs bedrooms, with a kind of windowed closet for the hired help, a young woman named Florence Sexton who was paid two dollars a week, plus room and board. Dwight Eisenhower, with the head for numbers that would make him such a formidable military logistician, calculated that the house had 818 square feet for a family of eight,[10] plus the hired girl, and reckoned that in later years many of his offices, as a five-star general and as president, were roomier. Still it was a step up, and—perhaps more important—the property was big enough for David to keep a horse, Old Dick, and to give each of his boys a little plot in which to raise vegetables. There was also a big vegetable garden for the family's needs (the produce to be eaten fresh or canned) as well as a flower garden of her own for Ida. The boys were encouraged to sell what they raised on their own little plots to the neighbors. Young Dwight specialized in corn and cucumbers; he sold the corn for twenty-five cents a dozen, and stoutly resisted, as he would do all through his life, any attempt to haggle him down; Edgar would remember with embarrassment, to the end of his life, having to hawk his vegetables to neighbors.

Even with young Florence Sexton putting in the twelve- to fourteen-hour day that was the lot of hired girls at the turn of the century, there was still plenty of work for the boys, especially since the new house had enough land to serve as a small farm. There was, of course, Old Dick to be looked after, and the buggy and harness that went with him to be cleaned, not to speak of chickens to feed, cows to milk, stalls to be mucked, and bales of hay and water buckets to carry. Water still had to be carried into the kitchen for washing clothes and for baths; the washing machine and the clothes wringer had to be turned by hand; firewood and kindling had to be chopped and split—there were probably very few minutes from sunrise to sundown when the Eisenhower boys, like most children in that day and age, were not usefully employed. When they were not in school, one of them would harness the horse to the buggy and take their father's lunch over to the creamery, with a beating guaranteed if it arrived late or cold.

Soon the house became even more crowded, when old Jacob Eisenhower and his wife decided to give up their own home and move in with David. Why Jacob decided on this move late in his life is unclear—Dwight, in later years, insisted, rather unconvincingly, that it was because his grandparents had lost their longtime housekeeper; but most biographers think it was more likely that Jacob had impoverished himself by his generous gifts to his children and was no longer able to run his farm himself. Whatever the reason, a small wing was built onto the house, and Jacob settled in. His grandsons could not help remarking that he and David still talked German together, as they had in the old days when David was a child, back in Pennsylvania, though none of the Eisenhower boys seem to have picked up any degree of fluency in the language.

Their schooling was, by today's standards, narrowly focused, with the emphasis on rote learning, but intense and serious. Abilene might then have been in some respects a rural backwater, but it had a school system which was, however primitively housed, designed to drum book learning into the heads of its children, and in which any child who wanted to learn could go far. Like most cow towns, Abilene had made a rapid transition from a gambling hell to a serious and God-fearing community. If the first wave of women to settle there had been whores, those in the second wave were schoolteachers. People in the West took their schools seriously, and if there was one thing they were agreed on, it was the need to give their children a better education than they themselves had received. As Dwight himself would put it in later life, "Abilene folk believed in education and its value." That belief was even stronger in David and Ida than in most—

both of them had attended college, Ida in an age when it was unusual for a young woman to do so, and both were fiercely determined that their children would get a college education too, and if possible go out and make a success of themselves in the wider world beyond Abilene.

That they succeeded is obvious. All the Eisenhower boys were encouraged to become good learners at home, pushed hard by both parents, who were, it seemed, naturally gifted teachers. From early childhood on the boys participated in daily Bible readings, and they had memorized much of the Bible by the time they were in school. Dwight, despite what would seem to have been a notable lack of conventional religious feeling, perhaps because his home was so permeated with religion during his childhood, would still be able to recite large chunks of the Bible from memory as an adult. All the brothers succeeded in life: Arthur, the eldest, would grow up to become a banker; Edgar, a wealthy and successful lawyer, by far the richest of the Eisenhower brothers; Dwight, of course, grew up to attend West Point and become the most famous of them; Roy became a pharmacist; Earl a small-town newspaperman; and Milton an educator and public servant whose fame and reputation were —eventually—eclipsed only by Dwight's. This was, in short, by any standards a family of achievers. Many years later, when Dwight Eisenhower returned from the war to visit his hometown, he had one of his famous outbursts of temper at the sight of a sign outside the house where they had all grown up marking it as General Eisenhower's birthplace. He wanted it changed at once to one that marked it as the birthplace of "the Eisenhower brothers," and so it was.[11] Just as Dwight never got over the guilt he felt at Earl's losing the sight in one eye, he never ceased to admire the success of Arthur, Edgar, and Milton. There was intense competition among the Eisenhower boys, but they also formed a lifelong mutual admiration society.

Much has been made of Dwight's "epic" fights in school, particularly after he moved up to high school in 1904, but some of these stories resemble Parson Weems's about the young George Washington, especially the one about young Dwight's two-hour fistfight with an older bully. What is certain is that while Dwight was afraid of bullies when he first entered school—and also possibly influenced by Ida's dislike of fighting and her warnings against it—thanks in large part to the support and encouragement of his older brothers (Edgar in particular was a notorious scrapper), he soon developed a reputation as a boy who could look after himself, and wasn't afraid of a bare-knuckle fight when it couldn't be avoided. And of course, boys being boys, it often couldn't.

His elementary school had been primitive—it had no indoor plumbing, and writing paper was considered a luxury—and his high school was not much better, since it was crammed into one floor of the ramshackle city hall, along with the jail, the volunteer fire department, and whatever other municipal services Abilene provided. Separation of church and state was not the charged subject it has now become, and one room of the school was appointed as a chapel, where the students attended daily services.

For boys brought up with no sisters, the presence of girls must have been intriguing and full of surprises. Almost seventy years later, Dwight would still be able to remember the names of the girls in his class, so they must have had a considerable effect on him. "The south side boasted more than our share of pretty girls—Gladys Harding, Ruby Norman, the four Curry sisters, and Winnie Williams." Both Gladys and Ruby seem to have had a serious interest in the young Dwight, but by his own admission he was something less than a ladies' man at this stage of his life, and still less of a dancer (something Kay Summersby would remark on almost four decades later). Then too, in the age before the invention of the motion picture and the automobile, there was not much opportunity for boys to be alone with girls; and the Eisenhower brothers had no money to spend on dating and did have a full schedule of backbreaking chores, in part intended, no doubt, to keep them out of trouble.

At a time when not much else was thought to be important, Dwight was good at mathematics and at writing a clear sentence—both excellent skills for a future general. Perhaps more significant, he developed early on an interest in history, which he indulged in at home, since little or no attention was paid to the subject at school. Ida had a remarkably solid little library of historical works, and Dwight spent so many hours reading her books that she put them under lock and key, since he was falling behind in his chores—an unforgivable sin in the Eisenhower household. In a rare moment of disobedience, Dwight managed to locate the key; and henceforth, when his mother was working in her garden or out shopping, he would unlock the closet and plunge back into the history books that fascinated him.

He was particularly interested in the history of the ancient world, and found that he had an extraordinary facility for remembering dates, names, and places, as well as a vivid curiosity about the great battles of the Greeks and the Romans. Not many boys in the Abilene school system would have known the name and date of the battle of Arbela, but Dwight did, and moreover knew how it was fought and why. "The battles of Marathon, Zama, Salamis and Cannae became as familiar to me as the games (and battles) I enjoyed with my brothers and friends in the school yard," he would write later; and when he was very young his hero was Hannibal, the great Carthaginian general, surely an unusual choice for a boy from Abilene.

Equally unusual was his astute reckoning that since all accounts of the Carthaginians' history which have survived were written by their enemies, and since everything we know about Hannibal was written by the Romans he had so often defeated, Hannibal's greatness must indeed have been remarkable. "For a great man to come down through history with his only biographers in the opposite camp is a considerable achievement,"[12] Dwight wrote, a thought that may have crossed his mind again when he read the memoirs and published diaries of the British generals after World War II. He also liked Hannibal, because Hannibal was the underdog in the struggle against Rome, and this reasoning also led him to an early admiration for Robert E. Lee, despite the strongly pro-Union sentiments of the Eisenhower family.

What his father, his mother, and his grandfather Jacob thought of this early interest in war is hard to know. It is hard to imagine that they would have been pleased, given their firm belief in pacifism; and Ida, whose books were, after all, responsible for this turn of events, seems to have been mystified by the fact that although young Dwight was willing to read about the politics and culture of the past, he always came back to the great commanders and the crucial battles, which he read again and again until he knew what there was to know by heart.

As an adult he had to hold this side of himself sharply in check, since very often instructors in military history at West Point and senior officers in both the U.S. Army and the British army had their dates and facts wrong when they talked about the great battles of the past. People tended to think of Ike—and many still do—as something of a country bumpkin, uneducated as compared with Winston Churchill, say; but in fact Dwight Eisenhower, like Churchill, was above all self-educated, a product of reading, rather than school learning. Churchill, like so many others, soon discovered that Eisenhower could more than hold his own on the subject of history, and had, moreover, not only strong opinions, but a truly amazing memory for facts. This side of himself, in later life, Dwight was at some pains to hide, once he realized that the one thing most people don't forgive easily is being corrected in public when they're wrong. Besides, it was part of Dwight Eisenhower's genius that he never wanted to appear "to know more than the other fellow," or embarrass anyone if it could be avoided. He kept his knowledge to himself, and so people were all the more astonished when they realized how much he knew. This was clearly not a side of himself which he revealed to Generals Bernard Montgomery and Alan Brooke; and that perhaps explains the tone of pained superiority with which they wrote about Eisenhower.

Ida's library must indeed have been remarkable. Dwight was soon reading deeply in the history of the Civil War and the Revolutionary War—he conceived a great and enduring admiration for George Washington, particularly "his stamina and patience in adversity . . . and then his indomitable courage, daring and capacity for self-sacrifice," qualities which Dwight himself would show when his turn came to command an army. Washington had human qualities that Dwight "frankly idolized"; but Hannibal, Caesar, and Scipio interested him just as much, and he came to realize that all the admirable things in human history—from the time of Plato and Aristotle to the Declaration of Independence—would have been meaningless, and come to nothing, had not soldiers been willing to die for them, and had not great military leaders risen to protect and preserve them.

So great was Dwight's love of history that in the 1909 yearbook of

Abilene High School (*The Helianthus*—Latin for "sunflower") his classmate Cecilia Curry, charged with predicting the future of each member of the class, described Dwight as "our best historian," and predicted that while his brother Edgar would become a three-term president of the United States, Dwight would become a "professor of history at Yale."[13]

There was, in fact, a side of Dwight Eisenhower that would have liked nothing better, but this is something most people writing about him have ignored, as have most of those who served with him. Dwight would be a master at concealing himself behind his chosen mask of tongue-tied, good-natured affability—something which he had already learned to do as a child, and which may have come naturally to a boy with stronger older brothers, and weaker younger ones.

Athletics were not then the central feature of high school life as they are now. In the first place, boys had chores to do, to which athletics came second; and in the second place, in a poor community there was not a lot of money to spend on games. Then too, by the time they reached high school, boys began to drop out, in a world where a twelve-year-old who was strong enough could be earning good money, and as a result there were far more girls than boys in school. Still, Dwight was a natural athlete and a fiercely competitive football and baseball player, as were his older brothers, and it would rankle him for the rest of his life that Edgar was voted the best football player of his class. In those days the students were expected to provide for sports. "Each participant provided his own baseball shoes and bats," as Dwight recalled, but in order to meet the larger needs of competition, he helped to organize the Abilene High School Athletic Association, to which every student was supposed to contribute twenty-five cents. Dwight would eventually become its president, and he wrote the association's constitution, seeking to ensure its permanence. By 1908 the new Abilene high school, a landmark of civic pride, had been completed, and the students and faculty had moved out of their cramped quarters in city hall. They missed a certain degree of excitement connected with their old quarters—the presence of the volunteer fire department had meant constant alarms and activity, and at one point an inmate had attempted to blow up the wall of the jail in an escape plot—but Dwight seemed satisfied with both the school's new building and his own position in it. He was an athletic star, popular, tall, and good-looking, and would make his first (and last) appearance onstage in a school production of *The Merchant of Venice.* He played Launcelot Gobbo, servant to Shylock, and won praise as "the best amateur humorous character seen on the Abilene stage in this generation and gave an impression that many professionals fail to reach."[14] (Connoisseurs of American military history may recall that another victorious general and two-term president, Ulysses S. Grant, made his unique Shakespearean debut onstage when he was a second lieutenant in Mexico, playing Desdemona in *Othello*, in a cast which included future Confederate general James B. Longstreet.)

Not only was Dwight good at athletics, but he had a taste for the outdoors; and while he was still at school he formed some friendships that Ida could hardly have approved of. One of them was with a man named Dudley, who lived across the street, and who had been a deputy back in the rip-roaring days when the gunfighter Wild Bill Hickok had briefly been the town marshal of Abilene. Dudley was, as one might imagine, a good shot with pistol, and often took young Dwight to the unromantically named Mud River, where he, the town marshal Henny Engle, and the Wells Fargo agent Mr. Gish did some informal target shooting, in which Dwight was sometimes allowed to join.

At the end of his life, Dwight could still remember, with a military eye for detail, that each man—for these were men who were used to going armed—carried his revolver a different way. "Gish wore his in a shoulder holster under his left arm. Henny Engle used a conventional holster on his right side. Mr. Dudley slipped his revolver inside his belt, the barrel pointing towards his left foot and the grip handy to his right hand. . . . They were all above average in marksmanship and at least two had personal experience in gunfights."[15]

Dwight himself, it seemed, had a natural gift for marksmanship, and became a formidable pistol shot. This would stand him in good stead at West Point and in the Army.

Even more exciting was Dwight's friendship with his "hero" Bob Davis, a tall, soft-spoken hunter, fisherman, and guide, another link to the wilder days when Abilene had still been on the frontier. Davis made his living netting catfish (illegally) and trapping muskrats and mink, and deeply impressed Dwight by his ability to bring down two ducks with two quick shots from his old double-barreled shotgun. He taught Dwight the arts of woodcraft—how to survive and find his way in the woods—and although Davis was illiterate and could not even sign his own name, he also taught the boy the even more important skill of playing poker. Davis had a natural ability to calculate percentages, and a gift for teaching. Camping out in the woods, he taught Dwight how to win at poker. "Often, he would pick up part of the pack and snap it across my fingers to underscore the classic lesson that in a two-handed game one does not draw to a four-card straight or a four-card flush against the man who has openers."[16] He also instructed Dwight in the rudiments of fly-fishing and how to cook over an open campfire—skills that Dwight kept up almost to the end of his life.

It comes as a surprise to learn that only late in Dwight's years in high school did he finally come to be known as Ike, to the great annoyance of his mother, who had gone to such trouble to change his first name from David to Dwight. Apparently, all the Eisenhower children were called Ike—so he became, briefly, "little Ike," in contrast to his brother Edgar, who was known as "big Ike." Ike seems to have been a convenient way of shortening the name Eisenhower, which was admittedly a bit of a mouthful, rather than a specific nickname for Dwight. That said, the older Eisenhower boys outgrew the name, or perhaps simply left it behind them when they went to college, whereas it stuck with Dwight for the rest of his life.

On graduation from high school Ike had supposed a little vaguely that he would follow his brother Edgar to the University of Michigan, but two things stood in his way. One was the simple question of money—determined as David Eisenhower might be for his sons to go to college, his means were limited—and the other was, to a degree, the kind of young man Ike had become. Unlike his brothers Edgar and Milton, he did not have a natural appetite for study, and at the same time his friendships with people like Mr. Dudley and Bob Davis, and his love of the outdoors life, must have made it doubtful that life on a college campus would necessarily satisfy him. He was more active and physically restless than his brothers, somebody who would not, on the face of things, be easily confined in a library or an office.

While he waited for his chance to join his brother at the University of Michigan, he plunged into work—first on a nearby farm, then for a company that made galvanized steel grain bins, and finally in the creamery where his father worked, manhandling 300 blocks of ice a day, and then, by way of contrast, working in the creamery's furnace room. It was rough, hard physical labor, but Ike seems to have enjoyed it, perhaps because it kept his mind off the burning question of what his next move might be.

It seemed doubtful, with one boy already in college, and two more approaching college age, that David Eisenhower could do much to pay for Ike's further education, and no matter how many hours Ike himself worked, he was unlikely to accumulate enough money to pay for college. At the same time, like all the Eisenhower boys, Ike was determined to get out of Abilene. Fond as he would be of his hometown in later life, he recognized that it was a backwater for a young man with ambition. However much he admired his father—and there was no question that Ike did—there was no getting around the fact that David Eisenhower, despite a college education and an aptitude for engineering, was stuck in a low-paying, menial job, and that anybody who stayed on in Abilene would be more than likely to end up in a similar position. There were no great opportunities in Abilene—you farmed, or you ran a shop, or you took a job dealing with machinery of one kind or another. Apart from a thin upper crust at the top—bankers, professional men (lawyers and doctors), owners of the bigger commercial establishments—Abilene was still a small town in which most young men expected to marry a local girl and do pretty much what their father had done, if they were lucky. Ike's time working on a farm, in the feed box business, and at the creamery was enough to convince him of the supreme importance of getting out while he still could, if he hadn't already known that.

Considering that the nearest bodies of water were the aptly named Mud River and the Smoky Hill River, and that Abilene is planted pretty much in the middle of the prairie—there is about as much dry land in every direction as exists anywhere in the world—it seems ironic that Ike's first choice as a means of escape was the Navy.

CHAPTER 4

“Where Else Could You Get a College Education Without Cost?”

What put a career in the U.S. Navy into Ike’s head was his friendship with a young man about his own age, Everett E. Hazlett, known as Swede to his friends because of his blond hair. Swede differed from Ike in a good many ways, most importantly because he lived on the right side of the tracks, on the north side of Abilene, where his father was a prosperous physician and pharmacist (it was not uncommon in those days for a doctor to play both roles, both prescribing and preparing medications).

Swede Hazlett was a tall, gangling youth, good-looking, to judge from early photographs, with a shock of bright platinum hair, a firm chin, and rather soulful eyes. He seems to have formed a friendship with Ike at Abilene High School, when Ike was a football star, and it is possible that

Ike, who was then very much a big man on campus, took a hand in preventing Hazlett from being bullied. Swede was a big fellow himself, who did not on the face of things seem likely to be bulldozed by other young men, but he had grown up in a more genteel atmosphere than a lot of his fellow students, and may have lacked the Eisenhower boys' skill with their fists, and of course he lacked their mutual support against bullies. That, at any rate, is the way Ike remembered it; and the two of them became close friends and would remain so until Hazlett's death, when Ike, then president of the United States, attended Swede's funeral. Ike's correspondence over the years with Swede Hazlett is remarkable—long, hand-typed letters, deeply personal and perhaps as revealing as any Ike ever wrote. No matter how crammed his schedule was, as a general or as president, he found time to write to Swede.

Swede left Abilene High School a year early, somewhat to Ike's dismay; went to a military school in Wisconsin; then secured an appointment to the United States Naval Academy at Annapolis, but failed the examination and returned to Abilene to cram for a second try at it. Swede's return coincided with the period when Ike was working at the creamery in the furnace room, and trying to figure out a way of going to college without asking his father for any help. He had hoped to follow his brother Edgar to the University of Michigan, but that seemed increasingly unlikely—his father couldn't afford to send two boys there, and Ike couldn't possibly save up enough money to pay for it himself. Swede apparently managed to convince Ike that an appointment to Annapolis was his best way out of Abilene, a prospect made more attractive by the fact that the two young friends would be classmates there.

Up to that point, Ike had shown no interest in a seafaring life; his heroes had been generals, and his favorite subject for reading had been land battles. Ike had never seen the sea—he had not even seen a river more imposing than the narrow, sluggish Mud River or the only slightly more impressive Smoky Hill River. Swede may have helped convince Ike that none of this mattered. Swede himself, who loved horses and was a good rider, was no John Paul Jones. He had not initially felt the call of the sea at all; in fact he had wanted, much against his family's wishes, to go to West

Point and join the cavalry, but having failed to secure an appointment to the United States Military Academy, he tried out for Annapolis as a second choice. In any case, he seemed to feel that if the only way out of Abilene for good was by ship, it was worth taking, and Ike apparently decided the same.

Once this opportunity to escape was clear to Ike, he moved with remarkable speed and an uncanny political sense for a boy his age. The Hazlett family had clout, but David Eisenhower did not, so it was up to Ike to find his own way to Annapolis. Ike wrote to his congressman, who eventually wrote back advising him to write to Senator Bristow. While awaiting the senator's reply, Ike set out to get a letter of recommendation from almost everyone who mattered in Abilene. As things turned out, Bristow was unable to give Ike an appointment to Annapolis, but he was impressed enough by Ike's recommendations to advise him to take the competitive examination in Topeka, in three weeks' time. Swede and Ike coached each other diligently for three weeks, and Ike's gift for memorization paid off.

In those days the same examination was given for West Point and Annapolis, and those who took it could apply to one or the other, or both. Without giving the matter much thought, Ike applied for both academies, and came in second in the examination. Since the young man who came in first had applied only for West Point, Ike was the top applicant for Annapolis. Fate then intervened. Senator Bristow showed the results of the examination to an old friend, who thought the young Eisenhower would be a good choice for Bristow's vacancy at West Point, and the young man who came in first failed the physical examination, so Bristow promptly wrote to Ike offering him an appointment to the United States Military Academy for the student year 1911.

Apparently, Ike's fleeting interest in the sea evaporated overnight, unregretted. Swede Hazlett went on to Annapolis, and eventually became a naval officer, while Ike went to West Point, with results with which we are familiar.

He had a year in which to prepare himself, which he seems to have spent studying hard, working, and playing football; and of course his mother and father had a year in which to get used to the idea of a son of theirs becoming a professional army officer. Ike would later admit that his mother found his choice of a profession "rather wicked," and even his father, though pleased that Ike would get a good education at no cost, was still sufficiently a Mennonite to be quietly appalled.

Ike himself, it must be said, suffered from no guilt or reservations. West Point and the Army would take him out of Abilene, and give him a chance for a career on a wider stage—more than that, he did not need to know. When he left Abilene by train in June 1911, carrying a single small suitcase, his mother saw him off with the stony detachment of a Roman matron—"It is your choice," she told him—then went home and for the first time in her married life, was seen to be in tears, "bawling . . . like a baby,"[1] her son Milton observed with astonishment. She was crying not so much at Dwight's departure as at his decision to become a soldier, which went against her very deepest beliefs, and which she doubtless also thought would be hard to explain to her fellow Jehovah's Witnesses.

For a boy from Kansas, the sight of the Hudson River must have been almost as amazing a sight as West Point itself. In 1911, the countryside around West Point was still all farms and great estates, green and lush in the early summer, and heavily wooded. The Hudson was no Mud River—at West Point it was wide and deep enough to have permitted a fleet of British warships to sail up from New York City and attack the fort during the Revolutionary War, despite the massive chain which had been drawn across the river to stop them. Broad, stately, and magnificent, the Hudson was symbolic of an older America. Walking from the Highland Falls railway station the short distance to the United States Military Academy in the humid, sweltering heat ("West Point . . . could export heat without loss,"[2] Ike later commented), suitcase in hand and five dollars in his pocket, Dwight D. Eisenhower must have had ample time to reflect on the enormous step he was taking. It does not seem to have daunted him—although he described his first day at West Point as "calculated chaos," he kept his wits about him by reminding himself sensibly, "Where else could you get a college education without cost?"[3]

Then, as now, everything was done on the double; inexplicable orders were given at the top of their voices by furious, fiercely uniformed cadets; there was a din of massed meaningless "shouts and barks" as the new cadets—plebes—rushed back and forth trying to collect the bits and pieces of their uniforms and discard their civilian clothes. A lifetime later, Ike could still remember the day clearly: "We were all harassed and, at times, resentful. Here we were, the cream of the crop, shouted at all day long by self-important upperclassmen, telling us to run here and run there; pick up our clothes; bring in that bedding; put our shoulders back, keep our eyes up, and to keep running, running, running. No one was allowed to do anything at ordinary quick-time; everything was on the double. I suppose if any time had been provided to sit down and think for a moment, most of the 285 of us would have taken the next train out."[4]

Needless to say, Ike did not take or consider this option, and by the end of the day he had a reward of sorts. In the evening, the entire class was assembled and sworn in as cadets of the United States Military Academy. "Whatever had gone before," Ike recalled, "this was a supreme moment. The day had been one of confusion and a heroic brand of rapid adjustment. But when we raised our right hands and repeated the official oath, there was no confusion. A feeling came over me that the expression 'The United States of America' would now and henceforth mean something different than it ever had before. From here on it would be the nation I was serving, not myself. Suddenly the flag itself meant something. . . . Across half a century, I can look back and see a rawboned, gawky Kansas boy from the farm country, earnestly repeating the words that would make him a cadet."[5]

"It would be the nation I was serving, not myself." These, though Ike might not have realized it fully at the time, were the words that would define his future better than any others. Many years later, when his young bride, Mamie, burst into tears because he was ordered away on temporary duty with the National Guard shortly after their marriage, he would put his arm around her and tell her gently, "My duty will always come first." It was not a lesson she found easy to accept, then or later, but she came to recognize it as bedrock in Dwight Eisenhower's character.

Even the dullest and least romantic of the 285 cadets who took that oath in June 1911, must have been at least somewhat stirred by the solemnity of the moment, but in Ike the words struck deep into his heart and soul, changing him overnight from a civilian into a soldier, and giving him both a sense of purpose and the prospect of membership in an organization he could revere and admire. Perhaps without fully realizing it then, as he took the oath he was deliberately choosing a life that would differ in every respect from his father's—secular, not religious; as a soldier, not a pacifist; and as a small cog in the power structure of the United States government, not an outsider. An officer in the United States Army might be many things, but he would never be a loner as David Eisenhower was.

There is a photograph of Ike on a visit home to Abilene, sitting on the steps of the porch, taken in 1926, fifteen years later. The six brothers are all there, showing a distinct family resemblance. David, their father, is wearing a dark suit, a white shirt, and a bow tie; Ida is wearing a long-skirted light summer dress. Ike is seated in front of his family, between his mother and his father, but lower down; his brothers, like their father, are all dressed rather formally, in neat suits, but Ike stands out from the rest of them in full uniform—an officer's tunic, flared riding breeches, gleaming riding boots—visibly part of a very different world from theirs, "serving the nation," as he had sworn to do at the end of his first day at West Point.

The West Point that Ike entered was still, by current standards, a small and very homogeneous place. After the Civil War a halfhearted attempt had been made to introduce black cadets, but was met with sharp hostility by both the cadet corps and the faculty—a hostility that, to his great embarrassment, not even so distinguished a graduate as Ulysses S. Grant was able to overcome, despite considerable prodding from Frederick Douglass. Yet while the racial divide remained unbridged, and had even widened since Grant's era, and although the cadet corps was still predominantly Anglo-Saxon, the United States Military Academy at least offered cadets a chance to mix with young men from all over the country—southerners, boys from the big cities, farm boys, Texans, New Englanders. West Point brought them all together. They were mostly middle class, but many of them were poor (among the poorest were two future five-star generals, Ike and Omar N. Bradley) and a very few were rich; and they offered, if nothing else, a cross section of young Americans more varied than you might have found in most colleges and universities of the time. For a boy from Abilene, Kansas, that alone was an exciting change. At any rate, despite the harassment from upperclassmen and the discomfort of "Beast Barracks," Ike seems to have felt no trace of homesickness from the very first day, and had no regrets at mailing his civilian clothes home, despite the fact that cadets then spent two years at West Point before receiving their first leave.

Above all, West Point itself was then, as it is today, a deeply impressive, moving, and even solemn place; steeped in history; rich with the memory of countless American heroes; still retaining the commanding sense of the fortress it had once been. No one can say what Ike would have made of his life had he followed his brother Edgar to Ann Arbor and attended the University of Michigan, as he had briefly hoped to do, or if he had accompanied his friend Swede Hazlett to Annapolis; but it is safe to guess that neither place would have had the same emotional impact on him as West Point did, when he and his fellow plebes assembled, awkward, clumsy, and newly uniformed on the Plain, for a ceremony that went back to 1800. At any rate, fifty-seven years later, when he was near death, Ike still talked about his years as a cadet with unconcealed emotion to a visitor who was also a West Pointer and a general.

Ike was awed by the solemnity of the swearing-in ceremony, but on the other hand he was not overawed by West Point. He had every confidence he would succeed there (perhaps more confidence than was truly justified). He was a natural—and enthusiastic—competitive athlete, in an institution where being athletic counts for a lot; and he was tall, strong, and fit, thanks to hard physical labor. Bob Davis, the old trapper and hunter who taught him poker, had also given Ike a good knowledge of field craft, and had made him a superb marksman with rifle, shotgun, and pistol; and like most boys who grew up on a farm, Ike was inured to discomfort and used to horses, at a time when horsemanship was still a vital part of the West Point curriculum. His education was neither better nor worse than that of most of his fellow plebes; indeed, he had come to the one academic institution in the United States where his early passion for military history would be an advantage. Finally, having grown up in a family with six boys, two of them older than himself, Ike was used to the rough-and-tumble of life, and not afraid of a fistfight—he did not suppose that the bullying, threats, and harsh discipline the senior cadets inflicted by tradition and by right on the hapless plebes could be any tougher than what he had habitually received at home from his older brothers, and he was right.

Roommates were paired alphabetically, so it was ironic that in an institution which presented such a varied cross section of American youth Ike's first roommate (or "wife," as a roommate was known at West Point)[6] was also from a small town in Kansas, as was his second. The first roommate, unable to bear the harsh discipline and the systematic persecution of life as a plebe, broke down in tears and quit early on, despite Ike's earnest and well-intentioned attempts to change his mind. Ike was struck by the contrast between the young man's having left for West Point as a small-town hero, and the prospect of his return as an abject failure, and perhaps congratulated himself on the fact that his own departure had been relatively low-key; but it does not ever seem to have seriously occurred to him that he was likely to fail. Indeed, from the very beginning, Ike's self-confidence was striking—despite West Point's draconic disciplinary code, he constantly skated on thin ice, cheerfully taking risks that would give him, early on, a reputation for being something between a class clown and a rebel against all forms of authority. His big smile, his evident good nature, and a certain reputation for rough horseplay and practical jokes made him popular, and quickly brought him, not always favorably, to the attention of the authorities.

Reading about Ike's early days at West Point makes one wonder if there was a wild streak in him that he would, later in his career, be at some pains to suppress—perhaps a reflection of the rather less cheerful wild streak that made his father leave Ida behind after the store failed and go off by himself to work for the railroad in Denison, Texas. In any event, Ike set out to rack up a remarkable (though not necessarily record-breaking) number of demerits, without seemingly being much troubled by them, even though demerits were used to calculate a cadet's position in his class, which, after graduation, would determine his all-important seniority in the Army.

Ike was not just high-spirited; in all likelihood, like many another small-town boy, he was kicking up his heels at being away from home, and above all at no longer being subject to his father's rather dour view of life, or his mother's intensely religious sense of right and wrong. As for the demerits, he may have taken some consolation from the fact that Generals Ulysses S. Grant, J. E. B. Stuart, and George Armstrong Custer and many other famous graduates had built up impressive numbers of them and still gone on to distinguished military careers. The Army took discipline seriously, but this is not to say that a certain wild streak was not regarded, then and now, as a good thing in a young officer, provided it involved nothing dishonorable and he was smart enough to get away with it.

In Ike's case it was evident from the beginning that his high spirits were matched by an extreme concern for honor: in all the things that mattered most in a cadet—telling the truth; taking full responsibility for one's own actions; never cheating or shirking a duty or behaving in any way shamefully or so as to jeopardize the reputation of the United States Military Academy—Eisenhower was, and would remain, beyond reproach.

The stories of Ike's high jinks at West Point have often been told (and no doubt exaggerated), not least by Ike himself; and like most such tales, they don't sound nearly as funny as they no doubt seemed at the time. Ordered to appear before a cadet corporal wearing "dress coats," Ike and a friend elected to take the order literally, and turned up stark naked except for their uniform dress jacket (then, as now, a tailcoat cut straight across the waist in front), and were severely "braced" for it, to the uproarious laughter of their fellow cadets. Ike soon got into the habit of "going over the wall" at night to get a sandwich or an ice cream cone in one of the nearby towns (an infraction for which he could have been expelled) and smoking cigarettes (then strictly forbidden, for reasons not of health but of class—"gentlemen" smoked a cigar or a pipe, not cigarettes); and he exhibited a disdain for the authorities that terrified his roommate. Although warned to dance more sedately, he persisted in "whirling" around the floor, and received a severe punishment, which he neither resented nor seemed to care much about. In one six-month period, he was disciplined fifteen times, among other things for being late for chapel, for having failed to make a "reasonable effort" to have his room ready for inspection on time, for being late to breakfast, and for failing to wear his gymnasium belt—he would eventually stand 125th in a class of 162 in discipline. This is by no means a record among famous figures (Edgar Allan Poe, George Pickett, and George Armstrong Custer did worse at West Point), but it was still remarkable for a young man who had been strictly brought up and was not, by nature, a hell-raiser or a malcontent. The explanation seems to have been that Ike was bored by the curriculum and had already come to the conclusion that while he might scrape by academically, he would never excel or become a "tenth boner," as the highest ranking students were known. His real passion was for athletics, and here he suffered what was to be, for many years, his greatest disappointment.

To begin with, when Ike arrived at West Point he was five feet eleven inches tall and weighed 152 pounds—he was fit and wiry, but very light compared with the men who were taken seriously on the football field; and his style of hitting, while it had been appreciated in Abilene, did not please the baseball coach at West Point.

Ike did his best to learn to hit the West Point way, and subjected himself to a fierce regimen of exercise, running, and eating that eventually brought his weight up to 174 pounds, most of it solid muscle. That and his "love for hard bodily contact" gave him a crack at making the varsity team, and wearing the coveted A on his football sweater, but then bad fortune struck. Just before the Army-Navy game, he played in a game against Tufts, twisted his knee badly, and was briefly hospitalized. Shortly after his release, he reported to the riding hall and injured himself again in what was called "monkey drill"—cadets were supposed to swing off their horse at a canter, run alongside it, then vault over the horse's back to land on their feet on the other side as it jumped over a hurdle. The shock of landing was more than his injured knee could take, and Eisenhower was taken back to hospital with torn tendons, ligaments, and cartilage. He was in great pain, and in the age before microsurgery there was nothing much the doctors could do but put his leg in a cast and hope for the best.

To Ike's dismay, the injury was permanent—he would never play football or baseball again. He kept himself fit with calisthenics and gymnastics, but his hopes for athletic success, even fame, which had once meant so much to him, were ended.

Paradoxically, the end of Ike's career as an active football player did nothing to diminish his belief that football produced "leadership qualities"; nor would it diminish his tendency, in later life, to promote officers with a good record on the football field at West Point. The duke of Wellington remarked famously, "The battle of Waterloo was won on the playing fields of Eton." Similarly, Ike's respect for football as the proving ground for military leadership was deeply entrenched, and the roster of those who served in high command in the U.S. Army is studded with football players from West Point, two of whom, Patton and Bradley, would serve under his command in World War II. Eisenhower, to his intense disappointment, never became a football player; but over fifty years after his injury he could still write that "victory comes through hard—almost slavish—work, team play, self confidence, and an enthusiasm that amounts to dedication." Whatever else he might be, Ike would prove to be the ultimate team player, and once he reached high command this was above all the capacity that everybody respected in him, even those who had doubts about his military expertise.

The connection between sport and war has always been evident to Anglo-Saxons on both sides of the Atlantic, though puzzling to others. Even in World War II (and after it), the first question asked of a candidate for a commission in the British armed services was invariably, "Do you play cricket?" A lack of enthusiasm for games was a serious liability, even in the Royal Air Force, where the connection between cricket and flying an aircraft was by no means immediately apparent. That the British military valued cricket—a game of skill which is highly, even ritualistically formal, and in which fair play, individual style, and above all the art of making the difficult look easy are the most important qualities—while the American military valued football, a game in which violent physical contact and team spirit matter most, says much about the differences between the two armies Ike would eventually command in battle. Perhaps more important, an interest in sports seems to have played a very small role in the selection of officers for the German armed forces. For the

Germans, the military was a profession to be mastered—skill and enthusiasm on the playing field were not thought to be reliable indicators of a talent for soldiering.

Even the inability to play could not keep Ike off the football field. He became coach of West Point's junior varsity, made up of men who were "not quite good enough to make the first team," and here he first put into practice his own passion for team spirit, and his skill at motivating men to do more than their best—more, in fact, than they believed they were capable of—and succeeded in sending some of his squad on to the first team. It was not the same as playing football himself, but it was perhaps better training for his future profession.

Poker was perhaps of equal importance to a future general; and thanks to Bob Davis, the old hunter who had befriended him in Abilene, Ike was a devoted and skilled poker player. This was, in his words, "his favorite indoor sport," and he indulged in it to the full at West Point, so much so that he kept a careful account book of his winnings, which would be paid off after graduation.

The fact that Ike's disciplinary and academic record at West Point was mediocre is something of a puzzle—it even puzzled Ike when, much later in life, he came to write about his years as a cadet. He described his attitude then as "nonchalance," but one suspects that it was more complicated. From the very beginning, Ike seems to have been sure that he could do well enough at West Point to graduate, but not to be at the top of his class, or anywhere near it. He was never going to equal the record of Robert E. Lee (class of 1829), for example—second in his class academically, and the first cadet in the history of the United States Military Academy to graduate without a single disciplinary demerit—or the athletic record of George S. Patton, Jr. (class of 1909), who not only was a football player but became the first American to compete in the modern pentathlon (horsemanship, running, swimming, fencing, and marksmanship), at the 1912 Olympics.

Ike's attitude at West Point seems less like nonchalance than fatalism, coupled with a concern less for standing out than for joining in—acceptance by his fellow cadets meant more to him than a high academic rating. Then too, he had clearly set his heart on being an athletic star, and when that possibility was removed, there was an inevitable decrease in his interest and enthusiasm. On the other hand, Ike was already exhibiting two personality traits that are unusual in the young, but that would stand him in good stead in his future career: patience and a certain degree of humility, or at the very least a marked reluctance to push himself forward. He was in every way the opposite of the flamboyant, fierce-tempered George Patton, who graduated from West Point three years before Ike became a cadet. Ike went out of his way to hide his light under a bushel.

This is not to say that Ike was without ambition, or that he had a placid temper. Like the good poker player he was, he learned early on to hide his feelings. He worked hard to conceal his ambition behind an easygoing pose and an attitude toward his own disciplinary demerits worthy of Peck's Bad Boy. He struggled hard to keep his temper under control, mostly successfully, though those who knew him well, even as a young man, came to recognize and dread the hard flat stare, the reddening of his face, and the pulsing of the arteries in his temples, particularly when he felt that he had been insulted, or when he saw somebody else humiliated or unjustly punished. Perhaps what he most disliked about West Point was the mindless bullying of the plebes by the upperclassmen, then as now a hallowed tradition of the academy.

One incident, which took place when he had become an upperclassman and a corporal himself, reveals much about the kind of young man Ike was, and the kind of officer he would be. A plebe from his own home state accidentally ran into him while running to comply with an order from another upperclassman. Ike "reacted with a bellow of indignation and mock astonishment," determined to "crawl" the offender—i.e., to humiliate and harass him. "Mr. Dumgard," he shouted, "what is your PCS?* You look like a barber."

The dejected plebe replied softly, "I *was* a barber, sir."

Flushing with embarrassment, Ike walked off, and said to his roommate, "I'm never going to crawl another Plebe as long as I live. . . . I've just done something that was stupid and unforgivable. I managed to make a man

* A West Point abbreviation meaning "previous condition of servitude," i.e., former employment.

ashamed of the work he did to earn a living."[7] And during his remaining three years at West Point, despite the social pressure from fellow upperclassmen to do so, Ike stuck to his guns—he would not humiliate a man.

Nor would he be humiliated. Midway through Ike's cadetship, an instructor in mathematics warned the class that the next day he would give them a problem in integral calculus so difficult that he would also provide the correct approach to the problem, and the answer. Ike, who described himself as a "lazy student" who relied on luck, paid no attention during the long explanation, which the instructor was clearly reciting by rote, figuring that since there were twelve cadets in the class, the odds against his being called on the next day were eleven to one.

As it happened, the next day Ike's luck ran out—he was the one the instructor called on. Since he had hardly listened to a word of the previous day's explanation, Ike hadn't a clue how to begin. He did remember the answer to the problem, so he wrote that in a corner of the blackboard, then tried to find some way of getting there by himself. After several false starts, he stumbled on a simpler approach to the problem that, to his amazement, yielded the correct answer. He went back to his seat, and when called on to recite, got up, went to the blackboard, and went through his solution to the problem step by step.

To Ike's surprise and chagrin, the instructor upbraided him for having paid no attention to the previous day's explanation of the right way to solve the problem, and for having remembered the answer and produced a meaningless solution to reach it. Ike faced not one but two penalties—a zero, which he could ill afford, with his academic record; and a disciplinary demerit for having failed to pay attention when the officer had recited the way to solve the problem. Worse still, there was the underlying accusation of having come very close to cheating—the worst offense for a cadet.

His fellow cadets in the class were alarmed to see all the signs of an approaching outburst of temper on Ike's part—he was "red-necked and angry"—despite the fact that insubordination toward an officer was grounds for dismissal, when Major J. Franklin Bell, the associate professor of mathematics, who had been quietly sitting in on the class, spoke up, and said, "Just a minute, Captain."

Fortunately for Ike, Bell's interruption shut him up and prevented an explosion of the famous Eisenhower temper. Bell asked the captain to let Ike go through his explanation again, and listened with every sign of patience and even admiration. When he had heard Ike out, he told the captain that Eisenhower's solution to the problem was "more logical and easier than the one we've been using," and that it would be incorporated in the curriculum henceforth.[8]

There are two things worth noting about this anecdote. One is that Ike was willing to argue back to an officer, even at the risk of losing his temper and being dismissed. The other, less obvious, is that Ike, despite his poor academic record, had instinctively produced a shorter, simpler solution to the problem than the existing one.

This would be one of Ike's greatest strengths as a senior officer—his ability to produce a simpler solution than anybody else's to a difficult problem. Many of the disagreements between Ike and Montgomery in World War II can be traced back to the talent that had so impressed Major Bell in the classroom at West Point. Napoleon once defined the most important quality in a general (he was, of course, thinking of himself) as *le coup d'oeil de génie*—the quick glance of genius—by which he meant the ability to cut through complexity and see the core of a problem and its solution instantly, by instinct rather than by analysis. Napoleon himself exemplified this form of genius—with one quick glance at the battlefield or at a map he could guess where the enemy's weak spot was and work out in his mind how to use the terrain to concentrate his forces against it, at the *Schwerpunkt* so beloved of German generals of the two world wars. So long as Napoleon possessed that genius, he won victories, and only when it began to desert him in 1812 did he start on the long losing streak that would end at Waterloo. Lee had it too—never more clearly than at Fredericksburg and Chancellorsville, though it deserted him, tragically for the Confederacy, at Gettysburg—and Grant, much as he detested Napoleon, had it throughout his career as a general.

Although Ike spent as little time at West Point reading *Le Mémorial de Sainte Hélène* as Grant had, he showed, early on, just the quality Napoleon most admired in generals, apart from luck—the ability to get to the heart of a matter in a quick burst of intuition, rather than working one's way toward it by conventional formulas or traditional methods. In the young Eisenhower, like the young Bonaparte, this was often mistaken by others for laziness, impatience, lack of preparation, or reliance on luck. Like the young Bonaparte, too, the young Eisenhower tended to blame himself for just those defects for which his teachers criticized him—it would take many, many years before he realized that they were in fact strengths, and developed confidence in his own way of tackling military problems and in the simple clarity of his vision. Unlike Napoleon, who was a general at the age of twenty-four and who was forty-six when he lost at Waterloo, Ike did not receive his first star until he was fifty-one, a long preparation that might have caused bitterness in another, lesser man.

Even at West Point Ike liked to be thought of as easygoing, good-natured, and affable—the kind of cadet who spent more time learning to play bridge than studying, and who picked up the habit of smoking cigarettes, which was strictly against the rules—though in fact his academic record was nothing like as bad as he would later pretend it had been, and he would graduate sixty-first in a class of 164. He absorbed things quickly, and he might easily have done far better had he not been determined never to be thought of as a bookworm—indeed, when he was accused of "bookworming" his way to the top of his class in English, he pledged not to read a single book in the subject outside class for the rest of the term, and stuck to his pledge, not surprisingly dropping to the bottom of the class when he might have been near the top.

Confident from the beginning that he could measure up to West Point's standards, Ike was willing to drift along, without making an effort to excel. Perhaps West Point did not challenge him enough; perhaps he was simply more concerned with being a well-rounded cadet than with graduating near the top of the list academically; perhaps his ambivalence about the discipline, the spit and polish, and the rigid inflexibility of the academy's traditions and regulations held him back from making a total commitment to military life—but for whatever mixture of reasons he was content to coast toward graduation at his own pace.

As it turned out, he came very close to not getting into the Army at all, for once again his bad knee continued to trouble him. After examining his medical records, Colonel Shaw, the head of the medical department, called Ike in to tell him that he might find it necessary to recommend that while he "be graduated and receive a diploma, [he] not be commissioned in the Army."

The United States Army was still small in 1915—it ranked in size behind that of Bulgaria—and there was no shortage of second lieutenants. Considerable care was taken not to commission anybody whose physical condition might pose a problem, and perhaps even lead to premature retirement with a disability pension. Service in the Army was a two-way street—the Army looked after its own, but you were expected to give a lifetime's service to it in exchange.

Typically, Ike managed to convince Colonel Shaw that he didn't much care whether he received a commission or not; this was probably the best way to prevent Shaw from dropping the ax. It must have taken some doing, since Ike had given four years of his life to earn a commission as a second lieutenant and would have had a very hard time telling his family and his friends in Kansas that he had failed. But he was equal to the occasion, explaining to Shaw that he had been thinking about going to live in Argentina, where there were great opportunities, and perhaps working as a gaucho for a while.

Puzzled—most cadets would have been horrified at the prospect of not getting their commission, and would have begged, pleaded, or blustered—Shaw dismissed Ike, promising "to think the matter over." After a few days he called Ike back. Shaw had partially relented: if Eisenhower would apply for service in the coast artillery, Shaw would feel justified in recommending him for a commission.

Eisenhower, however, politely replied that he had no interest at all in serving in the coast artillery, having failed to realize until he saw the expression on Colonel Shaw's face that Shaw had once served in the coast artillery himself.

At the time, the coast artillery was by far the most sedentary and pointless branch of the United States Army, still manning obsolete heavy guns in forts, many of which had first been established before the Civil War, in the event that an enemy fleet might someday turn up within range off

America's shores. The big guns, and the ponderous machinery that moved them, were meticulously maintained but seldom fired, even for training purposes, since the enormous shells were so expensive. There was, in fact, nothing for the coast artillery to do, and there had been nothing to do for generations, except wait patiently until the War Department finally realized how useless the guns were and disbanded it—each battery was a monument to many pointless lifetimes of spit and polish, and paint. Not surprisingly, the prospect of spending the next eight years overseeing the maintenance of these useless weapons daunted Ike.

Shaw once again dismissed Ike, who claimed to have spent the next few days writing away for travel brochures to South America, in the expectation of life as a gaucho. In later life, he worried that he may have offended the colonel by dismissing so cavalierly the chance to serve in the coast artillery; but he felt that Shaw, in the West Point tradition, had been asking for an instant reply, not encouraging a discussion about his future. Whether this was true or not, Shaw was apparently impressed by the fact that Ike knew his own mind and had spoken up so frankly. He called Ike back once more, and this time said that on further reflection, since the injury had been caused by a riding accident, he had decided that if Ike would agree not to make any request for mounted service, he would recommend him for a commission.

That was all right with Ike. He had no desire to join the cavalry, and not much more interest in the field artillery, and agreed to put the infantry down as his first choice. On that basis Colonel Shaw, who apparently combined good judgment with the patience of Job, gave his blessing to Ike's becoming a second lieutenant in the U.S. Army after graduation.

Among the officers responsible for instructing the cadets there was a considerable difference of opinion on how much Ike had benefited from his four years at West Point, ranging from the rather dour hope that he would serve under a "strict" commanding officer to the surprisingly prophetic judgment "born to command."[9] But in some ways, Ike was far better prepared for his future career than he appeared to be on paper. He had a deep interest in tactics, and he had spent a good deal of his spare time roaming the network of forts around West Point, to see what the battles of the Revolutionary War described in his textbooks must have been like on the ground. The field craft he had learned from Bob Davis would always play a role in Ike's view of war, and even as a cadet he had a remarkable ability to see beyond the two dimensions of a map the actual landscape of battle.

Like so many things about Ike, this was a side of himself that he tended to keep hidden. Another was the extensive knowledge of ancient warfare that he had gained from reading the books in his mother's library—knowledge that would, one day in the future, deeply impress no less a personage than Winston Churchill, also an omnivorous reader of military history. Much of this Ike kept from his fellow cadets, as well as from many of his instructors, partly out of a desire no doubt to be thought of as "base, common, and popular" (to use Pistol's words to the disguised King Harry, in *Henry V*), rather than brainy; and partly because West Point, then as now, did not encourage independent thinking or inquiry into military tactics on the part of cadets, who were merely expected to memorize the correct answers to questions that had been put to many generations of West Pointers before them.

Roaming around in the woods in the Hudson Valley with a map and a history of the Revolutionary War was not what most cadets did during those few moments when they were not parading, playing sports, or polishing their boots, buttons, and equipment. Yet somehow, without having seemed to make much of an effort at it, Ike had managed to acquire a good education for his future profession at West Point, and despite his congenital high spirits he struck those who had known him before he left Abilene for West Point as more mature and responsible—still good fun to be around, but with a serious streak that many found surprising.

And perhaps this serious streak was also appealing. Ike's social life at West Point had been minimal, restricted both by the regulations of cadet life governing leave, and by his own lack of money; but he had the natural instincts of a healthy young man of college age, and he wanted to have a good time, which included girls, of course. Money was a perennial difficulty—the only cadet in the class who was definitely poorer than Ike was Omar Bradley—but there were also other problems. The East Coast in general and New York City in particular were foreign territory to a boy from Abilene. Girls were more sophisticated, were harder to meet, and had higher expectations of what constituted an acceptable date than Ike's high school girl friends back home; and apart from the occasional strictly supervised dance, there was no easy way to meet them in any case. Ike's skill as a dancer was—and would remain—fairly limited, and on the rare occasions when he let himself go on the dance floor, he got into trouble with the authorities, whose views on dancing were Victorian.

He seems to have had no trouble finding a few girlfriends in his senior year, though not much can be made of it all except to conclude that writing letters to young women was something less than Ike's strongest talent—the love letter would never be his natural form of communication. In this area, practice would not perfect him; expressing his feelings was not something that came easily to him, and perhaps even recognizing them was difficult, so determined was he to avoid sentimentality in all its forms. Although hampered by lack of money, he made several trips to New York City, where he soon discovered that the Astor Hotel extended credit to West Point cadets, who did not have to pay until after graduation; and he attended a few Broadway shows with a date—about at the limit of what he could afford, or even beyond it. None of this was particularly serious, either to Ike or to the young women—a penniless West Point cadet from Kansas would hardly have been regarded as a catch by a New York girl, or her family. Ike was tall, handsome, and good company, but if he felt anything approaching strong feelings for any of the girls he wrote to, they are not reflected in his letters to them. To one he wrote, rather offhandedly, "Remember that it is good to write once in a while, just to keep in practice"; to another he confessed, "The whole purpose of this letter was to ask you to write to me, and since I got up the courage to do it, I reckon I'd better stop,"[10] then went on to analyze in detail his roommate's triumphs at the Army-Navy game. These were not letters calculated to turn up the heat of a woman's feelings, or even his own.

On his first furlough home, in his second year at West Point, Ike had discovered that he was something of a small-town celebrity in his stiffly starched white summer dress uniform, so much so that he quickly gave up wearing it; but in midtown Manhattan West Point cadets were a familiar sight, and nothing to get excited about. It's hard to escape the feeling that after four years as a cadet, Ike approached graduation in a somewhat confused and uncertain state. He had no real girlfriend to confide in—however difficult confiding might be for him. His closest friend was his roommate P. A. Hodgson (at West Point a roommate was called a roomie or, as noted earlier, a "wife"), star of the Army-Navy game, but it was improbable that they would be posted to the same place. And Ike's mother, however sympathetic she might be, was deeply ambivalent about the profession he had chosen for himself. West Point had been the center of his life for four years, but now he was leaving it for the bigger and much more challenging world of the U.S. Army.

It was perhaps typical of Ike's career at West Point that he came close to missing his own graduation, having been sent to the infirmary with acute influenza six days before it was to take place.

Ike only barely managed to persuade the medical officer to let him go in time to stand, pale and shivering, on June 12, 1915, with his classmates and hear the secretary of war's short speech. How much of it Ike heard we do not know—he did not refer to it in his memoirs, so perhaps it made little or no impression on him. Had it impressed him, he might have reflected that while the greatest war in human history had been raging for nearly a year in Europe, consuming at a rapid rate the lives of many, many thousands of British, French, Austro-Hungarian, Belgian, German, and Russian second lieutenants, no mention of it was made in the secretary's speech to the graduates of the United States Military Academy.

Across the Atlantic, Armageddon was in full progress. Mired in the trenches on the western front, the combatant nations were already contemplating casualties beyond anything they had ever imagined possible, with no end in sight; and on the eastern front the death of millions of Russian soldiers had already deeply eroded the Russian people's confidence in its monarchy and the Army's faith in its own officer corps.

The United States, with an Army that barely exceeded 100,000 men, most of them spread out across the country in tiny (and useless) forts dating from the Indian wars, remained a negligible military power, and President Woodrow Wilson intended to keep it that way.

CHAPTER 5

Second Lieutenant Eisenhower

Although the rest of the world was at war in 1915, the United States was at peace, and, it seemed, determined to remain so. The only chance for any kind of combat in which a young officer might have an occasion to distinguish himself was along the border with Mexico. There, a combination of inept diplomacy on the part of the United States and the inherent instability of Mexican politics had pushed the charismatic Pancho Villa (part ambitious politician, part bandit, part flamboyant guerrilla leader, and always a ruthless and gifted master of hit-and-run tactics) into a series of mounted raids, some of them bloody, on American border towns.

Ike's thoughts had been turned toward more exotic places than the Mexican border—he submitted his preference for service in the Philippines on graduating from West Point, and confidently expected to be assigned there, since in normal times hardly anybody wanted to go. Why the Philippines was on his mind is hard to know—no doubt service there offered a chance for adventure, and this was, at the same time, about as far away from Abilene as it was possible to go, and about as different as it was possible for a place to be. Perhaps an echo of his plan to become a gaucho if he didn't receive a commission on graduation, it seems to have been a romantic impulse in the Kipling tradition. Much as Americans disliked the idea of being a colonial power, the Philippines played, on a smaller scale, something of the same exotic role in the imagination of young Americans in uniform as India did for so long to the British, and it is curious that Ike's wish would be granted twenty years later, in a form he could hardly have imagined.

In the meantime, however, Ike had to go home and wait there for his commission, which was indefinitely delayed. In those days the president still signed an officer's commission, and Woodrow Wilson apparently had more important things to worry about than dealing with the paperwork required to create new second lieutenants for an army he was determined not to use.

For Ike this was an unusual period of forced idleness—he received a second lieutenant's modest pay, but there was nothing he was supposed to do for it except sit at home and wait patiently for the arrival of his signed commission and his orders. Until these were in his possession, he was not even officially a second lieutenant; he was a fledgling second lieutenant in limbo. Fortunately, this difference, while crucial to Ike, was hardly noticeable to people in Abilene, who were mightily impressed by the sight of him walking around town in his crisp summer uniform.

No doubt the sight of Ike in his new uniform was striking in a town not accustomed to military glamour; but despite attempts by many of his biographers to paint a picture of him as a twentieth-century Escamillo in *Carmen*, fatally attractive to (and attracted by) women, lounging around in the evenings on the front porches of Abilene, his buttons gleaming in the twilight, Ike seemed to have spent his time rather quietly waiting for his commission. After a few dates with girls he had known in high school, he pursued—without, judging from his letters, any success—an old flame, the beautiful Gladys Harding, whom he had also dated briefly when she visited New York City while he was a cadet. Ike's feelings for Gladys were strong—apparently rather stronger than hers for him—and it was widely assumed by many in Abilene, though not by Gladys or her father, that the two would marry. Gladys had aspirations to a career as a concert pianist, and her father was the wealthy owner of a shipping company, so her marriage to a penniless second lieutenant from the wrong side of the tracks was a prospect Mr. Harding probably discouraged as much as he could.

How serious the relationship was is hard to say. According to Carlo D'Este, Ike and Gladys continued to correspond with each other until Glady's death in 1959; this argues for a substantial degree of affection and friendship—and when Ike visited Abilene in 1953 as president, she ran out to intercept his motorcade and he greeted her warmly, to the alarm of the Secret Service and the unconcealed annoyance of Mrs. Eisenhower. Still, however strong Ike's feelings were, he cannot have been unaware of how difficult it would be for him as a second lieutenant to support Gladys in the manner to which she was accustomed, or how unlikely it was that she would be willing to leave behind her family and her hopes of a career to go and live on an army post in the Philippines. No doubt, as is often the way, there was an element of fantasy on both sides, making the relationship that much more intense.

In any event, Gladys eventually left to rejoin her touring company, and in August Ike's commission and orders finally arrived. Throughout his life, if there was one characteristic that stood out beyond all others, it was his congenital optimism. So great in fact had Ike's optimism been on the subject of the Philippines that he had purchased only tropical gear from the uniform tailor.

Cadets were docked about fifteen dollars a month from their pay during their four years at West Point. They could then draw from this money to outfit themselves with uniforms after graduation. Since tropical khaki was much cheaper than the many different uniforms required for service in the United States, Ike had a healthy balance left in his account, which he was entitled to withdraw, and did; and it was on this that he had been living while he was on leave in Abilene.[1]

It should not be thought that he was living high on the hog—the amount of money was small, "several hundred dollars," he remembered years later—but unfortunately for him it suddenly became significant when his orders finally arrived. With its usual genius for disrupting the lives of its personnel, the Army was sending him not to the Philippines, but to Texas. This meant that he would require not just tropical khakis, but an officer's full outfit, including OD (olive drab) and blue dress uniforms, an OD and a blue dress overcoat, and full uniform evening dress.

Ike had no option but to throw himself upon the mercy of Mr. Springe, of Springe Uniform Tailors, in Fort Leavenworth, Kansas, and ask for credit, which Mr. Springe, experienced as he must have been in the ways of second lieutenants, agreed to. Already owing money to his father, Ike therefore began his military career deeply in debt, as well as serving many thousands of miles from where he had counted on going. He had assumed that he would be on his way to Manila, but instead he ended up at dull Fort Sam Houston, in San Antonio, Texas, less than 750 miles from home.

His first concern was to get a start on repaying his debt to Mr. Springe. In this he was helped by receiving, in dribs and drabs, his winnings from poker games during his years at West Point—cadets were not supposed to gamble for money (and few of them had any ready money with which to gamble), so it was usual to keep a record of gambling debts and settle them after graduation. Even though this must have given Ike a certain satisfaction—he was always proud of his ability at card games—it cannot have compensated for being assigned to San Antonio, a very pleasant city, but decidedly peaceful, and not likely to be attacked by Pancho Villa.

By contrast, the flamboyant young cavalryman George S. Patton would shortly be making a name for himself chasing *Villistas* south of the Rio Grande under the command of Brigadier General John J. Pershing, and would eventually succeed in hitting one of them himself in a shoot-out. Patton brought three corpses back, tied to the hood of his car like hunting trophies, to present to Pershing—exactly the kind of action which Ike had hoped for, and which would elude him throughout his career.

In the meantime, he plodded through the stultifying daily routine of a second lieutenant's duties with an infantry regiment, determined to make the best he could of it. There seems to have come over Ike a kind of sea change—not a sudden infusion of blazing ambition, or excitement with his military career, but a solid, serious determination to devote himself to his profession, as if he had finally accepted the reality of being an officer in the Army. He had hoped for something more exotic, but if it was to be Fort Sam Houston and the Nineteenth Infantry Regiment, so be it.

He was still capable of high spirits and horsing around, however. He made the acquaintance of his commanding officer one evening when he bet a more senior second lieutenant that he could climb overhand up one of the guy wires that secured the flagpole on the parade ground to the very top and touch the pole, without using his feet or his legs on the wire. The flagpole was nearly sixty feet high, and when one of the regiment's senior second lieutenants angrily challenged Ike's ability to pull the feat off, Ike, who happened to be carrying five dollars (a considerable sum for a second lieutenant), bet him, handed the bill to a stakeholder, took off his uniform tunic, and started up the wire. As it happened, Ike had been good at rope climbing at West Point, and was confident of winning the bet, but just before he reached the top, a voice from below ordered him to come down immediately. With a shock, Ike realized it was none other than the commanding officer of the Nineteenth Infantry, Colonel Millard Waltz, who was a veteran of the Spanish-American War and was what would be described in a later conflict as a Colonel Blimp.

Having heard on the grapevine that the colonel was a betting man, Ike respectfully requested permission to make his way to the top and win the bet before coming down, but Waltz would have none of it. Fifty-two years later, Ike could still recall Waltz's exact words, as he bellowed up from the darkening parade ground: "DO AS I SAY AND DO IT RIGHT NOW! GET DOWN HERE!"[2]

Ike slid down, and stood there looking (and no doubt feeling) sheepish, as Waltz raked him over the coals, calling him "foolhardy, undignified, untrustworthy, undependable, and ignorant." Having attracted the unwelcome attention of his commanding officer, Ike was then confronted by the officer against whom he had bet, who maintained that Ike had lost. Ike's view was that the bet had been interrupted, and was therefore null and void, but tempers flared so high between the two young officers that they had to be separated by their friends before they went "behind the barracks" "to finish the discussion with fists"—a serious offense.

In the end calm was restored, but clearly, despite his good intentions, Ike was still given to the kind of high jinks that puzzled or infuriated his superiors, as well as to reckless betting—and to outbursts of temper, which, in a junior officer, could have serious consequences. His confidence in his own ability to win at bets and poker was not altogether misplaced—between the money coming from what he had won at West Point and his poker winnings at the Infantry Club in Fort Sam Houston, Ike was able to pay off his debt to Springe Uniform Tailors in record time (no doubt to Mr. Springe's surprise), and throughout his early years in the Army he seems to have relied on cards to supplement his tiny income. There was no doubt that at cards, Ike had not only skill but luck. He was also shortly to disprove the old adage, "Lucky at cards, unlucky in love."

As a second lieutenant Ike didn't have much money to spare for dating girls, and the competition among young officers at Fort Sam Houston for eligible young women was fierce, but he had made the acquaintance of Mamie Geneva Doud while on duty as officer of the day. Lulu Harris, an attractive woman married to a major (Ike describes her as "popular with all the second lieutenants of the post," leaving one to wonder just what that meant), called over to him to come and meet somebody. Ike—for whom the duty of officer of the day was a novelty, and who, like most second lieutenants, relished the idea of walking around the post as an authority figure, resplendent in "olive drab with campaign hat and blouse and side-arm" and an "Officer of the Day" brassard on his sleeve—came over to the little group around the ebullient Mrs. Harris stiffly, and heard her say to the young woman standing beside her, "Humph! The woman-hater of the post." Apparently this was enough to intrigue the young woman, whom Ike describes as "vivacious and attractive . . . smaller than average, saucy in the look about her face and in her whole attitude."[3]

The Douds, it transpired, were friends of the Harrises from Denver, who spent the winters in the warmer climate of San Antonio and had stopped at Fort Sam Houston in their "large car" (the car almost as much as the girl caught Ike's eye) to pay a call on Lulu. Mamie Doud's sauciness was sufficient to persuade Ike—who was anxious to show he wasn't a "woman-hater"—to invite her to accompany him on his rounds. To his surprise, she accepted readily; although there was no way he could have known it, she hated walking, and she was wearing a brand-new pair of expensive, elegant lace-up beige boots, a purchase she had made on a trip to New York City, that were very uncomfortable. Whatever spark was lit between the two of them, it was strong enough to overcome Mamie Doud's aching feet and her desire not to get her fashionable new boots dirty as she made the rounds of the guard posts with Ike.

Ike felt it appropriate to warn her about how to behave. "Now, Miss Doud," he told her sternly, "this in an army post, and the men in the barracks are not expecting ladies. I suggest you keep your eyes to the front."

"Of course," she commented later, "I immediately looked both left and right."[4] Coming as he did from a family in which "saucy" behavior and flirting were pretty much unknown, Ike must have felt a certain gravitational pull toward Miss Doud, though he can hardly have failed to be aware that her background was very different from his own—the Doud family's luxurious and impressive touring car would have been enough to tell him that, not to speak of the fact that Mamie had made her debut in San Antonio the year before. Debutantes were not a feature of life in Abilene, nor were showy cars and shopping trips to New York City. The Douds were, by the Eisenhowers' standards, conspicuously well off. Mamie's father, John S. Doud, was a wealthy meatpacker in Denver, with a gift for making money and a taste for travel and adventure, who unashamedly loved spoiling his beloved children. He was always referred to as "Pupah" within the family, and even when Mamie was grown up he still sometimes began his letters to her, "My Dear Little Puddy." Mr. Doud was the first person in Denver to build a "recreation room" in his house, in the days when that was a new phenomenon, with "a piano, a pool table and a Victrola,"[5] and every Sunday evening the Douds gave a sumptuous buffet dinner for friends and neighbors. When the Douds made their annual winter trip to San Antonio, where they rented a mansion in a fashionable neighborhood,

Mamie at the time of her debut.

they always sent two of their maids ahead of them by train a few days before their own departure to make sure everything was properly unpacked and in place for their arrival, including Mrs. Doud's gleaming black electric brougham. Mamie Doud had been to the most refined finishing school in Denver and was not only described by those who knew her "as one of Denver's most captivating belles,"[6] but clearly shared this opinion of herself. It would be difficult to imagine a family that less resembled the Eisenhowers, or whose values were more different.

Of course opposites attract, and that was no doubt a part of what Ike and Mamie saw in the other. Tall, athletic, serious, Ike was very different from most of the boys Mamie Doud had known, while she—spoiled, fun-loving, indulged, fashion-conscious, and naturally flirtatious—was in every way the opposite of him. In later life almost everybody who met Mamie would describe her as intensely "feminine." Wherever she lived, her dressing room was a riot of pink and peach flounces, mirrors, silver-framed photographs of her loved ones, and rows of crystal and gold-topped perfume bottles and atomizers. In photographs taken of her in her debutante year (1914–1915), her expression is strikingly coquettish; her hat is an amazing confection of silk and long feathers, almost as if a brightly colored bird of paradise were seated on her head; an enormous black lace ruff frames her face; and there is a good deal more makeup around her eyes than would have been considered normal at the time in San Antonio, let alone Abilene. In fact, for an eighteen-year-old from Denver, Colorado, she looks surprisingly like a sophisticated and elegant *parisienne*, with an expression that is at once frank and beguiling—one can easily see why young Second Lieutenant Eisenhower was attracted by those eyes.

However strong the attraction may have been on both sides, she managed to keep Ike hanging for a while. As a striking young debutante with a wealthy father, she did not lack for suitors. Ike called "two or three times a day" for a date, only to find that she was always busy, and finally took to walking over to the Douds' house to sit on the porch chatting with Mamie's parents until she finally came home, rather as if he were already part of the family. That may have been a shrewd move on Ike's part, and shows early proof of both patience and a gift for strategy. The Douds got to know the young lieutenant, and liked him, while he got a chance to see how they lived—for the Douds' rented home on McCullough Street was symbolic of a whole different lifestyle from that of the Eisenhowers. It had a high, columned facade facing the street, curved porches with beautifully carved white balustrades, striped awnings, and what appears to have been a glassed-in dome on the roof. The house has a certain Gatsby look to it—or at any rate looks rather like those grand southern houses on the porches of which Gatsby courted Daisy before the war—an impression enhanced by Mrs. Doud's electric brougham parked in front of it, a vehicle which cannot have been common in San Antonio in 1915.

Eventually, warned by her father to stop playing games with the young lieutenant, Mamie accepted an invitation to dinner at the St. Anthony Hotel, followed by the theater, which she remembered as a "vaudeville show" at the Majestic. Once the ice was broken, they went out together frequently, usually to an inexpensive Mexican restaurant called the Original on the banks of the San Antonio River (a pretty place for young lovers then, as now), and a vaudeville show afterward at the Orpheum, which was very popular with the officers from Fort Sam Houston. Ike remembered that they always ate chili, tamales, or enchiladas, and that dinner for two cost him about "$1.25 including tip," which was the most he could afford—he was still paying off the debt for his uniforms. He paid thirty dollars a month for his board, and six dollars a month for laundry, which didn't leave him much for dates with Mamie.[7]

Introduced to the game of craps by two young officers who had come to Fort Sam Houston to take the physical examination for the Aviation Section of the Signal Corps, as the infant Air Corps was then known, Ike, with his usual skill at games of chance, promptly won fifty dollars, and was finally able to pay off the last of his debt to Mr. Springe. He was also interested to learn from the two would-be aviators that fliers got an automatic 50 percent raise in pay—a very welcome piece of news to a penurious infantry second lieutenant. Ike promptly put in his application for transfer to the Aviation Section.

Through the winter, he continued to see Mamie, who eventually dropped all her other admirers. Her parents raised no objection. On one of his frequent business trips, "Pupah" wrote to his "Dear Little Girl" promising her a new dress, and added, "Next month if you still wish to 'horse back' and the Lieut wants you to ride, perhaps we can manage a 'habit' too," so clearly Ike was accepted by the entire family as Mamie's *cavalier servente*. The two spent as much time together as they could—often Ike and Mamie would take the jitney down to the "old Original" restaurant, eat a Mexican meal, then see a movie, and then sit on the porch of the Doud house and talk, sometimes so late into the night that Ike had to walk back to Fort Sam Houston—though that was no hardship, as the balmy San Antonio nights made walking a pleasure, especially "to a young man in love." For there was no doubt about the strength of the young people's feelings for each other, and they became engaged on Valentine's Day, 1916, when Ike gave her his West Point class ring, which, however, her mother refused to let Mamie wear until her father gave his consent to the engagement.

Mr. Doud was away on business—he traveled so much that some biographers of Eisenhower conclude that Doud must have had a mistress or two, though no proof seems to exist[8]—and did not return until March, when he gave his rather cautious approval of the engagement. He worried that Mamie, who had been brought up in the lap of luxury, might find it difficult to live on a second lieutenant's pay—he pointed out to her that he did not propose to give her an allowance once she was married—and thought that nineteen was a little too young for her to marry, so the wedding date was fixed for November, when she would be twenty.

The only hitch came when Ike announced at dinner that his application for transferring to the Aviation Service had been accepted, and that he was to report to the post hospital for his physical examination. In Ike's words, "the news of my good fortune was greeted with a large chunk of silence." The fact that his pay of $146.50 a month would be increased by 50 percent if he became an aviator did not dissipate "the chilliness in the atmosphere." Mr. Doud said sternly that flying was a "dangerous experiment" and a risky novelty, and that it was "irresponsible" of Ike to take it up just as he was preparing to marry. If he went ahead with it, Mr. Doud said, he and Mrs. Doud "would have to withdraw their consent." Glum and miserable, Ike took two days to make up his mind, then returned to the Douds to tell them that he was "ready to give up aviation." This came as a great relief to the Douds, since Mamie, always open to daring new ideas, completely sympathized with Ike's need for adventure and a new challenge and had been "raising quite a fuss" at home.[9]

Ike's disappointment did not last long. As always, his natural optimism prevailed. If he was going to stay in the infantry, he would make the best of it. He decided then and there that he would "perform every duty given me in the Army to the best of my ability and do the best I could to make a creditable record, no matter what the nature of the duty."

In the meantime, determined to save as much money as he could for marriage, he gave up buying ready-made cigarettes, which cost a dollar a carton in those days, and took to rolling his own, at which he soon became expert.

. . .

Mamie was soon obliged to return to Denver with her family, leaving a disconsolate Ike to court her by correspondence. Soon, however, his own quiet life as a junior officer in the Nineteenth Infantry Regiment was disrupted. The relationship between the United States and Mexico, never calm or satisfactory to either country, suddenly heated up again, when President Woodrow Wilson recognized the "constitutionalist" government of General Carranza, thus effectively taking sides in a civil war, and turning Carranza's opponents, including Pancho Villa, against the United States. The regular Army of the United States was still so small that it was necessary to call up a number of regiments of the National Guard to serve in the border region, and regular officers were assigned to assist in training them and handling their numerous administrative problems—for the National Guard was, for the most part, poorly equipped and trained, and notoriously weak in administration. Ike left Fort Sam Houston to live under canvas at Camp Wilson, where the National Guard was being assembled, as "inspector instructor" of the Seventh Illinois Infantry Regiment, composed mostly of Irish-Americans from Chicago, few of whom were pleased to have been plucked from civilian life and sent to Texas.

The regiment's commander was genial, elderly Colonel Daniel Moriarty, who acted more like a ward boss than a commanding officer, and was not overly disturbed by the boisterous behavior of his men, who had a vigorous and long-established reputation for drinking and fistfights. Colonel Moriarty had a politician's instinctive distrust of his own officers, and kept his power by a strategy of divide and conquer when it came to his subordinates, to whom he gave as little authority as possible—with the result that the Seventh Illinois was even more chaotic than most National Guard regiments. As a regular, young Lieutenant Eisenhower was no threat to the colonel, who soon gladly handed over to him full responsibility for training and administration—Ike, in his own words, "wrote all his orders, prepared his reports and other official papers for his signature, and became the power behind the Irishman's throne."[10] This was, in fact, the first instance of an ability which would eventually carry Ike to the top of his profession. Whatever his other merits, older men invariably liked and trusted him, and were happy to let him do their work for them, giving him broad authority and trusting him to know the facts and make sensible decisions in their name. Ike was, from the beginning, an officer who always did his homework, could be relied on in any situation, deferred in public to his seniors, and knew how to make them look good however little time they put in at their desks. It must have taken a good deal of work on Ike's part to make Colonel Moriarty look good, but he was equal to the task.

If Ike felt disappointment at being deprived of the opportunity to play the *beau sabreur* on the Rio Grande, he managed to conceal it. Already, his skill at the job he was doing, militarily unglamorous as the job might have been, was enough to make it unlikely that he would be commanding an infantry platoon, even had the Nineteenth Infantry or the Seventh Illinois National Guard been sent into combat—and this was not imminent or even probable.

When nineteen Americans were removed from a train near Chihuahua and shot to death by Pancho Villa's forces, an American "punitive expedition" was organized at the request of President Wilson. This expedition crossed the border into Mexico and in a hot, sharp little battle killed 200 of Villa's men, with a loss of seven American soldiers and eight American civilians. It seemed to many people that the United States was drifting toward war with Mexico or with Germany, or with both—for the Germans' decision to starve out the British by waging submarine warfare against shipping in the Atlantic led, inevitably, to the sinking of an increasing number of American flag merchant ships. A rigid determination to pursue a policy of brute force even when that seems likely to bring about fatal consequences for itself had always been a hallmark of Germany's diplomacy and strategy, and now Germany continued to increase its submarine fleet in the Atlantic despite the Americans' strong protests. In 1916 the failure of the British offensive on the Somme (which began with 60,000 British casualties on the first day and dragged on from July through November to become one of the bloodiest and most futile battles of the war, costing the British more than 432,000 casualties)[11] and the stalemate at Verdun (where nearly 500,000 French and German soldiers were killed in an area less than three miles square), made it increasingly evident to the Allied leaders that only the entrance of America into the war with unlimited numbers of fresh troops could defeat Germany. Although the Germans could draw the same conclusion—as many of them did—they nevertheless embarked on a naval strategy that may have been the only sure way of bringing a reluctant United States into the war against them.

One way or the other, it seemed to most people that the United States would soon be in the war. Mamie had been nervous enough at the prospect of Ike's going to Mexico, which "seemed to her like going around the world," but with all the talk going on of war with Germany she became more nervous still, and persuaded Ike that they should move up the wedding date. From Denver, she wrote to her parents, who were traveling in the East, while in San Antonio Ike set about the more difficult task of getting twenty days leave. The Army, though unprepared for war in terms of manpower, weapons, and training, was already "on a wartime footing," which meant that all leave was canceled, except for emergencies.

Ike's request did not impress the colonel—in his view, marriage did not constitute an emergency—but he sent it on to the department commander General Funston, in the expectation that it would be turned down. Instead, to his surprise, Ike was ordered to report to department headquarters to explain himself to the general. He reported as ordered, in his best uniform, with boots and brass gleaming, and was left to cool his heels in the office of the general's adjutant, who told him that it was a poor time for a second lieutenant to be getting married. After some time he was sent into the office of the general's chief of staff, who was equally pessimistic, and who warned him to prepare himself for a long wait.

But Ike's luck—and the workings of the old Army—saved the day. During the preceding autumn, Mr. Peabody, the head of one of the military academies in San Antonio, had asked Ike to coach the school's football team, making a tempting offer of $150 a season. In fact Ike had been sorely tempted, but his feeling was that coaching a school football team was incompatible with the dignity of an officer, and he had felt obliged to refuse.

Some days later, General Funston had appeared in the officers' club and asked if Mr. Eisenhower was in the room. When Ike stood up and introduced himself, the general invited him over to the bar for a drink—a most unusual invitation from a general to a second lieutenant. Mr. Peabody, it turned out, was a friend of the general's, and had told him about Ike's refusal.

"Mr. Peabody tells me that he would like you to coach his team at the academy," General Funston said.

"Yes, Sir," Ike replied.

"It would please me and it would be good for the Army if you would accept this offer."

"Yes, Sir."[12]

And that was that. Ike, who as a West Pointer knew that a general's wishes are as good as a command, coached Mr. Peabody's team successfully, and in the way of the Army his obedience was about to be rewarded. When he was finally ushered into the general's office, Funston looked up from his desk, recognized Ike, and gave him ten days' leave to get married.

The Douds were not quite so easily persuaded—they were afraid that people would think the wedding date had been moved forward because the bride was pregnant[13]—but in the end they gave in to Mamie, as they usually did. The wedding would be held on July 1 rather than November; plans for a church ceremony were canceled; and Mamie's wedding dress would be bought off the rack instead of custom-made.

Ike, whose leave papers came through only at the last moment, had no money, as usual, and gave an early and convincing demonstration of his persuasive skills as well as his capacity for initiative—he managed to arrange for an overdraft from the Lockwood National Bank of San Antonio on a Sunday (in those days young Army officers were regarded as a good credit risk), and even persuaded the manager of Hertzberg's, a jeweler in San Antonio, to open the store on a Sunday so he could buy a wedding ring on credit. "So with a new ring," he commented, "new debts, ten days, and high hopes, I started on my journey." He also brought with him the good news that he had just been promoted to first lieutenant. This promotion added a very welcome twenty dollars a month to his pay, so babysitting for Colonel Moriarty and the Seventh Illinois National Guard had paid off.

Ike's train was delayed by a flash flood in northern Texas, costing him one of his ten precious days of leave—in those days, the train journey from San Antonio to Denver took fifty-six hours, even if everything went right, and since he and Mamie were counting on a honeymoon and a quick visit to Ike's parents in Abilene on the way back to San Antonio, every minute mattered.

They were married on July 1, 1916, at noon, in front of the fireplace in the music room of the Doud home. Ike was handsome in his dress-white uniform (he had refused to sit down before the ceremony for fear of ruining the razor-sharp creases in his trousers); Mamie looked tiny beside him, in the off-the-rack wedding dress of white lace, wearing rather less eye shadow than in her formal portraits. (One of these portraits, taken before the wedding, shows her wearing a long-sleeved white dress; an enormous wide-brimmed, beribboned white hat with the edge of the brim pulled low over one eye; a huge white-lace ruff that frames her face and almost reaches her mouth; and more eye shadow than most people would have thought proper in the United States in 1916. Her expression is frankly sultry—no other word will do—and her general look is something like what Marlene Dietrich aimed at many years later, in Hollywood, in the days after she had made *The Blue Angel*. The portrait is fetching, but also startlingly provocative for a young woman who had only recently graduated from Miss Wolcott's finishing school, and it is hard to imagine that Miss Wolcott would have approved.)

They spent their two-day honeymoon in the nearby mountain resort of Eldorado Springs, then returned to the Doud home in Denver and took the afternoon train for Abilene, arriving there about four o'clock in the morning—a hot, sultry, and exhausting trip in the days before air-conditioning. They were picked up at the station by Ike's father—Mamie was astonished to see him standing on the platform in his shirtsleeves, since her beloved Pupah would never have left the house without wearing a jacket—and taken home for an early breakfast of fried chicken, since Ike's mother was determined they should not board the train for San Antonio hungry. Ike later commented that two of his brothers, Earl and Milton, were home and became friends of his bride instantly. What is certain is that Mamie was surprised at the modest size of the Eisenhowers' home, and at the fact that Ida Eisenhower did all the cooking and cleaning herself, as well as disconcerted to discover that the Eisenhowers did not permit alcohol, smoking, or card games in their home. Austere, withdrawn, and rigidly moral, Ike's parents must have seemed a little forbidding to a young woman brought up in the affluent and fun-loving Doud family—one hopes that Mamie had the good sense to forgo her eye shadow for the trip to Abilene! In later years she "would hang out of the upstairs bedroom window to smoke her cigarettes, hoping she wouldn't get caught," and she must have boarded the train for San Antonio with a certain sense of relief.

Married officers' quarters at Fort Sam Houston must also have come as a shock to her, however hard Ike had worked to prepare her for the experience of being an army wife. Even these days, base housing is a constant source of complaint; and in the days before World War I life on an army post was not very different from what it had been in the time of U. S. Grant, with the addition of electricity and indoor plumbing. Ike may have been content with his quarters, but it cannot have been an easy transition for a young woman who had been brought up in a large house and had never even made her own bed to take up married life in two small rooms and a bathroom in what amounted to a barracks building, with the absolute minimum of privacy, let alone to adjust to a rigidly graded world in which everybody's position was defined by rank and seniority.

The formalities of Army life had to be observed—an officer arriving at a new post had to "call" on each of his superior officers within twenty-four hours of his arrival—if he was married, accompanied by his wife, who must be wearing gloves and a hat. The time spent on each call and the order in which the calls should be made was prescribed, as was the fact that the calls would be returned over the next two weeks, as well as the etiquette of "leaving cards" if the person called on was not at home. Mamie, as might be expected, had good manners, but she did not take this kind of thing seriously, and the Army did—more important, so did Ike, who as a West Pointer had been taught the niceties of Army social life as thoroughly as Army drill.

With the help of a wedding present from the Douds, the Eisenhowers were able to buy a secondhand car, and to rent a piano for Mamie, who regarded it "as a necessity"; and between favorite objects and carpets brought from home and wedding presents from Ike's fellow officers, she did her best to make their Spartan quarters comfortable. Rather surprisingly, Mamie had been taught sewing, which she put to good use making curtains; but it had not been thought necessary for a daughter of John Doud's to learn how to cook. Indeed, Mamie's mother, according to Mamie's granddaughter Susan Eisenhower, had advised her, "If you don't learn to cook, nobody will ask you to do it."[14]

Mamie had taken this advice to heart, and as a result, the Eisenhowers were obliged to take their meals across the street at the officers' mess, where, Ike complained, the tapioca pudding and the mashed potatoes were indistinguishable, until that stretched their budget to the breaking point. Ike on the other hand enjoyed cooking, the elements of which he had picked up from Bob Davis, the old guide who had taught him to hunt, shoot, and look out for himself in the woods in Abilene; and he managed to teach Mamie how to a make a few basic dishes—"pot roasts, steaks and chicken," as he remembered it years later. Eventually, they could invite friends home for dinner, and it became worthwhile to dip into what remained of their wedding present nest egg for a tiny refrigerator.

Mamie was shortly to get a more severe lesson of Army life, when Ike was made provost marshal of the post, responsible for keeping order—no easy task with so many of untrained soldiers, and the normal conflict between the regulars and the National Guard. Ike was out late every night, patrolling "the bars and other establishments of dim repute" of San Antonio, which were not in short supply, and was obliged to leave Mamie alone, which cannot have been easy for a nineteen-year-old former debutante who until then "had never even spent a night alone in a house." The duty of provost marshal was more dangerous than Ike liked to admit—one night a drunken National Guard officer fired at him twice with a pistol at close range, and had to be subdued by Ike's corporal—and "there were times when [he was] frightened for Mamie," alone in a camp full of men, many of whom were poorly disciplined. He obtained a Colt .45 automatic pistol for her, and carefully taught how her how to cock it and shoot it, not easy for a young woman with small hands to do. Ike was determined that she should be able to defend herself if anyone broke in, but when he simulated a break-in to test her ability to use the pistol, Mamie had hidden it away behind her piano so securely that he decided "she couldn't have gotten it out in a week," and made up his mind that rather than rely on her ability with a handgun he would simply have to work harder to make the camp a safe place.

Less than a month after their wedding, Ike was sent back to Camp Wilson to continue coaching the National Guard, and Mamie, accustomed to being looked after, complained because he was leaving her. Ike heard her out—he was as calm and reasonable as she was volatile, and he hated any kind of scene or emotional display—then put his arm around her and with an expression that made it clear how serious his words were told her that much as he loved her, she must understand that his "duty would always come first," a remark that could not have been better calculated to chill a romantic and distraught young woman to the bone.

It was a watershed in their marriage—much as she hated it, Mamie came to accept that Ike had meant just what he said, and that it was a deep and unalterable part of his nature. Duty came first, and always would come first, and he was not about to pretty it up for her, or pretend that he was open to compromise on the subject. Eventually, she would come to admire Ike for his devotion to duty, but proud as she was of him, the fact that he could and did choose the call of duty over her was still capable of causing Mamie pain nearly half a century later.

They were, in so many ways, opposites. She was flighty, vain, fun-loving, talkative, wore her heart on her sleeve, was addicted to emotional scenes and outbursts ("tantrums," as Ike called them), was totally uninterested in the outdoors or sports, and was capable of spending one day a week in bed because she had been told it was good for her complexion. He was firmly devoted to duty; was devoid of vanity; loved sports, particularly football, and the outdoors; hated any kind of a scene or disagreement with her; and remained a lifelong early riser who was in constant motion from the moment he got up in the morning. If the marriage worked, it was because in some way she needed a protective father figure, while he needed somebody with whom he could laugh and relax, a woman who would take him seriously, but never too seriously—a combination which, fortunately, came naturally to Mamie, for with her firm sense of independence and fierce loyalty, she was at once amused and impressed by Ike. She was proud too. No matter how difficult things might be between them, she was determined, from the very beginning, not to go running home for comfort, to stand on her own two feet, and to make the marriage work. Making the marriage work meant, among other things, making sure that Ike's career worked, for in the small world of the peacetime Army (and, far in the future, in the even smaller world of general officers), a man's wife was an important element in his promotability. Few things could slow or prevent an officer's rise through the ranks as surely as marrying the wrong woman; and Mamie, despite an upbringing that can hardly have prepared her for the loneliness and rigors of Army life, set about becoming an army wife almost from the first day of their marriage, and succeeded to an extraordinary degree.

Ike's homily on duty had been, as it happened, timely. By the beginning of 1917 the troubles with Mexico had cooled down sufficiently that the National Guard regiments, to their great relief, could be sent home. No sooner had Ike returned from Camp Wilson to Fort Sam Houston than the German government announced its intention of resuming "unrestricted" submarine warfare. This prompted President Wilson to warn Germany that if it did so, it would bring the United States into the war. Once again the Germans ignored the warning and went ahead with their plans. Shortly, American flag merchant ships were being sunk by U-boats, and in April the United States declared war on Germany.

With a regular Army of less than 150,000 men, the United States was, for the moment, in no position to engage meaningfully with the enemy in a war where armies numbered in the millions, and in which the British had just lost almost three times the entire strength of the United States Army in a single, pointless five-month battle that took place in a few square miles of mud for no significant gain. Nor was the regular U.S. Army equipped for battle on the western front—it lacked heavy artillery, field artillery, machine guns, tanks (the importance of which was just becoming evident after their first use by the British in the Battle of Cambrai), and aircraft, as well as skill and experience in combining all these things effectively in battle on a vast scale. This was no secret. Field Marshal von Hindenburg and his chief of staff General Ludendorff, who had replaced the luckless General von Falkenhayn after his failure to bleed the French Army to death at Verdun, had supported the German Navy's demand for the resumption of unrestricted submarine warfare because they believed they could win the war before any significant number of American troops arrived in France, and because in any case they doubted America's ability to create an effective mass army quickly—after all, the United States had not produced a major field army since the Civil War, over fifty years ago.

The German strategy was ruthless and simple—drive Russia out of the war, at once freeing 1 million or more German troops to be moved from the East to the West; starve the British of food and supplies by sinking their shipping; then, in one huge hammer blow, with overwhelming superiority in numbers, split the British and the French armies and drive the British back to the Channel ports before the Americans had arrived in large numbers or were ready for battle. The February Revolution in Russia more or less guaranteed the first part of the plan, and the next move was up to the Kriegsmarine and its U-boat commanders. The Americans' declaration of war was unfortunate, but to the cold, clear, practical minds of the German general staff—and the somewhat less clear but always optimistic mind of the kaiser—it was a gamble well worth taking for the chance of a decisive German victory in 1917.

Of course, like all plans, however ambitious and carefully prepared, it depended on certain assumptions. One of them was that the German Navy could actually bring Britain to its knees by means of unrestricted submarine warfare, and this was to prove unfounded—the adoption of the convoy system for merchant ships, the development of underwater listening devices and depth charges, and above all the Admiralty's ability to break the radio code by which every German submarine reported its position to Berlin once a day would prevent the U-boats from succeeding, close as they came to it. The other assumption—that the United States could not produce a mass army quickly enough to matter on the European battlefield—was invalidated by the hard work and professionalism of the officers whose job it would be to train millions of Americans to fight and to lead them in battle, just such young officers as Lieutenant Eisenhower, now back with Company F of the Nineteenth Infantry in San Antonio.

Unfortunately for the Eisenhowers, the plan for expanding the Army rapidly was, as Ike explained with the clarity of a good staff officer, "to draw cadres of officers and men from each regular regiment to form either one or two more regiments around them with recruits." Thus the Nineteenth Infantry Regiment was to provide a nucleus of regular officers and NCOs to form a new regiment, the Fifty-Seventh; and Ike, who had already proved his ability as a trainer of men and an administrator with the Seventh Illinois National Guard, was picked to help form the Fifty-Seventh. This was a "crushing blow" to both Ike and Mamie. In the peacetime regular Army, rather like the British army, a man's regiment was his home, and most officers expected to stay with the same regiment until they retired or were lucky enough to be selected for staff school and a whole different and more ambitious career path. Ike's friends, as well as Mamie's, were all officers of the Nineteenth, or these officers' wives; indeed it was the only part of the Army that either of them really knew, except for Ike's brush with the National Guard. Then too, the Fifty-Seventh would form at Camp Wilson, under canvas, and so Ike and Mamie would once again be separated. As Ike put it, however, there was "no time to moan over disappointments . . . our outlook had to be forward,"[15] though he does not record whether Mamie took the news as philosophically as he did.

Ike's job in the newly formed cadre was regimental supply officer—a position of considerable responsibility for an officer with less than two years of service. His new commanding officer, Colonel D. J. Baker, told him to expect 3,000 recruits, who were already on their way to Camp Wilson and would be arriving in a couple of days with nothing more than their first issue of Army clothing. Ike was to ensure that they got a good hot meal on arrival, and "to provide them thereafter with shelter, food, supplies, and anything else that men needed to subsist and train." This was a tall order, but Ike was fortunate enough to have at his disposal one of those essential figures of Army life, an experienced and resourceful old supply sergeant, who knew how (and where) to borrow trucks and collect enough tents and food to meet the men's basic needs. Shortly after the men arrived, they were followed by a group of newly commissioned officers who had just graduated from a short course at Fort Leavenworth. On paper, the Fifty-Seventh now had its full complement of enlisted men, NCOs, and officers, but it was still completely untrained and lacking every possible kind of supplies and equipment, from rifles to tent pegs. To add to the confusion, the regiment was suddenly moved from Camp Wilson, which was at least near San Antonio and had been more or less laid out along military lines by the National Guard, to Leon Springs, twenty miles away, a barren piece of government property that was dusty, windswept, deserted, and totally without amenities. "There was not a building or a shelter on it of any kind," Ike remembered, and it had only one well to supply the needs of 3,000 men.

Ike and his sergeant did their best to scrounge what they needed, haunting "the quartermaster, ordnance, engineer, and medical services"; throwing themselves on the mercies of old friends and making new ones; "constantly pleading the case of our new Infantry regiment." Gradually, and with every possible difficulty, the small cadre of West Pointers and experienced old NCOs, the latter by definition wise in the ways of the Army, transformed the Fifty-Seventh into a trained infantry regiment, one with which Ike looked forward to going overseas.

It is one of the paradoxes of military life that war—which diplomacy and politics are meant to prevent and which almost every civilian fears—must be in fact the summit of a professional soldier's ambition. Despite the famous slogan of the U.S. Air Force Strategic Air Command during the cold war—"Peace Is Our Profession"—nobody has ever taken this concept seriously as a reason for joining the armed forces; it is a public relations gesture toward civilians, not a motto to inspire warriors. An argument can be made (and often is) that regular armies dislike war because it tends to upset their careful routines; bring into the army undisciplined, headline-seeking, overenthusiastic amateurs (best exemplified by such people as T. E. Lawrence in Britain and Theodore Roosevelt in the United States); promote the wrong people to high rank; and spoil equipment that has been lovingly cared for over many years (navies feel even more strongly about damage to ships that sailors have spent their professional lives maintaining in spick-and-span condition). There is of course some truth to this; still, most career soldiers spend a lifetime preparing to fight, and do not want to miss the opportunity when it arises. Besides, war provides rapid promotion (as senior officers get killed—or, more frequently, are revealed to be incompetent—those below them in seniority have to be promoted to replace them), as well as medals, decorations, and the chance to make a reputation. "The paths of glory lead but to the grave" may be true enough, but soldiers have ever been inspired to follow those paths, and doubtless always will. Finally, even for the most peace-loving of soldiers, with no particular urge to kill, war is what military life is about, the purpose of the entire thing. A lifetime of making sure that your men's rifles are spotlessly clean and their bayonets brightly polished is not the same as leading them into battle with bayonets fixed, rifles loaded and cocked, and "the light of battle in their eyes," as Field Marshal Montgomery liked to say he looked for in his troops, to the intense annoyance of the troops themselves and of his British and American colleagues in high command.

It would have been unnatural for Ike not share this ambition—he was a West Pointer, a professional, a regular soldier, trained to lead men in combat, and more than usually well-read in the triumphs, glories, and tragedies of military history. As a boy he had combed through his mother's books for stories of Hannibal, Caesar, Pericles, Themistocles, Militiades, and Leonidas; steeped himself in the character and the campaigns of Gustavus Adolphus, Frederick the Great, Washington, Napoleon, Lee, and Grant; memorized Washington's farewell address and his speech to the mutinous officers of his army at Newburgh; imbibed, despite his mother's pacifism, the heady, inspiring accounts of bloody battles, noble sacrifice, and heroism—Hannibal crossing the Alps; Washington crossing the Delaware; the young Napoleon Bonaparte leading his men toward the enemy's guns across the bridge at Arcole; Lee, hat in hand, riding up and down among the survivors of Pickett's charge on the third day of Gettysburg, in full view of the enemy, telling them over and over again that it was his fault and not theirs. "The battles of Marathon, Zama, Salamis, and Cannae," Ike would later write, "became as familiar to me as the games (and battles) I enjoyed with my brothers and friends in the school yard." Long before he reached West Point Ike was steeped in what Stendhal called "the glories and the miseries of war" to a degree that was rare for an American boy, let alone one from a Mennonite background. There are many men, even not a few generals, who are unmoved by the romance of war, but Ike was not one of them. He longed for glory.

For the moment, however, it escaped him. Although Colonel Baker was not as much of a Colonel Blimp as Colonel Waltz had been, Baker was elderly, dyspeptic, and fussy about his food. After several disappointments about his meals, Baker gave Ike the unwelcome job of mess officer. Since breakfast was the meal about which Baker complained most, Ike and his friend Captain Walton Walker (the regimental adjutant, who would go on to become a general in World War II) decided to go out dove shooting on horseback every morning at four, to provide the colonel with a breakfast delicacy he enjoyed. They learned to shoot at least half a dozen doves, since Baker frequently asked if there were any left over for lunch, and together with one of the more experienced army cooks, they experimented on various ways of cooking them—wrapped in strips of bacon and broiled, "stewed with mild vegetables," or made into a dove pie, which turned out to be a winner—and managed to alternate this from time to time with eggs and bacon or tiny lamb chops to keep Baker from getting bored with his breakfast.[16] This was apparently sufficient to keep the colonel in a good mood for the rest of the day. Shooting doves (or anything else) from horseback was no easy feat. Once, when Walker took his usual horse from the picket line and found that it was lame, he chose another only to discover that it wasn't used to having a shotgun being fired right above its head and began to buck. With each buck, Walker's shotgun went off, as he grasped desperately for the reins and tried to hold on to the shotgun at the same time, while Ike and the orderly who accompanied them flung themselves from their horses as shot whistled past their heads. Like being fired at by the drunken National Guard officer it was surely a dangerous experience, but not the kind of action that Ike was hoping for.

He was struck by lightning while giving a lecture on supply in an open field. And he was nearly killed when Mamie, who had never driven anything but her mother's electric brougham, decided to drive down to see Ike in their newly acquired car, only to discover that while she could start it, and steer it, she had no idea how to stop it. She telephoned ahead to say she was coming, and left early to avoid traffic, because she wanted the whole road to herself, and as Ike would remark, "I must say she needed it." When she reached the camp he was waiting for her, and she steered right at him, shouting, "Ike! Get on! Get on quickly! I don't know how to stop this thing!" He managed to get out of the way, then leap onto the running board, open the door, and get in to take over from her. He then spent the rest of the day teaching her to drive.

Other than this kind of excitement, Ike spent his time trying to equip the regiment down to the last item, in the belief that once it was fully supplied, it would be sent overseas to fight. Eventually, thanks to the Herculean efforts of him and his sergeant, the Fifty-Seventh Infantry was equipped with everything it needed except for the new pattern of entrenching shovels, and the carriers by which they were attached to the men's packs.

When a huge packing case arrived at the supply tent, Ike exulted, until the case was opened and was found to contain the old pattern of shovel. He had the cover and the metal straps replaced and shipped the case back to the ordnance department, only to receive months later a bill for $22.04 from the War Department, claiming that nineteen items had been missing from the returned case. Ike wrote back that nothing had been removed from the case, enclosing affidavits to that effect from several officers and men who had witnessed the opening and closing of the case; but eventually, despite many months of correspondence, he was obliged to pay the $22.04 anyway—a lesson in the remorselessness of Army bureaucracy, very similar to that which young Ulysses S. Grant had experienced during his brief service as a quartermaster during the Mexican-American War, which just goes to show that some things about the Army never change.[17] A further lesson in the ways of the Army was that the moment the Fifty-Seventh Infantry was ready for combat Ike was promoted to captain and assigned to train candidates for commission at Fort Oglethorpe, Georgia. Since he had set his heart on going into action soon with the Fifty-Seventh, he asked his colonel to propose another officer for the assignment, and the colonel, presumably still under the spell of Ike's breakfasts, agreed. The War Department, acting promptly for once, disapproved the request. Ironically, it was irrelevant in the end (the War Department, as Ike would discover, rather specialized in irony, when it came to the desires of junior officers)—the Fifty-Seventh, by now one of the best-trained and best-equipped infantry regiments in the Army (except for its missing new-pattern entrenching shovels and carriers), would be assigned to garrison duty in Houston, Texas, for the duration of the war.

Ike's new posting was "distressing" in more ways than one. Because it was "field duty," Mamie, who was pregnant, could not accompany Ike to Georgia. At just the moment when she most needed Ike's support, he would be away while she remained alone in their quarters at Fort Sam Houston. It was his hope to be posted to combat duty straight from Fort Oglethorpe, so she might not see him again before he went to France.

As it turned out, that too was not in the cards. At Fort Oglethorpe, Ike went into the field and helped to design and dig replicas of what candidates for commission might expect in France—trenches, dugouts, barbed wire, shell holes. There, Ike taught infantry tactics and helped in "weeding out the weak and the inept," experiencing all the discomforts of war without the danger, and there he was informed by a telegram (which took three days to reach him) that Mamie had given him a son, named Doud Dwight.

It cannot have been an easy time for Mamie. Fort Sam Houston had no maternity facilities, and although her mother came from Denver to help out, conditions were primitive. She was taken to the post infirmary "on a mule-drawn wooden wagon that made the rounds of the post," and placed in a small windowless room furnished with nothing more than a table and a chair. Mamie later commented, "It is lucky that babies came easily to me," but the experience can hardly have been anything but a difficult ordeal for a young woman without the support of her husband on an army post in which childbirth was, at best, a distraction and an inconvenience to those in command. Mamie chose the name Doud Dwight herself, but almost immediately called the baby "little Ike," a nickname that became shortened to Ikey and then to Ikky.

Once the course at Fort Oglethorpe was up and functioning, the War

Department immediately closed it down, and Ike was ordered to proceed to Fort Leavenworth to instruct "provisional second lieutenants," who had been commissioned without receiving infantry training. (Connoisseurs of the War Department at work in crises will note that in Georgia Ike had been training young men who would not be commissioned as second lieutenants unless they passed the trench warfare course, whereas in Kansas he would be instructing young men who had already been commissioned as second lieutenants without any infantry training at all. Obviously, two different and entirely contradictory policies for commissioning second lieutenants were being followed at the same time.)

In a rare display of compassion the War Department gave Ike three days to see his wife and new son on the way to Fort Leavenworth. Mamie was, by now, surrounded by her family, the Douds having moved from Denver to support her; but she was suffering from pneumonia which, in the days before antibiotics, was very often a fatal disease. "You either came through or you didn't" was how she later summed up her medical treatment, and she was obliged to return to Denver to recuperate, while Ike went on to his new assignment in Fort Leavenworth.

Before leaving he had reported in at Fort Sam Houston and learned that a machine-gun battalion was being formed there to go overseas, under the command of an old friend from the Nineteenth Infantry, Gilbert Allen, now a lieutenant colonel. Ike immediately volunteered to join the new battalion—his study of the infantry tactics of the war had made it clear to him that the machine gun was now the "queen of weapons," as Napoleon called the artillery, in the trench warfare of the western front. This lesson had not yet been learned by the British generals—nor by the U.S. Army, which, despite the genius of Sam Colt and John Browning, possessed so few machine guns that it was forced to borrow them for the American Expeditionary Force from the French, whose Chauchat light machine gun was widely recognized on both sides as one of the least efficient and most unreliable automatic weapons of this or any other war.

Germany, for its part, had a superbly effective water-cooled machine gun, the Maxim (invented by an American, Hiram Maxim, who, after being cold-shouldered by the War Department, had impartially sold the right to manufacture his gun to both the British and the German armies), often mounted on a sled so that it could be pulled quickly from shell crater to shell crater by its crew—and the Germans did not think 1,000 machine guns for a mile of front an excessive number. Deadly accurate at more than 1,000 yards, placed in reinforced concrete bunkers and in shell craters so as to enfilade advancing troops, and moved continually to avoid being hit by artillery fire, German machine guns dominated the battlefield. On the first day of the First Battle of the Somme, a mixed British and Canadian brigade of the Twenty-Ninth Division—to take a single, but typical and pathetic, example—was cut down by machine-gun fire before most of the men had reached their own front line; the few survivors were massacred in the cross fire from machine guns hidden in shell craters as they tried to make their way, bayonets fixed, toward the German front line through barbed wire. In minutes, a battalion of the Newfoundlanders "lost 272 killed and 438 wounded out of 750 who deployed, a casualty rate of 90 per cent." Not a single man reached the German front line.* [18]

Ike had spent his evenings assiduously studying the tactics of the war—he was not the kind of man who would try to teach what he didn't understand himself—and Colonel Allen wanted him badly. No doubt it did not escape Ike's attention too that his friend Allen had been promoted to lieutenant colonel while Ike was still a captain—promotion came quickly to officers put in command of a unit that was going overseas to fight—but Ike was more than willing to stay a captain if he could get into the action.

It did no good. The War Department sent a "curt reply" ordering Captain Eisenhower to proceed "at once" to Forth Leavenworth.

Once there, Ike was summoned to appear before his new commanding officer, who read him a letter from the adjutant general noting with disap-

* Siegfried Sassoon's bitter lines in "The General" express this kind of warfare perfectly:

> "Good morning; good morning!" the general said
> When we met him last week on our way to the line.
> Now the soldiers he smiled at are most of them dead.
> And we're cursing his staff for incompetent swine.
> "He's a cheery old card," grunted Harry to Jack,
> As they slogged up to Arras with rifle and pack.
> But he did for them both by his plan of attack.

proval that Captain Eisenhower had applied for combat duty overseas several times, and reminding him that it was up to the War Department to decide where he could serve most usefully, and up to him to obey orders.

Since the colonel reinforced this reprimand from Washington with one of his own, Ike for once "reverted to the old red-necked cadet" he had been, and furiously defended himself. He was asking nothing, he said, "except to go to battle."

That calmed the colonel, who said, "Well, I think you're right," but it did nothing to mollify Ike's opinion of the War Department or the opinion apparently held there that he was too good an instructor to let go. Through the winter of 1917, he remained assistant instructor for Company Q, consisting of young provisional second lieutenants, most of whom would go straight overseas into combat.* "Whenever I had convinced myself that my superiors, through bureaucratic oversights and insistence on tradition, had doomed me to run-of-the-mill assignments," Ike wrote later, "I found no better cure than to blow off steam in private and then settle down to the job at hand."

That would almost always be his method for dealing with the frustrations of his military career, and he would pay a severe price for it in his later years, when the combination of heavy smoking and suppressing his anger finally caught up with him. In the early months of 1918 his discontent was at its highest—he seemed stuck in his role of instructor. He was good at his job, and he knew it, and he did his best to make things interesting for his pupils; but for himself, he feared that he was sinking into a dull routine. Many members of his class at West Point were already fighting in France—that he was becoming a skilled instructor and had acquired "a mastery of military paperwork" was no consolation. He did not think there would be a future in the Army for "anyone who had been denied the opportunity to fight."

* One of these young officers praised Ike to his parents, writing home that he had "put more fight into us in three days than we got all the previous time we were here. . . . He knows his job, is enthusiastic, can tell us what he wants us to do, and is pretty human, though wickedly harsh and abrupt. He has given us wonderful bayonet drills." Another, less enthusiastic student of Ike's was F. Scott Fitzgerald, who slept through Ike's lectures and disliked him intensely.

His hopes of leaving "the Zone of the Interior," as the United States was strangely called by the Army, were raised sharply when he was ordered to join the Sixty-Fifth Engineers at Camp Meade (later Fort Meade), Maryland. This was the parent group for the formation of an American tank corps for overseas duty—a chance for combat, and for mastering a new weapon.

The tank had been Britain's secret weapon, intended as the means of breaking through the German lines on the western front to open up the way for the cavalry, which would—at last—be able to deploy behind the German trench system on open ground, cut the German communications lines and transport network at crucial points, and bring the war to an end. This touching belief in the importance of cavalry was a deeply held conviction of Field Marshal Sir Douglas Haig, commander of the British Expeditionary Force, and had resulted in keeping tens of thousands of cavalrymen and their horses in readiness behind the British lines, thus adding to Britain's difficulties of supply the immense quantities of forage and bedding that were necessary for horses in the field.

Several people have claimed to be the father of the tank, including Winston Churchill, who put the resources of the Admiralty, backed by his own restless energy, to work on it when the War Office showed an initial lack of interest, bordering on outright hostility. Because Churchill was first lord of the Admiralty at the time, the machine was developed by the Royal Navy and originally known by a naval name: "landship." The word "tank" was Churchill's suggestion for a code name disguising the nature of the invention from the curious. Nobody disputes that it was based on an American invention, the Holt Caterpillar Tractor, which, as early as 1914, had been suggested by its manufacturer as a way to breach the trench system that had, to all intents and purposes, turned the war into a static, bloody stalemate. By 1916 the first British tank, a prototype known somewhat paradoxically as "Mother," was ready for trials. An impressive monster, it was shaped like a rhomboid, weighed about twenty-eight tons, moved at two miles an hour (somewhat less than the average walking pace of an infantryman with rifle and full pack), and could be armed with a six-pounder cannon or four machine guns.[19]

Those who volunteered as tank crews faced great discomfort and hazards. The early tanks were stupefyingly noisy, and little thought had been given to the crews' safety or to ventilation: the interior was hot; smoke-filled; stinking of hot oil, gasoline, and, when the guns were fired, burned gunpowder and fumes. Vision was limited to small slits in the armor plate, and was so poor that someone seriously proposed having each tank guided by an officer on horseback who would pound on the hull with the pommel of his sword to indicate whether to turn left or right.

The early tanks could not be penetrated by German machine-gun fire, but hits from German field artillery could blow off a tread, immobilizing the tank, or, if the hull was hit, send the rivets that held the steel plates together flying and ricocheting through the interior of the tank, lethal chunks of steel—heated red-hot by the impact—that could shred the occupants.

The initial tactical use of the tank was severely compromised by the fact that Sir Douglas Haig had neglected to let either the French or the British army commanders in on the secret—in fact General Sir Henry Rawlinson, commander of the Fourth Army, in whose sector the tanks were first to be used, had never even seen one until a special demonstration was arranged for him a few days before the attack he was planning. Rawlinson's reaction was cautious. "He noted that two of the six tanks on show broke down and that an officer fainted while driving. He thought the crews green and in need of much practice. Nevertheless, overall he pronounced himself 'rather favorably impressed.'"[20] Even coming from an Englishman, and therefore a master of understatement, this was faint praise indeed; and in the end, the debut of the tanks on the battlefield more than justified Rawlinson's caution (as well as his opinion that the cavalrymen were sharpening their sabers in vain). Many tanks broke down, got stuck in shell holes, or became lost; after an initial panic the Germans brought up armor-piercing ammunition; and the "lanes" that had been left clear of advance artillery bombardment for the tanks to advance through were used by the Germans to slaughter the accompanying troops with unobstructed machine-gun fire. Still, as everybody including the American observers could see, the new weapon had immense potential, and in subsequent battles it gradually began to prove its worth.

The French pursued—naturally—an entirely different policy with regard to tanks. Rather than building behemoths like the British tanks, they settled on a tiny two-man tank, relatively light and much faster. The Renault tank, rather than opening the way for cavalry, was in fact designed to *replace* cavalry—it was "an armored, mechanical horse," as one British observer remarked—and to travel rapidly on the roads once the breakthrough had occurred, as one might expect from anything designed and built by M. Renault of the famous automobile firm. After much trial and error, these two approaches would eventually lead to two different types of armored fighting vehicles: "Mother" was in fact the ancestor of the heavy tank, the modern "main battlefield tank" now exemplified by the M1-A1 Abrams or the British Chieftain; and the Renault "tanklet" was the progenitor of fast-moving light tanks like the Stuart and the Chafee in World War II or the Bradley today.

For the moment, having displayed as little interest in the Holt Caterpillar Tractor as it had in Hiram Maxim's machine gun, the War Department of the world's greatest industrial power possessed no tanks at all, and was arranging to have British and French tanks shipped to the United States for evaluation and training purposes. Ike's job was to undertake the equipping and organization of the 301st Tank Battalion, Heavy, which was to go into action in British Mark VI tanks—the latest version of "Mother."

Like the rest of the 301st, Ike, now promoted to the temporary rank of major, eagerly awaited the arrival of these "juggernauts of combat . . . an irresistible force that would end the war." He would later write, "The men dreamed of overwhelming assault on enemy lines, rolling effortlessly over wire entanglements and trenches, demolishing gun nests with their fire, and terrifying the foe into quick and abject surrender," pretty much the same optimistic expectation that Sir Douglas Haig had for the tank, but without the mass of cavalry following behind. Whatever the men may have been dreaming of, this clearly represents Ike's dreams, or at least his hopes, now that he was at last on the verge of leading men into combat.

No sooner had Ike received word that the 301st would be sailing from New York in a matter of weeks, to be supplied with its tanks once it was in France—and, best news of all, that he would be in command—than his hopes were dashed. His "organizational ability" had been so impressive that he was directed to take the remainder of the troops who would not be going overseas to Gettysburg, Pennsylvania, and establish a new training site from scratch, Camp Colt.

Even Ike, who was generally an optimist, had to admit that his "mood was black." And with good reason—he had trained the 301st, arranged for the embarkation, and then been left behind. The War Department, capricious as ever, eliminated the Sixty-Fifth Engineers altogether, and replaced it with a new entity, the Tank Corps, under the command of a veteran of the Boxer Rebellion, who remained in Washington while Ike struggled to build a camp in Gettysburg. This was his first dose of serious, independent responsibility, and he carried it off brilliantly. Starting with no more than 500 men, Camp Colt would quickly grow to become a major training facility with over 10,000 men and 600 officers. Ike improvised training methods by securing naval cannon and machine guns to mount on flatbed trucks so that the men could learn to fire while in motion, using the base of Big Round Top, one of the salient features of the Battle of Gettysburg, as a backstop, and made them learn how to strip and reassemble the machine guns blindfolded. He instituted a "development battalion"—the only one in the country at a camp run by anyone below the rank of brigadier general—to deal with recruits who had learning disabilities or problems, rather than discharging them; and so successful was it that twenty-one "problem soldiers" went on to be commissioned as officers, and many became NCOs, and even managed to close down a hotel in Gettysburg that had been serving liquor to soldiers in uniform against his orders, despite a personal plea from the local congressman. Ike, never a slow starter when it came to learning about how the Army worked, had his first direct experience of the effect of politics on military command, and came out the winner. Still, it was not combat, and although Ike was proud of the job he was doing, he chafed at being in Pennsylvania while American soldiers were at last beginning to fight in large numbers in France. His only consolation was that Mamie and the baby had joined him in Gettysburg, and for the first time he was able to come home at night to the pleasure of being a father and a family man.

. . .

The reason Camp Colt's population was swelling so rapidly was that the British and the American armies in France had negotiated what became known as the "Abbeville agreement," which for the immediate future limited the Americans to shipping overseas only infantry and machine-gun battalions. This was partly a response to acute shipping shortages—the German submarine campaign was at its zenith—and also to the Allies' desperate need to put more men in the trenches. As German reinforcements began to arrive from the East in increasing strength with the collapse of Russia into revolution and civil war, what the French and the British wanted from America in a hurry—though nobody was about to say so frankly—was cannon fodder, not partially trained tank and artillery units without their weapons. Since Ike's tank troopers couldn't be sent to France, they stayed for the moment, under canvas, at Camp Colt.

Though it was not clear at the time, in France the last act of the war was being played out. The all-or-nothing strategy that Hindenburg and Ludendorff had proposed to the kaiser was reaching its logical conclusion. They had played the Russian card, with success—Russia had collapsed, and massive German forces had been freed to move from the eastern front to the West, as predicted. They had played the submarine warfare card, admittedly with mixed results. Now, before the bulk of the American army arrived in France, they had no choice left but to play their last card, the great attack that would drive the British to the sea. Ironically—or, some thought, subtly—they chose to name it the *Kaiserschlacht* ("kaiser's battle"), thus saddling Wilhelm with the responsibility in the mind of the German public for a battle over which he had no control, and the necessity of which he had accepted only with deep reluctance.

It began with a sustained barrage of more than 4,000 heavy guns, and was followed by an attack of 106 divisions across a fifty-mile front that drove the outnumbered British back twelve miles, costing them 40,000 casualties and over 600 guns. A week later the British army had been pushed back almost fifty miles, and of Haig's sixty divisions almost half were shattered. The French were under heavy attack too. The Germans had accomplished what the Allies had failed to achieve since 1914, despite several million casualties—the breakthrough. When Haig and Marshal Pétain—who disliked each other intensely—finally and reluctantly met to discuss the situation, Pétain, then, as in 1940, pessimistic and deeply Anglophobic, revealed somberly that he and his government were determined "to cover Paris at all costs,"[21] i.e., even if it meant abandoning contact with the right flank of the British army.

Maintaining contact between the French and the British armies had been the keystone of the Allies' strategy since 1914—the notion that the French might retreat and leave the British with their right flank "in the air" was the worst nightmare of the British army and government. Pétain's words were almost as much of a shock to the British as the scale of the German attack itself. It was the great crisis of the war. The British war cabinet grimly discussed the possibility of withdrawing the BEF "to the Channel ports"—a foretaste of events in May 1940, when the loss of contact between the French and the British armies precipitated the retreat of the British army to Dunkirk, and then home to England.

On April 11, Sir Douglas Haig released his famous "order of the day" to the retreating British army, which ended: "Many amongst us are now tired. To those I would say that victory will belong to the side which holds out longest. . . . There is no other course open to us but to fight it out! Every position must be held to the last man; there must be no retirement. With our backs to the wall, and believing in the justice of our cause, each one of us must fight on to the end."[22]

Despite more than 300,000 casualties, the British eventually formed a line and held it; and despite Pétain's pessimism, the French army, though badly mauled, managed to hold its own, stopping one of the most powerful German attacks of the war on the Chemin des Dames. "*La bataille, la bataille, il n'y a que ça qui compte!*"[23] the ever-ebullient Marshal Foch cried, as he took supreme command of all the Allied armies at last, steadied the French, persuaded General Pershing to throw his untested American troops into combat, and reassured the British about their right flank. By the end of June, the German breakthrough had been contained. Von Hindenburg and Ludendorff had come within a hair's breadth of winning their gamble, but it had failed. Worse still, their secret support of the Russian revolutionaries, undertaken to destabilize the czar, had succeeded only too well, producing a Bolshevik government in Petrograd whose ideas had already begun to spread among German troops returning from the eastern front and German prisoners of war in Russia. The exhausted Germans, reeling from the losses of the *Kaiserschlacht*—and from the fact that despite astronomical casualties it had failed to produce the promised victory for Germany—were now infected by the political contagion from the East, which they themselves had helped to create.

Through all this, Ike continued to train his men at Camp Colt. Perhaps as a gesture of goodwill toward the American Tank Corps, three tanks finally arrived from France in the summer of 1918—not the British Mark V's Ike had been waiting for, but tiny seven-ton Renaults, which could be armed with either a machine gun or a diminutive one-pound cannon, neither of which had been sent. Two experienced British tank officers also arrived as advisers, and Ike got his first opportunity to develop the good relationship with the British that in the future would be one of his major skills. From them he first heard the name of Winston Churchill, who, he thought, "sounded like a good chap."[24]

Eventually, once the Abbeville agreement had been fulfilled, tank soldiers began to be sent overseas in small batches, and Ike again risked the ire of the War Department by requesting to go with them. This his colonel would not agree to—as usual, Ike was doing his job too well—but he did promise to let Ike go to France in command of the November draft and, perhaps to sweeten the pot, recommended him for promotion to lieutenant colonel, which would come through on October 14, 1918, on his twenty-eighth birthday.*

This promotion coincided with the arrival of another, final disaster of the war: the "Spanish flu," an influenza pandemic that would very soon sweep away millions of people worldwide, soldiers and civilians alike. New recruits brought the disease with them to Camp Colt, complaining of aches and pains, headaches, and high fevers, and by the day after their

* If this seems young for even a temporary lieutenant colonel, it is worth bearing in mind that another West Point graduate, George Armstrong Custer, was made a major general in 1865 at age twenty-six.

arrival men started to die—175 in the first week. There were no coffins; a mortuary tent had to be set up and mass graves prepared. Inevitably, the disease quickly spread to civilians, and Ike was in terror of Mamie's or the baby's catching it. There was no known treatment, but a doctor from the National Guard tried using a strong antiseptic spray on the throat and in the nostrils once a day, and perhaps as a result Ike and his family survived. Losses were catastrophic, however, as they were around the globe. Somewhere between 25 million and 50 million people died, dwarfing the number of deaths in the war, and nowhere was the death rate higher than in the crowded confines of barracks and tented camps.

Ike remained busy preparing the November draft that he was to command for embarkation to France, and, like the rest of the world, was astonished to learn that at eleven in the morning on November 11, "the eleventh hour of the eleventh day of the eleventh month," an armistice had at last been signed with Germany.

The war was over, just as Ike was about to join it.

CHAPTER 6

"The Greatest Disappointment and Disaster in My Life"

Ike beside a Renault tank.

Ike can hardly be blamed for feeling a certain bitterness that he had been cheated of his chance to prove himself as a soldier, but it was eclipsed by Mamie's grief at the news that her beloved sister had died of a kidney infection two days before the end of the war. Mamie and Ikky returned to Denver for the funeral—a trip that turned into a nightmare of inconvenience and discomfort as the whole country ground to a halt to celebrate the armistice. Ike was obliged to remain at Camp Colt, while the War Department went about the business of turning an Army of several million men back into a tiny peacetime force of regulars.

Despite the tactical success of the American Tank Corps in the battle of Saint-Mihiel and the Meuse-Argonne offensive, its survival was questionable. Tanks were seen as an innovation, firmly resisted both by the cavalry generals, who were reluctant to give up the horse, and by the infantry generals who thought of the tank as a specialized weapon for trench warfare—something which everyone agreed had been an aberration, unlikely to be repeated.

Hardly anything is more difficult for a professional soldier to deal with than winding down a thriving establishment. "No human enterprise goes flat so instantly," Ike wrote, "as an Army training camp. Everything that sustains morale—peril to the country, imminent combat, zeal for victory, a sense of importance—disappears. The only thing that counts for a citizen soldier is his date of discharge."[1] Disciplinary problems mushroomed; training programs were discontinued, since there was no point in teaching tank warfare to men who would soon be going home to civilian life; and everybody except the regulars was determined to get out of uniform as quickly as possible.

Fortunately, Ike was able to keep his men busy restoring the camp area to what it had once been—the site of Pickett's famous attack on the third day of Gettysburg. When that major task was completed he was to take nearly 6,000 men by train to Camp Dix, in New Jersey, together with all Camp Colt's equipment (including the three Renault tanks) and records; assist in discharging them, a lengthy and nit-picking process; then take the regulars, a remnant of about 200, to Fort Benning, Georgia.

Even forty-five years later, Ike still remembered that train trip as one of the worst experiences of a lifetime. The trip took four days, "each a year long," as the rickety, bare, unheated, uncomfortable troop train, obliged by War Department regulations to seek out the cheapest route possible, "meandered tortuously through the eastern United States."[2] Ike set up a field kitchen in one of the cars, so the men could at least be fed, and bought candles, since there was no electric lighting. It was so cold that nobody—not even Ike—took his uniform or overcoat off for four days and nights; and by the time the men arrived in Fort Benning, together with the only three working tanks in the United States, they looked like a defeated army, rather than part of a victorious one.

Their welcome at Fort Benning was not warm, either. They were merely temporary visitors, since the War Department still had not decided where to put what remained of the infant Tank Corps, and soon they were on the move again, this time back to Camp Meade, Maryland. Here, to his disappointment, Ike was billeted in bachelor officers' quarters (BOQ), since Camp Meade was not able to house families. Unhappy as Ike was at this continued separation from Mamie and their son, he was about to embark on an unexpected and unusual adventure.

Although the automobile had long since begun to replace the horse permanently, particularly in urban areas, the United States still lacked anything resembling a coherent system of paved roads. An automobile trip of any length was an adventure; nor were the machines themselves reliable even at the best of times—least of all their tires, which were subject to frequent blowouts and punctures. Not only were there few decent roads, there were few garages and still fewer filling stations, which were, in any case, still a novelty. Prodded in part by enthusiastic automobilers and in part by a desire to see whether the automobile and truck could replace the horse and the mule in an Army that remained fond of both animals, the War Department decided to send a military truck convoy from coast to coast. This was something that had never been done before—indeed, many people believed it could not be done; and in view of the fact that most roads outside the big cities were merely graded dirt, which turned into rutted pits of mud in the rain, such doubt was reasonable. There were no road maps, nor was there even any assurance that roads as such existed in large parts of the country. Ike had a lifelong love for anything automotive—largely unrequited, because during most of his career he couldn't afford the kind of car he would have liked to own—and he volunteered to go on the trip as an "observer." For once, the War Department agreed, and Ike and a friend, Major Sereno Brett, joined the transcontinental convoy as it set off for San Francisco, California, on July 7, 1919, from a point just south of the White House grounds (the "zero milestone" marking the occasion still exists), after a formal ceremony and speeches.[3]

The convoy intended to use the "Lincoln Highway" as much as possible, though this was more an idea than a functional thoroughfare. Automobile enthusiasts who believed there ought to be a transcontinental highway had set up an organization that built "seedling miles," short sections of paved road, in order to encourage communities to pave a stretch of road through their area, and promised to provide enough cement for a sixteen-foot-wide roadway if the grading and drainage were done to the standards of the Lincoln Highway Association. In 1919 the seedling miles were few and far between, and mostly unconnected, even in the East. Some notion of what the roads were like can be gleaned from the fact that on the first day out of Washington, D.C., the convoy "traveled forty-six miles in seven and a quarter hours," with three breakdowns of vehicles. On the second day, it traveled "sixty-two miles in ten and a half hours," on roads that were mostly paved, and in good weather. On the third day, the convoy reached Bedford, Pennsylvania, where it was greeted by a crowd of more than 2,000 people, a band, a banquet, and street dancing, having covered, according to Ike's calculations, 165 miles in twenty-nine hours on the road, at a dizzying average speed of under six miles an hour.[4]

Ike was delighted that Mamie and the Doud family met the convoy at South Platte, Nebraska, and drove with it for the next three or four days—it was the first time he had seen Mamie since leaving Gettysburg—but there was still a lot of the country left to cross, and the convoy didn't arrive in San Francisco, where it received a tumultuous greeting, medals, a speech from the governor, dinner, and a dance, until September 6, having taken sixty-one days to cross the country.* Ike was granted four weeks of leave. He spent some of it in Denver, and some—ironically—at a hotel in Lawton, Oklahoma, near Fort Sill (not normally thought of as a vacation spot), where he and Mamie were stranded when torrential rains turned the roads into a bog, making it impossible for them to drive any farther. If Ike had not already believed in the value of paved highways, the week he spent in Lawton would no doubt have convinced him.

* Many years later Ike would remark that this experience, combined with his first view of Hitler's autobahns in 1945, gave him the idea of building the federal highway system, starting with 41,000 miles of superhighway. He regarded this as one of the major domestic accomplishments of his presidency.

. . .

When he returned to Camp Meade, it was to find that the officers of the Tank Corps who had fought in France had arrived home. Among them was a man with whom he was to have a long friendship and, ultimately, a tangled professional experience in the next world war, Colonel George S. Patton, Jr.

He and Patton hit it off from the beginning, partly because they were both true believers in the cause of tanks. Patton, Ike noted, was "tall, straight, and soldierly looking," his military bearing spoiled only by "a high, squeaking voice" that seemed ill-suited to a warrior. It would have been easy enough for a lesser man than Ike to have felt some jealousy toward Patton, who had not only fought in France but had won—though only after much lobbying on his own behalf—a Distinguished Service Cross, as well as a Distinguished Service Medal, four battle stars, and the French Croix de Guerre. Patton had been seriously wounded in combat in the Argonne, where he had advanced toward the German trenches carrying a walking stick and a pistol, against massed machine-gun fire that killed five of the six men who were with him. On top of that, he was rich, kept a string of polo ponies, and was socially well-connected, out of the Army and in it—his sister even had an affair with General John Pershing. Patton was, in fact, in many respects completely the opposite of Ike, flamboyant where Ike was self-effacing, relentlessly pulling every string he could for promotion and decorations, a martinet when it came to his soldiers' appearance, and, in the view of many, a swaggering bully and a snob. It was to Ike's credit that he somehow saw the serious soldier behind the pose, and understood that Patton was a thoughtful and well-read student of war.

Ike may have understood too that courage did not come easily to Patton, and did not come without a cost. There are those who are naturally brave—Ike's total calm and lack of concern when a National Guard officer shot at him twice with a pistol at point-blank range in San Antonio suggests that he may have been in this category—and those who have to screw themselves up to act courageously by a supreme effort of will. Patton was in the latter category; he was terrified of showing fear or acting like a coward, and determined to overcome, at whatever cost to himself, what he perceived as his own weakness. In later years, during World War II, Patton's exaggerated response when he confronted soldiers whose nerve had broken, and the almost embarrassing tough-guy quality of his speeches to his men (which deeply offended his rival, General Omar N. Bradley, and were used to such good effect by George C. Scott when he played the role of Patton in the movie), were both the outcome of Patton's obsessive and relentless concern with his own bravery, and his constant need to dramatize and display his own toughness. These were character traits that his British colleagues—with their national predilection for understatement and (occasionally false) modesty—often found mystifying.*

Both men were fiercely ambitious, but Ike did his best to conceal his ambition, whereas Patton wore his on his sleeve. Unlike Ike, Patton was eccentric, erratic, vain, deeply emotional, and a full-fledged military romantic, in love with the whole idea of glory, capable of writing, as Ike would surely not have been, of his beloved cavalry, "You must be: a horse master; a scholar; a high minded gentleman; a cold-blooded hero; a hot-blooded savage."[5] Such words—and sentiments—came easily to Patton, who saw himself (and wanted others to see him) as a cavalier, a swashbuckling hero on horseback, a student of war history and war poetry; and who at times seriously believed himself to be the reincarnation of great warriors of the past. Perhaps no soldier has ever had a more romantic view of war, and, at the same time, a better understanding of its hard practicalities, than Patton.

The curious combination of a high-pitched voice and a scowling aggressive manner also concealed a certain unexpected gentleness in Patton's character, at any rate toward those who were close to him. The Army had finally relented, allowing post commanders to assign wartime barracks as quarters for married officers and their families; and Patton did everything

* Although much has been made (and continues to be made) of the rivalry between Patton and Montgomery in World War II, the two men in fact admired each other as generals, and also as consummate military showmen, brilliant performers in the roles they had created for themselves. Both understood—and loved—the theatrics of command, and, like their opponent Rommel, dressed for the part. Monty wore baggy corduroys, a gray pullover, and a black beret; Rommel had his trademark checked desert scarf and motorcyclist's goggles; Patton had his polished helmet, his nickel-plated ivory-handled Colt .45 single-action revolver, and his many stars—in one photograph he glitters like a Christmas tree, with a total of twenty polished silver stars!

possible to help Ike prepare for the arrival of Mamie and Ikky, and once they were there, showed them the more gentle, kindly side of his nature, which he was at such pains to hide from most people. The Eisenhowers and the Pattons were next-door neighbors—they had been assigned adjoining dilapidated wooden barracks buildings—and although Ike and Mamie had to do most of the work of making their quarters livable, whereas the Pattons, who between them had serious money, were able to leave the work to builders, decorators, and their own servants, there seems to have been no strain between the two families over the difference in wealth. Ike was a lieutenant colonel, commanding officer of the "heavy" brigade, equipped at last with the Mark VIII tank, the latest version of the original British "Mother," while Patton was a colonel, and commanding officer of the "light" brigade, equipped with the small French Renault tanks. Both of them were only too conscious of the fact that their ranks were temporary, and that these ranks would almost certainly be lowered now that the war was over; and both were enthusiastic about the possibilities of the tank as an independent weapon of war—an enthusiasm which put them sharply at odds with the War Department's view that the tank was merely a means of preparing the way for the infantry.

The conventional view—shared by the War Office, in London—was that the tank's only purpose was to advance at the same pace as the infantry, but about fifty yards ahead, eliminating machine-gun nests and clearing a path for the troops through the wire. Given this understanding of the tank's role in war, a tank did not need to move any faster than an infantryman could walk over rough ground, of course; nor was its range of any great concern.

Young officers who had some experience of tank warfare, like Patton, and those who had given some thought to the potential of the tank, like Ike, had very different, more ambitious, and unconventional ideas about its role. They were not alone. Captain Basil Liddell Hart in Britain, the prophet of tank warfare, and such visionaries as Charles De Gaulle in France and Erwin Rommel in Germany, would all come to the same conclusion, and each of these three would write a highly controversial and influential book on the subject. Their theme was the same: the war of the future would be one of total mobilization and swift movement; and the tank, moving rapidly, would be the chief instrument, breaking through the enemy's forces before he could form a line (and thus making a repetition of the static trench warfare of what was then starting to be called the Great War impossible), spreading out behind the enemy in fast-moving columns that would cut his communications and lines of supply, creating the conditions for victory which the infantry could then exploit. This would not be the tank as it was at present, of course, conceived of as a lumbering, slow-moving support for the infantry—the tank of the future would travel at twenty or thirty miles an hour, and move 100 or 200 miles in a day, "with surprise, brutality, and speed," to quote De Gaulle. What they foresaw would later be called blitzkrieg, "lightning war," in which masses of tanks and aircraft would range far and wide, defeating whole armies, even whole countries, before there was time to fight a conventional major set-piece battle.

This was not a vision that most generals shared, or were even prepared to tolerate, not in Britain, France, or Germany, and certainly not in the United States. It was a radical concept of war that would appeal to the imagination of Adolf Hitler, but in the 1920s Hitler was still merely a fanatic right-wing demagogue, leader of one among many fringe groups in the lurid chaos of postwar German politics, and even the most enthusiastic and starry-eyed believers in what would eventually come to be called armored warfare had to face the fact that the tanks they were describing simply did not exist. It was one thing to envision such tanks heavily armed and armored, maneuverable and swift, leapfrogging behind the enemy 100 or 200 miles a day, supported by fleets of aircraft, and able to communicate with each other by radio; but meantime the real tanks were still the huge, unreliable Mark VIII, which, even when fitted with a Liberty aircraft engine, was hard put to keep up with the infantry, and the tiny Renault, which bogged down easily in trenches, lacked the power to climb steep slopes and in combat had often been guided by an officer walking beside it and rapping on the hull with his cane, as Patton had done under fire in France.

Neither Ike nor Patton was by any standards a wild-eyed visionary.

These ideas were in the air after the war among junior officers in every major industrialized country, but most senior officers and military bureaucrats in the victorious Allied nations took the view that the war had been won by massed infantry and artillery; that the bayonet, the hand grenade, and the machine-gun represented the summit of military technology; and that there was thus no need to contemplate a change, since these tactics had proved successful in the end. In Germany senior officers and military bureaucrats believed that since it was clearly the politicians, not the army, who had lost the war, there was nothing to be learned by the army from Germany's defeat except to insist on a different kind of politician—the German army had been treacherously stabbed in the back on the home front, a belief which swiftly took on the most ineradicable and sinister of meanings.*

Ike and Patton were certainly conscious that their ideas about tanks were controversial and unlikely to be welcomed by their superiors or by the War Department. Without drawing too much attention to themselves, they experimented with tanks, and spent their spare time taking a tank completely to pieces like a pair of shade-tree mechanics, right down to the last nut and bolt, then painstakingly reassembled it to understand how each component worked. Ike's fascination with the internal combustion engine and the automobile was certainly a factor in his enthusiasm for the tank—after all, he had helped to prove that automobiles and trucks could cross the United States—while Patton's combat experience gave him a practical sense of what the tank could do, without, however, in any way diminishing his belief in the value of the horse as an instrument of war. Both men made a study of J. Walter Christie's revolutionary designs for tank drive and suspension—Patton even made a trip at his own expense to Christie's

* According to Sir John Wheeler-Bennett, the phrase "stabbed in the back" was in fact accidentally suggested to General Ludendorff by a British general at a dinner party in Berlin after the war. Ludendorff had been explaining, in vaporous Teutonic generalities, how it was that the German army, while undefeated in the field in the West and victorious in the East, had been forced to surrender by betrayal, chaos, and revolution at home. The Englishman, exasperated by the torrent of words, sought to put an end to them, and, interrupting, asked, "You mean that you were stabbed in the back?" Ludendorff reacted with amazement—it was exactly the phrase he had been searching for, and the *Dolchstoss*, the stab in the back, soon became a core belief of the German right wing, at first attributed to the Social Democrats, then to the communists and Jews, and finally, in Hitler's version, to the Jews alone.

workshop in Hoboken, New Jersey, a mecca for tank enthusiasts. These designs promised a tank that could travel at highway speeds and cross even the roughest terrain without bogging down.*[6]

Ike and Patton were hands-on tank tacticians—theirs was the American way, as opposed to sitting down and writing a book like Liddell Hart, De Gaulle, or Rommel. Together they devised ways for the big Mark VIII to tow two or three Renault tanks behind it over rough terrain, and they were nearly decapitated when one of the cables snapped. Trying to decide which was the best machine gun with which to arm tanks, they fired so many rounds from a Browning .30 caliber that it overheated and began to "cook" the round in the chamber just as they went forward to examine the target, spraying them with machine-gun fire while they worked their way around the fire back to the gun so that Patton could flip the belt over, jamming the feed. "We had acted like a couple of recruits," Ike later wrote ruefully, but if nothing else a bond was formed between the two men, over those months, that Ike would always bear in mind whenever Patton got into a serious scrape with the press or the politicians during World War II.

The bond was reinforced when they finally managed to provoke the War Department's wrath. They had "analyzed the tactical problems used at Leavenworth in Command and General Staff School courses," added tank forces into each one of the problems, and proved, on paper, that the side with "a complement of our dream tanks" always won. Ike wrote a paper on this for the infantry military journal, while Patton wrote another for the cavalry journal. That was enough to bring down official disapproval on their heads, and put an end to further experimentation with tanks. Both of them were told that their ideas were wrong, dangerous, and in contradiction of U.S. military doctrine, and warned that if they continued they would be court-martialed. Unlike Billy Mitchell, whose unorthodox ideas about airpower helped to bring about his court-martial, Ike and Patton were not prepared to become military martyrs. Soon further proof

* Christie's design for tank suspension was at first adopted, then largely ignored and forgotten by the War Department, rejected by the War Office in Britain, and finally sold to the Soviet Union, where it was embraced with revolutionary enthusiasm, and would serve as the basis for the Russian T-34, probably the most advanced—and influential—tank design of World War Two, and on which all modern tanks have since been based.

of official disfavor came. They had put in a lot of time training a Tank Corps rifle team to enter in the marksmanship competition at Camp Perry, only to be told at the last minute that there would be no Tank Corps rifle team, since the Tank Corps was merely part of the infantry. This was a sign of things to come. In 1920 the Tank Corps was finally abolished and made a part of the infantry. Patton would return to his first love, the cavalry, while Ike would search elsewhere for a new assignment.

Professional disappointment aside, Ike's period with the Tank Corps at Camp Meade seems to have been one of the happiest times of his career. He was busy, he had the sense of being on the cutting edge of things, he was effectively commanding a brigade, and above all he and Mamie and little Ikky were together as a family at last. They went to a great deal of trouble to improve the quarters provided for them by the Army, waxing floors, painting, and installing modern conveniences; and Ike, a Kansas boy who loved to work on the land and grow things, put in a garden, planted a lawn, and had a white picket fence built to make the old barracks look like home. Though they had very little money, the Eisenhowers entertained constantly, their house becoming a second home for young officers who were tank zealots; and they adopted "the Doud tradition of a Sunday buffet open house," which became known as "Club Eisenhower."[7] Mamie loved entertaining, and was good at it, not just with Ike's fellow officers, but with the occasional bigwig who came from Washington to see for himself what the Tank Corps was up to. Their little boy, now age three, was adored by both his parents. When Ike came home at the end of the day, he would pick Ikky up and carry him on his shoulders, and he loved to take Ikky to parades, football games, and noisy tank displays. The younger officers had a tiny Tank Corps uniform made up for the child, and dubbed him the mascot of the corps. When Mamie's parents came to pay a visit in the autumn of 1920, they were relieved and delighted to see how happy Ike and Mamie were, and what a "harmonious" life they were living. The trauma of continual separations and the rough living at Fort Sam Houston in San Antonio were behind them. Mamie sent home for her "tea wagon," a sure sign that she was settling into her role as an army wife; and she described for her family Ike's surprise thirtieth birthday party, at which his officers presented him with a silver cigarette case with "a small tank engraved on the inside and the words 'Col. Ike a friend indeed.' "

The Pattons, nearby, often entertained too, on a more lavish scale. At one of their formal Sunday "dinners"—a meal we would now call "lunch"—Ike made the acquaintance of Brigadier General Fox Conner, former operations officer on the staff of General Pershing in France, and, as Ike put it, "one of the Army 'brains.' " Conner was that rare creature, a general officer who loved reading, studied ancient history and warfare, recognized talent when he saw it, and tried to get talented officers promoted to positions where they could learn more and make a difference. Conner wanted to see the tank training schools Ike and Patton had created, so they took him for an afternoon tour. He found himself a chair in one of the machine shops, and began to ask questions about tanks, directing most of them at Ike, whom he recognized as the intellectual leader of the two—the brains, as it were, of tank warfare. Conner stayed until dark, listening, as Ike explained the theory and practice of tank warfare as he and Patton had developed it, then remarked on what an interesting afternoon it had been, and went home. Ike had no idea that he had just met—and impressed—one of the men who was to have "a tremendous influence on [his] life."

In the meantime "the greatest disappointment and disaster" of Ike's life was about to strike him. Feeling that they could, at last, afford a part-time maid, the Eisenhowers hired a local girl. She was "pleasant and efficient"—but what she did not mention was that she had recently recovered from a mild bout of scarlet fever. Little Ikky fell ill while Mamie was out doing her Christmas shopping. She concluded that he had a stomach ailment, and called the doctor. He agreed, but after he had gone the child's temperature began to soar. In the morning, he had to be taken to the hospital, and his condition began to deteriorate. Since Army doctors had little or no experience of pediatric medicine, Ikky was eventually moved to Johns Hopkins, in Baltimore, where he was diagnosed with scarlet fever and placed in quarantine. Meningitis set in—in those days, before antibiotics, there was no cure—and the child died in Ike's arms on January 2, 1921

(Mamie had been sent home to bed with a cold so bad that she was thought to have pneumonia). Many years later, at the end of her life, talking to her granddaughter Susan Eisenhower, she said, "We never talked about it. I never asked him because it was something that hurt so badly."[8]

Ike would blame himself for the rest of his life for having missed so much of his boy's life by his frequent absences from home, while Mamie blamed herself for having not been there when he died—and perhaps for having trusted the Army doctors, who could hardly be expected to know much about children's diseases. Even a lifetime later, forty-five years after the boy's death, in a book that is for the most part very far from being soul-searching, *At Ease*: *Stories I Tell to Friends*, Ike's pain is still evident when he comes to Ikky's death. It was a blow that he had "never been able to forget completely," he wrote. "Today when I think of it, even now as I write of it, the keenness of our loss comes back to me as fresh and as terrible as it was in that long dark day soon after Christmas, 1920."[9]

Dealing with the bills from the doctors and nurses who had been able to do nothing for the child, Ike allowed himself a rare moment of bitter sarcasm, quite out of character, in writing to Mamie's parents: "Tell Mother that the nurse *must* have been good. She charged fifty dollars for 4 days."[10]

Returning to work helped dull Ike's grief, but Camp Meade must now have been difficult for him to bear. When General Conner sent word that he was going to Panama to take command of an infantry brigade, and hoped Ike would accompany him as his executive officer, Ike leaped at the chance. Ike's commanding general was less than enthusiastic—he did not think he could spare Ike—but he at last agreed to send Ike's application on to the War Department, though he predicted it would be turned down.

He was as good as his word, and absolutely correct in his prediction. As usual, when it came to a change in Ike's career, the War Department turned him down flat.

The next few months must have as been painful for Ike as they were for Mamie, stuck at Camp Meade, where everything reminded him of his son, knowing that the Tank Corps was now a dead end, reduced back to the rank of major, and with no bright prospect to excite and challenge him.

Then, in the winter of 1922, "orders came from out of the blue" for him to proceed at once to Panama. As usual, the Army was responding to the powerful pull of rank, influence, and old Army friendships—it was a perfect example of what is called, in the British armed services, the "Old Boys' Club" at work. General Conner was an old friend of General Pershing, who had just been made Army Chief of Staff, and had mentioned to Pershing that his request to have Major Eisenhower as his staff officer had been turned down by the War Department. Within days, Ike's orders had been issued, and the furniture packers from the Quartermaster Depot were at the door. For both Ike and Mamie it must have come as a great relief—whatever difficulties might lie ahead in Panama, it was at least a change, a novel, exotic world in which they could start again in new surroundings.

CHAPTER 7

The Education of a Soldier

One thing that Ike had not had so far during his career was a mentor. Now he was about to have three mentors in succession—Fox Conner, John Pershing, and Douglas MacArthur. Of the three, the least well known, Fox Conner, would have the greatest influence on him.

Even without the advantage of a mentor, for a young officer of thirty-one Ike had received valuable training. He might regret the fact that he had not experienced combat or commanded troops in battle, as Patton had, but few junior officers in the United States Army had had more responsibility thrust on them early in their career, or learned more about planning, logistics, and administration. Ike was a rara avis in military life, a disciplined, dedicated, efficient, hardworking officer who didn't appear to take himself too seriously, but who knew how to make things work by the book, and also knew, as every good old-time NCO did, how to get thing done "the Army way" when there wasn't time to go by the book. His efficiency, energy, and sheer competence were screened by what had by now become second nature to him, a genial, easygoing, relaxed style that put people at their ease. Unlike Patton, whose fierce, undisguised competitiveness and constant need to win, and whose rigid insistence on perfection down to the last small detail of dress, turnout, and appearance in his soldiers made him feared (and sometimes ridiculed), Ike managed to get things done smoothly by patience, example, and exhortation—the qualities that had made him, among other things, a highly valued football coach. Everybody in the United States armed forces claims to be a team player—that, indeed, is a rationale for the intense concentration on sports, particularly football, at the service academies—but Ike (unlike many of the senior officers he would work for, and others he would later command, who were, while giving lip service to team spirit, in fact prima donnas) was genuinely a team player. His talk about teamwork wasn't pious rhetoric—he was a true believer, and people sensed this, both the soldiers and junior officers below him and his superiors as well.

He could be tough too—as F. Scott Fitzgerald had discovered—with anybody who he thought wasn't doing his best. Ike wasn't a screamer like Patton, but when he bore down on somebody, the Eisenhower grin vanished instantly, his face flushed, his mouth become set in a rigid line, his jaw was thrust out, and his eyes were hard; and nobody doubted that he meant business. Stupidity sorely tried his patience; he was not, then or in the future, a man to suffer fools gladly.

Although Ike would develop a well-earned reputation for tact—and for soothing the ruffled feathers of difficult generals, admirals, and political figures—he was by no means the easygoing figure many of his critics have made him out to be. In particular, British generals tended to write Ike off as a "nice guy," sometimes even using this typically American phrase to describe him, as if he were the American equivalent of Field Marshal Sir Harold Alexander, the one British general whom everybody liked, from Churchill and Monty to Ike himself. But this was an illusion; in fact, Ike was tough and could be, in his bleaker moments, very harsh indeed. He axed people who didn't come up to his standards, even if they were old friends; and political figures who tried to impose their will on him, including such difficult and overbearing men as Churchill and De Gaulle, found that once Ike had made his mind up he was immovable. Very early on he learned the importance of being able to say no.

Also, if he had ever had any doubts on the matter, his son's death taught him not to take things for granted. Happiness could be taken away as quickly as it was given. The death of their son transformed both parents. As Ike's granddaughter-in-law Julie Nixon Eisenhower would write many years later, Ikky's death "closed a chapter in the marriage." At twenty-four, Mamie was suddenly matured by tragedy—not only the death of her boy, but that of her beloved sister. She somehow found in herself the strength not to slip into depression and grief, and instead did her best to take Ike's mind off the child's death—and learned the hard way that he dealt with that kind of thing by not talking about it and plunging deeper into his work. He feared, for both of them, what he called "the ragged edge of breakdown,"[1] and it is largely thanks to Mamie's courage and determination to put on a cheerful face that no such thing occurred.

Probably nothing could have been better for both of them than leaving Camp Meade, with its memories of Ikky and his death, and going to Panama, even though Panama was in those days by no means an easy or comfortable posting for a married couple. It represented a particularly courageous decision on Mamie's part, for she was pregnant. But she was as determined not to let Ike turn down on her behalf an overseas posting that he ardently desired as she was to accompany him. Nor was the Army solicitous about the comfort of junior officers and their wives when it came to transportation. Even Ike, not ordinarily a complainer, noted tersely about the voyage, "The accommodations were miserable." Packed in like sardines aboard the military transport *San Miguel*, the passengers were subjected to every possible discomfort, including a storm at sea. The Eisenhowers had been bumped from a roomy suite by the last-minute arrival onboard of a general, and instead were obliged to share a cramped, tiny, airless cabin, with a small couch and two narrow bunk beds, one above the other. Although Ike later remembered with pride that neither he nor Mamie was seasick, that does not correspond at all to her recollection of the voyage. "Suffering from morning-sickness, claustrophobia and seasickness," she could not bear to lie in her bunk, and instead huddled on the small couch, as close as she could get to the single porthole. The ship traveled at the leisurely pace of all military transports through an endless succession of tropical storms, pausing to dock at Puerto Rico and Port-au-Prince in suffocating heat and humidity. When the *San Miguel* finally arrived in Balboa, Mamie, longing for dry land, was almost floored by the hot, steamy air, while Ike was infuriated to discover that their Model T Ford had been inadequately strapped down, and had broken loose during one of the storms, suffering serious damage, in addition to which somebody had stolen the new distributor and carburetor that Ike had painstakingly fitted back at Camp Meade.

Panama had not changed much since the digging of the canal, the great impediments to its progress over the years having been the awful climate and the almost unique abundance of tropical parasites and diseases. Had the means for treating yellow fever and for preventing malaria not been discovered, the canal might never have been completed at all, but the country was still plagued by every biting insect, bat, and poisonous snake, and by tropical diseases that in the age before antibiotics and other modern treatments could be incurable and sometimes fatal. The country itself—much of which was inaccessible, owing to the rugged terrain, the heavy jungle, and the lack of roads—was awash with corruption, poverty, and the funereal consequences of politics played as a blood sport. Even Panama's neighbors, none of them models of civic virtue, looked in its direction with horror.

Ike and Mamie made their way alongside the canal on an old-fashioned train until it reached the point nearest Camp Gaillard, where they were obliged to disembark and "walk hundreds of yards in tropical heat across the Canal on one of the lock gates."[2] Ike seems, once again, to have been playing down the difficulties of the trip for a young woman who was pregnant and not physically adventurous at the best of times. Mamie remembered that the two banks were connected by a narrow catwalk over the lock and that it was necessary to make one's way across it holding on to a rope.

Nor was Camp Gaillard, when they finally arrived there on foot, a sight calculated to cheer Mamie up. Built by the French during the closing years of the nineteenth century, during their failed attempt to dig a canal, the houses had been "unoccupied for nearly a decade" and were in a state of decay, disrepair, and collapse. The Eisenhowers' quarters, although next to General Conner's, consisted basically of a ramshackle wooden hut on stilts, with a porch that opened directly onto the encroaching jungle. Torn screens and broken latticework covered the windows, and the interior was "infested with lizards, spiders, cockroaches, and snakes," as well as the ubiquitous rats (which were one of the reasons for building the houses on stilts in the first place) and the bats, which had been imported by the French in the mistaken belief (or hope) that they would eat the mosquitoes.

Ike seems to have taken their new home in his stride, but he does pause to note that "Mamie hated bats with a passion,"[3] as if the other vermin didn't bother her much. It is hard to imagine how the pampered daughter of John Doud coped with Camp Gaillard, or felt about it, but among the things she did note with displeasure were the bedbugs that infested their mattress, and that required the bed frame to be doused in kerosene and set on fire once a week; and the mildew that constantly covered her treasured belongings and furniture, no matter how vigorously or frequently they were cleaned. In those days the Canal Zone was not a neatly landscaped small replica of middle-class America, with golf courses, pools, and tennis courts, as it later became; nor was there much to attract officers and their wives in Panama City—indeed the attractions of Panama City presented one of the serious disciplinary problems for the United States Army, since they consisted of countless bars, tattoo parlors, gambling joints, and brothels. Prostitution was endemic, and, not surprisingly, so was venereal disease.

Eventually, and with great difficulty, Mamie adapted to life at Camp Gaillard, though she never became accustomed to the bats. The house was repaired and renovated, she cut her hair short for comfort in the heat—this was the style she would wear for the rest of her life—and she made friends with the other army wives on the post. The camp was poised on the edge of the famous Culebra Cut—essentially a wide, deep ditch—and large sections of the banks on either side collapsed regularly into the canal, keeping a small fleet of dredges busy. The view from the back of the Eisenhowers' quarters was solid jungle, constantly trying to grow its way around and into the house as if it were determined to reclaim the camp. The view from the front was across the canal, hardly an inspiring sight, since it consisted then of muddy, terraced earthworks on a gigantic scale—photographs of the Culebra Cut at the time make it look like a huge strip mine, with a muddy river at the bottom. Many years later, after World War II, flying over the Canal, Ike was startled to see that Camp Gaillard itself had collapsed into the Culebra Cut; and while he was living there, the fear that it might do so at any time must have added to the feeling of impermanency everybody seems to have had.

Mamie shopped at the post commissary, and—despite the alarming catwalk across the canal—occasionally accompanied the other wives on shopping expeditions in Panama City. She seems to have devoted a lot of energy to building a decent maternity hospital for enlisted men's wives—something which the War Department had apparently not thought of doing—and was determined to have her baby there. The arrival of the Douds on a visit, however, put an end to that idea, and it was decided she would return to Denver to have the baby. Ike remained at Camp Gaillard until just before she was due to have the baby, then joined her in Denver. Meanwhile he was a faithful correspondent, though his letters are more dutiful than passionate—his talent for writing love letters had not improved since his days at West Point—and were full of accounts of his continuing struggle against mildew and bat guano in the ceilings: hard as it is to believe, the rainy season in Panama made its normal level of discomfort seem idyllic by comparison. In the rainy season mildew covered Ike's boots and Sam Browne belt overnight, metal rusted, brass turned green, and the damp seeped through every wall and ceiling. "I will have it painted afterwards," Ike wrote, referring to the bathroom ceiling, then added plaintively, and no doubt correctly, " . . . I don't suppose it will ever dry."[4]

Ike arrived in Denver just in time for the birth of his son John, and returned to Panama three weeks later. Mamie, no doubt happy enough to be home in Denver again, far away from bats and mildew, stayed there another couple of months, then returned to Panama with the baby and a nursemaid, sailing from New Orleans aboard a banana boat, which her parents had supposed would be more comfortable than a military transport. This was apparently not the case, however. Shortly before docking, Mamie wrote to her parents complaining about her fellow passengers and the dirtiness of the ship, and adding, "This has been a terrible trip." Nor was her return to Ike as happy as she had expected. Ike was delighted with his new son, and pleased to have Mamie back, but he was single-mindedly engrossed in his work and with his relationship with General Conner, and Mamie found herself, once again, alone a great deal of the time.

Ike was away a lot, on inspection tours, and it would have been hard for Mamie not to feel abandoned, and frightened, too—having lost one child, she could hardly have failed to feel that the Canal Zone was not the ideal place to be bringing up the next. Army doctors had failed poor little Ikky, but at least Johns Hopkins had been near at hand, had they only brought the baby there sooner, whereas here, in the Canal Zone, there was no medical help to rely on except that of the Army. As Mamie wrote later, "I was down to skin and bones and hollow-eyed; so ill I'd have to walk all night long. The porch was screened on three sides and I would walk all night listening to mosquitoes buzz. I could hear the monkeys scream in the jungle, and I felt like screaming too."[5]

Photographs taken of Mamie in Camp Gaillard show very well that she was neither exaggerating nor guilty of self-dramatization—she looks emaciated, forlorn, wasted, like a concentration camp survivor, and totally stripped of her usual elegance and joie de vivre. Other wives thought that the marriage was breaking up, and perhaps it was, for Mamie finally managed to work up the energy to confront Ike and tell him that she and the baby were going home.

Her departure was surely one of the most difficult moments of what was not, by any stretch of the imagination, an easy marriage. Ike was heartbroken and resentful; while Mamie was deeply upset but determined to go. She must have been aware that he had found the most demanding of mistresses, a complete devotion to his career. For as Ike put it, having briefly noted how little there was for Mamie to do, "*my* tour of duty was one of the most interesting and constructive of my life. The main reason was the presence of one man, General Fox Conner."

Conner was "a tall, easygoing Mississippian," both a man of action and a widely read military historian, with "an extraordinary library," and what was clearly a great deal of charm. He was astute enough to perceive Ike's intelligence, and to know how to guide it toward increasing Ike's knowledge of his profession. He started by lending Ike historical novels, some of which Ike could remember forty years later, like *The Long Roll* by Mary Johnston; *The Exploits of Brigadier Gerard*, about the Napoleonic wars; and *The Crisis*, by Winston Churchill (not the English politician but an American best-selling novelist). Then, when Ike had fallen into the habit of reading these novels—there was little else to do at night in Camp Gaillard—Conner asked him if he wouldn't be interested in learning what the armies were actually doing in these periods.

Ike took the bait, and was soon devouring Conner's remarkable collection of military history, and sitting up well past midnight every night talking about the decisions and personalities of each period. Ike's dislike of military history—a reaction against the way it was taught by rote at West Point—gradually gave way as Conner quizzed and challenged him about each book he read, engaging his interest and curiosity. Ike, who had developed a particular aversion to the history of the American Civil War, now read memoirs and histories of the war until he knew every campaign by heart, while Conner questioned him, night after night, about the decisions that had been made and what the outcomes might have been had they been made differently, teaching him to think about strategy and about the entire process of making military decisions. He made Ike read Clausewitz's *On War* three times—no mean feat—until Ike could not only recite pages of it by heart but defend his own view of Clausewitz's maxims about waging war.

On reconnaissance trips, when they were out on horseback for days at a time, camping at night in the jungle, the two of them sat at the campfire talking about Napoleon's or Caesar's campaigns, or the Civil War, or going over the maps they were making of the terrain of the Canal Zone, and arguing how best to fight and defeat an enemy landing there, in the unlikely event that there should ever be one. Conner was something of a philosopher, as well as a student of military history, and two of his favorite quotations stuck in Ike's mind forever. One was "Always take your job seriously, never yourself," which became the standard by which Ike judged his own actions. The other was "All generalities are false, including this one,"[6] which he used time and time again to make people concentrate on the facts of a matter, not their opinions about it. Conner instilled in Ike a love of Shakespeare, particularly the soldiers that fill the plays, and encouraged him to read Plato, Tacitus, and Nietzsche. It was, Ike wrote, "a sort of graduate school . . . leavened by the comments and discourses of a man who was experienced in his knowledge of men and their conduct." Conner introduced Ike to the works of Polybius, Xenophon, and Vegetius, and encouraged him to delve deeply into the battles of Greek and Roman antiquity. He even went so far as to suggest that when the next world war began (for he was in no doubt that it would come, and believed, like Marshal Foch, that the Versailles Treaty was, in Foch's pregnant words, merely *une armistice de vingt ans*, a "twenty-year armistice"), Ike should hitch his way onto the star of Colonel George C. Marshall, who Conner thought was the officer most like himself in the United States Army.

It is hardly surprising that Ike's young wife—accustomed to being the center of attention from earliest childhood and the life of the party since adolescence, still deeply shocked and grieved by the death of her first child, and more or less stranded in the jungle, away from her family and her friends—should have decided that if Ike was so devoted to General Conner and wanted to spend more time talking to his commanding officer than to her, she might as well take the baby and the nanny, and go home to Denver. The wonder is that she had stuck it out as long as she did.

And it is perhaps no accident that in Ike's account of his time in Panama he devotes far more space to training his horse Blackie than to Mamie's feelings about being there, or her decision to go home. This is certainly the only part of *At Ease* in which Ike waxes rhapsodic, as he describes picking Blackie out of the corral to be his mount—a coal-black gelding of sixteen hands, who looked more like a mule than a horse, even in Ike's indulgent opinion, and seemed clumsy and short-necked—and gradually teaching him aids and tricks until Blackie became something of a local equine celebrity. Ike taught him to kneel when he wanted to dismount, to halt and move forward on voice commands, and even to climb and descend steps—a most difficult and unnatural thing for a horse to do. General Conner, an admirer of Ike's skill with horses, had the camp carpenters build a special flight of steps for Blackie to climb and descend at horse shows. Blackie even managed to win a ribbon in a horse show, despite a sneering comment about his breeding and appearance from the judge, an Army colonel. From his experience with Blackie, Ike derived a certain philosophy about training, an "enduring conviction that far too often we write off a backward child as hopeless, a clumsy animal as worthless, a worn-out field as beyond restoration."[7]

In the only photograph I have seen of Blackie, he doesn't look nearly as bad as Ike's description of him—he appears to have had something of a Roman nose, which is a matter of taste, and there is a very lively and intelligent expression in his eyes. But more striking is the fact that Mamie is mounted beside Ike on a much smaller horse, almost a pony, wearing an unbecoming round straw hat, and looking frail. There is a wan smile on her face, and Ike, towering over her on Blackie, looks down at her indulgently. Of course one can read too much into a snapshot.

Mamie told her granddaughter Susan that when she left Panama for Denver with the baby, John, Ike had implored her not to leave with words "that stayed with her for the rest of her life;[8] but whatever they may have been they didn't stop her from going, and Ike's fellow officers and their wives—including Mrs. Conner—not surprisingly drew the conclusion that the marriage was disintegrating. Though Ike passes over this episode, Mamie was rather more forthright about it. Home in Denver, cared for by her parents, she gradually recovered her health, and with it a more balanced view of the marriage. She does not say outright that she was contemplating divorce, but when one reads between the lines it seems likely that the thought had crossed her mind. At first she envied her women friends the stability and security of their marriages, with husbands who lived at home seven days a week and turned up on time for dinner; but then, as she began to feel stronger, she realized that the men they were married to did not seem to her altogether serious, and that Ike's strengths perhaps outweighed his single-minded dedication to his career. Some such summing up of the pluses and the minuses of being married to Ike went through her mind, at any rate, including, no doubt, the experiences they had shared together; and she drew from it the conclusion that she would have to adjust to the kind of man he was, and to the kind of life the Army offered—and, perhaps to her parents' surprise, she decided to return to Panama and Ike.

Although Mamie's view, as expressed in her conversations with Susan Eisenhower, can be summed up as an unconditional surrender—"With her return to Panama she threw all her efforts into keeping up with her husband and providing unstinting support"[9]—those who observed the Eisenhowers as a couple later might have expressed some doubts. Certainly, Mamie made more of an effort to join in the activities that Ike liked, such as riding and golf; and she seems to have learned from Mrs. Conner, apparently a perfect example of an army wife, just what it took to push a husband's career in the narrow world of the old peacetime Army; but it also seems likely that Mamie may have laid down some conditions of her own. She appears to have gone riding with Ike every morning from seven to nine o'clock; and she spent a good deal more time with Mrs. Conner, and playing bridge at the officer's club with the other wives, than she had spent before—but Ike also gave her a set of Minton china that she had coveted for some time and, more important, added a cook and a butler to their household staff so she could give gracious dinner parties a couple of times a week. She was able to boast to her parents that "she was gradually getting Ike housebroken again," and one can read into her letters a dawning awareness on Ike's part that he would have to make at least some concessions to Mamie's need to be treated like a belle and to have a little fun and glamour in her life. "He walked on tip-toes around her, and was always very anxious not to upset her," an observer would later comment, on seeing the Eisenhowers together, and it seems very likely that this first great crisis of their married life (there were to be several more) was the occasion when Ike first realized that not all the changes in their life together would have to be made by Mamie.

Both of them were dismayed when General Conner was ordered back to the United States—without him there was very little reason for Ike to be at Camp Gaillard—and even more dismayed when Ike's orders came through shortly after the Conners' departure, assigning him once again to Camp Meade, with its sad memories of little Ikky's death, and, worse still from a professional point of view, making Ike a football coach again instead of giving him the kind of serious job for which he had been preparing himself under Conner's guidance.

Ike had hoped to be assigned as a student to the Command and General Staff School, at Fort Leavenworth, where the Army trained officers who had been handpicked for future promotion. Nothing—not even his newfound interest in military history—could diminish Ike's passion for football, but he had come to feel, with a degree of pessimism that was rare for him, that he was at best a mediocre football coach, and also that football coaching in the Army was a dead-end job. Also, once the football season was over he would be put back in command of the same tank battalion which he had commanded two years earlier (and which was still equipped with the same old, obsolescent World War I tanks). He worked up his courage to request an interview with the Chief of Infantry—a fairly daring thing for a mere major to do—and asked to be sent to the prestigious Infantry School at Fort Benning instead, but was firmly told to go where he was ordered. A photograph exists of Ike at this time as a coach, wearing his football sweater with the big A, and scowling miserably into the camera, the very picture of a man unhappy with his lot.

It seems to have been at this low ebb in his fortunes that Ike actually discussed with Mamie the possibility of resigning from the Army, and received from her the very sound advice that he should do no such thing. She had seen enough of civilian life during her stay in Denver to be certain that Ike was not cut out for a business career. Her good sense was shortly to be confirmed by General Conner, now serving as deputy to the Army Chief of Staff in Washington, who sent Ike a cryptic telegram, indicating that the "Old Boys' Network" was hard at work on his behalf:

> NO MATTER WHAT ORDERS YOU RECEIVE FROM THE WAR DEPARTMENT, MAKE NO PROTEST ACCEPT THEM WITHOUT QUESTION [SIGNED CONNER][10]

When the orders arrived, Ike had even better reason to be grateful for Conner's message. Had it not been for the telegram Ike might well have resigned, or at least lost his temper, endangering his career, for he was posted to serve as a recruiting officer in Colorado—an assignment so humiliating that it would normally have been given only to an officer who had blotted his copybook rather badly in some way.

Conner, who seems to have been a master of Army paperwork on the order of Colonel "Jumbo" Trotter in Evelyn Waugh's *Officers and Gentlemen*, had instantly realized that as an infantryman Ike could not be transferred to any Army school without the approval of the Chief of Infantry, who, having just declined to give it, would certainly not be persuaded to change his mind by any mortal on earth. The solution was not to attack the Chief of Infantry frontally, but to outflank him: Ike would be transferred temporarily from the infantry to the Adjutant General's office (which controlled recruiting, among other things), and then selected by the Adjutant General, a friend of Conner's, to attend the Command and General Staff School at Fort Leavenworth as part of the Adjutant General's quota of officers, thus bypassing the Chief of Infantry altogether. QED. Ike not only was assigned where he wanted to go—admittedly by a circuitous route—but also learned a valuable lesson about the way the Army works, and the importance of not trying to overcome obstacles by butting against them headfirst. He was a little reluctant to take the crossed muskets of the infantry off the collar of his tunic, but consoled himself with the thought that he would eventually be pinning them back on again.

Ike quietly took up his recruiting duties—which were anything but demanding, since the Army was still being reduced, rather than expanded—at Fort Logan, Colorado, conveniently near Denver and the Doud family, to the delight of Mamie. One senses the deft hand of General Conner at work again in the choice of Ike's posting. Photographs of the Eisenhowers at this point in Ike's career mostly show Ike as a happy family man and proud father, sometimes dressed in the golf clothes of the period, plus fours, a tweed jacket, and a cloth cap—clearly, an interest in the sport that was to become something of an obsession for Ike had already taken hold. Another snapshot shows him smiling happily at the tiller of Mrs. Doud's old-fashioned electric brougham—automobiles of any kind or vintage always seemed to cheer Ike up. All the same, he was beginning to have second thoughts about Leavenworth. Those who were assigned there to the Command and General Staff School were not just the Army's intellectual elite, the ones who might one day expect their epaulets to carry a star; they had almost all graduated at or near the top of their class from one or more of the Army's other prestigious schools, such as the Infantry School that Ike had been prevented from attending. He felt himself unprepared for what he knew was a challenging and hugely competitive educational experience, and wrote to General Conner about his doubts. As usual, Conner responded quickly and with encouragement. "You may not know it," he wrote back, "but because of your three years work in Panama, you are far better trained and ready for Leavenworth than anybody I know. . . . You became so well acquainted with the technics and routine of preparing plans and orders for operations that included their logistics, that they will be second nature to you. You will feel no sense of inferiority."[11]

In short, if Ike relied on what Conner had taught him, he would do well—an opinion that contrasted sharply with that of an aide to the Chief of Infantry, who, by way of congratulating Ike, told him, "You will probably fail," and moreover added that attending the school would make him useless in the future as an infantry officer.

Staff schools are pretty much the same the world over; most of them are modeled, as one might expect, on the German army's staff school. Those who think of them as a form of postgraduate university education for army officers are mistaken. A staff school is highly specialized, and not designed to promote or encourage originality of thought or new ideas—indeed, the ideal of the German staff school was that if each officer-student in a class was presented with the same problem, all of them should come up with the same solution. For there is only one correct solution—the challenge is to present it in clear-cut steps and in language that cannot be misinterpreted or misunderstood even in the heat of battle, and to draw up an operational plan to respond to a specific strategic problem that covers every element of an army unit's movement, down to the smallest detail, in the right sequence, and taking into account every possible difficulty, from the weather to unexpected movements of the enemy. Units must arrive at the right place at the right time; their supplies must be brought forward to them on time and packed in the right order and the right way;* communications, logistics, artillery, and what is now called "command and control" must all be integrated seamlessly, so that everyone does what he is supposed to do when it is supposed to be done, without regard to what is happening (or going wrong) around him.

A military movement of any kind, especially an attack, presents an almost infinite number of variables, and the staff officer's job is to reduce all this potential chaos to a series of crisply written orders that will deliver the unit, with everything it needs to fight, exactly where it is needed and on time, and knowing exactly what it is supposed to do, despite what Clausewitz, perhaps the model of a staff officer, called "the fog of war." Staff officers, in fact, learn to write in a language that is all their own—a language of acronyms, abbreviations, and map references incomprehensible to the uninitiated; a language with no ambiguities, offering not the slightest possibility of misunderstanding; a language in which everything, however complicated, has been broken down meticulously into the basic elements that everybody can (and must) follow to the letter. (At Fort Leavenworth, this was so much the case that the language had a name: Leavenworthese.)

To the staff officer there are no shades of gray; everything is in black and white: advance, halt; left, right; take this road to that crossroads, and no other—for a road or a railway can only take so many vehicles or troops or horses per hour, and two units trying to use the same road at the same time will merely produce a traffic jam (and a perfect target for enemy artillery or aircraft). The antithesis of good staff work is muddle, and muddles lose battles and even wars.

* Connoisseurs of inept staff work will appreciate the following example. In 1879, when a British army in South Africa advanced against Cetewayo's Zulus, the reserve supplies of ammunition were packed for the tropics in tin-lined, steel-strapped cases, but without the special tool needed to cut the straps and open the cases. As a result, when the troops began to run out of the ammunition they carried in their cartridge pouches at the Battle of Isandhlwana and called for more, nobody could get at it. The troops, trained infantrymen armed with (empty) modern rifles, were slaughtered virtually to a man by Zulus wielding spears. (See Donald R. Morris, *The Washing of the Spears* [New York: Simon and Schuster, 1965].) Bad staff work can—and does—lead to death and disaster. In warfare, every detail counts.

Ike, it almost goes without saying, was born with a deep dislike of muddle. In writing orders, his language was crisp, decisive, authoritative—Leavenworthese came naturally to him, since General Conner had instructed Ike to write out the daily orders for the Canal Zone in the language of the Command and General Staff School. Conner had been right—when he and Ike had played war games as they slogged on horseback through the hills and jungle of the Canal Zone or sat by their campfire in the evenings "talking about the campaigns of the Civil War or Napoleon," he had taught Ike the skills that a good staff officer requires by making him treat every feature of the landscape as a military problem, constantly challenging him to think about how he "might able to meet with considerable force any enemy landing on our sector of the Canal." In short, Conner had taught Ike to look at the landscape in front of him not as a sightseer but as a tactician—How would I defend this or attack it?—and as a military planner: What would I need, and how would I get it there? This is not quite the same as Napoleon's famous *coup d'oeil de génie*, the quick glance of genius that the emperor boasted of bringing to the battlefield; but a part of that genius was Napoleon's own ability as a staff officer, his insistence on good staff work, and his tireless eye for the details of logistics and organization. Before he cast his "glance of genius" on the battlefield, Napoleon, at the height of his greatness, knew where each of his divisions was, what brigades and regiments it contained and how reliable they were, where his artillery was sited and how quickly it could be moved, how much he could expect of each of his marshals and generals, what roads were before and behind him—in short, he proved that on the battlefield, as elsewhere, "genius is the infinite capacity for taking pains," in the words of no less a military authority than Frederick the Great, a remark* that might almost serve as the motto for ambitious staff officers.

Mamie seems to have reacted with enthusiasm to the move. The course, though crucial to Ike's career, was only a year long, and there was no guessing where the Army might send him afterward, but at least Kansas was not Panama, with its many hazards and diseases. In the ten years of her mar-

* Later repeated in a slightly different form by Goethe.

riage to Ike, Mamie had so far followed him from San Antonio, Texas; to Gettsyburg, Pennsylvania; from there to Camp Meade in Maryland; then to Panama; from there back to Camp Meade; then to Denver, Colorado; and now to Fort Leavenworth, Kansas. Though emotionally high-strung and not physically strong, she had grown accustomed to the rootless life of an army wife, and by now had learned how to cope with the difficulties of moving and settling into new quarters. Her first task, her granddaughter Susan reported, was "to hang paintings, unroll rugs, and put out cigarette boxes and family photographs everywhere."[12] Ike's quarters at Otis Hall were roomy, basically two apartments joined together, and he set up what he called a "command post" on the top floor, which he shared with an old friend from his days in the Nineteenth Infantry, Leonard T. ("Gee") Gerow (who would go on to become a corps commander under Ike in World War II). It was the custom for students to form groups for studying a problem, ranging from four to eight officers, but Ike thought working in pairs was likely to be more productive and found that Gerow agreed.

Teaching at the Command and General Staff School was done by the case method, which is very much the way law is taught, and for which Harvard Law School is famous. Students were presented with a pamphlet, describing an enemy force in a particular geographical location, and asked what action should be taken. If that action was deemed correct, the student was then asked to draw up in full detail the plans, orders, and logistical support necessary to undertake it successfully. Ike and Gerow would work late into the night, with maps pinned to the walls and plans strewn over makeshift tables, to come up with successful solutions. Ike set himself a routine that limited him to two and a half hours of work a night, from seven to nine-thirty, but no more—he was determined then, as he would be in the future, not to be one of those who approached the next day "without fresh minds or an optimistic outlook." During World War II he would share this concern with General Montgomery, who never allowed any crisis, however serious, to keep him up past his nine-thirty bedtime (except, of course, for the presence of Winston Churchill, whose bedtime was two or three in the morning).

It was Mamie's job to see that Ike got to bed at nine-thirty, and to run the household smoothly so that he could concentrate on his "war gaming" with Gee Gerow, as well as to play hostess to his fellows students and the senior officers of Fort Leavenworth. As Susan Eisenhower points out in her book about Mamie, this was an Army institution, which in the future would come to be known, in a typical Army acronym, as PHT, for "pushing hubby through"—it was the wife's job to entertain, to make friends of the senior officers' wives, to give "modest supper and card parties." Mamie became a close friend of Gee Gerow's wife, Katie, "who did much to teach [her] the ropes, to show her how one could contentedly survive the nomadic life of the Army." Katie Gerow even gave Mamie decorating tips for livening up the drab quarters that the Army provided for married officers, in this case a bright splash of yellow, which Susan Eisenhower suggests may have been responsible for "Mamie's lifelong love of yellow roses and gladioli."

All the same, it must have been a lonely life for her, with Ike taking courses during the day and working until bedtime every night upstairs with Gee Gerow. By now Mamie must have been used to Ike's single-minded devotion to his work—not to speak of his indifference to bright splashes of color around him at home—but one guesses that it still gave her pain. On the other hand, like the gypsy life an Army career imposed on them, it was part of the bargain she had made with herself when she decided to return to Panama from Denver. Ike's determination to rise in his career was not something which she could blunt or from which she could distract him, so her best course, perhaps her only course, was to learn how to help him get where he so desperately wanted to go. It would be a long wait for both of them before this course paid off.

Ike graduated first in his class from the Command and General Staff School—a result that must have pleased General Conner—while Gerow took second place, surely justifying their initial decision to work together as a team of two. For Ike it was a triumph—proof at last, within the small, tight world of the peacetime regular Army, of his intellectual and professional abilities. With his graduation, however, he returned to the infantry, and was therefore once again under the command of the Chief of Infantry. He was promptly ordered to take command of an infantry battalion at Fort

Benning—not exactly an appointment that might have seemed appropriate for an officer who had just demonstrated his intellectual skills. The situation was complicated by the fact that he had been offered two other appointments: one to become an instructor at the school from which he had just graduated, a very considerable honor; and the other as a military instructor in the ROTC at a major northwestern university, with the "additional opportunity" of coaching the football team for an extra salary of $3,500 a year. Ike was tempted by both—teaching at the Command and General Staff School would be professionally and intellectually rewarding; and the extra $3,500 that he would make for coaching a football team made the ROTC job very hard to turn down—but once again the Chief of Infantry rapped Ike sharply on the knuckles by ordering him to proceed at once to Fort Benning, where, to his fury, he was told to coach the soldiers' football team in addition to his duties as commander of an infantry battalion, and of course at no additional salary!

Mamie and Ike liked Georgia—except during the summer months—and were, for once, pleased by the quarters they were given. But commanding an infantry battalion and coaching a soldiers' football team were not exactly what Ike had looked forward to as a reward for graduating first in his class from staff school, and he clearly felt that he was "marking time" (and perhaps being discriminated against, though he did not say so). In the winter of 1926–1927, Ike, still simmering about the treatment he had received, went about his duties grimly, though with his usual efficiency, while Mamie had to be sent to Walter Reed Hospital in Washington, D.C. with a "digestive ailment," which today would probably be considered a combination of stress, disappointment, midlife crisis, and fatigue at the job of being her husband's full-time cheerleader and helpmeet. While Mamie was in the hospital, their son John was sent to live in San Antonio with her scapegrace sister Mike, the Doud family "hell-raiser," whom Ike described, with guarded admiration, as "full of zip and rebellion"; and Ike remained at Fort Benning by himself.

Very fortunately, General Conner intervened once more at this point. Ike was ordered to report to no less a person than General John J. Pershing—"Black Jack," as he was known in the Army–former commander of the AEF in France, and now head of the Battle Monuments Commission. It was Pershing's task to create and "beautify" the American military cemeteries abroad, and his desire to have "a battlefield guide, a sort of Baedecker to the actions of Americans in the war," prepared for visitors. Ike, because of his success at Fort Leavenworth, and his tact with senior officers—for General Pershing was a notoriously difficult man—seemed like the right person for the job, but no sooner had he accepted and started work than he was selected as a student of the prestigious Army War College at Fort McNair, "the ambition of almost every officer" and almost a guarantee of eventual high command. Ike managed to resist the combined arguments of General Pershing's executive officer, the eccentrically named Xenophon Price, and Mamie, who yearned, like most Americans in the 1920s, for a chance to see Paris; but eventually he managed to eat his cake and have it too—he went to the War College, graduated, and then rejoined the Battle Monuments Commission, Conner, one imagines, having persuaded General Pershing to be patient (no easy task) and wait for Ike.

The Eisenhowers had moved once again, of course, this time from Fort Benning to Washington, D.C., where they rented an apartment in the Wyoming, a comfortable apartment building off Connecticut Avenue, near Dupont Circle. This was a move to a very different lifestyle—for the first time in their marriage they were not living on an army post, or in Army quarters, surrounded by other officers and their wives; and Mamie luxuriated in the ability to go shopping in department stores and lead what amounted to a normal civilian life for a change. Not that she was completely removed from the Army in her new home. The Wyoming must have had a special reputation in the War Department as a good place for married mid-level and senior officers in Washington to live, or some kind of "fix" with Quartermaster Corps, which looked after officers who were moving, since among its residents were two generals, several lesser officers, and eventually the Eisenhowers' old friends from Fort Leavenworth, Gee and Katie Gerow.

Ike, as usual, plunged into his work at once with a vengeance, absorbing huge quantities of information about the battles of the AEF, and breaking it down into terms that would make it understandable to civilians—a task that came naturally to him and would be of enormous consequences later, when those civilians included Roosevelt and Churchill. Ike did so well at this that even a taskmaster as demanding as Pershing was impressed enough to offer him the ultimate prize—a tour of duty in France, to examine the battlefields and the cemeteries.

Ike had not been certain that this what he wanted to do—as a graduate of the War College, he was offered a coveted job on the general staff, and this was where he felt he belonged—but Mamie was determined that they should see "the Old World, and travel," and having taken her to so many places where she didn't want to go, he felt obliged to give way to her this time. Of course the fact that General Pershing also thought he should go was not something Ike could ignore either, and he himself was attracted by the chance to walk over the battlefields and study them carefully. He had missed his chance to fight in France, but he would at least see where the fighting had been done, and learn from it. This knowledge of the terrain would indeed come in handy some sixteen years later.

The Eisenhowers left for France in August 1928, this time onboard the luxury liner *America*, accompanied by the Douds, who were taking their first trip to Europe, and who "would look after John while Mamie went apartment hunting and Ike settled into his new job."

The France they were sailing toward was in some ways a very different country from France today, and feelings between the French and the American people were also different. At no time since the days when Benjamin Franklin had represented the infant republic at the Court of Louis XVI had relations ever been closer or more cordial, or Americans more welcome and admired in France. The Americans had joined the war late—almost too late—but eventually they had come in huge numbers, and the French were still grateful. Inarticulate as he seemed to some, the remark with which General Pershing is popularly credited when he disembarked in France perfectly expressed the relationship between the two republics: "*Lafayette, nous voici.*"* American volunteer airmen had formed their own squadron in the French air force before their country had even joined

* It was actually spoken by Lieutenent Colonel Charles E. Stanton, but it seemed to be exactly what people thought Pershing ought to have said, so he got credit for it.

the war, the legendary Escadrille Lafayette; American soldiers and marines had fought alongside French troops, eventually under the supreme command of Marshal Ferdinand Foch; and President Woodrow Wilson had loyally—some might say blindly—supported French claims at the peace conference, including France's occupation of the Rhineland, the creation of Poland and Czechoslovakia as independent new states intended to hem in Germany to the east of France should Germany ever recover, huge German reparation payments, the breakup of the Austro-Hungarian Empire, and the reduction of the German army to no more than 100,000 men. One of the grander thoroughfares of Paris was quickly renamed Avenue Woodrow Wilson, whereas one could search the Paris street map in vain for an "Avenue Lloyd George," despite the death of 750,000 British troops.

On the American side, the sentiment toward France was best represented by Oliver Wendell Holmes's remark, "When good Americans die, they go to Paris," as well as the words of one of the most popular songs of 1917, "How you gonna keep 'em down on the farm, after they've seen Paree?" Paris in the 1920s was more than ever *la ville lumière* that drew Americans like moths to a flame, and seemed to represent everything that the staid young republic across the Atlantic, with Prohibition in full force and Puritanism still rampant, was not.

France was triumphantly victorious, a great world power, its army as admired as Napoleon's, its empire only slightly smaller than Britain's (the French flag flew over Syria, Lebanon, Algeria, Morocco, much of central Africa, most of Indochina, and innumerable islands all over the world), its culture and way of life envied by everyone, even the defeated Germans. Yet behind the glamour and the glittering facade all was not well. Nearly 1.4 million Frenchmen had been killed in World War I, and over 4 million wounded, in a country whose population was only 50 million and whose birthrate was declining. The Germans, defeated and to some degree dispersed by the peace treaty, had a population that was close to 90 million, if they could be united. The worst of the war had been fought on French soil, leaving nearly a quarter of the country a gaping wound that would not be healed for generations. The Americans had gone home as soon as possible after the war, and immediately disbanded their army; the British had, as usual, taken their army back across the Channel and reduced it to its tiny peacetime size, leaving the French alone to face the Germans, their border touching that of their enemy, now impotent and awash in grievances and fratricidal political strife, but animated by bitterness and a ferocious thirst for revenge.

As Germany regained strength—and nobody in France doubted that it would regain strength sooner or later—what were the chances that France's erstwhile allies would return? Would France be left alone in Europe to face a resurgent Germany, perhaps at best with the somewhat doubtful help of the Poles and the Czechs? The fact that the United States Senate had voted against ratification of the Treaty of Versailles chilled the French, as did increasing signs that the British were becoming more interested in appeasing Germany than defending France. Facing this unpalatable reality, France, hardly surprisingly, retained compulsory military service and began constructing the hugely expensive Maginot Line, designed specifically to keep the Germans back if they ever invaded again. These dark clouds were already forming dimly on the horizon, but in 1928 goodwill toward the United States was still widespread, and in a country where foreigners were generally despised, Americans were treated with rare, if occasionally amused and condescending, courtesy.

That is not to say that the French were any less infuriating than before or since. The Eisenhowers and the Douds disembarked to find that they had missed the boat train to Paris, where they did not arrive until three-thirty in the morning, after endless difficulties and red tape with their luggage, and with nobody from the travel company available to help them.[13] They experienced many times, to Ike's fury, the traditional French shrug which means "It's not my problem," or "So what do you want *me* to do about it?" Nor was there anybody to meet them at the station, so they were obliged to trust themselves to a Parisian taxi driver, who drove them the long way around the city looking for a place to stay. Finding an apartment to rent at a price they could afford turned out to be an even more difficult struggle, but Mamie seems to have taken charge of this. She found a comfortable hotel for herself, John, Ike, and her parents while she did so; then enlisted the help of the United States embassy; and finally found a pleasant and reasonably modern apartment on the Right Bank, on the Quai d'Auteuil, overlooking the Seine and the Pont Mirabeau. She also found a nearby school for John. For somebody who had spent the last ten years with the U.S. Army Quartermaster Corps looking after this side of life for her, and who spoke little or no French, it was a remarkable accomplishment.

Although Ike himself never learned French, he did learn, in his travels around what had been until ten years ago the western front, a certain admiration for the French. The great battlefields that the war cut across northern and eastern France from the English Channel to the Vosges, more brutally than the Culebra Cut had split Panama in two, testified to the awesome powers of resistance and endurance of France's citizen soldiers, as did their vast war cemeteries, above all those of Verdun, which were virtually a national shrine. Here rose the grim tower of the Ossuaire, in which the bone fragments of over 120,000 unidentified French soldiers were collected, visible through glass windows; here the French had fought and died, some 350,000 of them; here Pétain had issued his immortal promise, "*Ils ne passeront pas!*" "They shall not pass"; here the young Charles De Gaulle had fought and was wounded and taken prisoner; here a whole generation had been martyred—there was no other word to do them justice. Hundreds of thousands of French soldiers had marched up the narrow, muddy Voie Sacrée—the "sacred road," constantly shelled, which was the only access to Verdun—many to their death. Here von Falkenhayn's plan to "bleed the French Army to death" at a place France could not abandon for reasons of national history and prestige backfired, with fatal consequences for Germany. To the north of Verdun was the battlefield of the Somme, originally begun to relieve the pressure on Verdun, where the British army had fought and died in 1916 and 1917, with its own neat cemeteries, in which the rows of headstones seemed to stretch forever to the horizon, silent reminders of a hecatomb. There could have been no more eloquent monuments for a young staff officer to study than these vast French and British burial grounds, on battlefields where bones, unexploded shells, and the debris of war still surface to this day. From them Ike learned respect for his future allies, however infuriating their behavior might sometimes seem, and a degree of caution—and personal feeling—on the subject of casualties rare in a general. It would never be easy for him to send men to their death.

He traveled up and down the western front by car, stopping along the road "to join groups of workers who were eating the noonday lunch, their tools at their side, the shovels, the hoes, pickaxes laid down, they were invariably relaxed and hospitable." Ike joined them, and they always offered him something from their own lunch boxes, and he developed the habit of carrying a bottle of red wine in the car so as to be able to offer something back, and since he did not drink wine, a bottle of Évian water for himself. When there were no road crews with whom to eat lunch, Ike sat down at an outdoor table under the trees with his box lunch, at some little country auberge, and talked to the people, who were, he commented, "warm, jolly, and courteous."[14] His driver, who was also his interpreter, sat with him.

This too would one day distinguish Ike from many other American senior officers—the ease with which he got along with ordinary people; his unfailing courtesy; his affection for the French, over whose land, after all, two world wars would be fought; and his need to join in whatever fun was going on. There were great gaps of misunderstanding between him and the austere, autocratic De Gaulle, but it was De Gaulle who in his memoirs would write of Ike, during the most difficult days between France and America, in Algiers early in 1943, in words of sincere admiration, "He was a soldier. By nature and by profession, action seemed to him straight and simple. . . . It was the good fortune of the alliance that Dwight Eisenhower discovered in himself not only the caution necessary to confront these difficult problems, but also an attraction towards the wider horizons that history opened to his career. He knew how to be adroit and supple, but although he relied on these gifts, he was [also] capable of great daring."

Although they were no Puritans, Ike and Mamie seem to have lived rather quietly for Americans in Paris. Mamie, predictably, loved the shopping; Ike had a taste—it is hard to guess why—for the famous Musée Grevin, the Parisian equivalent of Madame Tussaud's in London, a wax museum, full of convincing characters from French history, and, more interesting, France's abundant store of crimes, executions, and murders. The Eisenhowers' apartment became, for American officers in Paris, "Club Eisenhower," just as their quarters had been at Camp Meade; and Ike took great pleasure in an occasional evening of poker with his colleagues—throughout his life his skill at card games never deserted him. They entertained, both in the apartment, where Mamie, with a maid and a cook at her disposal (and despite the language barrier) gave dinner parties; and at the Union Interalliée, where they hosted a number of well-attended dinner dances. Ike took his young son John to Verdun, and showed him the ruins of Fort Douaumont, the center of the battle, taking him below ground into the damp corridors, where the boy was startled to see a human skull grinning at him from a recess, and to the famous "trench of bayonets," perhaps the holiest shrine of French patriotism and sacrifice, where a company of French soldiers advancing to attack had been buried by a German shell, the tips of their bayonets still sticking above the ground in neat rows. The Eisenhower family took time off to visit Italy, Belgium, the south of France, Switzerland, and even, briefly, Germany. It was a period of their lives that must have seemed in many ways idyllic—they were together as a family, in those days the dollar went a long way in Europe, and Ike was happy in his work.

Perhaps most important of all was the fact that Ike's work brought him into close contact with General Pershing, whom Ike described as "reserved and even remote in manner,"[15] which was something of an understatement. Pershing seldom came to his office before one o'clock in the afternoon, and then worked, mostly in silence, until well past midnight. Despite his inarticulateness and eccentric work schedule, Pershing somehow managed to convey to Ike that he was pleased by his work on the guidebook, a fact which would have some significance in Ike's future career.

At the time, unlikely as it may seem, Pershing was being urged to run for the presidency in 1928, despite his extreme distaste for the White House and the political process in general. A man named McLaughlin, the editor of the *Army-Navy Journal*, seems to have been eager to play the role of kingmaker in Pershing's campaign, and encouraged the general to prepare the appropriate speeches; but the general's writing habits were, as Ike put it, "cautious and slow," and he soon left it to Major Eisenhower, whose work on the guidebook had impressed him so much, to become his speechwriter. This was not a task Ike enjoyed—then, as later, speeches were far from his forte—but it brought him into closer contact with Pershing. Ike would later remark that "in no case was I successful in producing anything he wanted," but the truth seems to have been that Pershing, however decisive he had been on the battlefield, was an infinitely fussy man who, while unable to write himself, was determined to question and if possible change every word that was written for him.

Pershing was also involved in writing his war memoirs, or rather turning his war diaries into a form of memoir, but he seemed to be making scant progress. He soon drafted Ike to help him; and Ike, though he made no claim to be a literary man, instantly perceived that writing a memoir in the form of a diary would prevent the reader from following the story of a battle in an understandable form, and sensibly suggested that Pershing use narrative for the more important parts instead. He even drafted narratives of the battles of Saint-Mihiel and the Argonne, to show Pershing what he had in mind. For once, the dour Pershing seemed impressed, but he remarked that in such matters "he always looked to one man to give him final advice." That man was Colonel George C. Marshall.

A few days later Colonel Marshall arrived, closeted himself with General Pershing, read the diaries and the drafts that Ike had written, and then came into Ike's office to give his opinion. Marshall was a tall, lean, austere-looking man, with piercing eyes, already famous in the Army for his fierce sense of duty and his sharp intelligence. He did not mince words. He liked what Ike had written, but he thought General Pershing should stick to his original idea. Ike explained why he thought the narrative form would be better for the battles, and Marshall said he could see that, but in the end he still thought General Pershing would be happier with leaving the diaries as they were.[16]

In the end, Pershing's diaries were published, and failed to make much of an impact—just as Ike had predicted, most reviewers (and readers) complained that they couldn't follow the battles from the general's diary entries. But what mattered most, though Ike could not have guessed it at the time, was that he had made a favorable impression on George C. Marshall.

Ike's intelligence; his patience; his hard work, and his willingness to let it go by the board; and his obvious devotion to Pershing, a notoriously difficult and uncommunicative boss, were apparent at once to Marshall. Marshall was never much impressed by flashy types like Patton, however many medals they wore, but he had an eye for hard workers with good minds, self-effacing men who would see a difficult job through to the end without thought of personal fame or promotion.

Marshall was something of a self-appointed talent scout for the Army that he knew the United States would some day need again. He kept a small black leather notebook in his pocket so that he could write in it the names of officers who had attracted his attention favorably; and he wrote down the name of Major Dwight D. Eisenhower.

CHAPTER 8

The MacArthur Years

The America to which the Eisenhowers returned at the end of Ike's tour of duty in France was at the dizzying height of the jazz age and the bull market. One month after they arrived back in Washington, and resumed residence at the Wyoming, the stock market crash brought all that to a sudden end. This is not to say that the Eisenhowers were directly affected by it—they had no significant investment in stocks; and regular Army officers, while they lived on modest salaries even in boom years, were also protected from the collapse of the market because they didn't lose their jobs.

Still, the collapse of the economy naturally had an effect on the armed services. Already reduced to the lowest possible numbers, and to making do with weapons, tanks, and ships that had been obsolescent in 1918, they saw their appropriations cut to the bone, and deeper. In truth, the other major powers were in no better straits. In Germany, the army remained reduced to 100,000 men by the terms of the peace treaty. In Great Britain all the armed services were starved of funds, also cut to the bone, in their case by the "Geddes Axe," as the ruthless cheeseparing policies of Sir Eric Geddes were called, leaving the British hard pressed to police an enormous—and increasingly fractious—empire, let alone contemplate fighting a major war. Only in France was a significant amount of money being spent on the armed forces; unfortunately, however, in the form of the continuing construction of the Maginot Line—a project that rivaled anything accomplished by the pharaohs in size and expense, and as the future would prove, soaked up money that would have been better spent on tanks and modern aircraft. In the Soviet Union and far-away Japan, on the other hand, vast amounts were being spent to build up modern armed forces, and it passed almost without notice that after the Treaty of Rapallo, in 1922, by the terms of which the two outcast nations of Europe agreed to collaborate, German officers were already busily training with tanks and aircraft deep in the heart of Soviet Russia, out of sight of the inspectors of the League of Nations and the victorious Allied powers, and beginning to fashion there the principles of the blitzkrieg that would astonish the world in 1939 and 1940.

Ike had hoped to command troops once again, but he was to be disappointed. Instead, he was assigned to the military staff of the assistant secretary for war, with the thankless task of exploring the readiness of American manufacturing to convert to military production. Not surprisingly, in the twelve years since 1917, American industry had lost interest in producing military equipment, and few plans existed for reconversion, since there was no war in sight. In the aftermath of the economic crash, this was also the last thing on anyone's mind—American industry had its hands full just trying to stay afloat as the economy plummeted and millions of workers were laid off. The War Department was equally hostile to the idea of preparing American business to tool up for war production in time of peace, with no war in sight, and even more bitterly hostile to any discussion of government control over wages, prices, and the cost of materials, without which no efficient industrial planning could take place. The only person who seemed to take Ike's study seriously was Bernard M. Baruch, the financial genius who had headed the War Industries Board in World War I. Baruch listened to the young major, and gave him good advice—and along the way, as was so often the case in Ike's life, developed a respect for Ike. Baruch was shrewd, wealthy, powerful, a close friend and supporter of Franklin D. Roosevelt, then governor of New York, as well as a close friend of Winston Churchill. Once again, as with Marshall, Ike had made a good impression on a man who would soon have the ear of the president of the United States, and who would not forget Ike's name. In the meantime, any attempt to prepare American industry for war was stopped dead by Army Chief of Staff General Summerall, who went so far as to "forbid any General Staff Officer to go into the office of the Assistant Secretary of War," presumably to prevent their being contaminated by Ike's radical ideas about government control of the economy and planned cooperation between industry and the armed forces.

A new wind was about to blow the War Department clean of such outdated and blimpish obstacles, however. In 1930 General Summerall was succeeded as Army Chief of Staff by General Douglas MacArthur. Very shortly General MacArthur took the opportunity of listening to Major Eisenhower's ideas about war production, refined and broadened by the wisdom and experience Bernard Baruch had offered Ike, and the new Army Chief of Staff, who loved "big thinking" and ambitious projects, voiced his approval and ordered Ike to draft a comprehensive plan for the wartime mobilization of American industry.

This was a big job, and though nobody could have known it at the time, in less than seven years it would grow and develop to become the master plan by which Franklin Roosevelt would transform America into the "arsenal of democracy," converting American industry almost overnight to the mass production of war matériel on a scale beyond anyone's imagination, from the most ordinary supplies and tools of warfare to weapons of such revolutionary scientific and technological sophistication that they would not only win the war, but bring about the equivalent of a second, and more dramatic, industrial revolution, changing the entire postwar world. When, in later years, Ike warned about the "military-industrial complex" he was speaking not only about something he understood but about something he himself had helped to create.

By 1932 his part in the work was completed, and as if by a seamless transition he moved into General MacArthur's offices as the great man's "personal military assistant," becoming MacArthur's éminence khaki. As Ike described the job, with typical modesty, he was "an amanuensis to draft statements, reports, and letters for his signature." For the next six years he would work on the most intimate professional terms with one of the strongest, most flamboyant, and most complex personalities in the history of the United States Army, a man who was widely criticized as vain, arrogant, and dangerously ambitious, and at the same time deeply respected as a hero and a military thinker. The experience would very nearly cost Ike his health, his marriage, and his military career—for the controversy that swirled around the expansive personality and outspoken ideas of Douglas MacArthur, even in peacetime, as well as the jealousy of his colleagues and the fear of him among politicians, inevitably attached itself to those who were close to him and served him loyally (those who in his opinion did not were swiftly banished from his presence).

The phrase "bigger than life" is often overused when it comes to public men, but in General MacArthur's case it seems to fill the bill. These days, over forty years after his death, it is difficult to describe the hold he maintained over the imagination of so many of his countrymen for so many decades. First of all, he was a national hero, a figure who to many Americans virtually incarnated patriotism; he not only "walked with kings" (and presidents) but even talked down to them. Except for the fact that they were both West Pointers, he and Ike could hardly have been more different. MacArthur was wealthy, socially and politically well connected, famous, glamorous, eccentric, deeply theatrical, patrician, a shameless old-fashioned snob, a military aristocrat, and a reckless hero. To say that he lacked the common touch, with soldiers or with civilians, that Ike so clearly possessed would be putting it mildly—MacArthur's language, both spoken and written, was high-flown, rhetorical, and outmoded. Anybody today watching

MacArthur's famous farewell address of 1951 ("Old soldiers never die, they just fade away") after he had been relieved of his command by President Truman, a speech given to a rare and remarkable joint session of the United States Congress and watched by a record television audience of 30 million people, would very likely find it an artificial, sentimental, overblown, even mawkish performance; indeed quite a few people thought so then. But many more did not. His speech was interrupted by thirty standing ovations, and countless grown men, many of them lifelong enemies of MacArthur, cried openly. One congressman shouted out, "We heard God speak here today, God in the flesh, the voice of God." From his apartment at the Waldorf Tower in New York, former President Herbert Hoover likened MacArthur to "a reincarnation of St. Paul."[1]

No other American military figure except George Washington rivaled MacArthur's fame and prestige. MacArthur's father, a distinguished Army general who won the Medal of Honor* during the Civil War, went on to become the virtual American proconsul of the Philippines. Douglas MacArthur's first memory was that of the sound of a bugle on an Army post. He was valedictorian of his class at West Point—even then he had a glorious voice, as well as the gifts and presence of a great actor—and only two other students in the history of the United States Military Academy ever surpassed his academic and athletic achievements there. He may also have been the only cadet in the history of the academy whose mother lived in the hotel at West Point during the entire four years of his cadetship. Mrs. MacArthur was a beautiful and formidable woman of fierce, unreconstructed Confederate leanings—her good-night words to the infant Douglas were "to grow up like Robert E. Lee"—although she was the widow of a Union hero. She remained at West Point so she could be with Douglas every day and could look out from the window of her room at Craney's Hotel every night to his window in the barracks, to see how late he stayed up studying. Within four years of his graduation, MacArthur was appointed aide-de-

* General Arthur MacArthur and General Douglas MacArthur were the first and for many years remained the only father and son to have each been awarded the Medal of Honor. In 2001, a posthumous Medal of Honor was awarded to Theodore Roosevelt for his service in Cuba in the Spanish-American War; his son Brigadier General Theodore Roosevelt, Jr., had been awarded a posthumous Medal of Honor for service at Utah Beach on D-day.

camp to President Theodore Roosevelt—throughout his career his mother, the widow of a lieutenant general, lobbied tirelessly, and often successfully for each of her son's appointments and promotions. In World War I he was appointed chief of staff of the famous Rainbow Division; was swiftly promoted to become the youngest brigadier general in the history of the United States Army; commanded a division in combat; was wounded and gassed in combat; was awarded the Distinguished Service Cross, the Distinguished Service Medal, the Silver Star, the Purple Heart, the French Legion of Honor and Croix de Guerre (as well as too many other foreign decorations to list); then returned home to receive a hero's welcome and appointment to the coveted post of superintendent of the United States Military Academy. In World War II, MacArthur would be awarded the Medal of Honor; after victory he would become the virtual American proconsul of Japan, as his father had been in the Philippines. Congress would twice try to promote him from the new rank of General of the Army—a five-star general—to the unique rank of General of the Armies: a proposed six-star general.

This was the man to whom Ike attached himself, with whatever doubts—for MacArthur, it almost goes without saying, was not one to brook arguments, differences of opinion, or disagreement from subordinates (or from his superiors, including two presidents of the United States). Like one of the more difficult Shakespearean kings, he had a majestic sense of self. During World War II, his staff would be handpicked to be so obsequious and devoted to polishing MacArthur's public image that they were held in ridicule and contempt by everyone, sometimes including MacArthur himself; and at no time was he an easy man to contradict. No less a judge than Clare Boothe Luce (who, as the wife of Henry Luce, doubtless knew whereof she spoke) wrote that MacArthur's "temperament was flawed by an egotism that demanded obedience not only to his orders, but to his ideas and his person as well—he plainly relished idolatry."[2] Ike might be tactful, but he was used to expressing his opinion. For a man of his temperament, working closely with Douglas MacArthur was likely to be a severe trial, and to produce the kind of hypertension that, even in those days, unlike smoking, was associated with cardiovascular problems.

Like most great men MacArthur could be charming when he chose to be, even disarmingly so. He set out to charm Ike, and Ike seems to have allowed himself to be charmed. Certainly he appreciated his chief's intelligence; his interest in the "big picture"; his imagination (and his extraordinary ability to arouse the imagination of others); his almost unique ability to navigate in the murky waters where politics and the military meet in Washington, for MacArthur seemed to know everyone who mattered, from the president down; and especially his willingness to listen to a new idea, adapt it to his own uses, and let a subordinate run with it, which is perhaps what most impressed Ike at first. Above all, what attracted Ike was being, at last, at the center of things. Even in peacetime, the Army Chief of Staff, who wore during his tenure the four stars of a full general, was a major figure in Washington, all the more so when he was Douglas MacArthur. Ike had worked for Pershing, but by 1928 Pershing (who disliked MacArthur intensely and thought he had been promoted too high and too fast for his own or the Army's good) was near the end of his career, whereas MacArthur, at the age of fifty-three, was just hitting his stride. What is more, MacArthur's view of his role was at once ambitious and expansive—unlike previous chiefs of staff, he was as interested in politics, diplomacy, and what we would now call the media as he was in military affairs. As Ike would remark, "Most of the senior officers I had known always drew a clean-cut line between the military and the political . . . but if General MacArthur ever recognized the existence of that line, he usually chose to ignore it."[3]

Ike's office was next to MacArthur's, separated from it only by a slatted door. Ike described his new boss as "decisive, personable. . . . On any subject he chose to discuss, his knowledge, always amazingly comprehensive, and largely accurate, poured out in a torrent of words."[4] He qualified this judgment by adding that "discuss" was perhaps not the correct word, since it suggested dialogue, and MacArthur's conversations were usually monologues. He also found it odd that everybody referred to MacArthur simply as "the general," as if there were no others, and that MacArthur himself was in the habit of referring to himself in the third person, as if he were the deity.

Rather like his nemesis "Black Jack" Pershing, MacArthur kept eccentric hours. Ike refers to his chief's long breaks for lunch, sometimes lasting three or fours hours at a time, tactfully refraining from pointing out that MacArthur's lengthy absences from his office were occasioned by the fact that he kept a beautiful Eurasian mistress in downtown Washington, in a suite at the Hotel Chastleton, "with an enormous wardrobe of tea gowns, kimonos, and black lace lingerie."[5] Her presence was kept carefully secret from MacArthur's mother, but not, alas, from columnists like Drew Pearson. Ike was no prude, but he found the general's erratic hours difficult to cope with, since they meant that he was frequently obliged to be in the office until late at night, making Mamie unhappy.

During the years when Ike worked at the War Department, the Eisenhowers led a fairly active social life, given that the country was in the depths of the Depression. Although they were always short of money—Ike's pay as a major was modest, and the checks Mamie received from her father from time to time for luxuries grew smaller as his meatpacking business began to feel the pinch—they did their best to keep up the kind of social life that meant so much to Mamie, and that was in those days very much a factor in making an officer's career. (PHT—push hubby through!—was still the order of the day, as it had been at Fort Leavenworth.) They were surrounded by old friends—Patton, reduced back to a major like Ike, commanded a squadron of cavalry at Fort Myer; General Conner visited Washington frequently; Ike's brother Milton was a rising federal bureaucrat; the Gerows, friends from their days in Fort Leavenworth, lived in the same building—and they gave small dinner parties, made easier by the fact that Mamie's beloved Pupah enabled them to have a maid. At one point they even managed to give a plush black-tie dinner party for Secretary of War Patrick J. Hurley at the Willard Hotel—a significant, and expensive, career push for Ike. Still, they were poor enough that every penny counted—literally, for Ike saved pennies carefully through the year to pay for Mamie's anniversary present. He also saved as much as he could from his weekly allowance of five dollars for cigarettes, carfare, and razor blades, to buy an addition to Mamie's prized silver tea set for every birthday. Mamie too did her best to economize—her best dress, worn for the

Armistice Ball of 1931, was four years old and painstakingly remade for the occasion. When their old Buick broke a rear axle, it was a financial catastrophe for the Eisenhowers; and it was a constant concern that Ike needed to be well turned out in a civilian suit, coat, and hat every day, for it was the custom then that officers serving in Washington, D.C., did not wear uniforms.

Washington was still, in Mamie's words, "an overgrown country town," in which most people got along on relatively low—if reliable—government salaries. The Eisenhowers seem to have been happy there, and no doubt they were, but part of the reason that Mamie worked so hard to entertain on Ike's behalf was a certain guilt that he regretted having accepted the appointment in Europe, when he could instead have accepted a position on the general staff, which would have almost certainly have meant a promotion to lieutenant colonel. It was, she felt, a sacrifice Ike had made for her, knowing how much she had wanted to see Europe; and however much they had both enjoyed it, there was no question that since his return his career prospects had remained dim. Here he was, a rapidly balding, middle-aged major, shuffling paperwork instead of commanding troops, and beginning to suffer from a series of maladies, including, but not restricted to, dysentery, indigestion, chronic back pain, hemorrhoids, bad teeth, and bursitis, none of which the Army doctors could help him with, and most of which we would recognize today as psychosomatic disorders, symptoms of stress and a midlife career crisis.

Working for a prima donna in khaki like General MacArthur didn't help, of course. MacArthur, a perfectionist with a natural gift for extracting the maximum amount of work from those around him, was not the ideal boss for a man with Ike's health problems, and Ike was inclined to describe his role to others as a kind of commissioned office clerk.

No doubt something of this de haut en bas attitude had seeped through to Ike in the days when he was still toiling away behind the slatted door, waiting for MacArthur to raise his voice and call him into his office, to answer a question or provide a piece of information for a visitor. Then, as later, Ike was much given to what Mamie called a "slow burn," in which his lips were compressed, his forehead was flushed, the veins in his temple were enlarged, and the back of his neck became bright red as he fought to control his temper; and MacArthur, with his air of invulnerable superiority, was just the man to bring this condition on. Indeed, more important people than Major Eisenhower often experienced it in the Chief of Staff's presence. Many years later, when they were both famous generals, a woman asked Ike if he knew MacArthur. In a rare moment of candor, Ike replied that he certainly did. "I studied dramatics for seven years under General MacArthur,"[6] he said.

Matters were not helped by the fact that there was a certain feline quality about MacArthur's treatment of those who worked for him. He not only played favorites with his staffers, thereby increasing their insecurity, but he also encouraged them to stay in the office as late as he did, fully aware that this was only too likely to cause trouble at home. Circumspect as Mamie was, it is possible, reading what she had to say to her granddaughter more than four decades later about Ike's years with MacArthur, to detect a certain degree of jealousy and resentment for all those evenings which Ike spent working late next to "the general"* or for the amount of attention Ike paid to him. Ike would not be the first American executive whose wife was envious of his boss, or even the first executive in uniform in that position.

Still, despite the downside, being MacArthur's aide put Ike in a unique position for a mere major. MacArthur was the reverse of secretive; on the contrary, he expected his aide to be familiar with everything he was doing—and so Ike was aware of everything there was to know about the U.S. Army. War plans, intelligence reports, political problems, promotions, plans for new weapons—everything passed across his desk before it reached MacArthur, and passed back across it promptly with MacArthur's crisp replies and orders, for the Chief of Staff did not dawdle over papers or encourage his staff to do so. Even Ike, who was himself an amazingly

* General MacArthur himself may have been a victim of this kind of resentment. According to his biographer William Manchester, MacArthur's Eurasian mistress grew tired of sitting alone in her hotel suite in the evenings with only the dog MacArthur had given her for company, and started to go out on the town with other men, beginning a downward spiral in the relationship between them that eventually compelled MacArthur to try to buy his love letters back and pay up to keep the affair secret from his mother, establishing the woman in a prosperous hairdressing shop as the price for silence.

accomplished military bureaucrat, admired MacArthur's speed at making up his mind, as well as his skill at handling problems and dealing with the endless flow of paperwork. "Unquestionably," he wrote, "the General's fluency and wealth of information came from his phenomenal memory, without parallel in my knowledge—reading through a draft of a speech or a paper once, he could immediately repeat whole chunks of it verbatim."

This is high praise, coming from a man whose acute memory and capacity for putting even the most complicated of operations down on paper crisply and succinctly impressed no less a judge of these things than Winston Churchill. Of course Ike would rather have been commanding troops like his friend Patton, but in the meantime he was perhaps the best-informed major in the U.S. Army.

It was a beleaguered Army, as America stumbled downward into the depths of the Depression in the 1930s, reduced now to less than the size of the Greek army, with regular pay cuts making it even harder for both officers and men to live. It was also an Army that was no longer held in any special respect by most Americans. The U.S. Navy and the Marines commanded a certain nostalgic affection, increased by the fact that they were seldom visible except in the few big naval ports; but without a war in sight, or an enemy to defend against, the Army seemed to most people useless. It was poorly armed, poorly trained, poorly paid, and poorly housed, and seemed to many merely a refuge for those who couldn't make it in civilian life, and were willing to accept pointless drill, harsh discipline, abusive sergeants, and plenty of spit and polish in exchange for three square meals a day, when it came to the enlisted men, and an exalted idea of their own importance, when it came to officers. "The war to end wars" had been fought and won—largely, it was believed, by drafted civilian soldiers—and since there were not going to be any more wars, why spend good money to maintain an army when half the country was going hungry? Even in the depths of the Depression it was hard to recruit soldiers. The regular Army has rarely, if ever, been a popular institution in the United States, and in the 1930s it was not only unpopular, but increasingly seen as useless, and also a potential weapon for right-wing reactionaries to use against the people—a fear exacerbated by the frequent use of the National Guard against strikers, and soon to be reinforced by the ill-judged actions of the Army Chief of Staff himself.

Washington had been spared much of the industrial strife that had spread throughout the rest of the country and was affecting people from workers in the Hollywood movie studios to big-city teachers. The fact that Washington was to all intents and purposes a southern city, where most of the menial jobs were performed by blacks, spared Washingtonians the kind of class warfare that was spreading rapidly through the rest of the country, as it became clear that the Hoover administration had no plans for dealing with the Depression. The administration was simply waiting for the business cycle to rise again and putting its faith in the American businessman; it was markedly reluctant to undertake any schemes of social progress, expensive relief efforts, or national work programs, which were viewed by President Hoover and those around him as utopian and socialistic. Ironically, the man who had become famous for his humanitarian relief efforts on behalf of Europeans now presided over a nation in which acute poverty and even starvation were becoming commonplace.

By 1932 it was clear enough that the Republicans were likely to lose the next presidential election (and might not regain the White House for generations) and that only vigorous, progressive social programs could prevent the country from slipping into chaos. Because we know the outcome, the victory of Franklin D. Roosevelt of course looks inevitable to us, but it did not seem so at the time—indeed the Democratic Convention in Chicago in July 1932 sweltered its way, in the days before air-conditioning, through four ballots spread over two days and a night before finally nominating Roosevelt, to the increasing distress of Louis Howe, his political adviser and amanuensis. Howe, irritated because the band kept playing "Anchors Aweigh" over and over again in tribute to Roosevelt's wartime service as assistant secretary of the navy, finally cried out, "For God's sake tell them to play something else—tell them to play, oh, 'Happy Days Are Here Again,'" thus accidentally giving Roosevelt the theme song that would last through the next thirteen years of his political career.[7]

. . .

General MacArthur's political views were, to put it bluntly, reactionary, though that was not unusual among senior officers in the American armed services; and he was especially devoted to Herbert Hoover, whom, like most conservatives, he saw as a victim of the Depression, rather than one of the causes. It was a source of particular anxiety to Hoover that a small army of "Bonus Marchers" had gathered in Washington. These were former soldiers, veterans of World War I, who had been promised a bonus for their war service by Congress almost a decade earlier. Congress had waffled over the payment of the bonus, however, caught between the veterans who wanted it paid right away and the administration, which didn't want to pay it at all, and had finally voted to pay the veterans a liberal bonus in 1945.

The veterans were, naturally, infuriated by this decision. Many were out of work, evicted from their homes, unable to support their families—they needed the money now; they felt they were entitled to it (and had been promised it); and a grassroots organization began to recruit veterans to march on Washington and make a public protest there. By July there were approximately 20,000 veterans encamped in Washington, most of them in a hastily erected shantytown across the Anacostia River, some in abandoned wartime office buildings near the Capitol. Many lived in tents and huts made of whatever material they could scavenge, and in both veterans' encampments there was a certain amount of order, such as comes instinctively to soldiers, in terms of cooking, latrines, and the like. Nevertheless, the fact remained that 20,000 people were squatting on public land and in government buildings in the heart of America's capital city, and even those who sympathized with the veterans were not pleased by their presence.

It is easier now to understand the effect the Bonus Marchers had on the Hoover administration than it was at the time. The reaction of Hoover's White House to the veterans was very similar to that of Nixon's White House toward the peace marchers in 1971, and just as deluded—the marchers were deemed to be untidy, unhygienic, unruly, insalubrious, and a potential source of revolutionary violence. Admittedly, in the long hot summer of 1932, open latrine trenches and communal cooking fires in downtown Washington were as unwelcome to Washingtonians as the marchers themselves, but like the peace marchers who opposed the war in

Vietnam forty-one years, later they posed no direct threat to the republic. Since the crash of 1929 those who were out of work and homeless had formed shantytowns all over the country—they had no other option—and it had long since become common to refer to these as Hoovervilles. Clearly, it did not help matters that the biggest Hooverville of all was now within sight of President Hoover's windows at the White House, but humiliating as that was for the president and his administration, it hardly threatened the government. With a modest degree of goodwill and common sense, the Bonus Marchers could have been placated, and removed by degrees from their encampments. There were, of course, hotheads among them, and a certain number of agitators and communists determined to exploit the political situation—it could hardly have been otherwise—but most of the veterans were simple, honest, puzzled men who had fought for their country and had a grievance they wanted addressed, not revolutionaries. This made it all the more distressing when the Hoover administration finally lost patience with them, or perhaps surrendered to its fears, and like Nicholas II calling in the Russian army to stop the hunger marchers in 1917 (with fatal results), ordered the U.S. Army to evacuate the Bonus Marchers from their encampments.

To do MacArthur justice, there could have been no question of his disobeying an order from the president, even had he disagreed with it, which he did not. He was duty-bound to carry out his orders—though a different kind of man might have argued that it was a mistake to use force against the protesters, and a potential public relations disaster to use the Army against Army veterans. But MacArthur, with his natural tendency toward self-dramatization, saw himself in the role of savior of the republic, and decided that he would take personal command of the operation, which was intended to be modest in scope—no more than 500 or 600 infantrymen—and which fell to the Army only because the Washington police department, still basically just a small-town southern police force at the time, had neither the training nor the manpower to carry it out.

Nobody saw all this more clearly than Ike. In the first place, his sympathies were basically with the Bonus Marchers, whom he correctly described as "quiet and orderly," and whose soldierly discipline under difficult circumstances he admired. More important, he argued forcefully that nothing required the Army Chief of Staff, a four-star general, to take active command of 500 soldiers for a "street-corner" matter—at best a lieutenant colonel's job—and that if it turned into a riot, as might easily be the case, it would be "highly inappropriate" to have the Army Chief of Staff directly involved.[8]

MacArthur thought otherwise. He saw what was happening as a challenge to federal authority in the District of Columbia, and warned Ike that there was "incipient revolution in the air."[9] He changed into a uniform and sent Ike home to do the same. When Ike returned, they went out into the street, fully uniformed in tailored OD tunics, Sam Browne belts, riding breeches, boots, and spurs, MacArthur displaying seven rows of ribbons on his left chest. In the street, to his dismay, Ike found his old friend George Patton with a squadron of cavalry from Fort Myer and a few tanks, forming up around the base of the Washington Monument. What had been a kind of police action was rapidly turning into a serious display of force.

The object of the exercise was to drive the veterans away from the area around the Capitol back to the other side of the Anacostia River, and this was accomplished with the use of Patton's cavalry and tanks, plus a brief barrage of tear gas. Despite this, while there was a good deal of booing and shouting at the soldiers, the crowd moved peacefully, though slowly and under protest, in the direction of the bridge,

Orders were shortly received from the secretary of war, in the name of the president, that under no circumstances were the troops to follow the Bonus Marchers across the bridge into the larger encampment on Anacostia Flats. Herbert Hoover might not have been the quickest or most politically astute of presidents, but he must have had at least some inkling that the Army's charging into the shantytown with fixed bayonets and the support of cavalry and tanks would play badly in the press. These orders General MacArthur ignored—he later said that he was too busy to listen to people bringing purported messages, although they included Colonel Wright, secretary of the general staff; and General Moseley, from the office of the assistant secretary of war, neither of them military nobodies, or likely to be the bearers of false messages.

He and Ike—who was by now appalled at the unfolding of exactly the consequences he had predicted—marched across the bridge at the head of the troops, whereupon the entire encampment went up in flames. Whether the fire was started by the soldiers or the veterans was never subsequently made clear, but Ike clearly spoke for a lot of people in remarking, "The whole scene was painful. . . . The veterans were ragged, ill-fed and felt themselves badly abused. To suddenly see the whole encampment going up in flames just added to the pity one had to feel for them."

Two babies were suffocated by tear gas and died, and a seven-year-old boy was bayoneted through the leg while trying to rescue his pet rabbit; but MacArthur, once again against Ike's advice, was driven back to the War Department and held a triumphant press conference, cannily praising the president for "shouldering the responsibility" of putting down a possible revolution, in which "the institutions of our government would have been threatened." Although hailed by the secretary of war as "the man of the hour," MacArthur found his reputation permanently affected. He would later claim that it was a "gesture of personal responsibility," and that he could not honorably have delegated the distasteful task to anybody else, but that was not the way Ike saw it, then or later, and Ike was beside him throughout the day.

In his biography of MacArthur, *American Caesar*, William Manchester wrote that a few days later, in Hyde Park, New York, Governor Roosevelt, the Democratic nominee, was resting before beginning his presidential campaign against Herbert Hoover, when he received a lengthy telephone call from Huey Long. Putting down the receiver at last, he turned to one of his political advisers, Rexford Tugwell, with a sigh, and remarked that Long was one of the two most dangerous men in the country. Tugwell asked Roosevelt who the other one was, thinking it was probably the rabble-rousing anti-Semitic radio priest Father Coughlin, but Roosevelt shook his head thoughtfully. "Oh, no," he said. "The other is Douglas MacArthur."[10]

This being the case, it is interesting that one of Roosevelt's first acts as president was to extend MacArthur's term as Army Chief of Staff by an additional (and unprecedented) year, until 1934. Some people have speculated that Roosevelt preferred to have MacArthur close at hand in Washington, where the president and his advisers could keep an eye on him, as opposed to some distant Army command where he might serve as a magnet for all sorts of trouble, and that is very likely the case. To use the words of a later Democratic president, Lyndon B. Johnson, it may have seemed safer "to have him inside the tent pissing out than outside the tent pissing in." Roosevelt and MacArthur knew each other well—in fact MacArthur's great-great-grandmother "became a common ancestor of Douglas MacArthur, Winston Churchill and Franklin D. Roosevelt . . . and MacArthur was a sixth cousin once removed of F. D. R."[11]—and disliked each other intensely without ever allowing themselves to show this dislike.

In many ways (except politics) they were alike, which perhaps explains it. Both were mama's boys: Roosevelt's mother's bedroom was between those of Franklin and Eleanor at Hyde Park (this says it all); and just as MacArthur's mother had accompanied her son to West Point, Roosevelt's mother accompanied him to Harvard and, until her death, to the White House. Both men were handsome, intelligent, ambitious, and well connected; each had tried to keep an injudicious love affair a secret from his mother (successfully in MacArthur's case; unsuccessfully in Roosevelt's); both were capable of great charm and equally great ruthlessness, as well as deceit; and both were courageous, naturally gifted actors, and determined to be the center of attention at all times. Only one of them was president, however, and neither of them ever forgot that fact.

Very fortunately, this did not matter much in 1933. War was far from Roosevelt's mind, and the problems of the Army therefore farther still. Military appropriations were relentlessly cut back; the development and production of new weapons were slowed to a standstill, even for the most basic needs, like the replacement of the Army's venerable bolt-action 1903 model Springfield rifle by the new semiautomatic Garand M–1; and military bases all across the country were shut down in the interests of economy. Ike was aware of every detail of MacArthur's epic battles to win funding for the Army. He accompanied the Chief of Staff to Congress, to the White House, to the Treasury. He was present when MacArthur received a rare rebuke from the president because he had pushed too hard for Army funding and insulted Roosevelt;* and when, himself insulted by senators who had asked him to justify the Army's expenditure on toilet paper, MacArthur replied, with great dignity, "I have risen as high in my profession as you gentlemen have in yours,"[12] and stalked from the room, only to be coaxed back by the apologetic senators. Few people knew more about Washington than Ike, a fact which he was clever at concealing, and which many people forgot or never learned, to their cost. Ike was never the simple soldier he made himself out to be, certainly not after his return to Washington in 1928 and his apprenticeship in military politics beside Douglas MacArthur—it would be like calling the private secretary of the pope a simple man of God. In later years, people as different as Senator Joseph McCarthy and Orval Faubus, the segregationist governor of Arkansas, learned this lesson in politics the hard way. (Richard Nixon, perhaps because he was himself so duplicitous by nature, found it hard to accept that Ike was his master when it came to Washington politics.) Ike might well have said of himself, in the words of Voltaire, *Nourri dans le sérail, j'en connais les détours* ("Suckled in the seraglio, I know the back ways") had he known French, and ever been tempted to be frank about his political skills. Ike had learned them from a master, as he was the first to admit.

The year 1933 was pivotal in more ways than one. Japan's rearmament and intransigence were increasing at a rapid pace, while in Germany, the hitherto unthinkable took place, as Adolf Hitler, maneuvering deftly among the more respectable German right-wing parties, accepted from President von Hindenburg the chancellorship he had coveted during so many years as a political outcast. As the SA storm troopers paraded past their Führer, now Germany's chancellor, by torchlight, few people had any

* MacArthur told Roosevelt, according to William Manchester, that when American boys died with a bayonet in their guts because they didn't have the proper weapons to fight with, he hoped they went with the president's name on their lips and not his. Roosevelt replied, with exemplary patience, "You must not talk that way to the president." Ike, at any rate, apparently learned something from this, since it would be echoed in a famous exchange between himself and Montgomery during World War II, before D-day. After an outburst from Monty, Ike put his hand on Monty's knee and said firmly but quietly, "You can't talk that way to me, Monty; I'm your boss."

doubt that Hitler would rearm Germany and attempt to make it a great power once again—though most failed to guess the speed with which he would do so, or his calculated recklessness as he overturned, one after the other, the restraints against German hegemony that had been the most important clauses of the Treaty of Versailles.

Preventing the United States from collapsing into chaos and taking steps to rebuild its crippled economy and self-confidence were foremost in the president's mind, naturally enough, and neither he nor his Army Chief of Staff had much time to worry about events in Germany. The general opinion, in any case, was that Hitler was a flash in the pan; that "decent German conservatives" had deliberately given him just enough rope to hang himself, and at the same time thwarted the ambitions of the even more dangerous German Communist Party; that Hitler's chancellorship was a subtle political masterstroke on the part of the Germans who really mattered—the bankers, the big businessmen, and the industrialists, who had him firmly in hand. He was their puppet, and they would dispose of him as soon as he had served his purpose and the German political situation was stabilized. In the meantime, the important thing was to be patient, to give the Germans the benefit of the doubt, and not to make too much of a fuss about the inevitable "excesses" of Hitler's followers as they sought their revenge against communists, anti-Nazis and, of course, Jews.

Some measure of this was fervently believed, in Washington, in London, in Paris, in Rome, even in Moscow, and, for that matter, even more improbably in Berlin, where the German conservatives who had brought Hitler to power were still congratulating themselves on their own cleverness. The only person of importance who did not believe it was Winston Churchill. Less than a month after Hitler came to power, Churchill, with profound melancholy, made an accurate prediction of the near future: "My mind turns across the narrow waters of the Channel and the North Sea, where great nations stand determined to defend their national glories or national existence with their lives. I think of Germany, with . . . its youth marching forward on the roads of the Reich singing their ancient songs, demanding to be conscripted into an army, eagerly seeking the most terrible weapons of war, burning to suffer and die for their fatherland."[13] A month later, he rose in the House of Commons to castigate in stinging terms, "the odious conditions" already developing in Hitler's Germany and the threat of "persecution and pogrom of the Jews."[14] These were opinions that very few people wanted to hear, in the House of Commons or elsewhere, and together with his demands for British rearmament, especially in the air, they convinced those on the right as well as those on the left that Churchill was an enemy of peace.

No country was more determined to ignore the danger signals emanating from Germany than the United States. Subsequent attempts to lay the lion's share of the blame for Hitler's reckless foreign policy on the British "appeasers" ignore the reality that they were not alone. Almost everybody wanted to "appease" Hitler; and in any case the British and the French, who, if it came to war, would have to do the fighting, could hardly oppose Germany without the support and encouragement of America, and when they looked toward Washington, nothing of the kind was forthcoming, or seemed likely to emerge.

Like many Americans, MacArthur looked toward the west when he sought danger, not toward the east. He was by birth and by conviction an American imperialist—the Roosevelt he admired was Theodore, not Franklin—and it was in the Pacific that he saw America's destiny, not in Europe, which he had come to dislike and distrust after his wartime experiences there. Besides, his father had won fame not only as the conqueror of the Philippines but as the military governor and the architect of the modern Philippines as a kind of American protectorate. For MacArthur it was evident that America must be the dominant power in the Pacific; that the United States must ultimately replace the British, French, and Dutch empires with a new and more dynamic kind of empire, in which the "natives" would be encouraged toward American ideals and accept American hegemony; and that America's real rival and enemy was aggressive, imperial Japan. Ike, by contrast, was already an instinctive "Atlanticist," perhaps because of his family's German roots, perhaps also because not having fought in France, he had not developed an antipathy toward the British officer class and the British Empire that this class dominated and ran, or the rooted distrust of France, and the Old World that many Americans felt.

MacArthur's remaining year as Army Chief of Staff was painful, as Roosevelt, with the deft political cunning for which he soon became famous, carefully undercut the position of the person he regarded as one of the two "most dangerous men in America," while all the time continuing to profess admiration and warm affection for him. Although MacArthur continued to attend social functions at the White House, and paid elaborate old-fashioned courtesies to the first lady, Eleanor Roosevelt, who found it difficult to hide her dislike of him, he was only too aware that the New Dealers, as they were already beginning to be known, viewed him with deep suspicion, hated him for his reactionary political views, and were afraid that he might harbor political ambitions which would bring him into open conflict with the administration—that he might become, in fact, the proverbial "man on a white horse"[15] in the event of a fascist putsch in America. In short, their feelings about General MacArthur were as paranoid as his about them.

His mood was not improved by the administration's constant effort to cut appropriations for the Army and for the National Guard—Roosevelt even canceled the contracts by which the Army had traditionally carried airmail, and which, in effect, helped to subsidize the Air Corps and train Army pilots in long-distance flying. Still less was his mood improved by an injudicious libel suit which he had brought against the columnist Drew Pearson for calling his treatment of the bonus marchers "dictatorial, insubordinate, disloyal, mutinous, and disrespectful of his superiors," and which MacArthur was eventually obliged to withdraw, in circumstances of considerable personal humiliation, when the story about his Eurasian mistress was leaked to Pearson.

MacArthur attempted to regain lost ground by using the Army to help implement one of the New Deal's favorite programs, the Civilian Conservation Corps (CCC), which put over 250,000 poor, unemployed young men to work in America's forests, planting trees, digging ditches, and clearing brush. MacArthur was certainly not in sympathy with this idea, but he loyally devoted his energy and his organizational genius to it, though it did him little good. The New Dealers suspected him of trying to turn the youthful portion of the nation's unemployed into an army reserve—such an idea had in fact been at the back of the general's mind—and to his chagrin many of the people whom he balefully suspected of disloyalty to himself, like Colonel George C. Marshall,* became stars in the eyes of the administration by their efficient administration of the CCC.

Roosevelt dealt MacArthur the final blow, with a plot of such subtlety that even MacArthur failed to see it coming. Pressed by MacArthur to extend his term by another year, and by MacArthur's enemies in the administration not to do so, Roosevelt decided that MacArthur should stay on until his successor was chosen, timing matters so carefully that the man MacArthur wanted to succeed him was eliminated because he was too close to retirement age, then picking instead Major General Malin Craig, a favorite of the aging General Pershing, and one of those whom MacArthur had suspected of being against him as long ago as when he was commanding a division in France.

The president quickly moved to smooth the general's ruffled feathers by awarding him another Distinguished Service Medal and cannily offering him an appointment that he could not refuse—Chief Military Adviser to the Commonwealth of the Philippines, where his father had served as military governor. Roosevelt thus managed to send MacArthur 12,000 miles away from Washington, but promised him that in the event of war, he was not to wait for orders, but to hurry home at once. "I want you to command my armies," the president told him warmly.[16]

MacArthur had been given to understand that he would retain his four-star rank at least until his arrival in the Philippines, and was therefore unprepared for a telegram he received in his compartment on the train for San Francisco, reducing him back to his permanent two-star rank as a major general. Ike, who was with MacArthur when he read the telegram, described the explosion that followed as "a . . . denunciation of politics, bad manners, bad judgment, broken promises, arrogance . . . and the way

* General MacArthur had refused to promote Colonel Marshall to brigadier general, despite—or because of—a personal letter asking him to do so from General Pershing. Neither Marshall nor Mrs. Marshall ever forgot or forgave this refusal, it inevitably colored the relationship between MacArthur and Marshall when the latter became MacArthur's boss.

the world had gone to hell."[17] The wound it opened would never heal. MacArthur's confidence in Roosevelt, such as it was—and in most Democratic politicians—was fatally undermined, and it would affect his relationship with Harry Truman ten years later. Whatever else he was, Douglas MacArthur was not a man to forgive or forget what he took to be a slight, and nobody knew that better than Ike.

This explains why Ike was on that train. He had no great desire to follow MacArthur to the Philippines nor was Mamie enthusiastic about another tropical posting, after her experience in Panama. But in many ways the United States Army was essentially feudal—after several years of working for MacArthur, Ike was regarded as "MacArthur's man," both by his colleagues and by MacArthur himself. The Army Chief of Staff had the power to make or break a man's career permanently, or at the very least for years to come. Not only had MacArthur, to spite Pershing, refused to promote Colonel Marshall to brigadier general, but he had appointed Marshall, perhaps the most promising officer in the Army, as an instructor to the Illinois National Guard—professional oblivion for Marshall, who now sat in Chicago, looking so "gray and drawn" that Mrs. Marshall worried about her husband's health. Ike, as a mere major, was well aware that MacArthur could make or break him; and MacArthur, who valued Ike, was clever enough to tempt him with glittering prospects—and of course to hint at what might happen if he didn't play ball. After all, if an officer as distinguished as Colonel George C. Marshall could be sent to Chicago to instruct the Illinois National Guard, there were plenty of other remote Army posts in which Major Eisenhower could look forward to years of coaching the football team until he reached retirement age. Although Roosevelt could easily outfox MacArthur when it came to politics in Washington (or nationally), nobody understood Army politics better than MacArthur.

As usual, Ike's greatest desire was "to serve again with troops," but that was not to be. MacArthur told him that "they had worked together for a long time and he didn't want to bring in somebody new," and Ike was impressed, even touched, by the apparent sincerity with which MacArthur confessed that he needed Ike and wanted him in the Philippines. Ike later admitted that MacArthur had "lowered the boom" on him, and he would speculate about whether this was an instance on his part of preferring "the known over the unknown," but the truth of the matter was that he was in no position to argue with the Army Chief of Staff. He consoled himself with the thought that at least "there were no Filipino football teams over there"[18] for him to coach.

Although in his memoirs Ike hardly even hints at Mamie's reaction to his decision to accompany MacArthur to the Philippines, it is clear that she was determined not to go with him, and that her refusal pained him deeply but did not deter him. Her dislike of the tropics was certainly one factor, but perhaps more important was the fact that she and Ike had built up a pleasant life for themselves in Washington, surrounded by friends, with John placed in a good school. Mamie had experienced one tour of duty in the tropics, and had no happy memories of it; she had also disliked sharing Ike with his commanding officer, and knew that MacArthur's demands on Ike's time would make Fox Conner's look modest. In addition, MacArthur was as jealous, as seductive, and as vain as any woman, and she and the general would, in effect, be competing for Ike's attention. She worried too about John's health in a tropical climate, and about her own. Their friends Katie and Gee Gerow had returned from the Philippines quite recently, and to Mamie's distress Katie was deathly ill. It was cancer, and it would very shortly kill Katie, but Mamie's granddaughter speculates that Mamie may have attributed Katie Gerow's illness and death to her stay in the Philippines—certainly, it increased her fear of going there with John.

The Eisenhowers were poor—and shortly to become poorer, as Roosevelt cut military salaries, low as they already were—but Mamie had, at last, the kind of settled life she had always wanted, surrounded by all her beloved possessions, photographs, and knickknacks. Ike played golf on his day off, badly, but with great enthusiasm; they gave cocktail parties, buying, before the repeal of Prohibition, "denatured alcohol" from a bootlegger in ten-gallon cans, which Ike siphoned off into bottles with John's help, and flavored with a recipe of his own so that it tasted like gin; on Sundays they slept late (neither of the Eisenhowers was a regular churchgoer, nor did either of them, then or later, pretend to be); played cards with their friends; and gave small supper parties at what they continued to call Club Eisenhower. It was a pleasant life, and snapshots of the Eisenhowers at the time reflect that—Ike, looking cheerful and faultlessly dressed in a three-piece suit; young John standing between two beaming generals outside the Wyoming; Mamie in her finest, ready to get into the old Buick and pay a series of formal calls on various Washington officials, a task which Ike hated but which Mamie loved, perhaps because it offered an opportunity to dress up, and to look at other people's homes. That she was unwilling to give all this up for life in Manila, with Ike at the beck and call of an autocrat like General MacArthur, is hardly surprising, but more surprising is the fact that she summoned up the courage to put her foot down—if Ike wanted to go to the Philippines, he would go without her!

Another factor—it would be more accurately called a burning issue—between them was that MacArthur had refused to set a limit on the duration of Ike's tour of duty. MacArthur himself behaved as if he didn't expect to return to the United States so long as Roosevelt was president, and the only concession he would make to Ike's misgivings about accompanying him to the Philippines as his aide for an indefinite period was to allow Ike "to pick one associate from the Regular Army" to join them. Ike chose Major James B. Ord, a West Point classmate of his who spoke fluent Spanish and whose "quickness of mind" he thought MacArthur would appreciate.

Still, it was not a happy situation. Ike wrote to his parents-in-law plaintively that "we'll be separated for many months at the very least. I hate the whole thought—and I know that I am going to be miserable. On the other hand Mamie is so badly frightened (both for John and herself) at the prospect of going out there—that I simply cannot urge her to go."[19] Urge her, in fact, he did, but to no avail—she was determined to stay.

Even professionally, Ike was not optimistic about the Philippines. First of all, although MacArthur had grandiose notions about the importance of his presence there, it seemed to most people that Roosevelt had managed very adroitly to move a potentially dangerous enemy thousands of miles away for an indefinite period, to a remote backwater where he would either be forgotten over the years or fail. Hardly anybody in the United States was enthusiastic about the Philippines, or even much interested. Unlike Panama, which was in any case much closer to home, the Philippines did not contain a major object of national importance like the canal, or even any valuable natural resources to awaken the interest of American businessmen. The Philippines constituted—to the embarrassment of the few Americans who thought about the subject at all—a colony, won from Spain during the Spanish-American War, after which a fierce guerrilla war took place, startlingly similar to that in Vietnam six decades later, against Filipino nationalist insurgents who saw no reason to accept rule from Washington in the place of rule from Madrid.

In the 1930s, a series of face-saving moves—which will seem familiar to those old enough to remember Britain's embarrassed rush to rid itself of its colonial empire in the 1950s—had been instituted to prepare the Philippines for eventual independence. The most important was that the Philippines was designated by Congress a self-ruled "commonwealth" for a period of ten years, under loosely defined American protection; at the end of this time, in 1946, it would become an independent country.

Manuel Quezon, who was to head the commonwealth government, had known and admired General MacArthur's father, and had personally requested MacArthur as his military adviser, with the task of building up a native Philippine defense force. Such a force would surely be one of the first and most important needs of an independent Philippines, given the outspoken and extravagant ambitions of nationalist Japan. But the fact remained that in the middle of the Depression nobody in the United States, up to and including the president, wanted to think about the possibility of a threat from Japan, and despite the political concessions and promises made to the Filipinos, the Philippines remained a miserably backward place, in which a wealthy, Spanish-speaking oligarchy ruled over an impoverished, poorly educated, landless rural population. It resembled not British India but rather the Dutch East Indies (now Indonesia), though without the latter's immense natural resources; and the notion that the United States might have to try to make it into a self-sufficient modern nation, if possible, or still worse, defend it from the Japanese, was not one which anybody in Washington wanted to contemplate.

Despite Roosevelt's enthusiasm about Quezon's request for MacArthur as his military adviser, the general got nothing he had been promised in the White House. He had supposed that he would arrive in the Philippines to take up his appointment as Army Chief of Staff with the prestige of four stars, but instead he disembarked there as a retired two-star general. He had also supposed that he would be bearing with him, as Roosevelt had suggested, a cornucopia of money and weapons with which to build up a Philippine army, navy, and air force, but this too failed to materialize. His two aides, Majors Eisenhower and Ord, had labored mightily to come up with a plan for building a Philippine armed force, but every assumption they made about money and equipment had to be halved, then halved again, then further reduced. They had drawn up plans for a conscript army based on one year of intensive training, at a cost of $25 million (or 50 million pesos) a year, including the purchase of modern weapons, but by the time they reached Manila they were rewriting their plan with a budget of less than 16 million pesos a year. This entailed almost eliminating from the Philippine army trained career officers, reducing the training period for recruits to six months, and relying for weapons entirely on what the United States was willing to hand over from obsolete stores at bargain-basement prices. Both Ike and Ord were only too well aware of just how obsolete these weapons were likely to be, and in what kind of condition, though neither of them could have guessed how slowly and grudgingly the weapons would be sent, or in what small quantities. They could not afford to include an air force in their plans, or even think about a naval contingent, although the Philippines consists of over 7,000 islands, with an almost unlimited (and undefended) coastline, and any sensible defense would inevitably depend on naval and air forces. Even using the most optimistic guesswork, they could see no way in which the Philippines could hope to defend itself before 1946, at the earliest.

Although General MacArthur was well aware of the impossibility of his task, his natural optimism, combined with his ego, persuaded him that his presence alone would be enough to overcome the lack of money—he had a Napoleonic belief in his own ability to outweigh every deficiency, and to overcome, by the sheer force of his personality and the magic of his name and reputation, every obstacle. Any other man would have been shaken by the circumstances under which he had arrived. Not only had he been shorn en route of two stars and his title as Army Chief of Staff—a great loss of face in the eyes of the Filipinos, he felt sure—but his beloved eighty-four-year-old mother, who had insisted on accompanying him, had died shortly after arriving in Manila, leaving him bereft. Still, even though he was plunged into mourning, MacArthur refused to give in to depression or pessimism. To the amazement of his staff, he discussed with President Quezon the possibility of his becoming commanding general of the Philippine army, once it was formed, with the rank of field marshal, and thus be in a position to pin his four stars back on his uniforms, despite the treachery of Franklin D. Roosevelt. Quezon was enthusiastic about the idea; and indeed the only person who argued against it was Ike,[20] who would also, when the time came, turn down for himself the offer of becoming a general in the Philippine army. (MacArthur got his way, and in 1937 donned the four stars and the unique gold-embroidered cap of a Philippine field marshal, specially designed for him. The cap would become his trademark in World War II, and he would continue to wear it throughout the rest of his career in uniform.)

The general's small military family was not, from the very beginning, a happy one. MacArthur took the air-conditioned, lavishly furnished six-room penthouse of the Manila Hotel as his residence, while Ike got a small room without air-conditioning in the same hotel. (Ord, with his family, rented a modest house in Manila.) A rift swiftly developed between MacArthur and his two aides, after he ordered them to draw up plans for an ambitious parade. Quezon, when he saw the plans, turned the idea down as being too expensive and elaborate, at which point MacArthur blamed the whole thing on Ike and Ord, telling Quezon it had been their idea. Not surprisingly, Ike and Ord felt that the general had double-crossed them, and were deeply resentful.

Although the American empire was very different from the British, there is a certain Evelyn Waugh quality to the relationship between MacArthur and his staff in Manila—a sense of people bearing the "white man's burden" in a tropical clime, with glittering titles and uniforms

General MacArthur and Ike, in the Philippines.

(MacArthur changed his uniform at least three times a day), and not all that much to do; a world in which spite and small hurts loomed large and bitter. It was also a world farther removed from home than is easy for us to imagine today, in the age of cell phones, satellite television news, broadcasts, and jet travel. To travel from New York to Manila involved an overnight train trip to Chicago, a change of trains there, then a three-day train trip from Chicago to San Francisco, a wait there to board a liner for the four-day voyage to Hawaii, and a much longer wait in Honolulu for the rare Army transport ship to load up for the long and tedious sea voyage to the Philippines.* Communications by radiophone were unreliable and expensive, and mail took about the same time to arrive as a person did. It did not help that MacArthur was a general without an army, and for the moment had nobody to order about but his two hapless aides. Matters improved slightly when he fell in love with Jean Marie Faircloth, a wealthy young woman from Murfreesboro, Tennessee, whom he had met on the

* Pan American Airways inaugurated a service by Clipper flying boat from San Francisco to Honolulu in 1936. This service shortened the journey by several days, but it was very expensive and still seen as a daring adventure.

ship on the way out, luckily for them both while his mother was confined to her stateroom by her illness. Jean Faircloth soon moved into the Manila Hotel to be close to the general, and would shortly become the second Mrs. MacArthur—his brief first marriage had earned his mother's disapproval and ended in a very messy divorce.

In the meantime, Ike returned to his early interest in aviation, which had been squelched by Mamie's father as too dangerous for a man approaching marriage. The more closely Ike studied the defense of the Philippines, the clearer it was to him that airpower was the key to the problem. Although MacArthur constantly argued, for public relations purposes, that he would throw the Japanese back from the beach no matter where they landed, a country consisting of 7,000 islands, and situated only just over 1,000 miles from Japan, was vulnerable to a landing, or a series of landings, anywhere; and in the absence of a serious road or railway system, or any proposal to build either of them, the Philippine army, once it existed, could hardly reach any landing place on the coast in time or in sufficient numbers to throw the Japanese back off the beach. The only way to defend the Philippines, as Ike saw clearly, was to have an air force capable of patrolling the seas around the islands to detect—and, if possible, destroy—the Japanese before they landed.

Some kind of air force was in any case necessary merely to raise the kind of army that MacArthur had in mind, since the lack of any modern transportation infrastructure meant that Philippine conscripts would be trained, and perform their military service, at one of nearly 100 small posts spread around the country, very few of them linked by road. The recruits would be pleased, it was thought, because the camps would be close to their local villages, and of course so were the local politicians; but to supervise such widely separated outposts, it would be necessary to reach them by airplane, so from the beginning Ike insisted that their construction must include a dirt landing strip and that a number of light planes would be required. Ike purchased a few training aircraft from the United States, and "borrowed two instructors from the Air Corps." Before long he not only had built up a "miniature air force" for training Filipino pilots, but had resumed flying lessons himself after a gap of nearly twenty years.

Of course, it made sense—first of all, he was right about the defense of the Philippines; second, he was correct that airplanes were the only practical way to visit the training camps and see with his own eyes what was going on there; and third, flying was the best way, indeed the only way, to get a sense of the country—but one cannot help feeling that it must also have been a relief for Ike to get away from MacArthur and take to the skies. Flying single-engine primary trainers without a radio over virtually trackless jungle and open sea, while dangerous, was at least a way not to have to think about whether he had made the right career decision by accompanying MacArthur to the Philippines, which was beginning to seem more and more like exile, or when, if ever, Mamie and John would join him there.

He could be proud of the fact that he had swiftly gained Quezon's confidence—indeed, Quezon relied on Ike's advice so much that he provided Ike with a private office in the Malacañan Palace, partly because MacArthur's eccentric working habits frequently made the general unavailable. But Ike quickly recognized that no matter how hard he and Ord worked at drawing up plans, the Philippine army could not prevent a Japanese invasion, and would at best be able to carry out a series of costly delaying actions, hoping to hold out long enough for a battle force to arrive from the United States. Whether such a force would be available when the time came, or would be sent, was not something he could predict.

Recognizing that the Philippines was unlikely to receive modern weapons, or long-range four-engine bombers, or the support of a naval fleet, MacArthur himself, despite the buoyant optimism he displayed for the benefit of journalists, celebrity visitors like Clare Booth Luce, and members of the Philippine government, had, with a typical stroke of eccentric genius, pinned his hopes on a fleet of fast motor torpedo boats to disrupt the Japanese fleet as it approached. He actually managed to procure three early Fairmile MTBs (the British equivalent of the American PT boat) from the Thornycroft boatyard in the United Kingdom—his plan was to harass the Japanese fleet on the cheap with aircraft and motor torpedo boats, and to use the Philippine army as a guerrilla force, which he estimated could pin down at least 500,000 Japanese soldiers in the rugged terrain and among the innumerable islands for as many years as it took before the United States came to the rescue. Except for the initial naval and air components, MacArthur's intention was to fight pretty much the same kind of war against the Japanese as General Giap would fight in Vietnam against the French and the United States some twenty years later. But of course the Filipinos, when the time came, had neither the ideology nor the trained cadres needed to carry out a successful guerrilla war on a large scale and over many years against a well-armed, and ruthless occupier; nor did they have a powerful friendly neighbor to maintain shipments of weapons and supplies across a common border. Realistically, the Philippines, over 8,000 miles from San Francisco and less than 500 miles from the nearest Japanese air bases on Formosa, was on its own. Defending it would require either a major American garrison, including a modern air force and the paved airfields and roads needed to support it, or a commitment to meet any Japanese threat to the islands with the full force of the Pacific Fleet from Pearl Harbor. Neither was likely to be forthcoming.

MacArthur's confidence in his ability to defeat a Japanese attack was in part a reflection of the common European and American attitude toward Japan in the 1920s and 1930s. Of course we know now how formidable the Japanese were, and how quickly they conquered the Philippines (not to speak of Malaya and the Dutch East Indies), but in the 1930s westerners still regarded the Japanese with undisguised contempt and ridicule—they were caricatured as short, bow-legged, comic figures, always wearing spectacles because their eyes were too weak to let them shoot straight; and their aircraft and ships were assumed to be merely cheap, poorly made copies of ours. The fact that the Japanese had soundly defeated imperial Russia in 1905 on land and on sea was long since forgotten, and the British, the Dutch, the Americans (and even the Filipinos), if they agreed on nothing else, were all confident that Japanese soldiers would not fight, and that Japan was merely, in words later used about the United States by Mao Zedong, a "paper tiger."

MacArthur's optimism, like Quezon's, was based on a severe underestimation of Japan's strength and willingness to fight, as well as a fatal misreading of the Japanese national character. Even Ike, who was more realistic than MacArthur, would later underestimate the astonishing speed of Japan's advance, when it took place, and therefore overestimate the amount of time the United States would have in which to supply and reinforce the Philippines.

Apart from his daily flying lesson, and an occasional golf game with Marion Huff, who was the attractive wife of MacArthur's naval aide and was also a devoted bridge partner, Ike seems to have led a fairly quiet life in Manila. His letters to Mamie were, as she remarked, "far from cheerful," and his spirits were certainly not improved by her letters, which made it evident that she was keeping up a busy social life in Washington, and didn't lack for escorts to take her to parties. There seems to have been a certain lack of communication between them, beyond what was inevitably caused by sheer distance and the slowness of mail—Mamie complained that she didn't know when Ike was coming home, and didn't even know what his rank and salary were; and it may well be that Ike was still simmering over the fact that Mamie had insisted on staying home. In any event, it gradually began to dawn on Mamie that Ike wasn't coming home in the foreseeable future, and that if she wanted to stay married to him, she had better start making plans to join him in Manila. The widowed Gee Gerow's advice to her—whether about her marriage or about life in Manila is hard to say—was, "Doll yourself up and knock their eyes out!"[21] Mamie set to work to do just that, buying a "trousseau" for the Philippines. She was heartbroken at giving up her beloved apartment, and still deeply afraid of the health risks in the Philippines for herself and John, but eventually—and regretfully—she packed her trunks and left. She and John traveled by train to San Francisco and boarded the Army transport *Grant* there, for the long voyage to the Philippines, a journey of almost a month, arriving there in October 1936, a month after the Eisenhowers' twentieth wedding anniversary and Ike's promotion, at last, to lieutenant colonel.

Ike was waiting at the dock to greet them, but it does not seem to have been a happy, or even a conventional, reunion. He apparently said to her, once they were in the taxi, "I gather I have grounds for divorce, if I want one."[22] Although Mamie's son and her granddaughter describe the manner in which he said this as "terse but jocular"—something of a contradiction in terms—and suppose that it refers to the fact that Mamie "had not sat at home" in Washington during the year of their separations, this seems doubtful. First, Ike knew his wife well enough to guess that she wouldn't have sat quietly at home for long in any circumstances—social life was all-important to Mamie—and second, it seems an odd way to have greeted her after urging her to join him for so long, unless he thought he had a reason for saying it. Whatever the truth of the matter—and it is pointless, and possibly tasteless, to speculate—relations between them seem to have been strained, and cannot have been improved either by Ike's accounts of how much he enjoyed playing golf with Marion Huff, the mere mention of whose name would soon make Mamie grit her teeth, or by the fact that they had been assigned a suite in the unair-conditioned wing of the Manila Hotel, where MacArthur had insisted they must stay. Life with two adults and a child in a small hotel suite in the tropics would have been enough to make poor Mamie long for the apartment she had given up in Washington, but on top of that Manila had most if not all of the features she had hated about Panama: bats, bugs of every size and type, suffocating heat and humidity, countless endemic diseases, the need to sleep beneath mosquito netting (difficult for Mamie, who suffered from claustrophobia), and torrential rain that produced mildew overnight. MacArthur's apprehensions about the tropics were, rather endearingly, almost as strong as Mamie's—his own doctor regularly inspected the hotel kitchen to ensure that everything the general ate was cleaned with boiled water, and that the ingredients were fresh. Ike's own health, Mamie remarked, seemed to have improved since his long list of symptoms had puzzled the doctors at Walter Reed Army Hospital in Washington, though he had shaved his head for the tropics and now appeared completely bald. In photographs he looks tanned and fit; and the extraction of a bad tooth, of all things—or perhaps all the golfing with Marion Huff—had cured his back problems completely.

The heat and the lack of air-conditioning in their hotel suite nearly drove Mamie crazy—she and Ike spent a lot of time at air-conditioned movies (still firmly segregated in those days, with the Filipinos downstairs and the Americans upstairs in luxurious armchairs on the balcony), and often had lunch in the suite, sitting in their underwear beneath the big revolving ceiling fans. Adding to Mamie's dislike of the Philippines was its propensity for earthquakes, sometimes severe enough to tip the furniture over on top of the bed, and create panic in the hotel. It did not help matters that Ike continued to play golf with Marion Huff (who was also a member of the little bridge club he had formed), and to sing her praises. No doubt Mamie regretted the fact that she was not the athletic, sporting type, and that her nerves and her health frayed in the tropical heat of Manila.

With John placed in a boarding school in Baguio, almost 200 miles away, and once the rainy season had put an end to Ike's golf games with Marion Huff, the relationship between Ike and Mamie seems to have settled down a bit, perhaps because she had come to appreciate the intensity of social life in Manila for the small group who constituted its American elite. Manila was an elegant city, wonderful for shopping, and there was no shortage of parties to go to. The Filipino upper class, then as now, was wealthy, and loved fun, clothes, and jewelry—Imelda Marcos and her famous collection of shoes would be closer to the rule than the exception—and social life in Manila was glamorous and glittering. Ike's friendship with President Quezon brought the Eisenhowers many invitations to the Malacañan Palace; General MacArthur, now married to Jean Faircloth, was a courteous host and put himself out to be charming to Mamie; and invitations poured in from the American high commissioner and elsewhere for dinner parties (often held outside under glowing Chinese lanterns), formal luncheons, cocktail parties, bridge parties, and dances. Busy buying glorious silks and brocades in a city where a Chinese dressmaker could turn them into haute couture overnight, Mamie began to relax into her role as the American equivalent of a memsahib, in a country where even Ike's meager pay made them seem rich. In a photograph of her in "Philippine national costume"—a floor-length, tightly clinging gown of beautiful silk, and delicately puffed sleeves embroidered with flowers—taken in the hallway of the Manila Hotel on her way to a dinner party she looks thin and drawn, but happy.

Dark clouds were gathering, though. Ike agreed to another year in the Philippines, despite Mamie's desire to go home. Really, he had no choice.

MacArthur didn't want to lose him, and without MacArthur's blessing and approval he would find it difficult to get a decent appointment at home—besides, much as Ike wanted to get back to commanding troops, here he was learning to deal with responsibilities far beyond those of a mere lieutenant colonel. He and Ord were responsible for the military budget, however inadequate, of an entire nation; he was in daily contact with President Quezon, with whom he discussed not just military affairs but "taxes, education, and honesty in government"; he was involved in founding an infant air force, and even in naval matters; he had helped to plan and implement the training scheme and the training facilities for the Philippine army. MacArthur, for all his vanity and posing, had his good side—he never hesitated to let Ike get on with things, and never second-guessed him. The Philippine army, when it was ready, would be as much Ike's creation as MacArthur's, and if nobody else knew this, Manuel Quezon certainly did. Besides, Ike enjoyed the freedom—the flying lessons before the day's work began; the long flights to inspect remote, faraway army units; the ability to set his own schedule. All the same, he worried that in the event of another war he might miss his chance again, here in a kind of colonial backwater.

And even here, living in a tropical never-never land, at any rate for the tiny elite, the news was becoming rapidly more alarming to all but the most blindly optimistic. To the east, Japan's attempt to conquer China was causing waves of alarm to spread through Asia. One major Chinese city after another fell to the Japanese army, and the bloodshed reached a peak with the infamous slaughter of over 100,000 Chinese civilians in what would soon be known as the "rape of Nanking." Thousands of American missionaries and businessmen in China took refuge in the Philippines, each with his or her own story of Japanese brutality. In Europe, Hitler had occupied and remilitarized the Rhineland, and revealed the existence of a German air force, the Luftwaffe, in defiance of the explicit provisions of the Versailles Treaty, prompting Winston Churchill, in a grimly prescient speech to an indifferent House of Commons, to say,

> Herr Hitler has torn up treaties and has garrisoned the Rhineland. His troops are there, and there they are going to stay. . . . The

> creation of a line of forts opposite to the French frontier will enable the German troops to be economized on that line, and will enable the main forces to swing around through Belgium and Holland. That is for us a danger of the most serious kind.[23]

Alas, although that is exactly what would take place four years later, defeating the French army and sending the British army in headlong retreat toward Dunkirk and home, Prime Minister Stanley Baldwin was unmoved, as was the House of Commons. In Washington, no alarm was sounded; and even the French, who had the most to lose, seemed more concerned with domestic matters than with this sudden move on their own border. Only in Germany were Hitler's actions understood for what they were—the German generals who had, on the whole, been skeptical about the Führer, and alarmed by his plan to occupy the Rhineland, took note of the fact that nothing happened. Hitler had won a bloodless triumph—Germany's first victory since the early days of the *Kaiserschlacht* in 1918—and the generals moved closer to him, willing at last to consider that he might in fact be the genius his followers claimed him to be, and thereby taking their first fatal step toward their own subservience to him. As Göring's friend and fellow World War I fighter ace Lieutenant General Ernst Udet of the Luftwaffe would remark, "When the Allies did nothing, we all thought, well, maybe the *Führer* knows what he's doing after all, so let's see what he does next."*[24] What he did next, of course, was to seize Austria, then threaten Czechoslovakia, but by that time the generals were no longer in a position to object, even had they wanted to—like the rest of the German nation they would henceforth follow their Führer blindly.

Ike's mood was further darkened by the death of Jimmy Ord, whom he had chosen to accompany him to the Philippines. Ord, who was as easygoing as Ike was tense, and whom everybody liked, had decided to fly down to Baguio with a Filipino student pilot, despite Ike's advice to take a fully qualified American pilot instead. The aircraft had stalled and crashed

* Udet was a friend of the author's uncle, Sir Alexander Korda, having made his living doing flying stunts for movies in Germany in the 1920s, and it is from him that this remark comes.

while Ord was leaning out of the cockpit to drop a message to friends. Though the pilot was unharmed, and the aircraft was not seriously damaged, Ord's chest was crushed by the impact, and he died two hours later.

Ord's death not only shocked Ike, who felt personally responsible for bringing Ord out here with his wife and two young children, but seemed to make him in some way more than ever conscious that events were moving rapidly toward a major war, and that much as he wanted to command troops, his duty was to stay here and finish the work he had begun. Mamie's view of life in Manila had unexpectedly softened with time. She had come to appreciate, despite her dislike of the tropics, that service in the Philippines allowed them to live more luxuriously, and far more glamorously, than they could possibly hope to on a lieutenant colonel's pay at home—John was in a good school; at last they had money to spend on the little luxuries that meant so much to her; they were invited everywhere and treated with great respect by everyone up to and including the president. But Ord's death shook her newfound enthusiasm, which had never been all that deep to begin with, and Ike's decision to request the War Department's permission to stay on with MacArthur for a fourth year therefore pained her. She hid her feelings as best she could—the experience of 1935, when she had refused to accompany Ike to the Philippines, had apparently convinced her that from now on she must accept his career decisions without argument and make the best of them—though she must have hoped that the War Department would turn him down. But that was not to be. A replacement was sent out for Ord; Ike plunged into his work, increasingly aware that the ability of the Philippines to defend itself was still nowhere near sufficient; and Mamie, valiantly trying to make the best of things, bought herself a pair of golf shoes and started to learn golf—she was determined to replace Marion Huff as her husband's golfing partner.

Ike's concerns—which were not diminished by MacArthur's publicly stated conviction that the Japanese would never attack the Philippines—led to a succession of debilitating stomach pains for which he had to be hospitalized several times, and which would only much later be diagnosed as ileitis. As usual, overwork, worry, the self-imposed demand for perfection, and the long and irregular hours that MacArthur inflicted upon his staff when he was in the mood took their toll. The doctors called Ike's condition a partial blockage and prescribed a bland diet, but nothing did much good. In the end, what seems to have cured Ike was a trip home—he had a three-month stateside leave, and was determined to go home to make a plea for more weapons and money for the Philippine armed forces, while Mamie was going home for gynecological surgery, which she wanted to have performed by her own doctor. The Eisenhowers gave an elegant dinner party for seventy-five people in the Palm Court of the Manila Hotel to say good-bye to their friends—it was a lavish and conspicuous social success—and sailed for home in comfort with John on the ocean liner *President Coolidge*, at great expense—"over $2,000 in gold, more than four months of Ike's salary,"[25] Mamie noted. More than sixty people came to see them off, including General MacArthur himself, bearing a bottle of Scotch as a present, a rare honor, since he disliked farewells and was famously aloof with the members of his staff.

On the way home they stopped, among other places, at Hong Kong, and at Yokohama, where Ike, as a military officer, was sharply questioned by the Japanese authorities, to his great irritation. They journeyed on to Denver, to stay with the Douds, while Mamie had her surgery successfully in Pueblo; then, when she had recovered, they traveled on to Washington, where Ike met with a certain amount of indifference at the War Department: the attitude he encountered was, to quote him, that if the Philippines was so eager for its independence, the Filipinos "could jolly well look after their own defenses."[26] Ike had not been General MacArthur's aide for so many years without learning something from the great man, however. He made an appointment to see the Army Chief of Staff, General Malin Craig. Craig was no friend of MacArthur's—in fact, he had been the general chosen as Chief of Staff by Roosevelt instead of the man MacArthur wished to see succeed him, and was also an old Pershing man. But Craig was above all a professional soldier, who understood that while presidents (including Roosevelt) come and go, generals stay on forever, with their relationship to each other tightly prescribed by seniority going all the way back to their year of graduation and place in their class at West Point. MacArthur might have been sidetracked, but he had been a four-star general, was still a two-star general, and might one day return to Washington as almost anything at all. Craig would not normally have condescended to meet with a mere lieutenant colonel whose only claim to fame was as a football coach, but he nevertheless received Ike with the promptness and courtesy due to General MacArthur's representative, and heard Ike out. Ike's commonsense view was that the Philippines "was vital to U.S. interests," and that in the event of war with Japan would at the least provide a "delaying action." He wisely did not attempt to present to Craig MacArthur's more optimistic belief about defeating the enemy on the beaches in the Philippines—a delaying action in the Pacific was a more realistic premise, and exactly what the Chief of Staff was hoping for; in fact, the more delay, the better, given the lack of preparedness of the U.S. Army.

Craig obviously took away a good impression of Lieutenant Colonel Eisenhower, and at Ike's request made a few calls to those of his subordinates who had been unwilling to discuss Ike's shopping list. With the U.S. Army reduced to hardly more than 100,000 men, and an annual budget of less than $300 million, including the Air Corps, there was not much to spare for the Philippine army. But thanks to General Craig (and the "Old Boys' Network" of West Pointers), Ike managed to get a large quantity of excellent new Enfield .30'06 caliber rifles that had been stored in Cosmoline since 1918, when it was feared that it might not be possible to manufacture enough Springfields to arm the rapidly expanding Army. The Enfield was, in some ways, a better rifle than the Springfield, and far superior to anything the Japanese had. Craig's calls enabled Ike to scrounge what he could from the Signal, Quartermaster, Ordnance, and Medical corps, and made it possible for him to go to Wichita and buy aircraft, and from there to Connecticut to buy weapons and ammunition from the Winchester Arms Company. The only items on his shopping list that he couldn't get were infantry mortars, but the United States was woefully short of them.

As a result Ike returned to Manila in triumph, traveling in comfort aboard the Canadian Pacific Line's *Empress of Japan*, stopping in Kobe, Japan, where the Eisenhowers got another taste of Japanese authoritarianism; and in Shanghai, which the Japanese had just bombed. Perhaps as a result of his success at acquiring arms, Ike was reassigned by General MacArthur to "a new suite in the air conditioned wing of the Manila Hotel," with "a staggeringly beautiful view of Manila Bay from [the] windows."

Hardly anything could have pleased Mamie more than air-conditioning, then still a rare and expensive luxury, and she quickly resumed her busy social life. If anything, life for Europeans in Manila was even more hectic and glamorous than before, the approach of war in Europe and the increasingly menacing behavior of Japan making people determined to enjoy themselves while they still could. Both Ike and Mamie were disturbed by the number of society people they met in Manila who were pro-German, and by the virulence of arguments about politics, even in places as staid as the Army-Navy Club. Filipinos of Spanish blood tended to be pro-Nazi, because of Hitler's support of Franco in the Spanish civil war, and their attitude made for some uncomfortable evenings; at the same time, there was a sizable Jewish community in Manila, its numbers increased by German Jews who had left Hitler's Germany and made their way east. From them, Ike got a picture of the increasingly brutal persecution of the Jews that was taking place in Germany and would reach what seemed—mistakenly—a peak with *Kristallnacht* in 1938. That night, the government turned the SA brown shirts loose to smash the windows of Jewish shops all over Germany and burn down most of Germany's synagogues while the police and firemen looked on indifferently; and to murder, beat, or send to concentration camps on fictitious charges countless Germans whose only crime was to be Jewish.

Ike was well-informed about the unspeakable miseries and atrocities being inflicted on the Jews of Germany—unlike the vast majority of people in America, Britain, and France who didn't want to know. For this reason, he was offered the enormous salary of $60,000 a year for five years (almost ten times his Army salary), plus expenses, to resign from the Army and travel through Asia looking for countries that would accept significant numbers of Jewish refugees from Germany.[27] He was sorely tempted, both because of the money and because it seemed like the right thing to do; but as the bad news came pouring in from Europe—the Czech crisis, the Munich Conference, the escalating pitch of the Führer's verbal attacks against Poland—Ike felt more than ever compelled to remain in the Army. In Britain, trenches were being dug in the London parks; gas masks were being handed out to civilians; the Chamberlain government and the French, having reneged on their obligation to support Czechoslovakia, now fecklessly gave the Poles the military guarantee they had been seeking, thus virtually ensuring that if Hitler attacked Poland, as he constantly threatened to do, a world war would ensue, or, to be more exact, that the war of 1914–1918 would be resumed, with some reshuffling of the minor players.

General MacArthur seems to have convinced himself that war in Europe need not necessarily spread to the Pacific, and that the Japanese would never attack the Philippines, but he was perhaps the only person in Manila to take this rosy view of the future. Nevertheless, the parties went on: the Malacañan Palace and its gardens, like the rest of Manila, glittered with lights at night like a "fairy tale," and Mamie's life was a hectic succession of lunches, dinners, formal parties, and bridge games—she had apparently given up the golf lessons, on the theory that she and Ike would get along better if they did not compete with each other, and if he was left alone to enjoy the things he liked doing. It is also possible that the long trip home and back together had made Mamie less worried about Marion Huff than she had been, and no longer so determined to take Marion's place on the golf course.

In late spring, news came of a major change in the War Department—George C. Marshall was the new Army Chief of Staff. Marshall was a great friend of General Fox Conner, and had been impressed by the tact with which Ike had dealt with General Pershing's memoirs. For the first time, it seemed possible that Ike might be finally be pried loose from General MacArthur, and indeed he shortly received orders to proceed to Fort Lewis, outside Seattle, even though his tour of duty in the Philippines was not yet over and both President Quezon and General MacArthur wanted him to stay. Quezon offered Ike a $100,000 annuity to stay on, and MacArthur said that Ike was making the mistake of a lifetime, and that he could do more good in the Philippines than as a "mere lieutenant colonel" back in the States.

But Ike was determined to return. On September 3, he listened with earphones to a friend's shortwave radio and heard Neville Chamberlain sadly announce the fatal consequences of Germany's attack on Poland—words that Ike remembered twenty-seven years later:

> This morning the British ambassador in Berlin handed the German government a final note stating that, unless we heard from them by 11 o'clock that they were prepared at once to withdraw their troops from Poland, a state of war would exist between us. I have to tell you now that no such undertaking has been received, and that consequently this country is at war with Germany.[28]

Ike had no doubts now about where he belonged, and that was not at MacArthur's side in the Philippines, however satisfying it was to have President Quezon's friendship and respect. On December 12, 1939, Quezon gave a luncheon in honor of the departing Eisenhowers, and presented Ike with the Philippine Distinguished Service Cross, which he asked Mamie to pin on her husband's white dress uniform, with MacArthur beaming at them both like a happy uncle. "You helped him earn it," the president said to Mamie, and then presented her with magnificent table service. MacArthur praised Ike's "superior professional ability, unswerving loyalty, and unselfish devotion to duty."[29]

MacArthur came in person to see them off, and as their ship sailed from Manila Bay for home, the Philippine army band played on the shore, and two of Ike's pilots flew overhead to accompany the ship far out to sea. Mamie would remember feeling that their lives were about to change, but not even she could have guessed how far Ike was to go, or how quick his rise would be.[30]

Years later, when they were both five-star generals and supreme commanders, a woman asked General Douglas MacArthur if he knew General Eisenhower. MacArthur gave her a steely look and replied, "Best clerk I ever had."[31]

PART II

Command

CHAPTER 9

"As Soon as You Get a Promotion They Start Talking About Another One. Why Can't They Let a Guy Be Happy with What He Has?"

Fort Lewis had not been Ike's choice, and he and Mamie were both perplexed by his being ordered there. He had wanted to go back to San Antonio, Texas, to command troops, but as was so often the case with his career, there is a clear sense—beyond the normal confusion of Army life—of his being moved around from job to job in order to prepare him for something else. At this point, however, it was no longer General Fox Connor who was pulling the strings, but General George Marshall.

Marshall was not a man to play favorites, but despite his frosty manner, fierce objectivity, and ice-cold intelligence, he seems to have decided early on that Ike was destined to handle great responsibility, and apparently decided to put him through a series of tests to see if he would complain, and if he was up to it. No sooner had Ike arrived home than his orders were abruptly changed, keeping him in San Francisco, while Mamie shortly went on to Fort Lewis to unpack their belongings and settle into the house they had been assigned. As usual with Army moving, many of her most precious items arrived broken, and as usual their quarters were in a wretched state and required washing down and redecorating. This she did alone, while Ike stayed on in San Francisco to deal with the consequences of the Army's sudden growth, which had begun not a moment too soon.

Across the Atlantic, great events were taking place. On May 10, 1940, came the long-awaited German attack on France and the Low Countries. In rapid succession, Belgium and Holland fell, and the British army retreated to the beaches at Dunkirk and from there was ferried home in heroic circumstances that did not disguise a disastrous defeat, Paris was occupied, and the French army collapsed. In London there occurred perhaps the most significant event of all: Winston Churchill succeeded Neville Chamberlain as prime minister, and rallied his country with words that were only the grim truth—"I would say to the House, as I said to those who have joined this Government: 'I have nothing to offer but blood, toil, tears and sweat.' "[1]

Most of continental Europe was now in German hands; possession of the great French ports on the Atlantic coast gave the German navy an opportunity to strangle Britain's vital supply line by submarine warfare; and the airfields in northern France at last put the Luftwaffe within easy range of Great Britain. "What General Weygand called the Battle of France is over," Churchill told the House of Commons, in one of his greatest war speeches; "I expect that the Battle of Britain is about to begin."[2]

Spread through southern and central England, fewer than 2,000 young Royal Air Force fighter pilots would shortly fight the Luftwaffe for air control over the Channel, without which the expected German invasion could not take place; but nobody rated Britain's chances of surviving high. Even in the War Cabinet, Foreign Minister Lord Halifax, exceeding his authority, had already opened informal discussions with the Italian ambassador in London about the possibility of asking "Signor Mussolini" to explore what terms Hitler might offer Britain for peace, while the neutral ambassadors were universally skeptical about its survival, especially the U.S. Ambassador Joseph P. Kennedy, whose reports from London to Roosevelt were deeply pessimistic and tinged with anglophobia. Even Roosevelt, though profoundly conscious that Britain's surrender would constitute a threat for the United States and only too well aware of Ambassador Kennedy's defeatism, was chiefly concerned with obtaining the British government's agreement to send the Home Fleet to ports in the United States or Canada rather than surrender it to the Germans in the event that an invasion succeeded. This proposal prompted Churchill, normally respectful of the president's opinion, to note that no such agreement could be considered. "We are going to beat them," he wrote firmly.[3]

In the meantime, Ike was at Fort Ord, coping with the task of training thousands of new recruits, for selective service had just been enacted, and the U.S. Army was now expanding at the fastest rate in its history. Hundreds of thousands of bewildered draftees and uprooted National Guard units needed to be trained in a hurry, and Ike found himself temporarily assigned to the staff of General DeWitt, "whose army area comprised the entire West Coast and stretched northward into the interior, including the state of Minnesota."[4] Throughout this vast area all troops—including the National Guard divisions of several states, which were activated for only a three-week period—had to be moved by rail to southern California for training; camps had to be set up to receive them; and a railroad schedule had to be improvised to move them in an orderly fashion over great distances. Speed was required. There was no point in moving the National Guard troops to southern California for training if they arrived only in time to be deactivated and go home. Faced with this Herculean task, Ike confronted those members of DeWitt's staff who were simply immobilized by the extent (and irrationality) of the plan, improvised a plan of his own, and succeeded in turning chaos into order. Toward the end of this ordeal, he unexpectedly met the Army Chief of Staff, General Marshall, again, while Marshall was observing "a field exercise with an amphibious landing," then something of a military novelty, on a beach near San Francisco with General DeWitt. Marshall's only greeting was, "Have you learned to tie your own shoes again since coming back, Eisenhower?"

This was a reference, of course, to the life of luxury, waited on hand and foot by many servants, that still existed for American officers in the Philippines, with which Marshall himself was familiar as an "old Philippine hand," and to which, most stateside senior officers thought General MacArthur had completely succumbed, and become a kind of oriental potentate in khaki. "Yes, sir, I am capable of that chore, anyhow," Ike replied cheerfully.[5]

In fact, Ike had passed the first test with flying colors, and was able to proceed to Fort Lewis to rejoin Mamie and take up his duties as regimental executive of the Fifteenth Infantry Regiment of the Third Army—"troop duty for the first time since 1922," he exulted, and it cheered him up so much that he was able to write to his old West Point classmate Omar Bradley, "I'm having the time of my life." There is no doubt that he enjoyed his return to practical soldiering. Many of the men in his regiment had seen service in China, and the rest were volunteers—it was a "regular Army" regiment, through and through. They trained for combat over what Ike described as "some of the most difficult terrain in Washington State," and though they were woefully short of "trucks, machine guns, and mortars," by any standards they were being turned into a well-trained unit. Ike managed to get command of one battalion, on top of his duties as regimental executive officer, and, as he put it, sweated and froze with his men. In a return to his days with the famous cross-country motor convoy of 1919, he helped to take the regiment on a 1,000-mile "motor march" from Fort Lewis to Monterey, to test "control procedures, communications systems, and march discipline." He was more than ever convinced that he belonged with troops—"with them," he wrote, "I am always happy."[6]

He wrote to his old friend from West Point Mark W. Clark—whom he had last seen as a major, but whose undisguised ambition and acute sense of Army politics had always been sharper, or at any rate more obvious than Ike's, and who was now promoted to brigadier general and serving at General Headquarters as an aide to General McNair, the new Commander of Ground Forces, and in daily contact with such eminences as the Chief of Infantry—asking Clark to make sure he was able to continue working with troops, and would not be moved to another staff job. There is no doubt of Ike's sincerity, but Clark, at the hub of things in Washington, must have known that there was not the slightest chance Ike would be commanding an infantry battalion for long.

Mamie was happy herself, although now that she was "home"—or, at least, in the United States, for she still thought of Washington, D.C., a continent away, as her home—she was reconnected to her family and its problems. Her mother's health was failing, her father was deeply concerned, and both her parents were upset by her sister's impending divorce, in an age when divorce was still relatively uncommon. In the meantime, the Eisenhower's son John, though offered college tuition by Ike's brother Ed if he agreed to work in Ed's law firm after graduation, had made up his mind to go to West Point instead.

Ike was pleased, but full of dire warnings—a career in the Army would never make a man rich, the way a successful career in the law might, he told John, and the rigid seniority system made promotion glacially slow. Ike himself was the perfect example, he pointed out to John in a serious father-son conversation—the highest rank he could possibly expect to reach before retirement was that of full colonel, though admittedly that had always been the summit of his ambition. As usual in such cases, no amount of parental warnings could change John's mind, and he was sent off to a West Point cram school run by a renowned former West Point instructor in Washington, D.C., to prepare him for the competitive examination.

Far away, across the Atlantic, the blitz was beginning, as Göring, in one of the most fateful miscalculations of the war, switched his air attacks from the airfields and radar stations of RAF Fighter Command to more vulnerable but less vital targets: the big cities. The war was brought home to Americans in the pages of *Life* magazine and by the voice of radio reporters like Edward R. Murrow, describing the devastation of the air raids, the firestorm that threatened St. Paul's Cathedral, the sustained heroism of the fighter pilots and the firemen, and above all the calm courage of ordinary Londoners as they went about their lives in their bombed city, refusing to give way to panic. War did not seem so far away now, as Americans heard and read about London under fire. Some Americans actually watched British oil tankers torpedoed and set ablaze by German U-boats within sight of the glittering lights of the hotels at Miami Beach.

Stuck in Fort Lewis—and the small world of an Army base, with its rigid social rules, in which the wives were as rank-conscious as their husbands, or more so—Mamie yearned to return to a more cosmopolitan life in Washington, among her friends, both because she loved it and because she would be close to John. Her hopes were aroused by a veritable torrent of job offers arriving for Ike, as the Army expanded rapidly to prepare for a war that now began to seem inevitable, just as Fox Connor had predicted long ago in Panama. George Patton wrote that two armored divisions were being formed, and he expected to get command of one of them, in which case how would Ike like to command a regiment under him? Ike was sorely tempted—he liked Patton; he had been an early enthusiast of tanks; he would be in command of troops—but then his old study mate from the Fort Leavenworth Command and General Staff School, Gee Gerow, who had vaulted past Ike to become a brigadier general and Chief of the War Plans Division in the War Department, wrote to say that he needed Ike and wanted him assigned to War Plans right away.

Deeply impressed by the word "need"—which sounded to him very much like "the call of duty"—Ike wrote to his old friend to say that of course he would serve wherever he was assigned, whatever his personal preference might be. But no sooner had he written than he learned that General Thompson, the commander of the Third Division at Fort Lewis, had already requested him as chief of staff. Gee Gerow, hearing of this through the Army grapevine, regretfully withdrew his offer of a job at War Plans for the time being.

Mamie, true to her determination not to meddle in Ike's career decisions (and no doubt still conscious of the difficulties that had arisen between them when Ike turned down a staff job to go to Europe because of her), nevertheless must have been deeply disappointed that Ike had lost an opportunity to go to Washington, and that they were to stay on at Fort Lewis. Wisely, she did not press him on the subject. Ike spent only three months as General Thompson's chief of staff before he was moved up to become General Joyce's chief of staff, at the Ninth Army Corps, and promoted to colonel—the rank that had once been his highest ambition. Three months later he was ordered to San Antonio to become General Krueger's deputy chief of staff; and only two months afterward he was promoted again, to chief of staff of the Third Army.[7]

Much as he still wanted to command troops, between November of 1940 and August of 1941 Ike would rise from a divisional chief of staff to a corps chief of staff to an army chief of staff. (In civilian terms this was as if he had gone in only ten months from running the day-to-day operations of a corporation with 15,000 employees to one with 60,000 employees to one with 250,000 employees.) He was now responsible for the orders, training, and supply of an area stretching from New Mexico to Florida, and an army of nearly 250,000 officers and men.

As for Mamie, no sooner had she finished decorating her house in Fort Lewis and encouraged Ike to plant a flower and vegetable garden—apart from cooking, gardening was one of his only forms of relaxation—than she had to pack up their belongings again to move to new quarters in San Antonio. Admittedly, they both had fond memories of San Antonio, and it was closer to her parents than Seattle, but it was still a long way from Washington. One consolation was that as a full colonel Ike was now entitled to an executive assistant and a "striker." A striker, so named because in the nineteenth century his duties included striking his officer's tent for the day's march, was the American equivalent of the British army's batman: valet, chauffeur, messenger, and, when called on, cook. Ike chose Ernest "Tex" Lee as his executive assistant and Private Mickey McKeogh, a former bell captain at New York's Plaza Hotel, as his striker, and kept them all through the war. Both were devoted and resourceful, and for the moment, they were invaluable in coping with many of the problems that Mamie had hitherto had to solve by herself when it came to moving and setting up a new house.

This time, instead of the cramped, run-down quarters she and Ike had first shared in 1916, they were assigned a brick house on a tree-lined street, with five bedrooms, "spacious sleeping quarters and several bathrooms, and . . . a screened-in sleeping porch."[8] Once again, Mamie set herself to unpack their belongings, deal with what was broken, and decorate (and, unfortunately, fumigate) their new home.

Ike was seldom there. Even many years later, Mamie would still remember, with telling accuracy, that he was absent for six weeks at a stretch on maneuvers, and that out of eighteen evenings he was home for dinner on only two of them. He was, in fact, preparing for the largest war game in the history of the United States Army, in which his own expertise would be put to the most severe test of his career short of war itself.

The Louisiana maneuvers of 1941 would become an Army legend. They were the brainchild of General Marshall, who was determined to test the capabilities of the Army—and its commanders—on the largest possible scale. The Army had grown to 1.5 million men by the summer of 1941—fifteen times the size it had been in 1939—but Marshall was well aware that numbers did not tell the whole story. Too many of the troops were raw, untrained, and undisciplined; many of the National Guard units were poorly equipped, poorly trained, and unfamiliar with the demands of modern warfare; many of the senior officers were too old and rigidly devoted to a peacetime routine; not a single senior general officer on the active list "had even commanded a unit as large as a division" in action in World War I;* and nobody had any direct experience in the mysteries of combining airpower and tanks with infantry and artillery on the battlefield, or controlling them by radio.

More than a year after the German blitzkrieg had shattered the French army and forced the BEF to retreat to the Channel ports in just four weeks (succeeding in doing exactly what the German army of the kaiser had tried and failed to do for over four years, despite millions of causalities on both sides), the U.S. Army was still wrestling with the basics of modern warfare. The British army at least had the benefit of having suffered a major defeat, which is often helpful in weeding out incompetents and forcing the adoption of new techniques; but even defeat is not always enough to overcome time-encrusted traditions and the "Old Boys' Network," as its

* The one obvious exception was General MacArthur, sulking in faraway Manila like Achilles in his tent. MacArthur had commanded a division briefly in World War I and had been recalled to active duty by Roosevelt in July, at his own artfully orchestrated suggestion, to assume command of all American forces in the Far East and restored to four-star rank, while remaining Field Marshal of the Philippine army.

performance in North Africa against Rommel would demonstrate glaringly until the arrival in Cairo of General Montgomery. The U.S. Army was not yet even in the war—its defeats were still to come—hence General Marshall's decision to apply what amounted to a kind of shock treatment on a gigantic scale, by pitting two armies against each other, a total of nearly 500,000 men, in a war game that extended over more than a full state. The plans had been drawn up by General McNair and General Mark Clark, and were designed to put everything and everyone to the test of modern warfare. Even the location would be a test, for Louisiana in early September is hot, humid, swampy, and full of biting insects of all kinds; indeed, many of the participants would still talk in tones of awe about how rough the Louisiana maneuvers had been, when they were actually fighting in the South Pacific or Tunisia.

It was hoped that the maneuvers, apart from being a valuable training exercise, would also help to provide answers to what were still serious questions about modern warfare, many of which bedeviled the armies of all the major powers. Should armor be used in slashing, independent attacks in the enemy's rear, as the Germans had done in France with such astonishing success, or, more conventionally, used to support the infantry? (This was a question with which the British army was still wrestling ineffectively in North Africa, and its failure to resolve the issue would have a dramatic—and negative—effect on both British tactics and tank design.) Should aircraft be used independently to attack "strategic" targets, as the Air Corps wished to do, or as a kind of airborne artillery, directed from the ground? (This was a question which the Germans, though they had invented the Stuka dive-bomber and used it to brilliant effect in Poland and France, were never able to resolve satisfactorily, and which would still be causing the Allies enough problems as late as May 1944 to make Ike threaten to resign as supreme commander if he was not given control over the heavy bombers before D-day.) How could the immense amount of supplies required by a modern, swift-moving army on the march be brought forward quickly enough to prevent it from bogging down in the field? (This was a question that the U.S. Army solved better than anyone else, but nevertheless Patton's Third Army would be left stranded on the roads in the autumn of 1944 for lack of fuel, to his fury.) Still, even these questions, important as they were, paled before the most important one of all—who was fit for command, and who was not?

Of course no war game, however big or realistic, can test the resilience of troops under fire, or their willingness to kill and be killed; nor can anything duplicate the mental stress for commanders of making instant, life-and-death decisions on the battlefield, in the middle of the "fog of war." Still, as Ike later summed it up, "The beneficial results of that great maneuver were incalculable. It accustomed the troops to mass teamwork; it speeded up the process of eliminating the unfit; it brought to the specific attention of seniors certain of the younger men who were prepared to carry out the most difficult assignments in staff and command; and it developed among responsible leaders skill in the handling of large forces in the field."[9] As he pointed out, it also tested the ability of the Army to supply large numbers of troops on the move in the field with fuel, food, and ammunition, and confirmed the wisdom of the Army's decision to procure hitherto unthinkable numbers of four-wheel-drive trucks of all sizes.

Eisenhower's boss, the commander of Third Army, was Walter Kreuger, recently promoted to lieutenant general, a tough, no-nonsense soldier who had, unusually in an army dominated by West Pointers, risen through the ranks from private all the way up to three-star general (and would eventually win a fourth star). Kreuger was something of a martinet, in Patton's mold but without Patton's streak of craziness or his talent for self-promotion. He seems to have been at once hated and admired by his men—often a good sign in a general—and although he was sixty-two, old in the U.S. Army for a general, his energy and his physical courage were legendary. Kreuger's problem was that he had no experience in commanding large formations, and needed a staff officer who could look after the details while he got on the business of trying to lead 240,000 men in battle, simulated or not. He had requested Ike, and Ike did not fail him.

Kreuger's task, in the first phase of the maneuvers, beginning on September 15, was to defend southern Louisiana against an attack by the "Red" army (no political connotation attached to this choice of color, which is ordinarily used in military exercises to indicate the attacker) of

160,000 men, commanded by Lieutenant General Ben Lear. Lear began with a swift, blitzkrieglike attack with one armored corps and one motorized infantry corps, the armored thrust led by George Patton's Second Armored Division; but Kreuger's "Blue" army was able not only to defend its line but to surround Patton's tanks and compel General Lear to surrender on the fourth day.

Kreuger, that rarest of creatures, a self-effacing and modest victorious general, attributed much of "Blue" army's success to Ike, who had indeed played a major role in drawing up the plans, and above all seeing that they were carried out and supported in a timely and efficient manner. Significantly, Ike had made intelligent use of airpower in bad weather by sending aircraft to find Patton's tanks and pin down their exact route and position. Airpower usually played a largely theoretical role in such war games. Nobody was quite sure how effective air strikes would in fact be, despite the grandiose claims of the Air Corps generals, and the problem of linking infantry and armored commanders on the ground to pilots in the air had yet to be solved by equipping ground forces with improved radios and by "embedding" with the ground units "forward air controllers." These controllers were airmen who could translate the needs of infantry or tank commanders to pilots who might be several thousand feet above the battlefield, flying at 200 miles per hour, and perhaps miles away, and from whose perspective the ground ahead would look very different from the way it looked to somebody lying in a muddy ditch and trying to peer over the edge of it while under fire.

Despite his Army's victory, Ike was dismayed by the number of senior officers who failed to rise to the occasion, and who, to use Ike's phrase, had to be "weeded out." Too many officers were simply incapable of the kind of leadership that was required by the rapid pace of modern warfare, and bewildered by the constant movement of masses of heavy vehicles over primitive roads. Nobody hated getting rid of a man more than Ike did, and once he was in a position of high command overseas the British were always complaining that he was "too soft," but the truth seems to be that when it was necessary, he did it swiftly and without any second thoughts or regrets. Officers had to measure up to their responsibilities, and the sooner they learned that, the better for the whole Army. At least, he did not do it with the kind of malicious glee that sometimes appears in Montgomery's diaries at the expense of British generals he sacked (or wanted to sack), or those whom he thought Ike ought to sack.

Kreuger's benevolence toward Ike extended to letting him talk to the press—for the Louisiana maneuvers were big news, and many reporters for the top newspapers, magazines, and radio stations covered it. There was nothing secret about it—quite the contrary. After all, nobody could hide 500,000 men, and in fact it was part of General Marshall's plan to draw attention to the size and scope of what was going on, and to make people aware that the Army—their Army—was preparing itself for modern war. By and large, the reporters came away impressed, most of them fortunately unaware of the number of problems the maneuvers had exposed. Ike, though he had no training in public relations, turned out to have a natural way with reporters, who flocked to his tent. He was tall, handsome, good-natured, and good-humored; he had a natural gift for covering up an awkward moment with a funny story—and above all he was modest and sincere and clearly knew what he was talking about. He held a kind of open house for the press outside his tent—he called it "something of a cracker-barrel corner"[10]—with liquor for those who wanted it (large parts of Louisiana were still dry), a wealth of human interest stories, and an explanation for civilians of everything that was happening. As a result he got the lion's share of the credit for Third Army's victory, thus opening up a breach between himself and General Kreuger, although not all the reporters managed to spell his name right, one of them identifying him as "Colonel D. D. Ersenbeing."[11] One of the journalists who did get his name right was Drew Pearson, who gave Eisenhower the credit for Third Army's victory in his nationally syndicated column, no doubt further stoking the irritation of General Kreuger, for Pearson at the time was immensely popular, and read by everyone in Washington, including the president.

A week later, the Louisiana maneuvers continued, this time with Kreuger's forces playing the attacker while Lear's attempted to defend Shreveport against them. (Patton's Second Armored was moved to the attacking force and this time played a major role in the victory.) It took only five days for Kreuger to surround Lear and force him to surrender, despite the fact that the battle area was hit by the "tail end of a hurricane," turning roads to molasses and causing the rivers to flood.

Toward the end of the maneuvers Ike wrote to Gee Gerow that he was having a good time, but still wanted a command of his own; instead he was informed by Mark Clark, at a postmortem on the maneuvers, that he had been promoted to brigadier general.

He went home to San Antonio, where General Kreuger pinned the single silver stars on his uniform as Mamie looked on. It was, she said, "her proudest moment," though she was quickly amused to note how much more quickly things moved for her now that she had "a General tacked to [her] name," and was distressed to learn that it would cost $150 for the tailor to make all the necessary changes to Ike's uniforms now that he was a general.[12]

Neither Ike nor Mamie got much of a chance to enjoy Ike's new rank. Only eight weeks after receiving his star he was woken from a rare Sunday afternoon nap with the news of Pearl Harbor. Five days later "his direct-line phone to the War Department rang, and the voice of Colonel Walter Bedell Smith asked, 'Is that you, Ike?' It was. Thus Smith: 'The Chief says for you to hop a plane and get up here right away. Tell your boss that formal orders will come through later.' "[13]

After a difficult trip—his plane was grounded and he had to proceed on to Washington by train—Ike was picked up by his brother Milton, at whose home in Falls Church he would live until he knew what his plans were, and presented himself to General Marshall early on Sunday morning, December 14. Marshall was already on a full-time war schedule, in which weekends were no longer of any significance. Ike's feelings about the call to Washington were mixed, though he kept them to himself. He was pleased at having been summoned by the Army Chief of Staff in a moment of crisis, but deeply concerned that he would spend World War II, like World War I, at home, in a staff job, rather than in command of fighting forces overseas. He also suspected that he was here because of his knowledge of the Philippines, and feared being pigeonholed as an expert on one subject, or as the person most likely to be able to interpret what General MacArthur's messages really meant, and to extract the truth from the sonorous, oracular, grandiloquent phrases of MacArthur's communications. (No less an authority than the Commander in Chief in the Pacific, Admiral Nimitz, kept a picture of MacArthur above his desk as a reminder to himself "not to make Jovian pronouncements complete with thunderbolts.")[14]

This was only the fourth time Ike had met Marshall, and their previous meetings had lasted only for a couple of minutes each, but Ike in any case would hardly have expected an effusive greeting from a man as reserved as Marshall. "Without preamble or waste of time," Ike noted, Marshall outlined the general situation, as he saw it, in the western Pacific. He was brief and crisply factual, but the facts were not such as to arouse optimism. The Navy reported that the Pacific Fleet would be out of action for many months; Hawaii lay open to attack; the only remaining Allied capital ships in the western Pacific, the Royal Navy's HMS *Prince of Wales* and HMS *Repulse*, had both been sunk off the coast of Malaya by a Japanese air attack and Admiral Sir Tom Phillips had gone down with them; and Japanese forces had landed in strength in Malaya, the Dutch East Indies, and the Philippines. In the Philippines, Japanese aircraft had caught the B-17 bombers that had been sent there neatly lined up wingtip to wingtip at Clark Field (despite warnings cabled to MacArthur) and destroyed them, thus eliminating the last significant remnant of the Allies' airpower in the Pacific west of California.

"What should be our general line of action?" Marshall asked.[15]

Trying to keep what he—a skilled poker player—described as a poker face, Ike asked for a few hours in which to prepare an answer, and went off to borrow a desk and a typewriter from Gee Gerow. He quickly drew his general conclusions, writing them down on a lined legal pad, and then typed them up, trying to make them as concise as possible—for it was already clear to him that Marshall was not a man who wanted frills and flourishes or generalities, and that it should be his aim to reduce the answer to every question, however complex and far-reaching, to one sheet of paper.*

* In this he resembled Winston Churchill, who often instructed those to whom he sent queries and questions to reply "on one sheet of paper."

Ike's former mentor, General Fox Conner, had once told him that when war came again—and he had no doubt it would—we would be fighting it with allies, and victory would depend on a unified command. In Conner's view the man best qualified to exercise that command was George Marshall, who, he added, "was close to being a genius." Ike gave some thought to how Conner would have wanted his recommendations presented, and decided that his "answer should be short, emphatic, and based on reasoning in which I honestly believed." He faced the painful facts: no major reinforcements could reach the Philippines without the protection of the battleships of the Pacific Fleet, which still lay smoldering at Pearl Harbor. The first priority must therefore be to set up a secure base in Australia, and "to procure a line of communications leading to it," which meant moving instantly to save Hawaii, Fiji, New Zealand, and New Caledonia, even at the risk of further Japanese advances elsewhere. (This conclusion may seem obvious now, since we know that the plan succeeded, but nobody else at that time had reached it, stated it clearly, or proposed to make it America's first and most immediate priority.) In the meantime, every effort had to be made to supply the American and Philippine troops by air and submarine for as long as possible, "although the end result might be no more than postponement of the inevitable."[16]

When Ike went back in to present Marshall with his conclusions, he finished by saying, "General, it will be a long time before major reinforcements can go to the Philippines, longer than any garrison can hold out with any driblet assistance, if the enemy commits major forces to their reduction. But we must do everything for them that is humanly possible. The people of China, of the Philippines, of the Dutch East Indies will be watching us. They may excuse failure but they will not excuse abandonment."[17]

Marshall merely replied, "I agree with you. Do your best to save them," and sent Ike off to begin the process of building up a base in Australia, while getting anything he could to the Philippines. In effect, Marshall had asked Ike for his recommendations, then given him the task of carrying them out. Despite his other preoccupations, Marshall met with Ike once a day to hear his report, as Ike begged and borrowed from the Navy every ship of any kind he could to rush men and supplies to Australia, and even used the Navy's precious submarines to bring vital supplies of shell fuses to the American forces in the Philippines who were running out of them. From time to time it seemed to Ike that the enemy was the United States Navy, not the Japanese, and at one point he confided to his diary that "one thing that might help win this war is to get somebody to shoot [Admiral] King."[18] Eventually, he was obliged to beard the lion in his den by calling on the admiral, about whom one of his own daughters said, "He is the most even-tempered man in the Navy—he is always in a rage."[19] King began by saying "No!" to Ike without even looking up from his desk. He was famously rude, intolerant, difficult, exacting, and harsh—an admirer has written that his only "weaknesses were other men's wives, alcohol, and intolerance"—but he was given to sudden, unexpected moments of kindness and common sense, and despite the fact that he claimed not to trust officers "who didn't drink and enjoy the company of women," he ended by hearing Ike out and agreeing with him. Ike had simply made it clear that he wasn't afraid of the admiral, and went on presenting his request in a calm and reasonable way, thus acquiring King's respect, rarely extended to those in Army uniform. Indeed, when the subject of a joint Anglo-American landing in French North Africa came up in 1942—a landing that would rely almost completely on the United States Navy—King wouldn't hear of anybody commanding it but Ike.

Ike's position was deputy to Gee Gerow in the War Plans Department, but from the very start Marshall relied on Ike much as MacArthur had. One of Ike's first moves was to overrule the Navy, which after Pearl Harbor had wanted to recall all the supply ships on their way to the Philippines to United States ports, and send them on to Australia instead, guarded by the cruiser *Pensacola*. Ike worked closely with Marshall, drafting many of the Chief of Staff's most crucial messages, many of which dealt with the rapidly deteriorating situation in the Philippines. MacArthur's plan to "stop the enemy on the beaches" of the Philippines had failed from the first moment—at the first sight of the Japanese landing on the beaches, the Philippine soldiers threw down the newly issued Enfield rifles that Ike had procured for them, and ran.

Ike was, unsurprisingly, clearheaded on the subject of MacArthur. The Philippines was a lost cause, and in his opinion MacArthur should fight it out to the end, with whatever help could reach him—Ike, when the time came, would be against evacuating MacArthur from Bataan, and also against awarding him the Medal of Honor—and then surrender. He drafted Marshall's calm and fact-filled replies to MacArthur's complaints that no supplies were reaching him and that he and the Philippines were being abandoned. Ike also had a hand in drafting Roosevelt's reply to an angry, anguished plea from President Quezon—with the support, rather surprisingly, of MacArthur, who had swung from improbable optimism to describing the situation as "about to be a disastrous debacle"—to let the Philippines be granted immediate independence and be "neutralized," with the immediate removal of both American and Japanese troops.[20]

Ike's description of this as a "bombshell" was putting it mildly. It not only struck at the heart of Americans' assumptions about the Philippines—the Americans' attitude was essentially colonialist and assumed both loyalty and obedience from the Philippine government—but also brought home at last, even to optimists like the president, "the very somber picture of the Army's situation" and the possibility of a major defeat there. This picture was shortly to be confirmed by the news that Manila had fallen and that MacArthur had retreated to Bataan, where 15,000 American and 65,000 Filipino troops were trapped. In his reply to Quezon, Roosevelt rejected Quezon's proposal as unacceptable, and pledged that "so long as the flag of the United States flies on Filipino soil . . . it will be defended by our own men to the death," an answer which infuriated Quezon. Roosevelt's cable to MacArthur ordered him to fight on "so long as there remains any possibility of resistance," and said, even more uncompromisingly: "The duty and the necessity of resisting Japanese aggression to the last transcends any other obligation now facing us in the Philippines."[21] MacArthur's belief that he had been "betrayed" by Roosevelt, Marshall, and Ike, who had promised reinforcements and supplies and then withheld them or routed them elsewhere, is not borne out by the facts. True, the president made optimistic promises; and true, Marshall did his best to make the tone of his messages more encouraging than Ike's very much bleaker drafts. But in the absence of a fleet and an air force to protect them, no significant number of supply ships could have reached the Philippines in time to change the outcome, and it seems unlikely that so experienced a commander as MacArthur did not know this.

Of course the mild paranoia that tended to surface in MacArthur's mind at moments of crisis did not help smooth matters between Washington and Manila. MacArthur had always felt that what he called the "Chaumont crowd"—officers who had served with Pershing at AEF headquarters during World War I rather than fighting at the front, like himself—were his enemies, and the cause of his various disappointments and frustrations. Prominent among them, of course, was George C. Marshall. The Chaumont crowd envied MacArthur—they had hindered him in the field; had persuaded Pershing to award him the Distinguished Service Cross instead of the Medal of Honor he thought he deserved; had conspired with Roosevelt, "a man who would never tell the truth if a lie would suffice,"[22] to take away his four stars before he arrived in the Philippines; and were now, of course, withholding supplies from him. Ike had been MacArthur's aide long enough for MacArthur to recognize, in the messages from Roosevelt and Marshall, both Ike's characteristic crisp style and a close understanding of the realties of the military and political scene in the Philippines that could have come only from Ike; and this was enough to infuriate MacArthur and convince himself that Ike had joined "them" in betraying him. The more MacArthur thought about it, the more sense it all made—of course Ike had sent the *Pensacola* convoy to Australia instead of to the Philippines, toward which it had been on its way until the Navy got cold feet! It was the first step in the campaign against him, for in MacArthur's darker moments the enemy was always in Washington, not in Tokyo. He held Ike responsible for the shortage of supplies and modern weapons that was crippling both the Philippine army and the American forces in the Philippines. Why, he wanted to know, were aircraft being sent across the Atlantic to Britain instead of to the Philippines?

But there was no way to get aircraft to the Philippines except on carriers, and the Navy was not about to send its only two carriers in the Pacific into waters controlled by the Japanese navy and air force, to suffer the same fate as HMS *Repulse* and *Prince of Wales*. Besides, in Washington Ike was facing the problems of global war. There was only so much to go around; shipping was in short supply; the demands of the British and now the Russians for supplies and weapons of every kind were insistent and clamorous. "Arsenal of democracy" was a nice phrase, but it was impossible to convert American industry to war production overnight, and much of what was being produced had not been tested in combat and would prove to be of questionable value. And it was a question not only of supplies, aircraft, equipment, and ships—though these were all intractable problems, familiar to Ike from his work on the War Department's plans for organizing American industry for war production in the 1930s—but of men. There was hardly anyplace on earth where there was not a demand for American troops, even though the only place they were actually fighting was the Philippines. Men were needed to garrison the Aleutians, the Caribbean islands, and Iceland. A constant flow of men needed in the South Pacific went on their way via Australia. Huge numbers of men had to be trained—and new formations created—in anticipation of America's attack on Nazi Germany, wherever that might eventually take place. From Brazil, where bases on the coast needed to be built and secured for the task of attacking U-boats in the south Atlantic, to Burma, where the precarious and vital supply line to China, the famous Burma Road, had to be maintained and secured, the call for troops was endless. Islands nobody had ever heard of had to be garrisoned to protect the route to Australia; aircraft had to be flown across the Atlantic to Britain to begin America's contribution to the bombing of Germany; communications and supply routes had to be pushed northward from the Persian Gulf to make direct contact with the Soviet Union—America's greatest and most immediate problems were those of time and distance.

Ike, like the other members of the War Plans Department, worked late into the night, seven days a week; he joked that he had never seen his brother Milton's house in the daylight. Mamie's arrival a few days after Christmas cheered him up, though he saw very little of her. Mamie soon decided that the first thing Ike needed was a home. A friend of Milton Eisenhower's, an affable executive at CBS radio, Harry "Butch" Butcher, soon to become part of Ike's wartime "family," whom Ike and Mamie had known and liked when they were in Washington in 1930, was now a naval officer. Butcher's work seemed to involve knowing everyone and everything in wartime Washington, and he found them a small apartment in the Wardman Tower* (where he and Ruth Butcher lived in the CBS corporate apartment). The Eisenhowers' apartment had one bedroom and a kitchenette—something of a miracle in 1942, but then Butcher was a kind of miracle worker, a Mr. Fix-It who could find anything and whose gift for lubricating the working press would be invaluable to Ike in the future. Mamie went back to San Antonio to pack up their belongings once again, though this time most of their things would have to go into storage. All the same, cramped as the apartment was, and much as she hated packing again (she had no idea, fortunately, that it would be five years before most of the things were unpacked), Mamie was not displeased—Washington was after all where she had wanted to be ever since their return from the Philippines.

Ike was soon involved in matters even more important than the American debacle in the Philippines: the arrival in Washington of Winston Churchill, accompanied by most of the British chiefs of staff—Admiral Sir Dudley Pound; Air Chief Marshal Sir Charles Portal; and Field Marshal Sir John Dill, the former chief of the Imperial General Staff (CIGS), whom Churchill had just replaced with General Sir Alan Brooke, and whom he would shortly appoint as his military representative in Washington. Brooke had been left at home to mind the store, but otherwise it was an all-star British military cast. Each man was supported by his aides, and each, starting with the prime minister, was convinced that he knew how the war should

* The Wardman Tower was home to a number of major well-known figures in the Army and in Washington society—including, in the latter category, the famous hostess Perle Mesta. She would be immortalized in Irving Berlin's musical *Call Me Madam*, played by Ethel Merman, and especially in the song "The Hostess with the Mostest on the Ball." Mesta became a good friend of Mamie's, as she was of Harry Truman and practically everybody else who mattered in Washington.

Winston Churchill boards the new British battleship HMS *Duke of York* for his first visit to meet President Roosevelt, and pauses to stroke the ship's cat, to the amusement of the petty officers on his right.

be fought and won, and that Britain, having fought for two years, was the more experienced and sophisticated of the two English-speaking Allies. As they met in Washington, the German army was almost within sight of the Kremlin towers, so the Soviet Union's plans and strategy, such as they may have been, were academic, even had there been any means of learning them. The burning question, in fact, was whether the Russians could hang on, and few of those who gathered in Washington for the conference—which was code-named Arcadia—were optimistic on that subject. Even the normally optimistic prime minister rated the Russians' chances of survival at no better than fifty-fifty, and in a rare burst of American slang remarked to one of his military aides, "Moscow is a gone coon."[23]

The British were not entirely welcome guests in Washington. From the moment he heard the news about Pearl Harbor, the prime minister had pressed hard for what we would now call a summit meeting as soon as possible, despite advice from every side to wait until he was invited. The president had expressed cautious optimism on the subject—his favorite form of delay—but at this early stage in the alliance, Churchill's insistence eventually carried the day. The British had arrived like self-invited dinner guests at a household in crisis, with the kitchen on fire, the dining table not yet set, and the roast burned to cinders. America had been plunged into war unexpectedly, and was still in a state of chaos and shock, struggling to organize itself overnight. The Pacific Fleet lay sunk or burned at Pearl Harbor; the Japanese army was sweeping over the Philippines; attacks from the Japanese fleet were expected everywhere from California to Alaska. Besides all that, there was a universal wariness against being led by the wily British into plans that might be good for British imperial interests, but not necessarily useful for bringing the war to a quick and victorious end. Though disguised by his affability, courtesy, and congenital good humor, this caution was as strong in the mind of Roosevelt as it was in that of the most extreme isolationist in Congress—something which the British took a very long time to discover.

From the British point of view, the purpose of Arcadia was simple—to ensure that the war against Germany would be the Allies' first priority. Except for General MacArthur and Admiral King, however, nobody in America had ever argued seriously against this policy. Nazi Germany was infinitely the more dangerous of the two major enemies, and the United States could not take the risk that either the Soviet Union or the United Kingdom might be defeated, let alone both of them, leaving America to fight Germany alone. In the previous war, it had taken Great Britain, France, Russia, the United States, and Italy four years to defeat the Central Powers (Germany, the Austro-Hungarian Empire, and Turkey), and Hitler's Germany was dramatically more powerful than the kaiser's, as well as more ruthless and fanatic. Only a successful alliance could hope to defeat Germany, and two of the three Allies were already deeply and precariously engaged. In Russia the Germans had killed or captured millions of Russian soldiers, destroyed most of the Soviet air force on the ground, and reached the outskirts of Moscow and Leningrad. Whether Stalin, or that strongest of Russian allies "General Winter," could prevent a complete collapse of the Soviet Union in the very near future was the most serious question facing the participants in Arcadia, and surely the most difficult to answer.

As for the British, they had plunged from exuberant victory to dismal defeat in the Western Desert with the arrival of Rommel and the Deutsche Afrika Korps to shore up the collapsing Italian army. The vital port of Tobruk was surrounded, and its fall would be a strategic disaster and a terrible humiliation; the island of Malta, without air cover from which Allied shipping could not pass through the Mediterranean, was besieged by the Luftwaffe; the conquest of the Nile Delta and the Suez Canal seemed within Rommel's reach. The American military attaché in Cairo reported, accurately, if glumly, "With numerically superior forces, with tanks, planes, artillery, means of transport and reserves of every kind, the British army has twice failed to beat the Axis in Libya. . . . The granting of 'lend-lease' alone cannot ensure a victory."*[24]

Operating from French ports on the Atlantic, German U-boats were exacting an increasingly fearful toll of shipping. In the air, RAF Bomber Command, forced by unacceptable losses to bomb German cities by night, was only beginning to receive the first of the big modern bombers and new electronic navigational aids that would eventually turn it into such a fearsome weapon of destruction. The British had fought for two years and had very little to show for it, except the loss of much of their empire in the East to the Japanese, and the potential of a bitter and strategically disastrous defeat in the Western Desert. Whether in the worst case the United Kingdom could survive the impending loss of Malaya and Burma, as well as that of Egypt and the Suez Canal, was doubtful. Still, ties of blood and language apart, the brutal facts of geography made it clear that the shortest route to bring American forces into combat with Germany was across the Atlantic to Britain. The great ports of the United Kingdom were only 3,500 miles away across the North Atlantic, and a stream of ships was necessary in any case to provide the 20 million to 25 million tons of supplies a year

* Correct as, in this instance, the U.S. military attaché's comments were, it was unfortunate that Italian military intelligence had broken the American diplomatic cipher, so his messages were reaching General Rommel before they reached the War Department in Washington.

without which Britain could not hope to survive—and once American troops were in Britain in sufficient numbers, the coast of France was only eighteen miles from Dover Castle.

Shoring up its two principal allies and defeating Germany were therefore in America's best interests, even if doing so meant allowing the Japanese to advance within reach of Australia, the inevitable loss of the Philippines, and the delay of America's revenge for Pearl Harbor. On the other hand, deciding how, when, and above all where to bring America's might to bear against Nazi Germany was more difficult.

Even at this early stage, the difference between the British and the American position was embarrassingly evident. The Americans, including the president and General Marshall, envisaged a landing in France in 1942, even if it was limited in scope, both as the best way of coming to grips with the enemy and as perhaps the only way of relieving the enormous pressure on the Russians. The British, while cautiously polite, thought this was a dangerous fantasy. They had experienced combat with the German army in Norway, Belgium, France, North Africa, Greece, and Crete, and had no illusions on the subject. German troops were well trained, battle-hardened, highly motivated, well led, and equipped with excellent (and battle-tested) weapons. Their tanks, artillery, and machine guns would remain superior to those of the Allies throughout the war. There were more than enough German divisions in France to repel an Allied invasion without having to withdraw any from the eastern front or the Balkans. Also, the Germans would enjoy the advantage of interior lines, thus being able to reinforce quickly by rail and road, while the British and the Americans would have to rely on a precarious supply line across the Channel or the Bay of Biscay. Finally, the specialized landing craft, most of which had been designed in Britain and were to be built in quantity at American shipyards—the LSTs, LCTs, and the LCIs—simply did not yet exist, so success would depend on capturing intact, and holding, a major French port, and it seemed to the British unlikely that the German army would let this happen.

The prime minister trailed the possibilities of an American landing in French North Africa in front of the president, without at first eliciting much enthusiasm. It seemed to most of the Americans present that Casablanca was a long way from Berlin. Moreover, the United States recognized the collaborationist Vichy government of Marshal Pétain whereas Britain recognized that of General De Gaulle and the Free French "National Committee," whose forces had been fighting the Germans since the defeat of France in 1940.* There were difficulties in dealing with De Gaulle for Britons of every rank—General Brooke had commented in his diary on first meeting De Gaulle, "Not much impressed by him. . . . Whatever qualities he may have [are] marred by his overbearing manner, his 'megalomania' and his lack of cooperative spirit";[25] and Churchill, borrowing a phrase from his friend General Spears, complained, "The heaviest cross I bear is the Cross of Lorraine," that being the symbol of Free France. Still, the bond between Britain and Free France, as incarnated by De Gaulle, however thin it frayed, remained strong. When France fell and Britain was left alone to face Germany in the summer of 1940, De Gaulle had come to London and brought with him a small nucleus of Frenchmen who were determined to continue the fight. The British had not forgotten that. Pigheaded, obstinate, infuriating, and obsessed with France's grandeur (and his own) De Gaulle might be, but he had proved to be a loyal ally through many difficult and even terrible moments, like the sinking of the French fleet by the Royal Navy at Oran or the failure of the expedition to capture Dakar and the two great French battleships anchored there. Churchill recognized in him another great man on his own scale, had in fact whispered to him prophetically—at the last, dreadful meeting with the French government at Tours, when it was clear that France was defeated and would ask the Germans for an armistice—"*L'homme du destin*."[26] On De Gaulle's arrival in London, Churchill noted, "Under an impassive, imperturbable demeanour he seemed to me to have a remarkable capacity for feeling pain—I preserved the impression, in contact with this very tall, phlegmatic man: 'Here is the Constable of France.' "[27] In the austere, stubborn personality of De Gaulle the British recognized—whatever their reserva-

* On this subject there was already considerable acrimony between the Allies, though nothing to what was in store. Cordell Hull, the secretary of state, insisted on referring to De Gaulle and the National Committee as "the so-called Free French," to the great annoyance of the prime minister.

tions, feelings of guilt, and impatience—the essential spirit of the country they had admired, coveted, and fought for almost 900 years, only to end up as reluctant allies in two world wars. Much as the prime minister wanted to please the American president, he was not about to give up De Gaulle at the whim of American foreign policy.

The second purpose of Arcadia was to establish an effective link between the British and the American chiefs of staff, which the president was determined should be based in Washington. Here there was greater accord. Field Marshal Sir John Dill was appointed the head of the British military mission, and each of the British service chiefs would appoint a senior officer as a representative in Washington. Dill was a surprisingly happy choice for the post—he quickly developed a close friendship with General Marshall and the president, and he retained their trust until his death in 1944. (At Marshall's insistence Dill became the only non-American soldier ever to be buried in Arlington Cemetery, and his grave is marked by one of only two equestrian statues in the cemetery, the other being that of Major General Phillip Kearney, who was killed in the Civil War.)

Perhaps the most important decision was to appoint a supreme commander in the East. For this post, in a moment of great generosity and tact, Roosevelt suggested General Sir Archibald Wavell, former British Commander-in-Chief in the Middle East, who, when he had lost Churchill's confidence there, became Commander-in-Chief, India. Wavell was a man that perhaps only the British army could have produced: a warrior and a poet, at once a sensitive intellectual with refined literary and artistic tastes and a gifted commander of men. Churchill, when he became prime minister, learned that Wavell was an author, and sent at once for one of his books. He was puzzled when an aide presented him with an anthology of poetry, *Other Men's Flowers*, and the incident somehow set the tone of their relationship, as if they had gotten off on the wrong foot. Churchill admired Wavell; perhaps saw in him some of the same division that shaped his own character, between the man of action and the man of letters; and for that reason heaped on Wavell vast responsibilities that no one man could have shouldered successfully.

Everybody—including Rommel—admired Wavell. He possessed boundless courage, skill, a profound grasp of strategy, imperturbable calm under fire—even under fire from the prime minister—and, to use an old-fashioned word, gallantry. With his slender forces stretched from Libya to Iraq and from Greece to Abyssinia, there was no way he could have produced the kind of victories that Churchill demanded; yet in the end Wavell held the Middle East together calmly and thus paved the way for the later victories of General Montgomery—despite a constant barrage of intimidating messages from the prime minister, who interfered in even the smallest details.

Perhaps nobody but Wavell, with his detached and slightly melancholy view of events, would have undertaken to accept "supreme" command of ABDA, which was to include all American, British, Dutch, and Australian troops in the Far East, excluding India, where he was replaced by another unlucky Middle East Commander-in-Chief, General Sir Claude Auchinleck; and Australia and the western Pacific, which would shortly be split between General MacArthur and Admiral Nimitz (since no American admiral would agree to serve under a general, and MacArthur would not serve under an admiral).[28]

"Wavell never had a chance," Ike would write later, with sound professional judgment. Even the prime minister, not normally a pessimist, wrote of Wavell that "he would have to bear a load of defeat in a scene of confusion," summing up very well the situation in the Far East. Wavell would go to Java to take command of his widespread troops knowing full well that without a fleet or a functioning air force there was no chance of stopping the Japanese until they simply ran out of steam because of the sheer distances involved in their conquests,* though not before sweeping ABDA swiftly out of existence. In the meantime, his short-lived appointment set a valuable precedent—Allied forces (except for the Russians) would henceforth serve, in each major theater of war, under a single supreme commander. In short, what the Allies did not manage to do during World War I until their backs were to the wall in 1918, at which point they all

* The Japanese had apparently neglected to read Clausewitz, who believed that all attacks, however successful, eventually come to a halt as the attacker's forces become spread out and weakened, and as he outruns his own supply lines—e.g., Napoleon when he reached Moscow, or Ike in late 1944 when Monty had failed to free the port of Antwerp and Patton ran out of gas close to the German border.

agreed, when it was almost too late, to serve under the supreme command of Marshal Foch, they began by doing in January 1942. As might be expected, while Churchill hailed Roosevelt's move as "selfless," it was also shrewd and farsighted. If the president was willing to let American troops serve under a British supreme commander early in 1942, then the British could hardly object later on to serving under an American supreme commander when the time eventually came to invade Europe.

That future commander would later describe himself as having been "on the fringe of the conference,"[29] and since he was a mere brigadier general, that was no doubt true; but he did have a unique opportunity to observe the Allied leadership at work. The differences between the British and the American participants were significant, yet somehow joint policy in many crucial areas was eventually agreed on, and the complexity of turning even the simplest of high-level decisions into detailed military orders so as to ensure that they were carried out was enormous—and instructive. It was one thing to agree across the conference table that the United States would garrison Iceland immediately with one division, freeing the British troops there to fight in the Western Desert, but quite another to pick the appropriate unit, equip the troops with winter gear, and get them safely to Iceland, at a time when shipping losses to U-boats in the North Atlantic exceeded 500,000 tons a month. The vital task of moving American troops to Australia quickly was made possible only because the British agreed to transport them in the 80,000-ton RMS *Queen Mary*, with its sister ship RMS *Queen Elizabeth* the largest, fastest passenger liner in the world. The *Queen Mary* could take 15,000 troops at a time from the East Coast of the United States to Australia unescorted, at speeds so rapid that no U-boat, and very few surface vessels, could catch up with it.* Great decisions required a multitude of smaller ones lower down the chain of command to put them into effect, and a firm command structure to ensure that every detail went

* Churchill himself offered the ship to General Marshall during the Arcadia conference. When Marshall asked how many men she could carry, the prime minister replied that she could take 8,000 men with access to lifeboats and life rafts, or 15,000 if one ignored safety precautions altogether, and that he could only tell Marshall "what *we* should do," but had to leave that decision to him. Marshall chose to take the risk of loading 15,000 men, and both the *Mary* and the *Elizabeth* continued to take that number, or more, throughout the war.

smoothly—and to wring the maximum cooperation out of the Allies, and out of their proud, defiant, mutually hostile services, for the outcome of the war would depend on the ability of the Royal Navy, the U.S. Navy, the British and the American armies with their very different structures and traditions, the Royal Air Force and the U.S. Army Air Corps to work together effectively. Since cooperation between the American services—the U.S. Navy and the U.S. Army were hardly even on speaking terms—was already difficult, cooperation with the British armed services was likely to present even more severe problems, and this did not bode well for joint strategy. The details of this and many other thorny issues landed on Ike's desk during and after the conference—there was almost no period in his life when he had worked harder, and that did not escape Marshall's attention.

Given the volume of work and the amount of information that a worldwide war involved, General Marshall was determined to restructure the War Department more efficiently. He was also a strong believer in shifting senior officers out of staff positions before they grew "stale from overwork" and "burned out." He therefore assigned General Gerow to command the Twenty-Ninth Infantry Division, and replaced him with Ike, who now became briefly, in Mamie's words, "the No. 1 man at War Plans."[30] No doubt Ike would have much preferred to have been in Gee Gerow's shoes, commanding troops, but for the moment it was not to be. Soon afterward, Marshall decided to replace the War Department's Planning Section with a new structure, the Operations Division of the General Staff, a more streamlined organization intended to provide him with information in concentrated form and to implement his decisions rapidly and efficiently. Whoever commanded the Operations Division, or OPD, as Marshall's Command Post swiftly came to be called, would be, in effect, Marshall's deputy for operations, and he chose Ike.

Shortly thereafter, there occurred a scene that is central to Ike's legend. Talking to Ike one afternoon about the promotion of an officer, Marshall paused to give Ike "his philosophy" on the subject. In World War I, Marshall said, the staff had received promotions more quickly than officers in the field. He intended to reverse this practice. "The men who are going to get the promotions in this war are the commanders in the field," Ike quotes him as saying, "not the staff officers who clutter up all the administrative machinery in the War Department. . . . The field commanders carry the responsibility and I'm going to see to it that they're properly rewarded so far as promotion can provide a reward."

Marshall gave Ike a stern glare. "Take your case," he said. "I know that you were recommended by one general for division command and by another for corps command. That's all very well. I'm glad they have that opinion of you, but you are going to stay right here and fill your position, and that's that!"

Perhaps to underscore the point, he added: "While this may seem a sacrifice to you, that's the way it must be."

Normally Ike maintained his reserve around Marshall, who, for his part, only once slipped up and said "Ike," and then made up for it by saying "Eisenhower" half a dozen time in the next minute or two. This time, however, Ike was stung. Nobody was more eager to command troops in action, and he was still deeply resentful at not having been sent overseas in World War I and very conscious of the fact that he was probably going to have to repeat that disappointment now. It was enough to unleash, for once, the famous Eisenhower temper. "General," he blurted out angrily, "I'm interested in what you say, but I want you to know that I don't give a damn about your promotion plans as far as I'm concerned. I came into this office from the field and I'm trying to do my duty. I expect to do so as long as you want me here. If that locks me to a desk for the rest of the war, so be it!"[31]

This was as close as anybody ever came to insubordination toward General Marshall. Slightly awed by his own outburst, Ike got up and walked to the door—a long walk, since the Chief of Staff's office was enormous—and then went back to his desk, red-faced and feeling increasingly sheepish at having added to Marshall's concerns.

In some versions of this story, Ike recalls turning back briefly and seeing "a tiny smile . . . at the corner of General Marshall's face." But this seems unlikely—why would Ike pause and turn around?—and Marshall's tiny smile seems to have crept into the story only when Ike was already a retired five-star general and a former president of the United States, a lofty position from which it was surely possible for him to look back on the incident with hindsight and a sense of humor. One supposes that at the time, Ike more likely assumed he would be lucky to keep his one star.

There may have been more to Ike's anger than his desire to command troops in battle. Like many men, he was driven hardest by a secret weakness. His was ambition. He had always been at great pains to conceal it, sometimes even from Mamie. From time to time Fox Conner had seen through the pose of the affable, humble Ike who occasionally backed into the limelight by accident, but George Marshall was far too shrewd a judge of men not to recognize immediately that Ike was fiercely ambitious. Ike's moment of anger was, no doubt, as much because he realized that Marshall had seen through him as because Marshall proposed to keep him in Washington for the duration of the war.

In any event, if this was a test on Marshall's part, Ike passed it with flying colors. Three days later, he was startled to find on his desk a carbon copy of a letter from Marshall to Roosevelt recommending him for promotion to temporary major general. More important, Marshall—who had presumably told Roosevelt about his decision to promote troop commanders instead of staff officers—felt obliged to explain to the president in his letter, with a certain amount of jesuitical reasoning which was hardly Marshall's usual style, that Brigadier General Eisenhower was not really a staff officer "in the accepted sense of the word," but rather his "subordinate commander," responsible for "all dispositions of Army forces on a global scale," including the Air Corps. (An attempt to revise this letter and eliminate the word "commander" was made within the War Department bureaucracy, and was promptly rebuffed by a stiff reminder from General Marshall that he had drafted the letter himself, and that not a word of it was to be changed.)[32]

Ike and Mamie's pleasure at the promotion was overshadowed by the death of Ike's father. Ike had heard the news from Mamie while he was drafting a message from the president to Chiang Kai-shek, which he was to present to Roosevelt the next day, and it had pained him greatly that he was unable to attend the funeral. "War . . ." he wrote in his diary that evening, "has no time to indulge even the deepest and most sacred emotions." He wrote a moving eulogy to his father a few days later, shutting his door for thirty minutes to be alone and think about him, and expressing his regret "that it was always so difficult to let him know the great depth of my affection for him."[33]

He missed his father's funeral, but managed to find time to pay a quick visit with Mamie to West Point, to see John, and both his wife and his son noticed how exhausted Ike was. As for Mamie, she was preoccupied by yet another impending move—General Marshall wished to have those who worked most closely with him near at hand at all times, so Ike was assigned Quarters Number 7 at Fort Myer, a large, handsome, redbrick house on a tree-lined street close to Quarters Number 1, the home of the Army Chief of Staff, with a beautiful view of Washington. Ordinarily, Mamie would have been delighted—it was exactly the kind of house she had always wanted, and indeed the houses for general officers at Fort Myer were the best the Army had to offer. But moving the crates full of their possessions from San Antonio to Washington in wartime was a huge task, as was attempting to decorate the house with no help but that of Ike's orderly Mickey McKeogh, and with Ike hardly ever home. She persevered, and was soon able to give dinner parties, though Ike very often failed to arrive until dinner had already been served.

The spring of 1942 was perhaps the low point of the Allies' fortunes. The Philippines, the Dutch East Indies (now Indonesia), most of Burma, and all of Malaya were falling or had already fallen to the Japanese. The surrender of the American and Philippine forces at Corregidor was a national tragedy, perhaps more of a shock than Pearl Harbor. The fall of Singapore—the supposedly impregnable "fortress Singapore," where the Japanese took prisoner almost 85,000 British and Commonwealth troops,* the largest surrender in British military history—was the prelude to Japan's rapid conquest of the Dutch East Indies, with its huge supply of oil. At sea, the Allies' last two remaining modern ships in the southwestern Pacific—the Dutch heavy cruisers *De Ruyter* and *Java*—were sunk in the Battle of the Java

* Not the least of the tragedies, given the Japanese treatment of POWs, was the diversion to Singapore of an entire Australian division, which arrived there just in time to be captured.

Sea. In the Western Desert, Rommel neatly parried repeated British attacks, and the fall of Tobruk—unthinkable only a few months ago—seemed imminent. In Russia, although the German army had been unable to take Moscow and was forced to withdraw in subzero weather as Stalin poured fresh divisions from the East into the battle, Hitler had forbidden further retreat and insisted on holding a line through the winter by sheer willpower, against the advice of his generals. This was "the severest test in our existence," as the Führer later described it; but as soon as the *rasputitsa*, Russia's infamous spring thaw which turned roads to a quagmire, dried enough to a permit movement, the Germans would launch a massive offensive across a front hundreds of miles wide with over 1 million men, cutting deep into the heart of Russia all the way to the Volga and the Caucasus. Nowhere was there even a trace of good news. Anybody looking at a map of the world—and nobody looked at the map with more intense scrutiny than Hitler—could see that if only Rommel could take the Nile Delta and advance to the northeast toward Iran, and if the Germany army in Russia could cross the Caucasus and meet him, the Allies had as good as lost the war.

The U.S. Army was being raised from a planned strength of thirty infantry divisions and six armored divisions to a more ambitious planned strength of sixty infantry and ten armored divisions. This was a Herculean task, but as Churchill shrewdly noted, "What will harm us is for a vast United States Army of ten millions to be created which for at least two years while it was training would absorb all the available supplies and stand idle defending the American continent." Very clearly, some substantial portion of the U.S. Army, beyond the men being sent to the South Pacific via Australia, must take the offensive in 1942, whatever their state of training and equipment. It was General Marshall's job to decide where and when, and Ike's job to present him with alternative plans and to ensure that once a decision was made, it was carried out effectively.

The new major general was already appearing at the White House and meeting with the president. Although at first the White House usher wrote down his name in the official diary as "P. D. Eisenhauer,"[34] his correct name was beginning to be known in White House circles. When it was decided to send one of the presidential inner circle to Australia, to bolster the Australians' confidence in America's commitment to its defense and to improvise a fleet of blockade runners to get vital supplies to MacArthur, the former secretary of war Patrick J. Hurley, who was also a former New Dealer was sent to see Ike at the War Department to be briefed. To his surprise he was appointed a temporary brigadier general on the spot—Ike and Gee Gerow each actually took a star off their own uniforms and pinned them on Hurley's—and placed on a plane bound for Australia that same night.[35] Ike already knew Assistant Secretary of War John J. McCloy, the future high commissioner of Germany—about whom it was said, by no less an authority than Secretary of War Henry Stimson, that "nobody in this administration does a thing without first talking to McCloy." Ike had persuaded McCloy, after the Louisiana maneuvers in 1941, of the then revolutionary idea that every American division should have a Piper Cub and a pilot attached to it for artillery spotting; and McCloy was soon to become well known to Harry Hopkins, Roosevelt's closest and, at the time, most influential adviser, whom Ike described as "fanatically loyal to the President."[36] The "barefoot boy from Abilene" had already come a long way.

One of the things that surprised Ike most when he took over the OPD was the almost complete absence of an organized intelligence service. "Spying" was looked down on—the secretary of war had famously remarked, "Gentlemen don't read each other's mail"—and the Army's military attachés were chosen for their social skills and a healthy private income, since no public funds had been made available to meet the expenses of living in a foreign capital. As a result, the United States was forced to rely in part on such intelligence as the British passed on to their ally, with the obvious risk that much of it was slanted from the British point of view. The creation of General William Donovan's Office of Strategic Services (OSS) was intended to provide an American equivalent of both Britain's MI-6,* which dealt with foreign intelligence and espionage;

* MI-6 (also known as Secret Intelligence Service, the envy of Continental statesmen since the Napoleonic wars, and the ultimate repository of Britain's reputation as "perfidious Albion") achieved postwar fame as the employer of the fictional James Bond (whose creator, Ian Fleming, like Graham Greene, served in MI-6 in World War II), and was run during the war by the legendary and secretive Brigadier Sir Stuart Menzies, always referred to as "C." SOE was a more recent wartime invention of Churchill's, who appointed Labour MP Hugh Dalton to run it with the ringing order, "Set Europe ablaze!"

and Special Operations Executive (SOE), which was responsible for such activities as commando raids and the dropping of agents into occupied Europe, the support of underground movements, sabotage, and what are now called dirty tricks. The OSS rather overshadowed, by its glamour, the rapid growth of the Army's Intelligence Division, under General Strong, which was intended to provide the kind of detailed military and industrial intelligence and analysis that Ike felt the Army lacked.

It was not just intelligence about the enemy that was lacking, however—Ike needed intelligence about what American industry was doing, above all about what weapons and supplies were available and what was still to come; and this was almost as hard to come by as hard facts about Germany's and Japan's military production. The amount of detail Ike had to track down, analyze, and pass on to General Marshall in condensed, useful form is extraordinary. Even a quick glance through his memorandums to Marshall, and the messages he drafted for Marshall's signature, inspires awe at the breadth of his knowledge, his capacity to absorb detail, and his *Sitzfleisch*—best translated as the ability to sit at one's desk without a break until a job is done, however long it takes.

From early December 1941 to late June 1942 Ike delved deeply, up to eighteen hours a day, seven days a week, into every military, political, strategic, and logistic problem that faced the United States. The range is simply amazing, and constitutes a kind of crash course in modern generalship: from the exact number of rounds of ammunition available in the Philippines (his inquiries revealed that even in such basics as 50-caliber AP ammunition, the stock in the Philippines was 31 million rounds short of authorized supply; and in 30-caliber Ball, the standard Army rifle cartridge, the stock was over 18 million rounds short) to the correct pay scale for enlisting Chinese for guerrilla warfare. Ike swiftly advised on the fastest route for ferrying heavy B-17 and B-24 bombers to Australia (Brazil–Africa); drafted detailed instructions for General Wavell as supreme commander of ADBA in the Far East; attempted to unravel the many difficulties between the British, Americans, and Chinese in cooperating to fight the Japanese in Burma; and tried to deal with such headaches as the dive-bombers that had been shipped to Australia in crates, where they were to be reassembled and flown to the Philippines, only to find that the solenoids vital to their bomb-dropping mechanism had been thrown away along with the crates to which they had been stapled. This last subject produced an angry radio message from General MacArthur himself, and prompted the secretary of war, unfamiliar with aviation technology, to remark petulantly, "They have a name that sounds like a bad word—something like hemorrhoids."[37]

No problem was too big, too small, or too difficult to command Ike's attention, and in each case he reduced it to its essentials and did his best to solve it and, when necessary, explain it to Marshall. When MacArthur complained that his forces were running out of fuses for shells, Ike not only found them but persuaded the Navy to deliver them by submarine, and also pointed out that there were quantities of shells and fuses stored in the tunnels at Corregidor, unbeknownst to MacArthur and his staff. He (prophetically) drew up plans for landing American forces in French Morocco, after Churchill brought the subject up during the Arcadia conference, and complained to himself on his desk pad, "Tempers are short! There are lots of amateur strategists on the job—and prima donnas everywhere. I'd give anything to be back in the field."[38]

In April, against a background of absolute disaster in the Far East, and imminent disaster in the Libyan desert, General Marshall flew to Britain to confront the most important and opinionated "amateur strategist" of them all, and to see for himself what was happening. He returned having once again secured British agreement in principle to the importance of a cross-Channel landing, but dismayed to discover that the buildup of American forces in the United Kingdom was taking place at a rate so slow as to preclude any attempt in 1942. American officers in London seemed to have no idea of Marshall's ambitious plans for turning the United Kingdom into a base for offensive operations, "the greatest operating military base of all time."

Marshall returned to Washington determined to send somebody whose judgment he trusted over there to come up with a plan, somebody who could turn chaos into order, and who would win the confidence of the British. There can have been very little doubt in his mind who that person would be.

Ike requested permission to take with him "Major General Mark W. Clark, then chief of staff for General McNair, head of the ground forces," and one of Ike's old friends. Early in May the two major generals set off for London by air, via Maine, Newfoundland, Labrador, Greenland, Iceland, and Scotland.

Ike's job was to recommend to Marshall what needed to be done to get American forces in quantity to Britain, and who should command them.

He could not have guessed that this limited task would take him to war at last—and carry him onto the stage of world history in less than four weeks.

CHAPTER 10

London, 1942

Kay Summersby, outside Ike's London headquarters.

London in the summer of 1942 was a city at war. Huge areas of what was then still the world's largest city lay in ruins, and everywhere there were the marks of bombs and fires—rows of stately houses with one or two gone, like missing teeth; whole neighborhoods destroyed. Famous buildings—Buckingham Palace, the Houses of Parliament (walking over the still smoking ruins of his beloved House of Commons, Churchill had grumbled, "They *would* blow up the best bits!"), St. Paul's Cathedral—all bore the scars of bombs, and were shored up with scaffolding, while landmarks like Nelson's Column and Landseer's lions in Trafalgar Square were protected "for the duration" with piles of sandbags. The blackout was complete, and rigorously enforced, with the result that there had never been so many prostitutes in the streets since the nineteenth century, or so much street crime. Soldiers, sailors, and airmen of a dozen foreign nations, in exotic and unfamiliar uniforms, and speaking even more unfamiliar languages, thronged the pubs. The blitz was over, but the Luftwaffe still managed to bomb London frequently enough to keep people on edge at night when they heard the sirens, wondering whether or not to go down to the shelters. Nor was it just London. Other cities—Coventry, Portsmouth, Southampton, Birmingham—were even worse hit. Coventry had been hit so badly that for a brief time its name came to be used as a verb in both English and German to describe inflicting total destruction (as in, "We coventried Cologne last night").

The British had been at war for three years, and a noticeable shabbiness and fatigue had set in—clothing was strictly rationed, so people had a certain threadbare look; bathing was limited to four inches of hot water, to save fuel (many bathtubs had a line painted around the inside to mark the limit); the meat ration was a few ounces a week; the average citizen was allowed one egg a month; and fresh fruit of any kind was so rare as to be an unthinkable luxury. The British ate horrible substitutes for familiar dishes, like the notorious "Woolton pies," named after the minister for food, a sort of ghastly Cornish pasty with ingredients about which nobody cared to speculate; and variants of shepherd's pie or sausage that were treated with even greater suspicion. Cigarettes, candy, sugar, and "spirits" were rationed. Except for those of doctors and farmers, no private cars were in circulation. Everywhere, women were in uniform,* or doing men's civilian jobs, from farming to driving buses. The two most popular phrases in Great Britain were "carrying on" (when people were asked how they were doing, they replied, with resignation, "Oh, we're carrying on, you know") and "Mustn't grumble" (an alternative reply to the same question). Patient, stoic, good-humored, and brave, the British had been "carrying on" doggedly since September 1939, and opinion was divided as to how they would respond to the presence of a couple million American servicemen in their small, crowded island.

* The author's stepmother-to-be, Leila Hyde, first met his father, Vincent Korda, in 1942, when, as an ambulance driver in a Red Cross uniform temporarily assigned to be Lady Mountbatten's chauffeur, she knocked him down with Lady Mountbatten's car outside Claridge's Hotel in the blackout.

On the one hand, the British craved an ally—a real, powerful ally, in contrast to the governments in exile of France, Belgium, Holland, Norway, Czechoslovakia, Greece, and Poland. Of course, since June 22, 1941, the faraway Soviet Union had been an ally; except to those on the left, it did not seem all that much more appealing than Nazi Germany. On the other hand, the British resented America's "innocence," wealth, and attitude of moral superiority, especially since the Americans had joined the war late. When it might have made all the difference—in 1938, during the Sudetenland crisis; or in 1939, when Hitler attacked Poland; or in June 1940, when France was on the brink of surrender—America had refused to intervene. Now, when the war it had been unwilling to help prevent had reached the scale of a world cataclysm, it had been plunged into the conflict by the Japanese attack on Pearl Harbor, and already it was taking on the role of moral leadership, just as Woodrow Wilson had done in 1919, only to let the other Allies down with a thump. The Americans were big, noisy, loud, and well fed; their pay scale was enormous compared with that of British troops; their soldiers wore uniforms made of material almost as good as that of officers; and they seemed to have unlimited supplies of candy, gum, cigarettes, and that miracle of modern science and technology, nylon stockings. British children soon learned to stand by the side of the road when convoys of American troops drove past, shouting, "Got any gum, chum?" to the great annoyance of their parents. It would take an officer of supreme tact to prevent the relationship between the U.S. forces in Britain and the British from turning sour long before they saw combat together.

That such an officer might already have arrived in Britain was not immediately evident. Kay Summersby,* a young Anglo-Irish woman who was a former fashion model for Worth of Paris, recently separated from her husband—a wealthy publisher—and rather loosely "engaged" to a dashing but as yet, unfortunately, still married American captain named Richard Arnold, was a member of the Motor Transport Corps (MTC), a volunteer

* The author had the pleasure of talking briefly to Kay Summersby Morgan shortly before her death, and published her posthumous memoir, *Past Forgetting: My Love Affair with Dwight D. Eisenhower*, in 1975.

unit of uniformed drivers originally formed from post-debutantes, and was the first person in London to comment on Major General Eisenhower. The MTC drivers were smartly uniformed, bright, upper-class, and pretty, although most of them, like Kay, had done yeoman service during the worst of the blitz driving ambulances in London's East End, picking up the wounded, the dead, and the dying in the middle of air raids and firestorms. The drivers had worn boiler suits or corduroy slacks then, and gauze masks when they helped to put the dead into canvas body bags. Now, smartly uniformed again, Kay had the rather less exacting duty of driving visiting American senior officers, and having been at the hairdresser when drivers were being assigned to the members of a visiting party from Washington, she was disappointed to be given an unknown two-star general to drive.

She waited for two days while the American party, grounded in Scotland by the weather—a good introduction to the British climate—eventually decided to come down by train. The train was late too, and when the party finally arrived at Paddington Station to be met by the American ambassador, John G. Winant, they piled into the ambassador's car with him, so that Kay, by now furious, had to drive back to the American embassy in Grosvenor Square and sit there waiting in her olive drab Packard, until eventually, giving way to hunger, she went to the embassy canteen for a sandwich. When she came back, she was mortified to see that all the other cars had left and two American officers were standing beside the Packard, one of them very tall indeed and the other somewhat shorter, stockier, and fuming with impatience. She ran up to the car, saluted briskly, and asked if one of the two men was General Eisenhower, and when the shorter one indicated that he was, she opened the door for them and asked where they wanted to go.

"We'd like to go to Claridge's, please," Ike said, to Kay's surprise, since it was only two blocks away.

When she got back to headquarters an American asked her how she had liked "Ike."

"Ike?" she asked.

"General Eisenhower."

She said he seemed all right. The American said, "There are a lot of rumors about him. They say he's going to be in command here."[1]

That did not impress or convince Kay Summersby—major generals were small fry to the young women of the MTC—but the next day she was to spend over sixteen hours in the car with Ike; and the day after that, even longer. She soon learned that he was a chain-smoker, that he had unlimited curiosity, and, when not angry, an infectious grin and great charm. She also learned that while British major-generals were no big deal in London, this American major general was a very important man indeed, treated with deference by almost everybody. She experienced Ike's temper at close range, however, when she drove him to see Lieutenant-General Bernard Montgomery. "That son of a bitch!" Ike said, his face crimson with fury, as he got back into the car. It turned out that he had lit a cigarette as he talked to Montgomery, who told Ike sharply that he did not permit anybody to smoke in his presence, and to put the cigarette out immediately.[2]

She drove Ike and Major General Mark W. Clark to almost every American military installation in Britain, and it was apparent to her that they were not pleased by what they saw. When they had time to spare, she took them sightseeing. They went to Dover, where they could look through their binoculars at German-occupied France only eighteen miles away; to the ruins of the East End, where she had worked during the blitz; to an English pub, where they tasted warm English beer and switched gratefully to gin-and-tonics (without ice); and to lunch at the Connaught, where the hitherto unthinkable sight of two major generals and their driver eating together caused a distinct stir (and where Ike first began to call her Kay instead of Miss Summersby). Sitting opposite him for the first time, she remarked on his "brilliant blue eyes, sandy hair—but not very much of it—fair, ruddy complexion." She thought he had, for a general, "a nice face—not conventionally handsome, but strong, and . . . very American, certainly very appealing," and she succumbed immediately "to the grin that was to become so famous." She took them to see Warwick Court, where half of her mother's house had been blown away by a bomb; the Tower of London; the Guildhall; and, for some reason, Bryanston Court, to see the house where Mrs. Simpson had lived during the abdication crisis.

Ike, it turned out, had met the duke of Windsor, the former king in Washington, and had not been favorably impressed. "A shame the King lost sight of duty," he remarked crisply. Finally, after two days of delay caused by bad weather—another warning for the Americans of what was to come—she drove them to RAF Northolt, outside London, where they presented her with a box of chocolates, a gift as good as gold in London in 1942, and boarded an aircraft bound for Washington.[3]

She never expected to see either of them again.

On his return, Ike reported to Marshall. Ike admitted that he had not been impressed by Major General James E. Chaney, the United States Commander in the United Kingdom, or by anything else he had seen there of the American effort. The officers kept bankers' hours, wore civilian clothes, and saw themselves as a "liaison group" rather than the headquarters of an army in formation. What was needed was a real commander, one who realized he was there to fight—somebody who could put the entire American effort in the United Kingdom on a war footing as rapidly as possible. The officer in command of American forces in Europe, Ike thought, should be somebody totally familiar with the plans of the United States government, able to control a rapid buildup in Britain and to make the troops understand that they were there to fight the Germans—and the sooner, the better. Asked who this should be, Ike recommended Major General Joseph McNarney, but Marshall brushed that idea aside because McNarney had just been promoted to Deputy Army Chief of Staff. Instead, he ordered Ike to draw up a directive embodying the essentials for whoever was going to be the commanding general in the European theater of operations. Ike submitted the directive to Marshall on June 8, remarking that Marshall had better read it carefully, because it was likely to be a key document in waging war against Germany. "I certainly do want to read it," Marshall replied. "You may be the man who executes it. If that's the case, when can you leave?"[4] Three days later Marshall appointed him to command the European theater, with Mark Clark as his deputy. "I'm going to command the whole shebang,"[5] Ike told Mamie, when she asked him what his job would be.

. . .

Very shortly thereafter, following a meeting with Secretary of War Stimson, Ike was called to the White House (by this time the White House usher knew how to spell his name correctly), where Winston Churchill was once again a guest, to meet with the president and the prime minister. Ike had already made a favorable impression on Roosevelt, and he appears to have made an equally favorable one on Churchill, who was something of a connoisseur of generals.

The prime minister, together with his staff, had come to discuss the future operations in which Ike would now play a key role. Bolero was the administrative and supply buildup in the United Kingdom (a Herculean task in itself, demanding an immense increase in shipping, and a major Anglo-American assault on the German U-boats that were preying so successfully on Atlantic convoys). Round-Up was the full-scale invasion of France with no fewer than forty-eight divisions (eighteen of them British); the Americans were determined to accomplish Round-Up as soon as possible, but in no event later than September 1943. Sledgehammer was an alternative, smaller plan to capture and hold a bridgehead on the French coast, perhaps in Brittany, with no more than five divisions, as an "emergency measure" in case the Soviet Union showed signs of collapsing; or, again alternatively, to take advantage of "significant political changes in Germany." Finally, "Tube Alloys" was the code name for the atomic bomb.

The British were very doubtful that Round-Up would be possible even by September 1943, and were horrified at the very idea of Sledgehammer, which they thought was certain to fail without providing any meaningful help to the Russians. In the interests of the Allies' unity they kept these reservations to themselves, but with unfortunate consequences, since they left the Americans under the impression that they had agreed to both. This misunderstanding would lead to considerable subsequent rancor and accusations of bad faith.

The prime minister also brought up once again the subject of Gymnast, the full-scale invasion of French northwest Africa, and even presented the president with a long and persuasive paper on the subject, which he had written himself. The British were convinced that Gymnast at least offered a chance to get American forces into action in 1942 (and not by trying to storm ashore in France and put five divisions, of which at least two would very likely be British, on the rocky, inhospitable coast of Brittany, supplied precariously by sea, to confront up to twenty-five experienced German divisions).* And if Gymnast was successful it might force Rommel to defend himself on two fronts, and bring about significant changes in the Mediterranean. For their part, the Americans were still wary of becoming entangled in French colonial politics, or in the dispute between those French who were loyal to Pétain and those who followed De Gaulle. The Americans were made even more wary by Churchill's tendency, in discussing the Mediterranean, to launch himself out full sail on a flood tide of eloquence toward wider strategic horizons—landings in Sicily, or the Greek islands, or southern Italy, or, worse still, the Balkans; and plans to bring Turkey into the war on the side of the Allies.

This was Churchill's old stamping ground, it was here that he had deployed all his enthusiasm, energy, and formidable powers of persuasion during World War I to bring about the Gallipoli landings in 1915 that temporarily ended his political career and still haunted his reputation. By now, the British chiefs of staff merely rolled their eyes when listening to this kind of thing from the prime minister (Admiral Pound was said to nap until he heard the word "Navy," at which point he would wake up with a start), but the Americans were frankly alarmed—though America had not yet joined World War I when Gallipoli failed, it was still the one thing that most of them remembered about Churchill as a strategist. Even before the prime minister arrived in Washington, Secretary of War Stimson and General Marshall had been warning each other about the dangers of leaving the president alone with him, since, as they rightly guessed, Churchill would certainly try to sell Roosevelt on the merits of Gymnast. However, Roosevelt surprised them both by inviting his guest to his home in Hyde Park, New York, for a break during the conference, and returned half sold

* The wisdom of this was demonstrated two months later, perhaps not entirely by chance, when the British landed over 5,000 troops, most of them Canadian, to seize the French port of Dieppe in a kind of small-scale dress-rehearsal for Sledgehammer. Of the 5,000, more than 1,000 were killed and 2,000 captured. In the face of this disaster, enthusiasm for Sledgehammer declined rapidly.

on Gymnast, to their dismay. Indeed, Marshall was so taken aback by Roosevelt's sudden, if partial, conversion to Gymnast that despite his legendary self-control and respect for Roosevelt, when he heard about it he told the president that it was "such an overthrow of everything they had been planning for . . . that he refused to discuss it in any way, and he turned and went out of the room."

The conference was further marred by a moment of profound humiliation for Churchill. He learned from the president, who broke the news to him as gently as possible, that Tobruk, the port which was the key to any attack against the German and Italian armies in Libya, though heavily garrisoned by British and Commonwealth troops, had just fallen—over 33,000 officers and men had surrendered, a good many of them Australian, to the predictable outrage of the Australian government, which blamed the British.

This was an unmitigated disaster for the British. The fall of Tobruk left Rommel free to attack across the Egyptian frontier again, perhaps even to take Cairo and the Nile delta at last. Except possibly for the surrender of Singapore and the sinking of *Repulse* and *Prince of Wales*, it was the darkest and most humiliating moment of the war for Churchill, made worse of course by receiving the news from Roosevelt in the Oval Office. "Defeat is one thing," Churchill said glumly, "disgrace is another." But the president, with exemplary tact and generosity, merely asked what the United States could do to help, and offered to remove 300 brand-new Sherman tanks and 100 Grants from American armored units, which had just received them, and send them immediately to North Africa to replace British losses. The British Eighth Army, which had fallen back beyond the Egyptian frontier, had lost 1,188 tanks in seventeen days of continuous fighting, and had only 137 serviceable tanks left. This situation prompted the American ambassador to Egypt to cable the State Department in Washington the rather alarmist warning that he expected "Rommel to arrive at Cairo and Alexandria . . . within the next few days."[6]

The ambassador was not the only person expecting Rommel's arrival. From Cairo and Alexandria, the "gabardine swine," as Cairo's numerous staff officers in their well-tailored tropical uniforms were known, "got the wind up" (this was then a popular British phrase to describe fear), and fled north in their cars, with their luggage and their mistresses, to the safety of Jerusalem and the King David Hotel. But at the royal palace, the Gezira Sporting Club, the bar at Shepheard's Hotel, and Groppi's Restaurant in Cairo, King Farouk and the wealthy Egyptian elite openly celebrated the imminent arrival of the Germans, whose field guns could be heard in the distance above the clink of ice cubes and the thud of tennis balls on the courts of the Mena House Hotel. A curfew was instituted, officers were required to wear pistols at all times, and smoke eddied above all of official Cairo as British officials burned their confidential files—not an action calculated to convince the Egyptians that the British were preparing to make a last-ditch stand west of the canal.

What saved the day was the stubborn determination of British infantry to dig in and fight, and the exhaustion of Rommel's Afrika Korps, in which some of his armored divisions were reduced to as few as twenty-two tanks. The British were attacking the Italian infantry, which gave way, leaving the weakened German units to bear the brunt of the fighting. "Things are going badly for me at the moment," Rommel wrote glumly to his wife, apparently unaware of the panic in Cairo. "It's enough to make one weep."[7]

Above all, Clausewitz's iron laws of warfare were in operation, this time to Rommel's disadvantage—he was hundreds of miles east of his bases; his supply line was stretched to the breaking point over impossible roads or none; and the further he pursued the British, the weaker his attacking forces inevitably became.

In Washington, Ike, still immersed, despite British opposition, in planning the cross-Channel invasion for 1943, received the shocking news that his brother Roy had died, of a stroke. Once again, he was unable to attend a family funeral in Abilene. The Commandant of West Point gave John a weekend pass to say good-bye to his father, and Mamie wept at the sight of John, in uniform, giving Ike a crisp salute, soldier to soldier, at the end of the weekend. A few days later, Mamie walked Ike to his car and watched it drive away, taking him to the airfield where a plane waited for him. She had promised that she would stand beside the flagpole outside their quarters and wave as the plane flew over Fort Myer, but even the commander of the American forces in Europe couldn't persuade the pilot to battle high winds at a low altitude with a fully loaded aircraft, and all Mamie got to see was the plane receding in the distance as it climbed.

In accordance with harsh Army tradition, Mamie had to vacate Quarters Number 7, where they had lived for only two months, within a week of Ike's departure. She returned, thanks to the indefatigable Harry Butcher, to the Wardman Tower. Her larger possessions were once more crated up and moved into storage, since at first she would have to share the Butchers' small apartment with Harry's wife, Ruth, and their daughter—Harry himself would soon be joining Ike as his naval aide and general factotum.

Mamie was hurt but not surprised at being "ordered off the post." That was the Army way, and she was used to it. What was harder on her was not knowing how long Ike would be gone and not being able to reach him except by letter, as well as the discomfort of becoming a celebrity's wife—for Ike's new appointment was a big story, and since the journalists couldn't reach him, they went after Mamie, who had no help or advice from the Army.

Never at ease with reporters and news photographers, Mamie endured, rather than enjoyed, her sudden fame. She was suffering from a variety of ailments, many of them no doubt brought on by the continual moves, the uncertainty, and the separation from Ike, as well as the misery of a Washington summer heat wave in a small apartment without air-conditioning. She did volunteer work for the Red Cross and as a waitress in a canteen for soldiers, sailors, and marines, and fell back on the friendship of her fellow Army wives whose husbands were overseas; but there is no doubt that it was, for the time being, a lonely life, and she could not help feeling that the Army, having appointed her husband to high command, had simply abandoned her.

Ike arrived in Britain to assume his command on June 24, 1942. He was accompanied by his deputy, Mark Clark; by one of his aides, Ernest R. "Tex" Lee, a former insurance man and car salesman; and by Mickey McKeogh. He landed at Northolt aerodrome, outside London, to be greeted by a substantial party of brass hats, among them Vice-Admiral Lord Louis Mountbatten, the flamboyant British Chief of Combined Operations, whom Ike had met and liked on his first visit to London; Major General Carl "Tooey" Spaatz, commander of the fledgling U.S. Eighth Air Force; and Major General John C. H. Lee, an old friend of Ike's, who dealt with the vast and rapid buildup of the Army's logistics and supplies in Britain, and who, like many of Ike's old Army buddies, would prove something of a mixed blessing in the months ahead.

Ike's command was called the European Theater of Operations, U. S. Army (ETOUSA). Its headquarters were in swank Grosvenor Square, just a couple of blocks from the even swanker Claridge's Hotel, where he was billeted, and which he almost immediately disliked because it seemed too elegant, too expensive, and too stuffy a place for a soldier with simple tastes. Claridge's famous and much admired red-and-gold Chinese-patterned wallpaper in the bedroom of his suite reminded him of a funeral parlor or a brothel—he wasn't sure which—and he was deeply uncomfortable there. The staff of ETOUSA moved him hastily to the Dorchester Hotel on Park Lane, but this was scarcely less posh—though more in the style of traveling first-class on a transatlantic ocean liner, rather than the "stately home" look of Claridge's—and he could hardly imagine spending any substantial part of the war there.

Ike had presumably told Mamie at least once too often, when he was back in Washington after his first trip to London, how glamorous and what good company his British driver had been, since he now felt it necessary to reassure her rather self-defensively in his very first letter home, on June 26, 1942, that this time his driver was "an old time Britisher."[8] Still, it apparently occurred to him that there was somebody who would probably understand better than the ETOUSA staff what he was looking for in the way of accommodations, and could almost certainly find it for him. Coming back from an inspection trip a few days after his arrival, he landed at Hendon aerodrome, noticed Kay Summersby in the crowd waiting for him, and learned that she was now driving Tooey Spaatz. Ike had brought a basket of fresh fruit across the Atlantic for her, remembering that she had said how much she missed it in wartime London, but although he had asked Tex Lee to find Kay and have her pick up the fruit, Lee was still trying to track her down.

Now that Ike knew she was driving Spaatz, he called his Air Force commander as soon as he was back in his office and suggested a transfer for Kay. Spaatz protested at having Kay taken away from the Air Force, but it was no contest—they were both, for the moment, of equal rank, but Ike was the theater commander, and Spaatz knew better than to carry an argument with him too far, at that point in their relationship.

It turned out that the resourceful Tex Lee had in fact come close to finding Kay—he had discovered which car she was driving and left a note for her under the windshield wiper. However, since it just asked her to stop by 20 Grosvenor Square and see a "Colonel Lee," Kay had torn it up. Several American officers in ETOUSA had already tried that one on her, and anyway she did not think that General Spaatz, whom she described as having a "dour personality," despite the fact that his aides tried to arrange a cocktail party for him every evening at Air Force headquarters outside London, would approve of her calling on a stranger who had left a note on his Packard, particularly a mere colonel.

In any case, Kay dropped in to Ike's headquarters to pick up her basket of fruit, much to Tex Lee's relief, and to her surprise was taken right in to see Ike, who asked if she was willing to drive for him again. She said she'd love to, if it was all right with General Spaatz, and Ike gave one of the big grins which were his trademark and told her not to worry her head about Spaatz. Two days later she had joined Ike's staff.[9]

Prudently, he did not mention this change in drivers in his letters home—indeed, the subject of Kay does not reappear in Ike's correspondence with Mamie until March 2, 1943, almost eight months after she was moved from Spaatz's staff to Ike's. By that time Ike and Kay were in Algiers, and stories about Kay (as well as her photograph in *Life* magazine) had made it necessary for Ike to explain her unexpected presence on his staff to Mamie. What he came up with was not altogether convincing: "If anyone is banal and foolish enough," he wrote to her, in response to her letter about the story in *Life*, "to lift an eyebrow at an old duffer such as I am in connection with Waacs—Red Cross Workers—nurses and drivers—

you will know that I've no emotional involvements and will have none. Ordinarily I don't try to think of all the details surroundings my existence when I write to you."[10]

But in fact it was not WAACs or Red Cross workers or nurses who had upset Mamie; it was Kay's photograph in *Life*. The story described her as a "pretty Irish girl," and the photo showed her standing right behind Ike, tall and slim in her tailored uniform, with a very unmilitary coiffure of carefully tousled hair, meticulously penciled eyebrows, and full lipstick and makeup—not at all the usual look of an army driver, as Mamie, an Army wife, must have known very well. Given the number of drivers available in the combined American and British armies, it is hardly surprising that Mamie might have felt her husband could at least not have picked one who was young, pretty, social, and glamorous, with high cheekbones and great legs.

Ike's "official family," which was assembling in London, would not change throughout the war, despite Mamie's simmering resentment about Kay Summersby. At the top would be Ike's grouchy, workaholic, formidable chief of staff Brigadier General Walter Bedell Smith, whose nickname was Beetle. Smith was indispensable—everybody seemed to be afraid of him, as if he were the Lord High Executioner in Gilbert and Sullivan's *The Mikado*—and he performed two vital tasks: guarding Ike from people who would waste his time or whom he didn't want to see, and firing or disciplining people whom Ike didn't want to fire or discipline himself. Harry Butcher, officially Ike's naval aide, played a variety of roles. His main role was that of the man who could arrange or smooth over almost anything and deal with the press; but he was also court jester, golfing companion, the target of Ike's anger when Ike was in a bad mood and the butt of his jokes when he was in a good one. Tex Lee was Ike's principal aide, "who resembled a worried bloodhound" and soon developed an encyclopedic memory for the names and faces of everybody who mattered in the U.S., British, and Free French armed forces. Mickey McKeogh, a kind of Jeeves, looked after Ike's comfort and his uniforms, and occasionally did basic cooking when called on to do so.

Kay Summersby herself became Ike's driver; and very soon after joining his staff, she also became his social secretary, his companion, and the person charged with explaining to him the peculiarities of English behavior and the all-important complexities of the English class system. It fell to Kay, for example, to explain to the indignant general that at British dinner parties he could not smoke at the table until "the king's health" had been toasted at the end of the meal, and that when royalty was present Ike could not leave the party until after the royal personage had left, however busy and eager to get back to work he might be.[11]

The rumor that Ike and Kay were having an affair started early in their relationship, persisted during and after the war, and still vexes biographers. Kay wrote one book, shortly after the war, ignoring it; and another, shortly before her death, affirming it. Ike remained, perhaps wisely, entirely silent on the matter. Some people have argued that the fact that Kay was still married to (although perhaps separated from) her husband and at the same time "engaged" to an American officer when she and Ike met is proof that her relationship with her boss must have been innocent. But most adults would take the view that neither of those difficulties would necessarily prevent a woman from falling in love with another man, which seems to be what happened—though whether the relationship was consummated, is, of course, quite another question, and surely nobody's business. Also, war, with its dangers, excitement, deprivations, and stress, tends to produce intense, fleeting relationships and confused emotions. Many biographers, like Carlo D'Este, theorize that such an affair would have been impossible, given the fact that Ike lived surrounded by aides and guards, and that he shared his accommodations with Butcher most of the time; but students of human nature will take this theory with a grain of salt. When adults of the opposite sex are thrown into a tense and highly charged working atmosphere over a long period of time, they will always find a way to be alone with each other, however briefly, if that's what they want to do. Some writers, like Ike's granddaughter Susan Eisenhower, take the view that Ike would never have done such a thing to Mamie, and that Mamie would never have forgiven him if he had, and this may be so, but stranger things have happened in wartime. When men are away for five years at war, some do things they might never have done at home in peacetime, and wives have been known to forgive them when they return (Clytemnestra is a well-known exception).

The truth of the matter is that nobody knows, and prurient speculation is out of place. The only sensible verdict is the old-fashioned one of Scottish courts which lies somewhere between guilty and not guilty—not proven. What is certain is that from the very beginning Kay hero-worshipped Ike, that Ike was deeply (and obviously) attracted to her, and that throughout the war he treated her much more like a close friend than a driver. All this was bound to start tongues wagging, as Kay must have known even if Ike didn't. His letters and notes to her are not those that a general would normally write to a driver—how many generals would send a driver a handwritten note on a postcard from, of all places, the garden of Gethsemane, reading: "Good night! There are lots of things I could wish to say—you know them. Good night."*[12] This would admittedly not count for much as evidence in a divorce case, but it certainly suggests something beyond a purely formal military relationship.

Where there is smoke, there is fire, as we are always told, and there was an awful lot of smoke around General Eisenhower and Kay Summersby—enough to convince plenty of people that they were lovers, whether they were or not. The fact that Kay often accompanied Ike to lunches and dinners with both Roosevelt and Winston Churchill—at Ike's insistence—puts her in a very different class from most, perhaps all, other military chauffeurs; and as late as 1950 Churchill, temporarily annoyed with Ike, who was shortly to be appointed Supreme Allied Commander Europe, remarked that he "had not forgotten Eisenhower's 'horrible bad temper' at dinner in Marrakech in January 1944 because, as he [Churchill] later learned, the general's 'lady chauffeur had not been asked too.' "[13] Roosevelt, more easily charmed than the prime minister by a pretty woman, took her on a picnic in Algiers; called her "child," his usual way of flirting with attractive young women; asked her to sit beside him; and promised to make her an officer in the American WAACs, despite the fact that she was a British subject—hardly the normal behavior of a president toward a uni-

* Underlining is Ike's.

formed military driver. Perhaps the only people of consequence who snubbed Kay were King George VI, who was always petrified by the slightest hint of an improper relationship because of the misfortunes of his older brother, and who deliberately treated her like a chauffeur, which is to say a servant; and General Marshall, who considered it part of his job to telephone Mamie once a week, and was deeply suspicious of Kay Summersby. Whatever virtues Ike may have had, however—and he had many—discretion about his friendship with Kay was not one of them, and people can hardly be blamed then or now for drawing the logical conclusion.

The numerous efforts to suppress such rumors invariably backfired and made matters worse, as is usually the case. One example involved the two versions of the famous photograph of Ike and his staff on May 8, 1945, with Ike grinning and holding up the two pens with which he had just signed Germany's surrender. In the unretouched version, Kay (the only woman in the group) is standing close behind Ike's right shoulder, leaning toward him, an adoring look on her face. But in the "official" version, she has been carefully airbrushed out. In the former Soviet Union historical photographs were doctored and altered to suit the party's needs; in this case it is more likely that the photo was altered to spare the sensibilities of General Marshall and Mamie Eisenhower.

Much of the trouble may have been caused by sheer naïveté on Ike's part. If he was innocent and the relationship was harmless, as may have been the case, then he must have seen no reason for hiding his growing affection for Kay, however inappropriate it might have seemed to other people. Another factor was his fierce loyalty to the members of his small entourage, which led him to angrily rebut any criticism of them.

Then too, like Mamie, Ike had not anticipated becoming an overnight celebrity; nor was he comfortable with being one. He was astonished at the number of invitations he received, faintly guilty about turning most of them down, but also furious at the waste of time whenever he was persuaded against his better judgment to accept one of them. His popularity was an amazing phenomenon—so many people wanted to shake his hand at the American ambassador's Fourth of July reception that his right hand turned black-and-blue, his swollen fingers ached, and he was unable to write or even sign letters for days afterward. Unlike his old boss General MacArthur, Ike might have been called, in a different age, "the people's general." He was a traditional and beloved figure of American myth, the poor boy who has made good; he was (or appeared to be) a plainspoken midwesterner who didn't give himself airs or indulge in dramatics; he was friendly and outgoing—all in all exactly the kind of general in whose care ordinary Americans were happy to place their sons, and also exactly the kind of American Britons of all classes desperately wanted to believe existed now that the two countries were at last in the war together.

Whether intentionally or not, Marshall had chosen for high command in Europe a general whose popularity was instant, genuine, and enormous—an unexpected and formidable weapon in its own right, in fact—and as a result everything Ike did or said became a news story. In these circumstances, Ike was hounded by the press like any other celebrity, despite the efforts of his staff to shield him, and his friendship with his beautiful English driver, however innocent, was bound to cause a stir. Indeed it may be a measure of Ike's innocence that he didn't see why it should, and even at the end of the war, didn't seem to understand why Mamie might be wounded and angry about it.

That Ike paid no attention to what other people thought of this is yet another example of his stubbornness. So far as he was concerned, Kay was simply a member of his "military family," a bright, buoyant spirit who was able to make him forget, however briefly, the sheer daunting enormousness of his task and the loneliness of command; and in the few moments he could steal for any kind of recreation, she was his companion. Ike had always liked riding, and was good at it, having grown up as a country boy at the end of the age of the horse, having attended West Point when horsemanship was still an important part of the curriculum, and having spent his Army career wearing the tan riding breeches, shiny brown riding boots, and gleaming spurs that were then still a part of an officer's uniform, even in the infantry. He rode well, and with his tall, lean, broad-shouldered body, he looked good on a horse. As for Kay, she was the daughter of a British army officer, and much of her life had revolved around ponies and horses. She too, without being "horsey," rode well, knew something about horseflesh, and looked good on a horse. Far from being a timid and unwilling horsewoman like poor Mamie, she was a daring rider who had grown up fox hunting. Riding became one of the things they shared, another bond between them. There would soon be others.

Ike was working more than twelve hours a day or more at Grosvenor Square, chain-smoking, and averaging six hours of sleep a night. Once he had decided to curtail his social life he seldom went out except for his weekly luncheon with Winston Churchill and the occasional dinner at Chequers, the prime minister's official country house, which was over 500 years old and as cold and drafty as a barn (Ike called it "a damned icebox")—so cold in fact that the Prime Minister and Mrs. Churchill often dined wearing their overcoats. There, the prime minister kept Ike up every night until two or three in the morning talking strategy until Ike could hardly keep his eyes open.[14]

What Ike needed was a country place of his own, a place where he could have at least some respite, some privacy, companionship, and a feeling of being at home, instead of living between his hotel suite and his office. Kay and Butch seem to have gotten along without friction from the start, perhaps because Butch's many years of working for William S. Paley as an executive at CBS had given him a deep, unshockable sense of realism about human behavior, as well as an uncritical appreciation of pretty women with a good, bawdy sense of humor. The two of them managed to find a few suitable properties for rent, and Ike chose Telegraph Cottage in Kingston, less than half an hour from Grosvenor Square (in those days). It was a small, quaint Tudor house set in ten acres of dense woods and carefully cultivated lawns and gardens, with a winding drive and high hedges that ensured absolute privacy. In the English tradition, it had five bedrooms, only one bathroom, and no central heating, but a huge fireplace in the living room kept the downstairs warm. Ike moved there with undisguised pleasure and relief. In some ways, Telegraph Cottage made him even happier than his third star—he had been promoted to temporary lieutenant general. It is some measure of Churchill's growing regard for Ike that he insisted on having a bomb shelter built for him in the garden before he moved into the house.

A second thing Ike shared with Kay was a dog, Telek. There is some disagreement about whether Telek was Ike's gift to Kay or Kay's gift to Ike, but both of them referred to him as "our dog." Kay's coconspirator in the search for a dog was, improbably, Ike's ferocious, no-nonsense chief of staff, Bedell Smith, who turned out to be a knowledgeable dog lover. When Ike said that what he wanted was a Scottie (Smith had recommended a Dandie Dinmont), they managed to narrow the search down to two, and Ike made the final choice and picked the dog's name—"Tele" for Telegraph Cottage and "k" for Kay: "Two parts of my life that have made me very happy," he told her. The choice of a Scottie was not surprising, since Roosevelt's Scottie, Fala, was perhaps the most famous dog in the English-speaking world; but the name Telek was yet another indication that Ike's feelings for Kay were hardly the normal ones of a three-star general for his driver. Telek would accompany Ike and Kay everywhere, though after the war was over and Ike reached out toward the presidency, Telek became Kay's dog, and in his old age eventually moved with her to New York, a kind of living reminder of her years with Ike, and a familiar figure at El Morocco, 21, and the Stork Club.[15]

Despite these efforts to make his domestic front secure, the main focus of Ike's attention was always his job—first of all, to supervise the rapid growth of American forces in Great Britain; and second, and more important, to decide what to do with them. Ike lavished a good deal of time, energy, and tact on the first part. Unlike other American-area commanders—MacArthur in the South Pacific, or Stilwell in China—he and his forces were stationed in a country that had a long, tangled, intimate relationship with the United States and that was, at the time, an equal industrial and military power, with a rich history of glory, conquest, and empire. British sensitivities could not be ignored.

One of the first people whom Ike sought out, very wisely, was Brendan Bracken,* then Britain's minister of information. More important, Bracken

* Brendan Bracken, later Rt. Hon. the Viscount Bracken, PC, was a close and dear friend of the author's father and uncle, Sir Alexander Korda, as well as a friend and determined mentor and adviser to the author as a teenager and a young man, a kind of surrogate godfather, always available for caustic but well-intentioned advice.

was a wealthy, self-made young business tycoon who had devoted much of his enormous energy for the past twenty years to looking after Winston Churchill with ferocious devotion. Nothing in Churchill's business, political, or personal life was done without Bracken's knowledge and involvement, and in consequence many people believed him (incorrectly) to be Churchill's illegitimate son—a belief that Bracken, to the fury of Mrs. Churchill, did little to discourage. Bracken was pugnacious, shrewd, hot-tempered, and famously impatient (Ike remarked that Bracken's habit of swearing was intensified by "the rasping intensity of his voice").[16] With his pink face, bright blue eyes, and flaming red hair, he looked like what he was—not a man to be trifled with—but he had a deep core of common sense and decency, and a precise knowledge of just what the British would and would not put up with, an instinct that had saved Churchill from many a public relations scrape over the years. Ike compared Bracken to Harry

Brendan Bracken and Churchill.

Hopkins, and there was some truth in that comparison, though Bracken was a man of the right, not the left, as well as more worldly and social, with far more personal ambition than Hopkins, and no instinct to hide himself in the shadow of the great man he served. Although they made an odd couple, Ike and Bracken got along famously, and many of the steps that Ike took to make his troops pay attention to British sensitivities were suggested or encouraged by Bracken.

"Intensive programs were devised with Bracken's splendid organization to fit the newly arrived Americans into the highly complex life of a thickly populated area in such a way as to minimize trouble," Ike would write later. These included what we would now call widespread sensitivity training,[17] in which American troops received lectures about British history and life, were taken on visits to British families and tours of the bombed-out areas of the cities, and were carefully introduced to such national peculiarities as warm beer, driving on the left-hand side of the road, fish and chips, the need to stand up at the end of a play or motion picture when "God Save the King" was played, British football (soccer), and the national passion for tea. Ike, whose goodwill was spontaneous and instantly recognizable, put into effect a vast and carefully prepared program of mutual respect and awareness, and saw to it that this program was enforced rigorously throughout the American services in the European theater. Boastful behavior and anti-British slurs or jokes were severely punished—the higher the rank of the perpetrator, the more severe the punishment. Meanwhile Bracken did his best to persuade the British to be tolerant and hospitable toward their 2.5 million uniformed visitors. Of course there were "incidents" on both sides, involving all ranks from private soldiers to generals, but on the whole the program was remarkably successful, and without it, the two armies could never have fought side by side for the next three years.

Perhaps the only area in which Ike failed was relations between black American troops and white American troops, and British civilians. Here, he was undercut by racial issues in the U.S. armed forces, which were still strictly segregated. "Negro" or "colored" troops, as they were then referred to, served in separate units, often with white officers, and were at first relegated to noncombat, menial duty. The British were not without their own colonialist race prejudice against blacks, but that of the Americans surprised and dismayed them, and the occasional sight of a black American soldier dating a white woman led, inevitably, to brawls and near-riots, which the military police of both nations did their best to suppress with a liberal use of truncheons. After several stories about this kind of incident ran in the British press, the censors clamped down on the subject, until Ike heard about it, and with his usual common sense "at once revoked the order and told the pressman to write as they pleased." With that, interest in the subject began to die down, as he had predicted it would, though this development did nothing to change the reality of segregation in the Army. Like most senior officers of his time and generation, Ike held views on the subject that were not, by current standards, enlightened—he had two black sergeants on his staff as cook and cleaner—but at least he refused to let the censors have their way about hiding any mention of the subject.

Apart from training his troops and keeping them on good terms with the British, it was Ike's job to get them into action. From day one he pressed hard for Round-Up, but as the summer days of the "invasion season"—and the endless meetings about it—went by, he saw increasingly clearly that nothing could persuade the British that launching a full-scale invasion in 1942 was a good idea. Even a ten-day visit to London beginning July 16 by General Marshall himself, accompanied by Admiral King and Harry Hopkins, and bearing a personal plea for action from the president, could convince them. And enthusiasm for the task on the part of the British was essential, as Marshall eventually conceded wearily, since they would have to provide the lion's share of the invasion forces—there were nothing like enough American troops to provide more than a small percentage of the forces envisaged for Round-Up. Even had the British been enthusiastic, however, the difficulties of carrying out Round-Up in 1942 were overwhelming—beginning with a severe shortage of landing craft, which alone was enough to doom the prospect of an invasion.

Here, at least, the British were not part of the problem. Though many of the landing craft had been designed in Britain, mass production was supposed to take place in United States shipyards; and although the U.S. Navy insisted on running the program, nobody in the Navy was in fact much interested in it—naval officers, understandably, wanted to build (and command) warships, not seagoing barges with ramps for landing on beaches. In addition, there was fierce infighting between the commanders in the Pacific (Nimitz and MacArthur) and Eisenhower (and his naval commander Admiral Stark) for priority in landing craft.

By the last day of Marshall's visit to London, Sledgehammer was dead, and Round-Up was definitely postponed until 1943. Much as it pained Ike—and infuriated Marshall—the number of troops needed to invade Europe were simply not there, and neither were the vessels needed to transport them to French beaches. In the summer of 1944, when the invasion at last took place, Ike would have available for use in the United Kingdom seventeen British and Canadian divisions; twenty American divisions; one French and one Polish division; more than 7,000 vessels of all kinds, from battleships to Higgins boats; over 10,000 combat aircraft and 5,000 transport aircraft and gliders; and a total strength of 2,876,439 officers and men*—and yet the invasion was still considered a risky business, with no more than a fifty-fifty chance of success. In the summer of 1942, however, his American forces in Britain consisted of only one infantry division, one armored division, a single parachute battalion, and small detachments of the Air Force, still "only partially trained." Equipment was lacking; neither British nor American troops had yet been trained in the basics of carrying out an opposed landing (not just opposed, but opposed by the German army—i.e., fiercely opposed) from the sea onto hostile beaches; solutions to such problems as creating artificial harbors (the Dieppe raid had convinced planners that they could not count on seizing a port intact) or supplying fuel to the beaches in sufficient quantity to permit a breakout were still in the experimental stage—indeed, the problems themselves were new and baffling, not surprisingly, since never before in military history had anyone ever contemplated undertaking anything of this kind, or on this scale.

* These are Ike's own figures, from his *Crusade in Europe*.

. . .

The British had begun theoretical planning for the invasion of France as far back as 1940, at the urging of Winston Churchill, and by 1942 teams of experts had compiled data on every beach on the northern coast of France, at one point even going so far as to gather tens of thousands of snapshots from Britons who had vacationed at one time or another at French beach resorts. Hidden away in the nooks and crannies of the British Museum of Natural History, in Kensington, under conditions of the strictest secrecy, geologists painstakingly analyzed sand and shingle samples from beaches all along the northern coast of France; oceanographers compiled exact charts of tides and currents; fishermen were questioned about every bay, shoal, and sandbank; craftsmen, sculptors, and theater and film set designers and model makers constructed enormous scale models of the more promising beaches; and cartographers, armed with, among other things, the excellent maps of France's revered *Guide Michelin,* painstakingly traced every track, path, road, and railway leading inland from the beaches. Elsewhere, General Percy Hobart's men experimented with such unorthodox inventions as a bridge-laying tank, an amphibious tank, a flame-throwing tank (for burning out enemy pillboxes), a tank with a huge revolving flail in front of it that exploded land mines harmlessly as it moved forward; and many other ingenious devices. Still elsewhere engineers labored to design huge floating concrete harbors that could be towed to the French beaches and sunk there, tested out the idea of creating breakwaters by sinking old freighters, and attempted to develop a flexible pipeline that could be unrolled and laid on the sea bottom rapidly from England to France to carry enough fuel for an entire army as it broke out inland from the beaches.

No military event in history has ever been so meticulously planned for, or received the undivided attention for so long of so many scientists and scholars (boffins, as the British called them),* as well as talented ama-

* Michael Quinion, an expert on British slang, points out that there is a character in Dickens's *Our Mutual Friend* called Nicodemus Boffin, "a very odd-looking fellow," and that the name entered RAF slang in the 1930s for a "mad scientist" type.

teurs, artists, cranks, and self-proclaimed geniuses, as the invasion of Europe. In 1942, however, much of all this was still in the embryonic stage; nor had any decision been made on the crucial question of where to land. The Pas de Calais was a favorite of most of the planners, since it was, at its closest point, only eighteen miles from England, but many of the beaches there were sandy, and, as Dieppe would prove, unsuitable for bearing the weight of tanks. In addition, this was undoubtedly where the Germans expected the invasion. Brittany had the advantage of a major port, Cherbourg, if it could be seized undamaged, and a coast on the Atlantic, so that reinforcements could eventually be sent directly from the United States rather than via Britain; but it was a peninsula, and there seemed no good reason why the Germans could not pinch it off at its narrow end and trap the Allied invasion forces there.

A further problem was where you went to from Brittany, once you landed there. It was about as far as you could get from the heart of France or the German border and still be on French soil. Normandy was closer, and many of the beaches in Normandy seemed suitable for a landing—to the boffins at any rate, if not to the soldiers—despite worrying cliffs and high ground immediately behind the beaches. But anybody who had ever spent a summer holiday in Normandy knew that inland lay an intricate checkerboard of tidy fields, farms, and villages, each protected by the notorious *bocage*—dense, high, thick hedgerows—and solidly built, centuries-old stone walls, which would stop tanks and give the Germans unlimited opportunities for a stubborn defensive battle, exactly the kind of fighting at which the Germans were masters.

As a result of constant warnings on the subject from the Royal Air Force, British planning for the invasion had been based from the beginning on what was then the maximum range of fighter aircraft operating from airfields in southern England, since air cover over the beaches was essential. Indeed, it was the failure of the Luftwaffe to gain air control over southern England in the summer of 1940 that had doomed Fall Seelöwe, or Operation Sea Lion, as the plan for German invasion of Britain had been called. Since the maximum range of the Spitfire and the Hurricane in 1941 was 200 miles—they had been designed to defend metropolitan

Britain against air attack, not to go looking for trouble elsewhere—the invasion beaches could be no more than 200 miles from England. If possible, the distance should be even less than that, to increase the time the fighters could spend over the beaches. A half-circle with a 200-mile radius extending from the fighter fields in Kent to France therefore pretty much defined the rough limits within which the invasion could take place, but this still left planners several hundred miles of French coastline to examine—an enormous task.

"Fortitude," a particular hobbyhorse of Churchill's, would perhaps prove to be just as important. It was, if such a thing was possible, an even more closely guarded secret than Round-Up. "The truth," Churchill said of Fortitude, "must be surrounded by a bodyguard of lies." From the very beginning it was clear that preventing the Germans from learning the date and the place of the landings would not be enough—they must be systematically and patiently pointed in the wrong direction and tricked into believing that in fact they knew where the landings would take place. Just as a good magician fools the audience by making it look away from his hands at the very moment when he performs his trick, the British wanted the Germans to be looking hard in the wrong direction when the invasion took place. More than that, they wanted, if possible, to pull off the infinitely harder trick of making the Germans believe that the invasion, when it finally came, was merely a deceptive diversion, and that the real invasion would land later, elsewhere. This belief would persuade the Germans to hold back their armored divisions from attacking the invasion beaches while the Allied troops were still spread thin and the heavy equipment had not yet been unloaded. It was a tall order—perhaps the most delicate and difficult task of the war—but the success of Round-Up would ultimately depend on it.[18]

Still, it was one thing to draw up a plan, however sophisticated, for what would surely be the climactic and most important moment of the war and quite another to carry it out. In the meantime, bad news continued to flow in relentlessly from all over the world. Rommel seemed to be concentrating for another, and perhaps final, assault against the British in Egypt, while in the Soviet Union German forces had advanced so far in the south that a team of mountain troops climbed to the summit of Mount Elbrus and planted the swastika flag there. (Elbrus is the highest mountain in the Caucasus; from this peak on a clear day, it is possible to see Georgia and Iran. Far from being pleased by this exploit, according to Albert Speer, Hitler flew into a rage about "these mad mountaineers," climbing mountains when they should be concentrating on military targets.) In the Pacific, the Japanese were still advancing in all directions. Even a nonstrategist looking at a globe could see the possibility of any number of "nightmare scenarios."

In the days before shingles was understood as a viral disease, there was a popular belief, since disproved, that if the patches of rash around the waist or the chest of the patient finally met and formed an unbroken circle, the patient died. Geopolitically, the world situation in 1942 looked similar: if the German army in Russia could cross the Caucasus, break out into the Middle East, and meet up with Rommel's Deutsche Afrika Korps, the Mediterranean might be circled and closed at both ends, with the loss to the Allies of the Middle East oil fields and the Suez Canal—surely a fatal blow.

Worse still was the possibility, however remote it may now seem to us, that if the German army and the Japanese army met, the world would be effectively divided. After all, the Japanese had taken most of Burma and were fast approaching the Indian border, while the Germans were in the Caucasus, on the ancient land route to Asia. If the Japanese invaded India, might not the Indian masses break into revolt against the British, as "armchair strategists" in Berlin speculated? And if that happened, might not the German army and the Japanese army meet one day soon within sight of the Himalayas?

Hopes rose high in Germany among true believers that the Führer—"the greatest *Feldherr*," or military leader, "of all time," as Propaganda Minister Goebbels, with his trademark hyperbole, had made it customary by 1942 to describe Hitler—would succeed where Alexander the Great and Napoleon had failed. Accordingly, SS Reichsführer Himmler spurred on the racial researchers of SS Ahnenerbe, his personal "think tank" (to use a modern phrase) devoted to "the sphere, spirit, deed, and heritage of the Nordic Indo-German race," in the untiring effort to prove that the Japanese, despite their appearance, were Aryan. As the war ended in Europe, anthropologists and philologists of the Ahnenerbe would still be laboring (unsuccessfully) to prove that there was a connection between Nordic runes and Japanese ideograms.

Berlin in 1942 was awash with dreamers who hoped to rise to power when Germany and Japan conquered the world. Among them were Subhas Chandra Bose, a former member of the Indian Congress Party who was busying himself raising an SS unit, the Jai Hind,* from Indian army prisoners of war in Germany; and the fiercely anti-Semitic and anti-British grand mufti of Jerusalem, Haj Muhammed Amin al-Husseini, an intimate of Hitler and Eichmann, who was waiting in comfortable exile in Berlin to return to Palestine in the wake of the German army and preside over a fatwa, a jihad, and a bloody pogrom against the Zionists and Orthodox Jews in the Holy Land. The person least swept up in this global imperial fantasy was Hitler himself, who disliked the grand mufti and remarked of the Indians, "If we took India, the Indians . . . would not be slow to regret the good old days of English rule!"

Hitler, a land animal and a pragmatist, sought principally German hegemony in Europe, from Gibraltar to the Urals, from the toe of Italy to the northernmost point of Norway—a Europe dominated and ruled by Germany, in which the Slavs would be reduced back to the slavery from which their name came ("As for the ridiculous hundred million of Slavs," he remarked, "we will mould the best of them to the shape that suits us, and we will isolate the rest of them in their pig-sties. . . . At present, they can't read, and they ought to stay like that"),[19] while the Jews, of course, were to be exterminated altogether. He ridiculed the larger geopolitical dreams of his followers, but he understood that victory might require conquest on a larger scale than he intended to keep, and was not afraid

* Some of these unfortunates eventually made their way to Japan and were enrolled in the so-called Indian National Army (INA) organized by the Japanese and sent to fight in Burma against the British, the Americans, and the Indian army; others were sent, along with many other "foreign" units, to man the Westwall, the German fortifications along the coast of northern France, against the Allies' invasion of Europe.

of the possibility, which, as seen from the Reichskanzlei, looked promising indeed.

The only promising piece of news in London was that RAF Bomber Command had at last received the commander-in-chief it deserved in the person of the ruthless, outspoken, energetic Air Marshal Arthur ("Bert") Harris, soon to become better known as "Bomber Harris," and had finally outgrown its awkward stage and emerged as a formidable weapon of war with Harris's brainchild Operation Millennium.

On the night of May 30–31 Harris finally managed to bring off Millennium, his long-cherished ambition of a "1,000-bomber raid"—he first had to raid the Operational Training Units, the flying schools, and RAF Coastal Command to assemble enough aircraft and crews to do it—sending 1,046 RAF "heavies" by night against the cathedral city of Cologne. Guided by flare markers laid down by low-flying RAF "Pathfinder" Mosquitoes, the waves of bombers alternated between loads of high-explosive bombs and loads of incendiaries, with pauses between waves calculated to catch the German firefighters in the open when the next wave arrived. In less than three hours 1,455 tons of bombs had been dropped; 600 acres of Cologne had been destroyed; a firestorm had been created whose glow could be seen on the horizon by bombers at the tail end of the "bomber stream" as they crossed the Channel; 45,000 people had been made homeless; and one of Europe's most beautiful cities had been largely demolished. "They have sown the wind, and so they shall reap the whirlwind," Harris remarked with satisfaction.

Although it hardly seemed so at the time, the foundations were being laid in the late summer of 1942 for the eventual defeat of Germany—the vast drain of German manpower and matérial on the eastern front, the defeat of German forces in the Mediterranean, the relentless bombing of Germany's cities and industries, and, finally, the coup de grâce, the invasion of Europe and the opening of the "second front."

Despite the arguments of true believers—especially supporters of the Soviet Union, who, following the party line, demanded "a second front now," at any risk or cost; or the British and American "bomber barons," led by Harris and Spaatz, who believed they could defeat Germany all by themselves by bombing it day and night—no single strategy was enough to do the job. Germany would be defeated only through the relentless operation of all these strategies together—there was no panacea, or, as a later age would call it, magic bullet.

One other bright spot on the horizon in London was the appointment of Lieutenant-General Harold Alexander to replace Auchinleck as Middle East Commander-in-Chief. Everybody, British and American, liked and respected the elegant, charming, attractive, and tactful Alexander, an Old Harrovian like Churchill. Also, Lieutenant-General Bernard Montgomery (who had so angered Ike by telling him not to smoke in his presence) was appointed Commander of the British Eighth Army in Egypt. Hardly anybody liked Montgomery, but he was widely regarded as the most competent senior officer in the British army (and regarded himself as the most competent senior officer in *any* army). It was hoped that these two together, reinforced by the tanks which Roosevelt had offered Churchill on giving him the bad news of the fall of Tobruk, would manage to stop Rommel before he reached the Nile delta.

Their task was all the more important because General Marshall and the president had, at last, reluctantly accepted that Gymnast, now renamed Torch, was the only practical way of bringing American ground forces into action against the German army in 1942—the importance of which Roosevelt had emphasized to Marshall in a handwritten note before his departure. The president was deeply concerned about the Soviet Union, which seemed, once again, to be on the verge of collapse; and although he was tactful enough not to bring the matter up with Marshall, he was also anxious to have American troops in action against the Germans before the midterm elections in November. Marshall had firmly expressed the view that "Gymnast would be indecisive and a heavy drain on our resources," and had even suggested that unless the British agreed to undertake Round-Up no later than 1943, "we should turn to the Pacific and strike decisively against Japan with full strength and ample reserves, assuming a defensive attitude against Germany," a complete reversal of his previous position, and of American policy. This may have been a red herring, intended to stiffen the British spine regarding Round-Up and, if possible, Sledgehammer, but it certainly reflected the importance Marshall had always attributed to the cross-Channel landing. Now, in the face of the facts presented to him by Ike, and the strongly expressed doubts of the British chiefs of staff, Marshall accepted the inevitable. "The British staff and cabinet were unalterable," he later recalled. "Looked like the Russians were going to be destroyed," he added glumly (and inaccurately). The Combined Chiefs of Staff agreed in London, after interminable wrangling and horse-trading, to "complete all preparations . . . for operations against North and Northwest Africa"—a signal victory, the first, and perhaps the last, for Winston Churchill over the Americans' strategic plans.

It was now Ike's job to prepare the plans for what would be the largest and most ambitious American military action so far in the war; and a look at the map will show why this was no easy task. The troops that were beginning to arrive in the United Kingdom and undergo combat training for Round-Up or Sledgehammer would have to be reshipped to North Africa, together with all their equipment, across seas that were dominated by German U-boats, while other troops—together with heavier equipment, most of the air force, and a substantial naval force—would be sent directly across the Atlantic to North Africa. Since part of the invasion fleet would have to pass through the Strait of Gibraltar into the Mediterranean, there was almost no chance of keeping its passage a secret from the Spanish, who would certainly inform the Germans. The only possibility of a surprise was to keep the Germans guessing to the last minute about where it would land.

The distances involved were daunting, to begin with. Round-Up involved crossing the English Channel, only eighteen miles at its narrowest point, and had been done before successfully, in the opposite direction, by Julius Caesar and by William, duke of Normandy. But Torch would require the troops to be at sea for weeks before they were off-loaded into landing craft (of which there was an acute shortage) and transported to the beaches of north and northwest Africa. The northwestern landings would be on the Atlantic coast, where the surf was likely to be high, and all the landings would have to rely on air cover flying from the relatively few small escort carriers that the Navy could spare, and from Gibraltar, with its small airfield. There would be not one but three separate invasions, each with its own task force, spread across nearly 2,500 miles of coast—from Safi, in French Morocco, the easternmost point; to Algiers, the westernmost—posing huge and intricate problems of timing, communications, and command.

On July 30, Field Marshal Dill left Washington for London, carrying with him a "hastily dictated" letter from Marshall to Ike. Marshall wrote that he was sending Ike "a planning team" to advise him on shipping, transport, and supply, along with Air Force officers and a U.S. military attaché who had recently returned from North Africa. But Marshall reminded Ike that while he was "to take the bull by the horns," no firm decision had yet been reached by the president about who was to command Torch—Ike should proceed with the planning and consider himself the "Deputy for whoever is designated for supreme command."[20] To meet the president's "desire for an early date for Torch," planning would have to be completed by September 15, on which date the final decision would be made about whether to proceed or not. Since ninety days would be required to arrange for shipping and to make necessary modifications to the ships involved, all this would have to be sorted out with the British and put under way immediately.

In the meantime, Marshall informed Ike, the Air Force was already raising objections, which were soon to become familiar to him. Withdrawing bomber squadrons from the United Kingdom to support Torch, Commander of the USAAF General Henry H. ("Hap") Arnold predicted, "would produce a serious reduction in the power of the bombing attack against Germany, with the possible consequence [of] much heavier return blows against Great Britain." The Americans' daylight bombing of Germany had scarcely even begun, and there was no reason to suppose that the Luftwaffe, its offensive capacity already stretched paper-thin from the Volga River and the Caucasus to Egypt and Malta, would be able to renew the blitz against Britain, or even had any plans to do so; but these facts did not prevent Arnold from raising the same arguments that the "bomber barons" would make again and again over the next two years—left alone, the bombers would produce victory, and in the meantime attempts to use them for any other purpose but bombing Germany would merely prolong the war.

Arnold was staking the Air Force's claim to victory in Washington, just as firmly as "Bomber" Harris was doing in London, and it is to Ike's credit that he paid no attention. Unlike Britain (or Germany), where the air force was a separate service like the navy, the United States kept its Army Air Forces as part of the Army. This had many disadvantages in theory for American airmen, but it also meant that Ike, unlike his fellow generals in Britain or Germany—for whom the air force was a separate and mysterious entity, competing for men, money, and industrial capacity, and operating with an independent strategy of its own, hardly less detached than the navy from their own experience—knew something about airpower. He had learned to fly in the Philippines, and was largely responsible for such airpower as was there when the war broke out, and he had gone to West Point with many men who eventually became aviators; indeed, he had wanted to become an aviator himself, and would have, had it not been for Mamie's father.

Ike was careful to retain a degree of control over the Air Force units in his command, much as he liked—and trusted—Spaatz, and took a strong interest in air matters. One of his first visits to his troops in the United Kingdom was in connection with the initial offensive operation of the U.S. Army Air Forces against the Germans, when six American twin-engined Douglas "Bostons" joined a larger formation of RAF medium bombers in a daylight attack on Luftwaffe airfields in Holland, to mark July 4, 1942 (two of the American aircraft failed to return).[21] He supported Spaatz with great determination against the arguments of Churchill and the British air staff that the Americans' daylight bombing of Germany would prove to be a disastrous waste of men and resources, that losses would be prohibitive, and that American bomber crews should be retrained to attack by night like those of the RAF.* Ike was willing to let the Air Force pursue its bombing strategy, but not at the expense of the kind of air support he required for ground operations, on which he was always careful to insist.

On August 6 Marshall cabled two important decisions to Ike. The first was that the target date for the North African landings had been set for November 7, that being the earliest date for which the landing craft could be assembled and the modifications to the troop ships completed. President Roosevelt, though disappointed that the landings would take place three days after the elections, had agreed; and Prime Minister Churchill, though equally disappointed, had been so informed. The second was that the president had finally designated Eisenhower as the "Commander in Chief for the Torch operation," and that the British chiefs of staff had not only agreed to but recommended his appointment.

As usual, there was a political element in the president's choice, of which Ike was perfectly aware. Since the Americans, under pressure from the British, had at last reluctantly agreed to postpone Round-Up to 1943 and to abandon Sledgehammer altogether (Ike feared that the second decision might be "the blackest day in history"), the British, who had won on both points, felt it would be tactful to agree, and even suggest, that the commander of Torch should be an American.

The British also hoped that the appointment of an American general to command Torch might make the operation more acceptable to the French military commanders in North Africa, and Roosevelt, who, like many Americans, tended to overestimate the French people's admiration and affection for the United States, agreed. Torch was to be an Allied operation, but one in which it was deemed essential to downplay the British role for fear of offending French sensibilities—it was thought that the French

* There was no right or wrong in this argument. The British heavy bombers had been designed to carry very heavy bomb loads over great distances, and were therefore lightly armed to save weight—they could, therefore, only attack by night. American heavy bombers carried ten .50-caliber machine guns and heavy crew armor, which limited their bomb load and range but enabled them in close formation, and when supported by long-range fighter aircraft like the P-47 and P-51 (of which the RAF had none), to defend themselves against enemy fighters by day. The RAF specialized in massive night raids that destroyed whole towns, while the USAAF preferred to attack factories, transportation hubs, and oil refineries. Combat losses in the RAF Bomber Command and the U.S. Eighth Air Force were about equal in percentage terms.

might not open fire on American troops, or would, in any event, at least prefer surrendering to an American rather than a British commanding general.

Nevertheless, this was a moment of weighty significance, and not just for Ike's career. Ike, a newly minted American lieutenant general, would command not only the U.S. Army and Air Force elements in Torch, but elements of the U.S. Navy, as well as the Royal Navy, British army, and Royal Air Force units involved in the landings. All these services, many commanded by senior officers with vastly more combat experience, more seniority, and higher rank than Ike, and each fiercely jealous of its own independence and prerogatives, would be under his command. This was an extraordinary vote of confidence in his leadership on the part of the president, General Marshall, Admiral King, the British prime minister, and the British chiefs of staff, which would never have been possible without the good impression that Ike had made during the six weeks he had been in London. What is more, he would continue to be in command of ETOUSA, which included all American forces in the United Kingdom, as well as the preparation for Round-Up—an immense responsibility, to which was now added his first combat command, in faraway North Africa.

He was now, for better or worse, to use a favorite word of Churchill's, a "supremo," with direct political access to the prime minister and supreme command over a widespread international force. Nobody else in the U.S. armed forces, not even Douglas MacArthur, had a command anything like as ambitious and far-reaching as this, or as independent—or one with as many thorny diplomatic and political problems. Ike not only would draw up the plans for Torch but would carry them out. He not only would command British and American forces, but would be landing on French soil. Not since Foch, at the supreme moment of crisis in World War I, had so much power been handed by allied nations to one man.

It is worth noting that—again, unlike his old boss MacArthur—Ike was the least rank-conscious of generals. This made his task easier. He asked for and got his old friend George Patton, who was delighted to join him in London even though Patton was senior to Ike in age and permanent rank; he made Mark Clark his deputy, even though Clark had reached the rank of major general before him; and he reached out for Major General Charles W. Ryder, who was happy to serve under Ike, even though Ryder had himself been a highly decorated combat soldier and had won a battlefield promotion to lieutenant colonel at an early age during World War I. Men were happy to serve under Ike, even British admirals and generals who might easily have raised objections. His sincerity, his grasp of detail, and his lack of ceremony made it difficult, even impossible, to refuse him, and enabled him to assemble very rapidly a team whose members might quarrel and try to pull rank between themselves, but rarely, if ever, with Ike.

Though the planning for the cross-Channel invasion had been going on for two years, and every detail of the possible beaches was being studied by teams of experts, Torch, by contrast, was hastily improvised under tremendous pressure, not only in terms of time, but, even more crucially, in terms of the availability (what we would now call the window of opportunity) of shipping and naval vessels. Originally, Gymnast had been conceived as a landing to take Casablanca and occupy the Atlantic coast of Morocco, so that the Germans could not use it as a submarine base. In fact, no evidence existed that the Germans had ever based submarines there, or intended to (or that the French would let them), but this became firmly fixed in President Roosevelt's mind as a reason for invading North Africa, possibly because Winston Churchill planted it there at the Arcadia conference in Washington as a way of selling him on Gymnast.

Whatever the initial impulse may have been, Torch quickly expanded into a far more ambitious assault, in which two of the three French colonies in North Africa would be invaded simultaneously, at multiple points. The major focus of attention was no longer Casablanca, on the Atlantic, but Algiers, on the Mediterranean, from which it was hoped that an attack could be launched into Tunisia before the Germans could react. This attack would squeeze Rommel between two armies, while at the same time relieving the pressure on Malta, which was under constant air attacks by the Germans and Italians and was vital to the passage of Allied convoys through the Mediterranean to Alexandria and the Suez Canal—both serious British concerns.

The longer Ike looked at the map, the more tempted he was to land in strength as far east as Bône (later Annaba), almost on the Tunisian border, so as to come to grips with Rommel rapidly, even at the cost of dropping the Moroccan landings altogether; but just as the D-day planners were limited in their choice of landing sites by the combat range of the fighter aircraft available to them, so was he. In 1942 the USAAF was no better equipped with long-range fighters than the British, and in fact delivery of American fighter planes was lagging so far behind schedule that the first American squadrons arriving in Britain had to be supplied with British Spitfires (the pilots caused a mild flap at the highest levels in Washington when they said, in public, that they preferred these to the P-40s and the P-38s they were waiting for). This meant that apart from the carrier-based fighters, fighter cover would have to be provided from Gibraltar, and Bône was out of range for the aircraft that would be available, as well as exposed to attack from Luftwaffe bases in Libya, Sicily, and Sardinia.

A further objection was that the Combined Chiefs of Staff were concerned with securing the Casablanca-Oran railway, which, however rickety it might be, would be the best way of withdrawing troops from Algiers and Oran to the Atlantic, and from there home to the United States or Britain, if Torch went badly. Because Torch did not go badly—or at least was not an immediate disaster on the beaches—we are inclined today to underrate the dangers the Allies faced in North Africa, beginning with the possibility of sustained resistance by the French (and the embarrassment that it would cause) and including the very real possibility that Franco, alarmed by the American landings in French Morocco, next door to Spanish Morocco, might allow the Germans to move a substantial number of divisions from France through Spain to take Gibraltar—always a live issue for the Spanish. From Gibraltar to North Africa, the sea crossing is only eight miles wide, and it might then have been possible for the Germans to do to the Allies what they hoped to do to Rommel—trap them by attacking on both sides, and cutting them off from their supply line and their air cover.

In the end, as is usually the case in war, geography determined the course of events. The Western Task Force, consisting of American troops coming directly from the United States (under the command of Patton), would land on the beaches of French Morocco, at Safi, Fedela (later Mohammedia, just north of Casablanca), and Mehdia (close to Rabat and Port Lyautey, later Kenitra), and receive air cover from the escort carriers. British and American troops coming from the United Kingdom would pass through the Strait of Gibraltar and land at Oran (Center Task Force, under the command of Major General Lloyd R. Fredendall) and Algiers (Eastern Task Force, under the initial command of Major General Ryder, who would then hand over to the British Lieutenant-General Sir Kenneth A. N. Anderson once Algiers was secured, since the dour and outspoken Anderson, a Scot—Ike described him as "blunt, to the point of rudeness"[22]—was thought unlikely to win over French hearts and minds), with air cover from Gibraltar.

It was anticipated that Patton's greatest problem would be the long Atlantic swells breaking as high surf on the Moroccan beaches, which might swamp or capsize his landing craft. This was far from the only concern. An amphibious force of over 100,000 men (about three-quarters American and one-quarter British), carried on over 400 ships and protected by over 300 naval vessels, would constitute the largest operation of its kind to date in history. Special commando teams would be required to seize intact, if possible, the ports and port facilities of Oran and Algiers; paratroopers would have to be flown all the way from Britain, nearly 1,500 miles, to their drop zones; fighter aircraft and light bombers would have to be packed onto the runways of the airfield in Gibraltar—one of the world's trickiest airfields from which to take off or on which to land—like sardines in a can; troops had to be schooled to use weapons they had never seen before, like the new bazooka shoulder-fired antitank rocket launcher; supplies had to be provided for two armies with different weapons, calibers of ammunition, radio frequencies, and, above all, very different attitudes and ways of expression ("Two countries separated by a single language," as Oscar Wilde had put it). And all of this somehow had to be put onshore to face a French army of fourteen divisions (some of them made up of admittedly more or less ramshackle or exotic colonial troops, but others consisting of French regulars and French marines) and over 500 relatively modern aircraft, rapidly enough to turn east and attack the Germans before the Germans attacked them.

The most important question, and by far the most difficult one to answer, was political, not military: would the French fight back?

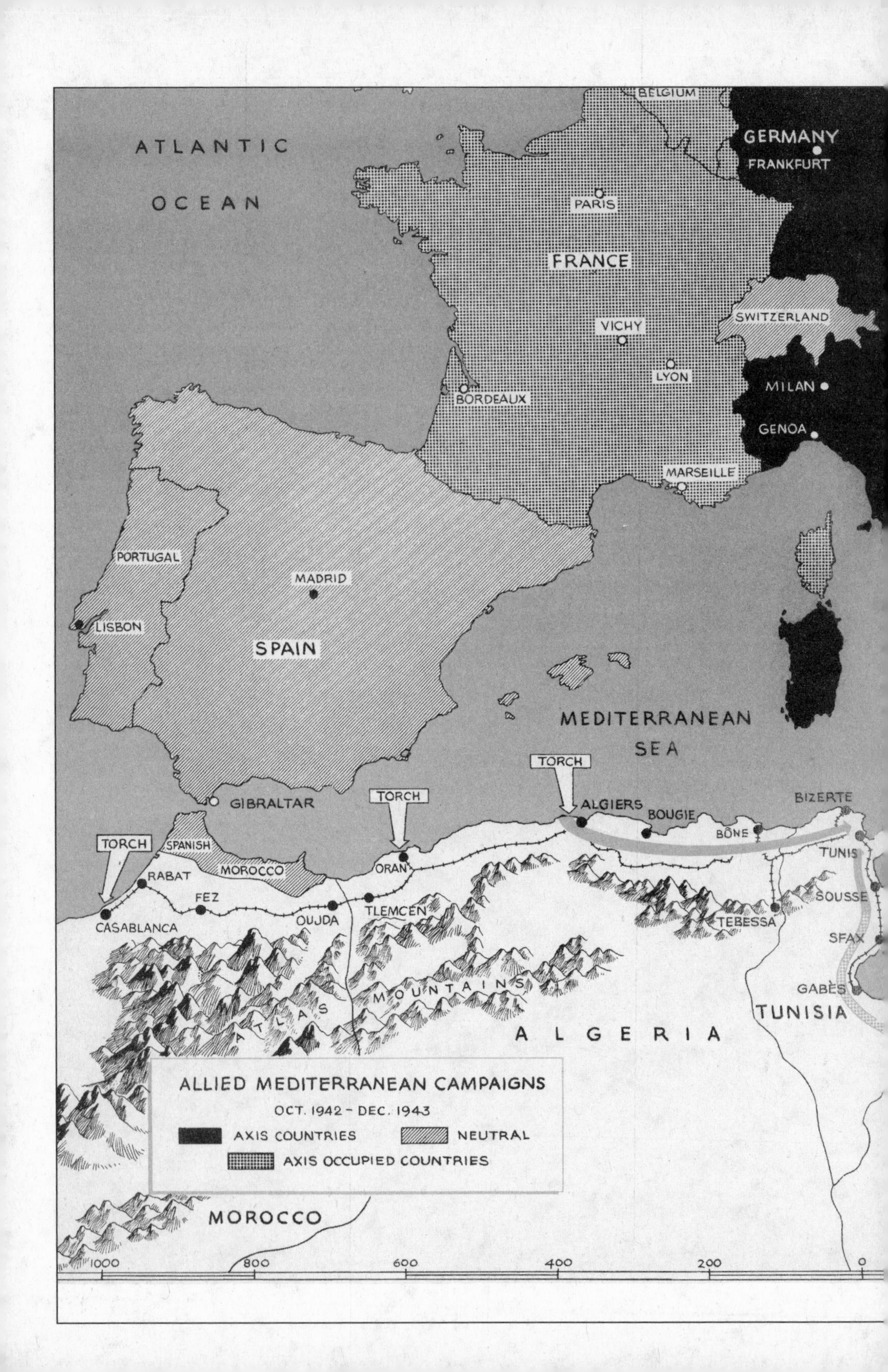

BELGIUM
ATLANTIC
OCEAN
GERMANY
FRANKFURT
PARIS
FRANCE
SWITZERLAND
VICHY
LYON
MILAN
BORDEAUX
GENOA
MARSEILLE
PORTUGAL
MADRID
LISBON
SPAIN
MEDITERRANEAN
SEA
TORCH
TORCH
GIBRALTAR
ALGIERS
BIZERTE
BOUGIE
BÔNE
TORCH
SPANISH
TUNIS
RABAT
MOROCCO
ORAN
SOUSSE
FEZ
TLEMCEN
CASABLANCA
OUJDA
TEBESSA
SFAX
MOUNTAINS
GABÈS
ATLAS
TUNISIA
ALGERIA
ALLIED MEDITERRANEAN CAMPAIGNS
OCT. 1942 - DEC. 1943
AXIS COUNTRIES
NEUTRAL
AXIS OCCUPIED COUNTRIES
MOROCCO
1000
800
600
400
200
0

PRAGUE
CZECHOSLOVAKIA
U.S.S.R.
ODESSA
MUNICH
VIENNA
AUSTRIA
BUDAPEST
HUNGARY
RUMANIA
BUCHAREST
BLACK SEA
BELGRADE
YUGOSLAVIA
BULGARIA
ITALY
SOFIA
ADRIATIC SEA
ROME
FOGGIA
TIRANA
ALBANIA
NAPLES
SALERNO
TARANTO
TURKEY
U.S.FIFTH ARMY
GREECE
BR. 1ST A.B. DIVISION
ATHENS
U.S. SEVENTH ARMY
BR. EIGHTH ARMY
MALTA
CRETE
MEDITERRANEAN SEA
DERNA
TRIPOLI
TOBRUK
BENGASI
MATRUH
MISURATA
BRITISH EIGHTH ARMY
FROM EL ALAMEIN
EL AGHEILA
EGYPT
LIBYA
MILES 200 400 600 800 1000

CHAPTER 11

Algiers

Patton, Ike, Monty.

To understand the difficulties that the Allies faced once they had decided on Torch—a better word than "difficulties" might be "muddle"—it is necessary to review briefly the calamitous history of France from June 1940 to the end of 1942. The terms of the armistice with Germany split France in two. Northern France, including of course Paris, was occupied by the Germans. Southern France, which came to be known as the Free Zone, remained unoccupied and was ruled by a French government headed by the aged Marshal Henri-Philippe Pétain, the hero of Verdun—"led" would attribute more energy to him than he possessed at his advanced age—from the provincial health resort of Vichy, whose famous spring water in its familiar bottle was then a mandatory accompaniment to all French meals.

The Vichy government, as it soon came to be known, was collaborationist in spirit, in the sense that most of its members, and certainly the marshal himself, disliked "perfidious Albion" more than they disliked the Germans; felt that since the Germans had won the war France's only hope was to collaborate with them; and, to some degree, also disliked communists, labor unions, Freemasons, and Jews. The members of the Vichy government were not all by any means "traitors," with some notable exceptions. Vichy was, for better or worse, the legitimate government of France, albeit a reactionary one that had signed a humiliating armistice after the defeat of the French army in 1940, and accommodated itself perhaps a bit more quickly than was seemly to a "new European order" ruled, for the moment, from Berlin. Those around Pétain were of the generation of the French bourgeoisie and upper class, predominantly Catholic, which believed that France had become too secular, that French workers were too well paid and had too many "rights," that Dreyfus had been guilty despite all the fuss made about him by the left, and that poor France had gotten its fingers singed once too often pulling Britain's chestnuts out of the fire. They were the kind of people who had said with a sad shrug in the late 1930s, *Mieux vaut Hitler que Staline* ("Better Hitler than Stalin") and even *Mieux vaut Hitler que Léon Blum*. (Blum was prime minister in the French leftist Front Populaire government.) In short, they were like the people in Britain who argued that there was no point spending government money to build public housing with bathrooms because the working class would only use the bathtub to store coal in; and if they were high enough on the social scale, they invited the German ambassador von Ribbentrop to dinner.

Distrust of England was not of course by any means limited to the right. Indeed, it cut across the entire French political spectrum, from left to right, not excluding De Gaulle and his followers in London; and it was a consequence of 900 years of warfare, colonial rivalry, and mutual contempt, as deeply ingrained as the popular belief in France of the efficacy of Britain's famous secret service, whose wiles and plots the French sought in every misfortune of *pauvre France*.

The Germans had not occupied, nor were they much interested in, the vast French overseas colonial empire. Thus Vichy France, itself something of a German colony as well as a rump state, still ruled, at any rate in principle, Tunisia, Algeria, Morocco, Syria, Lebanon, and what was then called French Indochina (Laos, Cambodia, and Vietnam),* as well as numerous other large and small colonies in equatorial Africa, the Caribbean, the Pacific, and Asia. The French fleet, one of the largest and most modern in the world—that part of it which the British had not sunk at anchor at Mers el-Kébir near Oran in July 1940 "in order to prevent [it] from falling into German hands"—was immobilized in Dakar, Alexandria, and Toulon; and most of its officers were still simmering with rage against their hereditary enemy the Royal Navy, rather than the Germans. The French colonial possessions in North Africa and the Middle East were effectively ruled for the time being by high civil servants, and by senior French army officers in whose eyes the chief threat to French colonial power had always been not faraway Germany but Britain, a rival colonial power forever eager, in French eyes, to take advantage of France in distress.

Defeat did nothing to change the mental habits formed over many generations—when French military men thought of Britain, they thought of the old enemy of Crécy and Agincourt, of Blenheim and Fontenoy, of Trafalgar and Waterloo, the country whose greatest heroes—Henry V, Marlborough (an ancestor of Winston Churchill), Wellington, and Nelson—had all fought against France, and whose most famous military exploits, learned by rote by generations of British schoolboys, were all at the expense of France.

The French military proconsuls who ruled North Africa, in the absence of any direct rule from the French government in Vichy, were at once deeply troubled by the sad state of France itself and all the more determined to keep alive the idea of a powerful, independent France in its colonial empire, pending the day of Germany's defeat and the reemergence of France as a great power. Their relationship with General De Gaulle and the Free French movement he led from London was prickly, and even poi-

* The Japanese occupied French Indochina in 1940 (as a kind of prelude to the attack on Pearl Harbor and in parallel with the Germans' invasion of France), when the French were in no position to defend it. Japan then ruled over it until 1945, maintaining in power France's puppet emperor Bao Dai and the Vichy French colonial and military administration, and thus producing in Southeast Asia a political situation analogous to that of Vichy.

sonous. The view of most of De Gaulle's colleagues was that the legitimate government of France had signed an armistice with Germany on June 17, 1940, and it was therefore, strictly speaking, the duty of every French soldier to lay down his arms and accept the terms of that armistice, whatever personal reservations he might have about it. When De Gaulle flew to London and proclaimed, in a radio broadcast to France on June 19, his own intention to continue to fight the Germans, he had placed himself, and his followers, in the position of deserters and mutineers. Events since then had done nothing to increase De Gaulle's popularity with his fellow generals—the British attack on the French fleet was bitterly resented, as was the fact that De Gaulle attracted to his movement a number of adherents who were "of the left," or Jewish, or socialist, in addition to the kind of officers who were thought of as troublemakers, headline-seekers, or adventurers. The growing resistance movement in the occupied zone of France—a movement largely organized and supplied from Britain—represented exactly the kind of thing conservative French officers were most afraid of: putting arms into the hands of civilians, many of them labor unionists and communists. Never mind that these arms were being used against the Germans for the moment; there might come a day when they would be used against the "forces of order" and the state itself. De Gaulle, from the point of view of those who opposed him, was a publicity-seeking, megalomanical junior general (only recently promoted to *général de brigade*, or major general) who had disobeyed his own government and was in the process of opening a Pandora's box of civil disorder and even civil war in France.

The Americans, who recognized Pétain's government and kept an ambassador in Vichy, and the British, who did not, were both conscious of the fact that landing in French colonial territory was likely to be taken by many Frenchmen as an act of war, and would very probably also provoke a violent reaction on the part of the Germans. There was no possible way to warn the French substantially in advance of the landings—to inform Vichy was the same as informing the Germans, and the Americans were equally concerned about leaks from Free French headquarters in London. There were no German forces in the French North African colonies, except for members of the German armistice commission, but nobody underestimated the speed with which Rommel, in Libya, or the Luftwaffe, close by in Sicily and Sardinia, would react when the Germans learned of the landings. Because Torch succeeded in the end, very few people today appreciate the enormous risks Ike was taking in his first experience of command. He was gambling that the French would allow him to get his forces ashore, and that despite primitive roads and railways, challenging terrain, inexperienced troops, and uncertain air cover, he could move them into Tunisia, and even take Tunis and the port of Bizerte before the Germans did. As things turned out, this was a gamble he would lose, and overcoming that loss would cost the Allies some 70,000 casualties, but it would also turn the United States forces into a battle-hardened army—and transform Ike from a skilled military bureaucrat into the toughest, most experienced, most formidable, and most realistic American commander since Ulysses S. Grant.

The problems of transporting the two armies were complicated not only by a critical shortage of shipping, but by the fact that the United States had counted rather optimistically on a cross-Channel landing in 1942 and had not only equipped the troops accordingly, but stored the vast quantities of equipment and supplies in the United Kingdom with a short sea crossing from England to France in mind, rather than a 1,500-mile sea journey to a climate and terrain very different from that of northern France. This alone constituted something of a logistics nightmare, in the age before the widespread use of the computer. Huge quantities of stores had to be located, repacked for shipping, and loaded on ships in the order in which they would be required by the troops. This was not accomplished easily or efficiently, as it turned out. "We should have paid more attention to 'red tape' and paperwork,"[1] Ike commented ruefully later, mildly rebuking his old pal General Lee, whose job this was. (As late as the winter of 1944, the "fighting generals" would still be complaining bitterly to Ike about Lee's shortcomings as a logistician.)

The political problems of landing in French territory not only required Ike to become something of a diplomat himself but made it necessary to supply him with a political adviser. In the autumn of 1942, Robert D. Murphy—a wily, charming, well-connected, French-speaking American diplomat, who had served in France before 1940 and then in Vichy and North Africa, and had won along the way the guarded confidence of right-wing Frenchmen and the somewhat less guarded confidence of Roosevelt—was sent to London disguised in uniform as "Lieutenant Colonel McGowan."[2] (This had been Ike's suggestion, on the sensible theory that nobody pays any attention to a mere lieutenant colonel.) Murphy was to brief Ike on the politics of North Africa, or at any rate on the politics of North Africa as Murphy and the president believed them to be, for Ike would soon be provided with a British political adviser as well, in the person of Harold Macmillan—heir to the great British publishing house of the same name, son-in-law of the duke of Devonshire, a staunch supporter of Churchill, and a future Conservative prime minister—to brief him on the politics of North Africa as seen from 10 Downing Street, a view which, unsurprisingly, differed sharply from that of the White House.

Murphy was not the conventional State Department striped-pants diplomat: he was a glib and resourceful intriguer and "operator," with an adventurous piratical streak (he had smuggled the gold from the Bank of France to Dakar aboard an American warship), plenty of Irish charm, and a well-honed gift for flattering important people. He was, in fact, exactly the kind of person who appealed to Roosevelt; and Roosevelt had more or less appointed Murphy on the spot as his de facto personal representative in North Africa, functioning independently of both the State Department and the War Department, and sometimes at cross-purposes with them. Murphy's energy and courage were beyond dispute, but his preference for right-wing politicians and military figures was curiously at odds with what one would have assumed the president's views might be, and was behind many of the more controversial political decisions Ike made in Algiers, and later in Italy. Murphy consistently overrated the good faith, judgment, and ability to deliver the goods of the Vichy government and conservative French military figures, and distrusted more "democratic elements" in French politics. (Later he would make the same mistake in Italy, placing far more faith in the Vatican, the king of Italy, and "conservative" businessmen and generals than in sincere antifascists whose aims were more in sympathy with American policy.) It was Murphy more than anyone else who fueled De Gaulle's suspicion of the Americans' motives in North Africa, the Middle East, and France, with unfortunate consequences both during and long after the war.

The purpose of Murphy's flying visit to London—he stayed only twenty-four hours—was to give Ike a preview of the situation in North Africa, and to describe from his personal knowledge the major French military leaders there, and this he did with uncharacteristic caution. He did not believe that the French would welcome American troops with open arms, still less British troops, and he predicted correctly that General August Paul Noguès, foreign minister and chief French military advisor to the sultan of Morocco, would be the most likely to fight, and the most determined fighter. Murphy had great confidence—unfortunately misplaced—in the ability of General Emmanuel Mast, in Algeria; and General Marie Émile Bethouart, in Casablanca, to win the other French generals over to the Allies' cause. Both generals believed—or at any rate told Murphy they believed—that if only General Henri-Honoré Giraud could be brought to Algiers, the whole of North Africa "would flame into revolt" against the Vichy government, and rally around the figure of Giraud.

Giraud's role as the "anti–De Gaulle" was to become an idée fixe of the president's, thanks to Murphy; and certainly there were, on the surface (and if one had not actually met Giraud), many reasons to believe in his capacity to unify and lead the French military away from Vichy and back to the Allies. Nobody believed in this more firmly than Giraud himself. He had put up a good fight in 1940, then managed to carry out a daring, dramatic escape from imprisonment in a German fortress in Saxony and make his way back to Vichy France—nobody doubted Giraud's courage, or his military credentials.* Almost as tall as De Gaulle, he was a handsome, dashing, impressive military figure, the classic *beau sabreur* of French military legend. Unfortunately for the Allies, he was also almost as stub-

* General Giraud's great gift was escaping from captivity. He had escaped from a German POW camp during World War I, making his way to England in a variety of improbable and exotic disguises. After escaping from the Königstein Fortress in 1942 he made his way through Germany despite a nationwide manhunt organized by the Gestapo and the Geheime Feldpolizei, and a price of 100,000 reichsmarks on his head, then talked his way across the Swiss border and from there to Vichy France. His fellow French generals admired his physical courage but did not think escaping in disguise was necessarily the most useful talent for a four-star general.

born and megalomaniacal as De Gaulle, but without De Gaulle's icy intelligence, acute political instincts, soaring historical imagination, or gift for inspirational, passionate patriotic rhetoric. General Sir Alan Brooke, a shrewd if jaundiced judge of men, wrote of Giraud, "He is no politician, [and] a very indifferent general," and remarked that he "had charm but no ability."[3] Others commented that Giraud's eyes reflected not so much a hard, military stare toward the far horizon as sheer stupidity. The truth seems to have been that Giraud was that most unfortunate of figures, a wooden and an unimaginative mediocrity thrust into a situation that required bold Napoleonic vision and daring, with which he was by no means equipped. Much as the other French generals might dislike De Gaulle, none of them turned out to be willing to defer to Giraud.

Murphy also believed that the cooperation of French senior officers in North Africa would be more easily secured if a high-ranking American general could be sent there to meet with them clandestinely, man to man. This was a delicate proposition: first, because if the meeting was discovered, the American would be interned, and the French officers involved would be arrested and sent back to Vichy to be court-martialed; and second, because the American would have to conceal from them at all costs that the invasion was in fact already at sea. On the other hand, the risks seemed to be outweighed by the possibility that the French, if approached in the right spirit, might agree to defy Vichy and refuse to resist.

Since it was out of the question for Ike himself to go, his deputy Mark Clark volunteered for the mission, and in mid-October was flown to Gibraltar. He was taken from there to Algeria in a British submarine, HMS *Seraph*, from which he and his small staff made their way to shore near Algiers by night in canvas kayaks, to be greeted by moonlight on the beach by Murphy, who said, "Welcome to French North Africa."[4] The French, when they arrived, were astonished at the fact that Clark, a major general, carried a carbine—French generals did not carry rifles like common soldiers. Despite Murphy's optimism, the secret talks were inconclusive. General Mast wanted to know the date of the Allies' landings, which Clark could not tell him, and insisted that only Marshal Pétain, or somebody who could speak for Pétain, could relieve him and his colleagues from their responsibility to defend French territory and French neutrality against any aggressor. Giraud's name did not seem to stir the French deeply, and a warning that the police might be on their way ended the talks prematurely—the French officers scattered; and Clark and those who accompanied him were obliged to hide out in a farmhouse. In trying to get back into the kayak to return to the submarine, Clark was dumped into the sea by the high, breaking surf and lost both his wallet and his trousers—a good lesson about the ability of the Mediterranean Sea to lose its picture-postcard image in a few hours. The mission accomplished very little except to suggest that even Murphy's cautious estimate of the French generals' willingness to stick their neck out on behalf of the Allies might be overoptimistic. Curiously enough, it did not shake Ike's confidence in Murphy, or even Mark Clark's—Murphy was the kind of man who had an explanation for everything.

The invasion went forward on schedule, as the convoys left the East Coast of the United States bound for Morocco, and left ports in Britain bound for Oran and Algiers. The landing date had been moved one day forward to November 8, and it was a measure of President Roosevelt's confidence in Ike's judgment that he had accepted a date four days later than the elections with good grace and without any visible sign of disappointment or regret, much as he had hoped the landings would take place in time to influence the voters. It also says much for Ike's professionalism and growing self-confidence that he had not tried to shave a few days off the schedule to satisfy the president—he knew to the day just how many days he needed, and that was that.

The Americans kept French sensibilities in mind, and few American armies can ever have been provided with more flags of all sizes: embroidered patches of the Stars and Stripes to sew on their uniforms, and American flag stencils so they could paint Old Glory conspicuously on their vehicles. Even the British vehicles were to be stenciled with a conspicuous white American star, to the great annoyance of the British. General Patton's troops would hit the beaches in Morocco with enough flags flying to satisfy an eighteenth-century painter of patriotic scenes—not that, in the end, it made much difference.

On the night of November 4 Ike and his staff flew from London to

Gibraltar in five B-17 Flying Fortresses. The journey, which was made at an altitude of a few hundred feet in rain and fog, was not without danger or mishaps—the arrival of Ike's aircraft in Gibraltar went unreported, so there were fears in London that he had been shot down; and the next day a B-17 carrying the rest of the staff was attacked by two German JU-88s, which the gunners only just managed to fight off. Even Ike, who was usually too indifferent to his surroundings to mention them, found Gibraltar "the most dismal setting we occupied during the war."[5] Quarters at the Rock, as Gibraltar was known without affection throughout the British armed forces, were in long, damp tunnels deep underground, carved out of the living rock, lit by a few dim lightbulbs, and ineffectively ventilated by noisy, ancient fans. Gibraltar had just two things to commend it: it was the only piece of Allied territory in the western Mediterranean; and it had a small—though treacherous—airfield, which was now packed with Spitfires, jammed in wingtip to wingtip, like sardines in a can, for want of any other place to fly from until airfields in North Africa were captured. The food and the living conditions were execrable, but there was no other place to establish a headquarters until such time as Ike could move to Algiers—if, of course, the landings were successful. Ike himself rated the odds at no better than fifty-fifty.

Even gloomier than the surroundings was the weather. Submarines reported that the seas were stormy and that there was heavy surf on the invasion beaches. In case of bad weather Ike had two options. One was to have the convoys steam in circles in the Atlantic and the Bay of Biscay until it cleared; but the obvious danger was that the invasion fleet would eventually be spotted by a German submarine or patrol plane, as well as that many of the ships would soon be running low on fuel. The second option was to gather the invasion fleet in the harbor at Gibraltar, but since everything in Gibraltar could be seen from Spain—the Gibraltar airfield itself was separated from Spanish territory by only a barbed-wire fence—that would effectively give the Germans advance notice of the invasion. Ike pondered these two options rather glumly, and decided on the second in the event that he had to postpone.

Fortunately, postponement proved unnecessary, but Ike's frame of mind was not improved by the fact that his radio communication with the invasion fleets and London was poor and sometimes nonexistent—whether the problem was geographic, or the quantity of signals simply overwhelmed the communications system, remained mysterious, but the effect was to cut him off in Gibraltar for three days.

On the day after his arrival at Gibraltar he wrote a hurried letter to Mamie telling her about his new title, "'Allied C-in-C' in this region,"[6] and two days later he dictated a longer one, explaining why it had been necessary to leak to the American press that he was coming home on leave as a blind to deceive the Germans and explain his absence from London—he was aware of how deeply she would be disappointed, and apologized for not being able to let her know the truth. He longed to write to her about where he was, what he was doing, and why, but "in war this can't be done," he pointed out, and censorship made "letter writing a terrible chore."

Two events occurred on November 7 to cause further dismay. The first was the sinking of the USS *Thomas* with a reinforced battalion of American troops on board. The *Thomas* had been torpedoed by a German submarine within a day of Algiers, with what was presumed—wrongly, as it turned out later—to be a considerable loss of life and the risk that the submarine would alert the Germans to the impending invasion.

The second event was the surprising arrival in Gibraltar of General Giraud, whom the resourceful Murphy had managed to smuggle out of what amounted to house arrest in Vichy France onto a small fishing boat. Once at sea, Giraud was transferred to a waiting British submarine—the ubiquitous HMS *Seraph*, placed for this one mission under the command of an American naval captain, a face-saving gesture devised by Murphy to allow Giraud to claim to his fellow Frenchmen that he had traveled in an American submarine rather than one belonging to the hated Royal Navy—and brought to Gibraltar. He was something less than a welcome guest, since his role in the invasion was still undecided, and Ike had been hoping to avoid a meeting with him.

Though still wearing civilian clothes, Giraud nevertheless immediately impressed Ike as "a gallant if bedraggled figure. . . . Well over six feet tall, erect, almost stiff in carriage, and abrupt in speech and mannerisms. . . . Very much the soldier."[7] It was not only Giraud's "carriage" that was stiff, unfortunately. He was suffering from "the grave misapprehension"—either because of an inflated view of his own importance or because Murphy, who had a gift for telling people what they wanted to hear, had accidentally planted it there in the effort to persuade Giraud to leave his family behind where the Germans could get at them and board an Allied submarine—that he was to be supreme commander of the entire invasion: that he would, in fact, replace Ike.

As the argument between the two men went on and on in what Ike called his "dungeon"—an interpreter was needed, since Giraud spoke almost no English, and Ike even less French—it gradually became clear to Ike that more was at stake than merely Giraud's amour propre, vanity, and wounded feelings, strong though these were, or even the fact that Giraud had four stars to Ike's three. Giraud, like his rival De Gaulle, was unable to think of *la France* as being anywhere but at the center of world affairs, and again like De Gaulle saw himself as representing, even symbolizing, his wounded nation. Personal feelings apart—and Giraud's personal feeling was that he must be in command of the British and American forces as well as the French—Giraud could not accept that he, or France, should be subordinated to an American general; nor did he think, in practical terms, that the French generals in North Africa would take him seriously unless he arrived there in the role of supreme commander (as it turned out he was correct about this). Over and over again he repeated woodenly, "General Giraud cannot accept a subordinate position. . . . His countrymen would not understand and his honor as a soldier would be tarnished."[8] (Like De Gaulle, Giraud was in the habit of referring to himself grandiloquently in the third person.)

To say that Ike was amazed would be putting it mildly, but he was also alarmed to realize how out of touch with reality Giraud was. The Frenchman seemed to have no idea of the role that airpower would play in the impending operation; and Giraud not only thought that landing in North Africa was a mistake but he tried to talk Ike into changing direction at the last minute and landing in southern France instead—where

Giraud's name, he believed, would ensure that the Allied armies would reach the demarcation line between Vichy France and occupied France before the Germans could move south in force. It was explained to Giraud that the troops were even now approaching their landing zones, and that Giraud's role, whatever he may have gleaned from Murphy, would merely be to act as a kind of figurehead and command those French soldiers who rallied to his name, but he remained indifferent to any opinion but his own. As the night wore on, Mark Clark was brought in to add his weight (and his partial knowledge of the French language) to the argument, but it made no difference, and at midnight Giraud finally rose and ended the discussion, saying by way of good-night, "Giraud will be a mere spectator in this affair."[9] He then retired with wounded dignity to sulk in his quarters, which were even less comfortable than Ike's.

The long comic-opera scene with General Giraud had one advantage—it distracted Ike from thinking about the landings. Even had he been able to communicate with the invasion fleets from Gibraltar, there was nothing more he could do at this point—the landings would take place as planned, and they would succeed or they would not.

He was fortunate in one respect. The Germans had noticed the huge amount of shipping passing through the Strait of Gibraltar, and the number of aircraft on the airfield at Gibraltar—they could hardly fail to have noticed that something significant was happening, given the number of ships involved, the ubiquity of German consular officials and agents around Gibraltar, and the increase in radio traffic at sea—but like Giraud they too thought first of the south of France, though there were also predictions that the conveys were intended to relieve Malta or to carry the Americans to Egypt. Apprised of the situation, Hitler drew the conclusion that the Allied convoys were making for Libya, to attack Rommel from behind, and he ordered the German submarine forces in the Mediterranean and the Luftwaffe to destroy the fleet as it passed through the narrow Straits of Sicily. As a result, what German forces were available to attack the invasion fleet were in the eastern Mediterranean, hundreds of miles away from the landing zones.

In the morning, Giraud woke in a less intransigent state of mind, or perhaps encouraged by the good news that was trickling in as the landings took place, and Ike succeeded in talking him into the less ambitious role the Allies had in mind for him.

Throughout the night garbled and scattered radio messages had reached Ike, but radio communications at Gibraltar were still ineffective—one lesson of Torch would be the need to provide invasion fleets with a headquarters or command ship and with more effective communications equipment. Although in Gibraltar he was only a relatively short distance from the landing zones in North Africa, in terms of his ability to control events and to receive reliable information Ike might just as well have stayed in Britain. Of course, to some degree this was inevitable. The operations were on a huge scale—and, as Ike described them, "an undertaking of the most desperate nature"—and spread out over an enormous chunk of geography. As Ike breakfasted in the dank corridors of the Rock, 35,000 American soldiers commanded by Patton were landing in French Morocco; 18,500 Americans under General Fredendall were landing on beaches east and west of Oran; and 20,000 American and British troops under General Ryder were landing on beaches to the east and west of Algiers. Meanwhile, at both Algiers and Oran, British Commandos and American Rangers were being landed in the port on destroyers to prevent the French from blowing up the port facilities, blocking the harbor, or scuttling their ships.[10]

The hope that the French would not oppose the landings, or that Giraud's name would have a pacifying effect on them, proved to be misplaced. In Morocco Patton found himself fiercely opposed, just as Murphy had predicted; nor was the situation improved when the naval vessel carrying him was forced to leave the beaches and put back out to sea to deal with a French naval attack, rendering Patton helpless for a time to command or reach his own troops—this was another argument in favor of a headquarters ship. Although the surf was manageable for the first landings in Morocco, it increased dramatically during the day, and many of the landing craft were swamped during subsequent landings. Also, some of the troops were landed on the wrong beaches, and this mistake added to their confusion as they came under heavy fire.

The landings on the beaches near Oran and Algiers took place without heavy opposition, but there was sporadic fighting as the troops made their way inland, and the French navy and air force both put up a stubborn, and in some cases successful, fight. The attempt to secure the airfield near Oran failed, since the paratroopers were dropped too far away from it and found themselves pinned down under heavy fire. The heaviest fighting, apart from that in Morocco, took place in the ports of Oran and Algiers. In the port of Algiers, one of the destroyers was sunk by French coastal batteries, and there was fierce fighting in the port itself. In the port of Oran the initial landings were at first "bloodily repulsed,"[11] two British destroyers were sunk by the French navy, and some of the Rangers were even obliged to surrender before French resistance was finally broken. In both ports the French managed to inflict a good deal more damage on the facilities than the Allies had anticipated.

By the night of November 8 it was clear, despite fragmentary communication, that the forces at Oran and Algiers were there to stay, but the unexpected strength of French resistance, particularly from the French navy and marines, worried Ike, who noted, "We are slowed up in the eastern sector when we should be getting toward Bône-Bizerte at once."

He was right to be worried. After all, the nub of the matter was that the success of Torch depended on Ike's ability to gather his forces once they were landed, turn them east quickly, and advance to take Bizerte and Tunis before the Germans did; and the stubborn resistance of the French, combined with the slowness of inexperienced troops coming under fire for the first time, was already beginning to cause concern. This concern was increased by the lack of reliable information from the Western Task Force in Morocco. Ike remained unable to reach either Patton or Rear Admiral H. K. Hewitt, the fleet commander, by radio; nor were their messages to him intelligible—and several light bombers sent to "gain contact" with American forces there were shot down by the French. Ike was finally obliged to send an American admiral off in a British destroyer to find out what had happened to the Western Task Force.[12]

Early on the morning of November 9, he sent Major General Mark Clark and General Giraud off to Algiers by plane—Royal Air Force and USAAF fighter and light bomber squadrons were already beginning to land on some of the military airfields in Algeria—in the hope that Giraud's presence and Clark's persuasiveness might put an end to French resistance. A broadcast by Giraud had little effect, partly because radio communication with Algiers was so poor that very few people could have heard it, and partly because Giraud did not have De Gaulle's gift of eloquence. But the one piece of news from Murphy that did get through in the other direction from Algiers to Gibraltar was astonishing: Admiral Jean-Louis-Xavier-François Darlan, "the commander in chief of the French fighting forces" and one of the most formidable and best-known members of Pétain's government, was actually in Algiers.

Though there has been speculation for over sixty years that Darlan knew about the invasion in advance, or was tipped off by Murphy, possibly at Ike's request, this remains unproved. Ostensibly, Darlan had come to Algiers to visit his beloved son, who had been stricken with polio, and his presence was explained as merely one of those unlikely and ironic coincidences.

Even in civilian clothes, Darlan was a force to be reckoned with. He was short and rotund, and his fleshy face had the kind of weary cynicism, intelligence, fierce energy, worldliness, and instinctive unscrupulousness of a Renaissance cardinal. With his firm chin, his thick neck, and his blunt, crooked nose, he resembled a French equivalent of such English historical figures as Cardinal Wolsey and Thomas Cromwell, physically powerful, plebeian in appearance, and unashamedly ambitious—even Darlan's smile was more knowing (and possibly mocking) than cheerful. His Anglophobia was unfeigned and deeply rooted (his great-grandfather had been killed at the Battle of Trafalgar), and his skill as a sailor, naval administrator, and political intriguer was beyond question. Darlan had been given command of the French navy by the socialist Léon Blum, despite Darlan's instinctive conservatism and his dislike of almost everything Blum's Front Populaire stood for; and when Paul Reynaud resigned in 1940 Pétain made

Darlan his minister of the navy. By 1941 Darlan had been appointed Pétain's successor, as well as minister for foreign affairs, defense, and the interior—so that Darlan, given Pétain's age and increasing senility, became the most powerful figure in the Vichy government. In 1942, as part of the fallout from his continuous power struggle with his rival Pierre Laval, Darlan was removed from these offices and appointed commander in chief of the armed forces (a post that gave him effective command not only of the French navy, but also of the army and air force) and high commissioner for North Africa. In the process Darlan came to be regarded in the English-speaking world as the arch collaborator and was almost as reviled as Laval.

In fact, Darlan was not an unthinking or a wholehearted collaborationist*—as a French naval officer he simply saw Britain as France's natural and permanent enemy. In 1940 the war had been lost, he thought (like many other Frenchmen), in large part because the United States had refused to join in despite Reynaud's desperate pleas for it to do so, and because Britain had sent a ridiculously small army relative to that of France, and then selfishly refused to commit the bulk of its fighter aircraft to the struggle in France at the moment of crisis. In short, France had been left in the lurch as usual by *les Anglo-Saxones*, and perfidious Albion had then withdrawn its troops from Dunkirk, abandoning France. Since the Germans had won the war, and were in the process of constructing a "new European order," France had no choice but to play its part in that order and wait for time and events to restore it to power in a Europe freed from the threat both of Soviet communism and of Anglo-Saxon meddling. If it was necessary, for a time, to bow down to the Führer and accept German hegemony, so be it. France had been humiliated before, and always had risen back to its rightful position. No doubt it would rise again. All that was needed was patience and guile, both of which Darlan possessed in abundance.

* This is not to say that he was an angel. As a senior member of Pétain's government, Darlan was responsible for anti-Semitic legislation in Vichy France and the French colonies, for allowing the deportation of foreign-born Jews in France to Germany (and from there to the killing camps in the East), and for helping the Germans keep up a flow of supplies from French ports in southern France across the Mediterranean to Rommel.

Darlan was a supreme realist, a Machiavelli in an admiral's blue uniform with gold braid; idealism as such did not form a part of his character. Even his love for the French navy, which nobody doubted, did not prevent him for looking at the French fleet unsentimentally as a bargaining chip, perhaps the last and most valuable one France possessed—hence his anger when the British sank part of it at Mers el-Kébir in 1940, and tried to seize one of its most modern battleships at Dakar. He would not under normal circumstances have allowed himself to be trapped in Algiers by the British and the Americans, had it not been for his son's illness; but once it was clear to him that this had happened, he naturally sought to make the most of the situation. Besides, unlike Laval, Darlan was not an ideological enthusiast for Nazism, or for anything else. It had not escaped his attention that Germany's victory no longer seemed quite as good a bet in November 1942 as it had in June 1940 and throughout most of 1941. The British were still stubbornly fighting; the Americans had joined the war; the German army had been defeated at El Alamein, in Egypt, and had advanced deep into the heart of Russia to the city of Stalingrad, on the Volga, but there, too, the Russians were still fighting. Darlan was not a man to back the wrong horse indefinitely, whatever Marshal Pétain might choose to do.

Ike's instinct to make contact with Darlan was a sound one, much as he underestimated the scandal it would create at home and in Britain. Winston Churchill had told him, before he left London for Gibraltar, "Kiss Darlan's arse if you have to, but get the fleet," and subsequently both Churchill and Stalin would defend his decision on the grounds of realpolitik, as would Roosevelt, though with great reluctance. Still, in hindsight, it is impossible not feel that the "Darlan deal," as it soon came to be known, was a reflection of Ike's inexperience as a combat commander and his precarious hold over the military operations of Torch, and the confusion and slow progress of the Allied landings when they met with unexpectedly strong French resistance.

Morocco worried Ike most, since despite poor communications it was clear enough that the fighting there was serious, both on land and at sea. Once Patton himself was onshore, he pushed his men hard, and focused

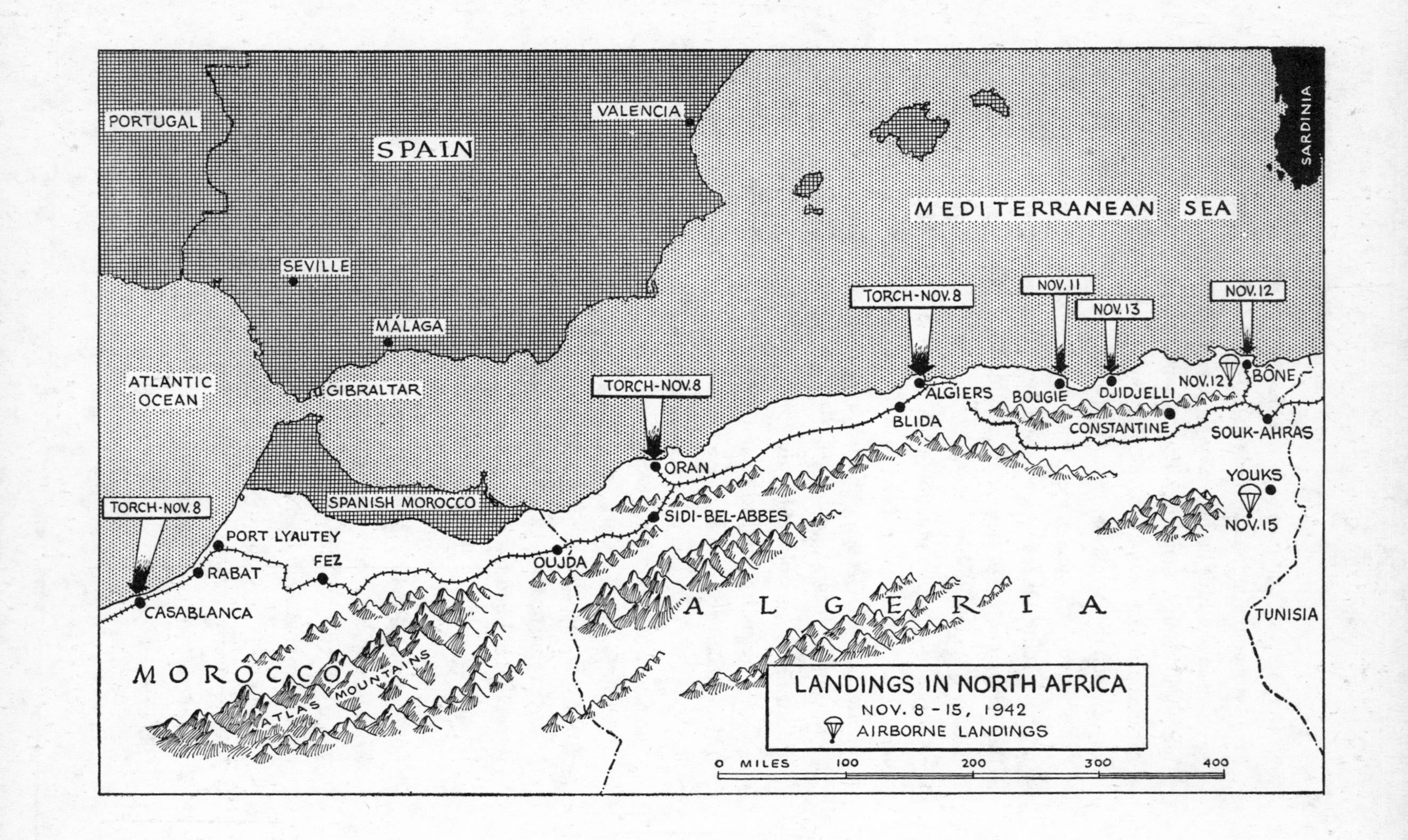
LANDINGS IN NORTH AFRICA
NOV. 8 - 15, 1942
AIRBORNE LANDINGS
PORTUGAL
SPAIN
VALENCIA
SEVILLE
MÁLAGA
GIBRALTAR
ATLANTIC OCEAN
MEDITERRANEAN SEA
SARDINIA
SPANISH MOROCCO
MOROCCO
ATLAS MOUNTAINS
ALGERIA
TUNISIA
TORCH-NOV.8
CASABLANCA
RABAT
PORT LYAUTEY
FEZ
OUJDA
TORCH-NOV.8
ORAN
SIDI-BEL-ABBES
TORCH-NOV.8
ALGIERS
BLIDA
NOV.11
BOUGIE
NOV.13
DJIDJELLI
CONSTANTINE
NOV.12
NOV.12
BÔNE
SOUK-AHRAS
YOUKS
NOV.15
0 MILES 100 200 300 400

them on getting off the beaches, but the fighting quickly escalated beyond anything that had been anticipated. Despite numerous attempts to sink it, and heavy casualties, the French modern battleship *Jean Bart*, though uncompleted and immobilized, continued to fire its enormous guns until it was finally put out of action by U.S. Navy dive-bombers with a heavy loss of life on board; and those French ships that could make it to sea fought courageously, and with some success. On land too, the French fought with courage worthy of a better cause. Despite Patton's bravura leadership, the Moroccan landings were plagued with problems, some of them of his own making. He had not paid enough attention to the logistic problems of loading the ships, with the result that vital supplies and equipment were buried under tons of inessentials. This was a reflection partly of Patton's lack of interest in logistics, and partly of the fact that he behaved as if the U.S. Navy were the enemy, rather than the Germans—he and Admiral Hewitt, though on the same ship, were hardly on speaking terms by the time they reached Morocco.[13]

Isolated 600 feet belowground on Gibraltar, Ike was in no position to exert military leadership, or correct the shortcomings of his commanders. This was, in any case, his first experience of real command, and it was a little like being thrown into the deep end of a pool to learn how to swim. Here he was, a man who had never led a company, a battalion, a brigade, or a division in combat, commanding the most ambitious amphibious operation in the history of warfare. One thing, at least, was becoming clear to him, as he started to receive signals from Algiers—no French commander would agree to a cease-fire, or join the Allied forces in fighting the Germans, without a legitimate order to do so, either from Pétain himself or from Darlan as commander in chief of the armed forces. Nobody doubted that Hitler would react swiftly to any breach by the French armed forces of the armistice agreement with Germany, and failure to resist the Anglo-American landings in North Africa would certainly constitute a major breach, one from which there could be no turning back.

Murphy had spent November 8 in a dramatic bargaining session with Darlan at the villa of General Alphonse Juin, commander of French ground forces in North Africa—at one point members of a pro-Allied resistance group that Murphy had organized placed Darlan and Juin under arrest; at another, French troops surrounded the villa with tanks, placed Murphy and his associate under arrest, and offered them the last cigarette that is traditional before execution by a firing squad. But by the morning of November 9, when Giraud and Clark left for Algiers in a Flying Fortress, resistance in the city was beginning to die down. This was not so in the rest of North Africa, however, and it was becoming increasingly apparent that the only man who could stop this resistance was Darlan. For the moment, he declined to do so, having received an order from Marshal Pétain in Vichy forbidding him to. Under the pressure of events—now he was a virtual prisoner of the Allies— Darlan eventually agreed to a cease-fire in Algiers, but he denied that he had any authority to do the same outside the city.[14]

The next day came the news that Hitler had ordered the German army to occupy the Free Zone of France, but for the moment this merely increased Darlan's leverage with the Allies—he was now the only senior member of the Vichy government who retained freedom of action, as well as the one man who could order the French naval commander in Toulon not to allow the Germans to seize the elements of the French fleet at anchor there. Darlan shrewdly used the German occupation of southern France to bypass Pétain. Since the Germans had now breached the armistice agreement, he decided, he was at liberty to ignore it himself; and he ordered fighting to cease throughout North Africa, which it promptly did—even in Morocco, where at the last minute Patton was spared the necessity of a full-scale battle for Casablanca, and soon found himself exchanging gallant toasts in champagne with his wily former adversary General Noguès. By the time Ike reached Algiers on November 13 all that was left to negotiate was Darlan's price, and to ascertain whether the French military leaders in Tunisia would follow him.

The price was the confirmation by the Allies of Darlan's role as French high commissioner of North Africa—"Not a pretty sight," as De Gaulle, in London, remarked acidly, and correctly. Giraud, who had been obliged to go into hiding to avoid arrest, was given command of all French troops in northwest Africa, something of a humiliating comedown for him, but he too recognized Darlan's legitimacy as the only person in North Africa authorized to speak for Pétain.

Unfortunately, the French commanders in Tunisia did not. The negotiations between Darlan and the Allies had taken too long, Darlan had confused them by changing his mind too often about his authority to declare a cease-fire, and they remained uncertain at heart about Pétain's wishes. It did not help matters that whereas Darlan and Giraud had seen the size and strength of the Allied forces with their own eyes, they had not—with the result that by November 9 the Germans were already beginning to land troops and aircraft in Tunisia, while the French forces there did nothing to inconvenience them. This was just what the "Darlan deal" was intended to prevent. By November 13, when Ike sat down with Darlan, substantial German forces were already in control of Bizerte and Tunis; Luftwaffe aircraft were landing on Tunisian airfields; and Luftwaffe Field Marshal Albert Kesselring, an aggressive and able soldier and airman (known to his men as "smiling Albert" not for his sense of humor but for his fixed, toothy grin of grim determination), had been given command of all Axis forces in the Mediterranean, with orders from Hitler to attack the Allied forces in Tunisia immediately. Kesselring, whose headquarters was in Rome, where he could keep an eye on the Duce and the Italian general staff, was mystified that the Allies had landed in Morocco and Algeria instead of Tunisia, which had been virtually undefended. It was nearly 400 miles from Algiers to Tunis, over miserable roads and across some of the most rugged and mountainous terrain in North Africa, and Kesselring was just the man to contest every square mile of it.

At home, the fallout from the "Darlan deal" was so bad that General Marshall found it prudent to withhold most of the press criticism from Ike, with the result that Mamie had a better sense of just what hot water her husband was in than he did. Having just promoted Mark Clark to a third star, Ike rather shrewdly allowed the new lieutenant general to announce over Radio Morocco his "pleasure" at the conclusion of the Darlan deal, but not surprisingly that failed to deflect a torrent of criticism aimed at Ike. On November 15 the *New York Times* pointed out the incongruity of radio broadcasts emanating from Allied-controlled Radio Morocco which ended with *Vive le maréchal!* and professed its "bewilderment" in thoughtful editorial. Others expressed their indignation more strongly. Many of those closest to the president were furious at Ike's making a deal with a notorious collaborationist—as was Wendell Willkie, the Republican presidential candidate in 1940; and it took all of Marshall's skill and his stiff upper lip to calm things down. In an effort to soothe his critics, the president announced on November 17 that the Darlan deal was a "temporary expedient" and asked for the abrogation of Vichy's anti-Semitic laws in North Africa; but on the same day there was a firestorm in Britain's House of Commons, in which Darlan was described as another Quisling (the head of the collaborationist government of German-occupied Norway), and the deal with him as a "bitter pill." Among the many ironies surrounding Ike's embrace of Darlan was that the story about the president's announcement appeared in the *New York Times* next to one about the death of Colonel Demas T. Craw, who had been machine-gunned by French troops in Morocco as he moved forward under fire for the second time to try to persuade the French commander to surrender (Patton said of him, "He was born for war," and praised his "superlative heroism"). Most of the American liberal press, and all of the British Press, railed against Ike's deal. The *Manchester Guardian* summed up the feelings of those who opposed the deal:

> It is not surprising that Darlan's position is causing widespread anxiety. His record is hardly to be distinguished from that of Laval. He has been responsible with his Vichy colleagues for all the persecutions in France, for allowing the Nazis to shoot hundreds of Frenchmen as hostages, for surrendering refugees, for hunting down the Jews.[15]

Under the circumstances, Ike was lucky to escape with his scalp, and he owed much of his good fortune to Marshall—and to Roosevelt's commonsense view that any deal, however distasteful, which ended the fighting between the American forces and the French and allowed the U.S. Army to move into Tunisia and begin fighting Germans was a step ahead.

In a gesture that was unusual for him, Marshall invited the eighteen members of the House and Senate military committees to meet with him in his office; asked for their support for Ike's decision; and arranged to meet privately with the press, to put the fire out. He cabled to Ike to say that he had discussed the matter with the president, and added encouragingly, "Do not worry about this, leave the worries to us and go ahead with your campaign." Nevertheless, despite Marshall's firm backing of his protégé, Ike would henceforth be on trial in the president's eyes until he produced a decisive victory in Tunisia.

It says much for the president's patience that he would have to wait six months for it. The worst thing about the Darlan deal, as it turned out, was that it didn't accomplish what Ike and Murphy had hoped for, and left the Anglo-American army struggling with the French instead of trying to reach Tunis before the Germans were there in strength. Ike's inexperience as a battlefield commander; the defects in supply, equipment, and training of his army; the failure of American and British commanders to cooperate with each other willingly; the geography and climate of Tunisia; and above all the amazing resilience of the Germany army and the Luftwaffe, even when outnumbered and at the end of a precarious supply line across the Mediterranean, all combined to produce exactly the kind of hard, slogging infantry battle, with high casualties and slow progress, that Ike had wanted to avoid. The opening weeks of the battle for Tunisia became a footrace, in which the Allied armies played the tortoise and the Germans the hare.

The "what ifs" of history are always fascinating, but futile. Perhaps had Patton been in command of the landings in Algeria, instead of those in Morocco, he might have seized the moment and driven his men straight into Tunisia—better yet, had Patton been Ike's deputy instead of the more cautious (and political) Mark Clark, he might have seen the vital importance of attacking the Germans before they had a chance to consolidate and begin to move west toward the Algerian border. Nothing could make up for the Allies' slow, flat-footed approach to the Tunisian battle. Admittedly, the unexpected strength of French resistance had come as an unpleasant surprise—though given Murphy's caution on the subject, it should not have—but all along the coastline from Algiers to Casablanca there now took place, after the initial landings, a process known in the British army as "regrouping," which basically means pausing after a major effort to dig in, count your supplies, set up your base and communications, polish your brass and your leather, brew a pot of tea, and wait for further orders. The German army did not regroup—German soldiers were trained to keep moving forward at all costs, and in this case they were moving forward into a vacuum.

Several other factors conspired to slow Ike down. One was the unwelcome pull of French politics and French colonialism. The French army was in North Africa not to fight the Germans (still less to fight the Allies), but to preserve the status quo in France's North African colonies pending France's recovery. This meant keeping the native (Arab-Berber) majority in each colony firmly under control, in Algeria and Tunisia as cheap agricultural labor for the European settlers (or *pieds-noirs* as they were known in France), in Morocco by a careful policy of divide and conquer among the warlike, colorful tribal subjects of the sultan of Morocco, many of whom had engaged in sporadic but bitter warfare with the French army and the French Foreign Legion throughout the 1920s and 1930s.

The arrival of American troops inevitably stimulated the Arabs' hopes for independence, and if there was one thing the French communicated loud and clear to Ike it was their fear of an Arab uprising. Ike himself was concerned about this—it was bad enough having to fight the French, without having to fight their Arab subjects as well—and was determined to secure his bases by garrisoning the French colonies with enough American troops to discourage rebellion or attacks from the desert tribesmen in Spanish Morocco. He even used this as a reason for making the Darlan deal, in a long cable to General Marshall in which he warned, "Without a strong French government we would be forced to undertake military occupation," and quoted Patton's estimate that in Morocco alone he would need 60,000 Allied troops "to keep the tribes pacified."[16] This was far in excess of the number of troops that France had needed to maintain order in Morocco; nor was there any convincing proof that "the tribes" were contemplating war. But concern about this kind of threat was diverting Ike from the whole purpose of the landings, which was to take Bizerte and Tunis swiftly before the Germans could, and then attack Rommel from behind as he retreated through Libya pursued (at a leisurely pace, unfortunately) by Montgomery. In the meantime, the only significant sign of the kind of trouble that Ike expected from the natives was as a result of Roosevelt's request for the abrogation of Vichy's anti-Semitic laws in North Africa. That request was praised throughout the West but caused consternation and anger among the Arabs, who regarded the Jews as their enemy.[17]

It was not just Ike's need to secure his bases and sort out American policy toward France and the Arab population that slowed things down. Nobody, it seemed, had had a clear picture of what Tunisia would be like in the winter, which was fast approaching—though a careful reading of *Baedeker's Guide to North Africa* might have helped matters. Much of North Africa, particularly Morocco, is desert, with oases, palm trees, camels, and snow-capped mountains in the distance, familiar to Americans because it was the colorful, exotic background for such movies as *The Road to Morocco* and *Beau Geste*; and Morocco had been for years the winter playground of the European rich and famous, including Winston Churchill, who called Marrakech "the loveliest place on earth." On the other hand, nobody had ever sought out Tunisia as a winter resort. Inland, away from the seacoast, it was harsh, forbidding country, freezing cold and windswept in the winter, with long periods of heavy rain that turned the entire landscape into a sea of thick, gluey mud, every bit as bad as the notorious *rasputitsa* of Russia's spring thaw. Tanks, vehicles—even those with four-wheel drive—men, mules, and parked aircraft would simply sink into the Tunisian mud, and soldiers eventually learned how to live in mud as if it were a hostile fifth element. Troops that had landed on the balmy shores of Morocco and Algeria wearing lightweight uniforms would shortly find themselves in need of a greatcoat, thick socks, and waterproof winter boots, and would curse "the brass" and the U.S. Army Quartermaster Corps, which had failed to provide them, more harshly than they cussed the Germans. Planning was no more reliable than supply: no one discovered until it was too late that the tunnels on the Moroccan railroad were too narrow by a few inches to permit rail transport of the American medium and heavy tanks landed there,[18] and as a result, tankers would initially end up confronting the fifty-eight-ton German Tiger with small, lightly armored reconnaissance tanks.

Throughout the Allied forces, British and American, progress was slowed by mistakes and oversights of this sort, many of which could no doubt have been solved by consulting the Free French officers in London, most of whom had spent substantial portions of their career in France's North African colonies. The responsibility for carrying out the initial thrust into Tunisia fell on Lieutenant-General Anderson, the dour Scot who took over command of the Eastern Task Force from the American Major General Ryder as soon as the troops were ashore in Algiers and it was no longer necessary to pretend to the French that they were all American. Anderson, whom Ike characterized rather mildly as "not a popular type," was nevertheless an aggressive commander, but the difficulties he faced were daunting. The distance from Algiers to Bizerte is close to 400 miles as the crow flies, with steep, jagged mountains coming right down to the sea, and at the time there were only poor roads and a single-track railroad in miserable condition. The French commanders in Tunisia were divided in their allegiance. The resident-general of Tunisia, Vice Admiral Jean-Pierre Estéva, remained loyal to Marshal Pétain and suspicious of Darlan's change of heart; the commander of the naval base at Bizerte, rendered confused and apathetic by the barrage of orders and counterorders from Vichy and Algiers, surrendered the base to the Germans without firing a shot; and the bey of Tunis, who ruled Tunisia, in theory, under French control, instantly pledged his loyalty to Berlin—while the army commander in Tunis, General Georges Barré, marched his troops, nearly 10,000 men, into the rugged mountainous country west of Tunis in the hope of joining up with the Allied forces; receiving ammunition, supplies, and tanks; and fighting the Germans.

Although great attention had been paid in planning Torch to give the French the impression that it was an entirely American operation, the troops on the far left of Ike's long line were predominantly—and unmistakably—British (as they would be on D-day), including a "floating reserve" still shipborne, one American and two British parachute regiments, and a mixed bag of British and American commandos. Now that French sensitivities could be ignored, thanks to Darlan, Anderson's forces would henceforth be known as the British First Army, and since they were the closest to the Tunisian border they would also be the first in the race for Tunis—indeed, they would serve as the pivot for the entire line of battle as American forces began to arrive from Oran and from Morocco. In the meantime, it was up to Anderson to get his scratch force across the frontier and into battle as quickly as possible, since he was under no illusions about how quickly and thoroughly the Germans would prepare their defenses in the forbidding terrain ahead, where every hill, gulley, and ravine offered innumerable opportunities for exactly the kind of stubborn and well-prepared defense in which the German army specialized. As everybody on the Allies' side was about to discover, troops in the German army didn't reach a point on the map, then sit down, light a cigarette, and wait for orders. Given half an hour, German troops would place an antitank gun and a couple of machine guns with a good field of fire exactly where they would do the most harm, lay a murderous minefield in front of them, then dig themselves in. They didn't need orders from an officer—a sergeant or a corporal would know what to do and make sure it was done quickly and efficiently, thanks to the harsh lessons taught by combat experience in Russia.

As a commander, Anderson was thorough and determined, rather than reckless and fast-moving in Patton's style. He was not showy; nor was he a flamboyant, colorful, interesting character like Montgomery; and his personality was marked by a degree of Calvinist pessimism unusual even among the Scots. Neither the forces available to Anderson, nor the attitude of the French, who were as apt to regard his troops as the enemy as much as those of the Germans, would be likely to encourage him. All the same, he set in motion an ambitious plan of attack for British First Army which had been intended to win him Tunis and Bizerte by November 12—as it happens, the day before both cities were actually taken by the Germans.

Opinions differ about whether a single strong thrust might have carried Anderson to Tunis before the Germans were fully organized, but it was never in the cards. Ike's plans called for a more conventional attack, and that suited Anderson, who was, as they say in Britain, a "belt and braces"

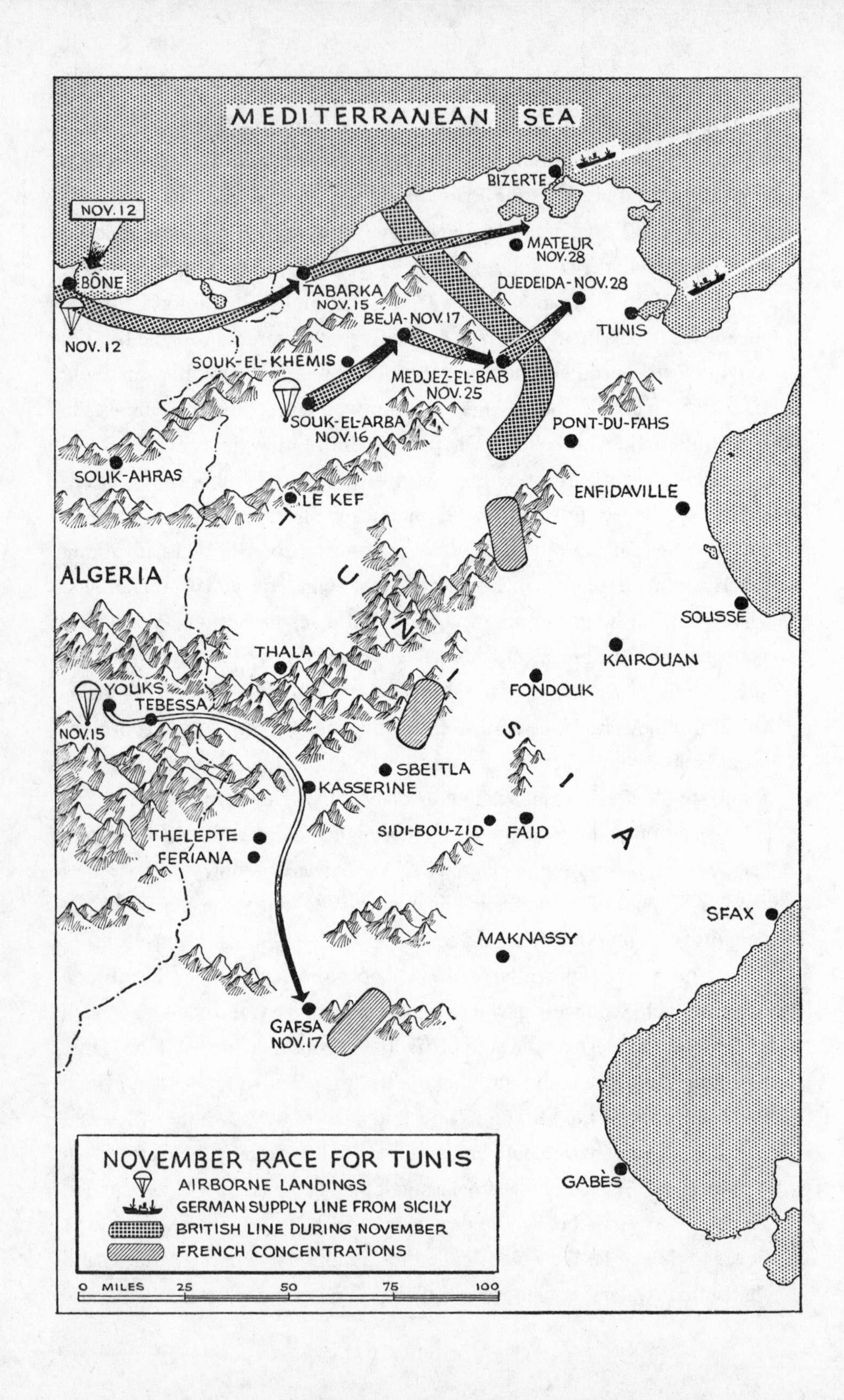

MEDITERRANEAN SEA
BIZERTE
NOV. 12
MATEUR
NOV. 28
BÔNE
TABARKA
NOV. 15
DJEDEIDA - NOV. 28
BEJA - NOV. 17
TUNIS
NOV. 12
SOUK-EL-KHEMIS
MEDJEZ-EL-BAB
NOV. 25
SOUK-EL-ARBA
NOV. 16
PONT-DU-FAHS
SOUK-AHRAS
LE KEF
ENFIDAVILLE
ALGERIA
T U N I S I A
SOUSSE
THALA
KAIROUAN
YOUKS
TEBESSA
FONDOUK
NOV. 15
SBEITLA
KASSERINE
SIDI-BOU-ZID
FAID
THELEPTE
FERIANA
SFAX
MAKNASSY
GAFSA
NOV. 17
NOVEMBER RACE FOR TUNIS
AIRBORNE LANDINGS
GERMAN SUPPLY LINE FROM SICILY
BRITISH LINE DURING NOVEMBER
FRENCH CONCENTRATIONS
GABÈS
0 MILES 25 50 75 100

man himself. Still, given the small number of troops available to him, and the fact that he had to rely for armor on the notoriously ineffective, slow-moving, temperamental British Valentine tanks, Anderson did his best to come to grips rapidly with the Germans, and very nearly succeeded. He began with a daring series of amphibious and airborne assaults to secure the coast road, landing a brigade of British infantry at the port of Bougie, about 120 miles east of Algiers; and using British commandos and a British airborne drop to seize the port of Bône, which was less than 100 miles from Tunis. Unfortunately, the troops sent on by sea from Bougie to capture the town of Djidjelli, only forty miles away, were swamped by heavy surf and unable get ashore—the Mediterranean once again demonstrating its ability to behave like the Bay of Biscay or the English Channel. As a result, fuel supplies could not reach the all-important airfield at Djidjelli, where the RAF Spitfires that had landed there sat idle for two days with empty fuel tanks, depriving Anderson of air cover. Kesselring was not the man to ignore that kind of opportunity, and the Luftwaffe pounded Bougie[19] unmercifully, sinking several ships and sending vital supplies to the bottom.

It was not an auspicious beginning, but four days later, as the commandos and paratroopers struggled inland and eastward from Bône—to which Kesselring's bombers had now switched their attention, gutting the town and almost completely destroying the vital port facilities—Anderson carried off not one but two airborne assaults, one British, one American, at a time when the use of paratroopers was still something of a novelty. On November 15 American paratroopers took the city of Tébessa, a key road and railway junction that commanded the important pass in the Western Dorsal Mountains, through which southern Tunisia could be reached. On November 16 British paratroopers seized Souk el Arba, less than sixty miles from Tunis. By November 17 British forces had made contact with General Barré's French troops in the hills overlooking Bizerte and Tunis, less than forty miles away, while a small force of American paratroopers had taken the city of Gafsa, within eighty miles of Gabès, one of the principal ports of southern Tunisia. Against the odds, Anderson had very nearly succeeded in securing Ike's main objectives by placing himself within striking distance of Bizerte and Tunis.

Reaching them was another question. Too much of Anderson's force "at the sharp end of the stick" consisted of lightly armed commandos and paratroopers spread out in "penny packets" all over a vast and forbidding landscape; he needed to bring up (and concentrate) the bulk of his forces quickly, as well as substantial reinforcements of infantry and armor from Algiers, and these arrived piecemeal, slowed down to a crawl by miserable roads, muddy tracks, poor maps, and appalling weather. It did not help that Ike had returned to Gibraltar, and was distracted by the intricate political maneuvers taking place in Algiers as Darlan sought to fortify the position he claimed as the legitimate representative, backed by the Allies, of Marshal Pétain, the only senior member of the Vichy government with freedom of action now that the marshal himself and the rest of his cabinet were for all practical purposes prisoners of the Germans. Algiers was awash with plots and factions—every French political group from left to right was present. Even the comte de Paris, the Bourbon pretender to the French throne, was there; indeed one of his aristocratic supporters was to play a key role in the rapidly unfolding drama over Darlan's legitimacy.

Impatient with the apparently insoluble communications difficulties of Gibraltar, where, in any case, he was out of touch with the battle, and tired of living in a damp tunnel underground, Ike insisted on moving his headquarters to Algiers immediately, despite protests from the communications staff that they couldn't get their equipment moved and set up in less than two weeks. Ignoring them—after all, they hadn't been able to make their equipment function in Gibraltar—he transferred his headquarters to Algiers on November 23, and took advantage of the journey to land at Oran on the way to inspect American troops. At Oran, he received a memorable example of North Africa's weather in late autumn—the mud was so thick that the only way his B-17 could be removed from the single paved runway so that other aircraft could land and take off was to have it towed by a tractor while soldiers walked alongside laying thick wood planks under the aircraft's wheels to prevent it from sinking into the mud.[20] And this was in "sunny Algeria," by the sea, not 750 miles farther east in the Tunisian mountains!

Nothing Ike saw once he arrived in Algiers was likely to lift his spirits. Though General Anderson was desperate for American forces, and had been pleading for them for days, the challenge of moving them forward rapidly on the inadequate railroad had proved insurmountable for the staff. The solution of moving them in half-tracks had been rejected out of hand because the journey would cost "half the useful life" of the tracks. Ike calmly cut this Gordian knot by ordering the staff to release the half-tracks, then checked into his hotel, where he was kept awake most of the night—as he would be for many nights to come—by the Luftwaffe, which was methodically bombing the port.

Still, one senses, reading *Crusade in Europe*, that Ike's spirits were lifted once he was out of his tunnel and installed at last on the same continent as his army. He seems to have welcomed the German bombs falling a quarter of a mile away from his hotel as a sign that he was "in action" at last, however modest the scale, and the incident of the half-tracks gave him a sense of the importance of his presence—nobody else had the ultimate authority to make things happen. After three days and nights of solving problems that nobody else could solve, he took off by car with Mark Clark to visit the battlefield.

Given the state of the roads in Tunisia, the fact that they went by car requires some explanation. The Allied air forces had performed a miracle in landing their aircraft on the relatively primitive French airfields, but much of their supplies and many of their ground crews had still to catch up with them. In addition, most of the radar and communications equipment had gone to the bottom when the ship carrying it was sunk during the German bombing of the port at Algiers that had kept Ike awake. The only paved, modern airfields in Tunisia were in German hands, so the Allies' fighter aircraft had a long flight from dirt fields in Algeria to reach the front, and very little time to patrol before they had to return home to refuel. Under these circumstances, despite the large number of Allied aircraft in North Africa, the Germans held, for the moment, air superiority over most of Tunisia. German ME-109s patrolled vigorously and in strength in the airspace to the west of Bizerte, while the supposedly obsolete JU-87 Stuka dive-bombers, which, with their fixed undercarriage and robust wings, were in fact ideal for operating off short, rough North African airfields, tormented Anderson's troops with constant attacks, and interdicted his already inadequate supply line. A lone B-17 carrying Ike would have been a sitting duck for German fighters.

Troops on the ground complained that they were not receiving any air cover, while airmen complained that they were being misused as a substitute for artillery. Like the ground forces, to many of whom the superiority of German tanks would come as a rude awakening, the airmen were also becoming aware that the P-40 and the early models of the P-47 and P-38 might not be as advanced as they had been told. One of them recalled seeing a dot in the sky behind him in his rearview mirror and watching it enlarge quickly into the unmistakable frontal silhouette of an ME-109G; he pushed his throttle all the way forward to full power, only to have the German aircraft pass by him effortlessly, so close that he could see the pilot's face, then climb away from him contemptuously into the distant clouds. At that precise moment, the American pilot decided "it was going to be a hell of lot longer war than I'd thought."

A lot of other people were having the same reaction, one of them being Ike, as he and Mark Clark made their way forward slowly by road, "long stretches of which practically disintegrated." All along the crumbling road men slogged through the thick mud while looking up at the sky fearfully for low-flying aircraft, and only "occasionally" would the aircraft turn out to be friendly. To either side of the road was "evidence of incessant and hard fighting."[21] They finally caught up with Anderson at Souk Ahras, about forty miles inland from Bône, and learned that his advance troops, though very thin on the ground, were already approaching Souk el Khemis and Béja, and in contact with General Barré's force, less than fifty miles from Tunis. None of this, however, was apparently enough to make the meeting with General Anderson a cheerful one. Barré's Frenchmen were poorly equipped and desperately short of ammunition, and therefore more of a liability for the moment than an advantage (they were also reluctant to take orders from a British general)—and the Germans were reinforcing through Tunis and Bizerte more quickly than Anderson was being reinforced from Algiers. Kesselring had just appointed Lieutenant General

Walther Nehring to command German and Italian forces in Tunisia; and Nehring, who had fought both on the Russian Front and with the Afrika Korps, was not a man to waste words, or time—he ordered every position held "to the last man," and he counterattacked vigorously. Nehring, like all generals who had served under Rommel, had learned from the master a kind of *Fingerspitzengefühl*, or sensitive instinct, for the moment when the enemy's attack was about to falter or weaken; and sensing that Anderson had outrun his supply line and that his mostly green troops were discouraged, he struck hard at British and American forces. By the time Ike was back in Algiers—the sight of the Allied forces advancing slowly along the road in Tunisia had convinced him that it was time to stop worrying about using large numbers of Americans troops in Morocco and Algeria to "protect" the army's rear and overawe the Arabs on behalf of the French, whatever Patton might think about the danger of the native "tribes" rising—Nehring had already attacked north and south of the road to Tunis with fresh troops newly arrived from Italy, including crack German paratroopers. A spirited attack by American tankers the next day caught the Germans by surprise and managed to reach a point just five miles from the outskirts of Tunis, momentarily shaking Nehring's nerve, but they were forced to withdraw after destroying a packed Luftwaffe airfield; and by Thanksgiving Day a stalemate had settled in—just the thing Ike had feared most.[22]

Recognizing that he would be in Algiers for some time—as it turned out, he was there longer than even *he* suspected—Ike sent Harry Butcher off to find him a house, and reconstituted his military "family." Mickey McKeogh arrived, with Telek, the Scottie that Ike shared with Kay Summersby, and yards of red tape were cut to facilitate the arrival of Kay herself. She was still essentially a British civilian in a fashionable paramilitary uniform, who required a passport and a visa to travel to Algeria, but Tex Lee obtained these in record time. Kay would later describe the house Butcher found for Ike as ugly and cramped, and Ike himself complained to Mamie about its enormous, freezing bathrooms—the largest, coldest rooms in the house—and its idiosyncratic French plumbing, of the kind that had puzzled and infuriated him during his tour in France in 1929. But "the villa," as everybody called it, had a view over the harbor and at least offered him some sense of privacy. He described it to Mamie in a letter home as "a big-looking but most uncomfortable place on a hillside overlooking the city . . . a sort of combination of Moorish-Spanish architecture with French habits controlling the inside. Awful."[23] In photographs, it doesn't look quite that bad—it looks just like what it was, the home of a nouveau riche French businessman, with comfortable furniture in white slipcovers, Oriental carpets, lots of Moorish tile work, and (as mentioned above) its view over the harbor, which was the fashionable place to live in Algiers. But those who read Ike's letters to Mamie with attention will quickly notice that he was always careful to paint his surroundings as grimly as he could, as if to emphasize that she had nothing to be jealous of, that she wouldn't have liked where he was living even had she been able to join him there, and that she was more comfortable and better off where she was. Ike does not exactly complain—he merely paints a dark picture for Mamie's benefit, a kind of reflexive marital gesture. In the same spirit he had played down his late nights with Winston Churchill in England and his "duty call on the king" in writing to her, even though both were examples of his very special status.

She herself was careful not to complain to him about her loneliness, her poor health, or her resentment at being cut adrift by the Army once he was overseas. There is a certain defensive quality in Mamie's biographies on the subject of her drinking; the biographers note that many of the Army wives who were her friends—and especially Harry Butcher's wife, Ruth—were heavy drinkers even by the standards of the day. Milton Eisenhower, Ike's "kid brother," who had become a well-connected Washington bureaucrat, later defended Mamie (rather tepidly) by saying that she "drank no more than anyone did in those days," but he apparently felt impelled to warn her at the time "not to drink too freely in front of strangers,"[24] a tactless piece of advice which she must surely have resented (and which makes it clear that Milton was his outspoken mother's son). Susan Eisenhower, Mamie's granddaughter, attributes Mamie's problems at the time to "Ménière's syndrome, a severe inner-ear disorder that is caused by pressure on a nerve and results in periodic attacks of dizziness . . . and vertigo,"[25] and so it may have been. But rather as with the attempts on the part of Ike's biographers to play down his relationship with Kay Summersby, the result is merely to make one suspect that matters were worse than they may in fact have been. Mamie's belief that she shouldn't be seen going out to parties while her husband was at war—and perhaps the memory of how angry he had been when she had tried to enjoy herself in Washington while he was alone in the Philippines—led her to stay at home with her few women friends, of course sparking off rumors that her drinking problem was so bad that she couldn't go out socially. Life in a small, unair-conditioned apartment in Washington, D.C., in wartime, with the company of a few military wives whose husbands were also overseas, cannot have been easy or much fun for Mamie, and surely cannot have been made easier by her strong sense of privacy, which made her fear the attention of the press; or by her consciousness of Ike's unique position and growing fame, which put her in the limelight everywhere she went—something she had hated and feared from the very beginning of Ike's rise. She knew too that her beloved son John would be going overseas as soon as he graduated from West Point, and that apart from playing cards with her friends, doing volunteer work with the Red Cross, and waitressing at the canteen for soldiers, sailors, and marines—which exposed her to the curiosity of total strangers—her life would go on like this until the war was over. She also knew her husband too well to imagine that he would come home for anything but the shortest visit until the war had been won. Of course it would be foolish to compare her unhappiness to the tidal wave of suffering that was spreading across the globe, and to do her credit, Mamie never tried to do this. But living with poor Ruth Butcher (whom Harry would dump as soon as he returned home from the war) while Ike dined with the British prime minister, hobnobbed with the king, and lived in a big house surrounded by devoted aides and servants cannot have been pleasant for a woman to whom marriage was the center of her life.

Throughout the first weeks of December 1942, the Germans strengthened their defenses around Bizerte and Tunis and carried out a series of attacks that pushed the Allied forces, hampered by divided command structures, lack of supplies, bad weather, and German air superiority, back from El Guettar in the south to Cape Serrat in the north along a front of more than 300 miles, producing a situation that resembled, in some ways, the western front of 1915 to 1918 on a smaller scale. The ambitious trench works of World War I were impossible to duplicate in the rocky Tunisian soil, but men dug themselves into squalid foxholes protected by barbed wire and mines and settled into static warfare, punctuated by fierce attacks on both sides in which the bayonet was as useful a weapon as the tank. General Nehring, whose momentary loss of nerve had alarmed the usually unflappable Field Marshal Kesselring, was replaced by Colonel General Hans-Jürgen von Arnim, a veteran commander from the eastern front, whose nerve was judged unlikely to falter under any circumstances, and the forces under his command were renamed the Fifth Panzer Army.

If the tactical picture looked gloomy seen from Algiers (or from London and Washington), the larger strategic picture was also cause for concern. Arnim's Fifth Panzer Army had a solid hold on the eastern half of Tunisia, its front line following almost exactly the heights of the Eastern Dorsal Mountains which run from north to south the length of the country.[26] To the south and east of Tunisia, General Montgomery's British Eighth Army, having defeated Rommel at the battle of El Alamein, on the Egyptian frontier, almost 1,500 miles away, was now pushing the Deutsche Afrika Korps back across Libya toward Tunisia, at what some of Monty's critics took to be a snail's pace. If Ike could not defeat Arnim before the latter was joined by Rommel, the German forces in Tunisia would be doubled, adding to the Fifth Panzer Army perhaps the most experienced and battle-hardened troops in the German army, under the overall command of its most daring and resourceful general. If anybody could make a good thing out of a defeat and a long retreat, it was Rommel. Once he reached Tunisia with his army, he would have the benefit of short interior lines, formidable natural barriers, and air cover from Luftwaffe bases in Sicily and Sardinia, as well as two deepwater harbors at Bizerte and Tunis, and two smaller, shallower ones at Sfax and Gabès, for his supply line. Of course in the long run he would be surrounded, holding out in a pocket, with his supply line in the hands of the never altogether reliable or enthusiastic Italian merchant marine; but he, or any competent and determined commander, could hold out for a very long time.* This would be particularly probable if Monty, advancing from the south; and Ike, attacking from the west, failed to cooperate, or failed to coordinate their attacks—something that must have seemed a very likely bet to those who had had any experience with Monty on either side of the war.

In the last week of December, Kay Summersby arrived. The troopship on which she had been sailing was torpedoed by a German submarine, and went to the bottom with both of her Vuitton trunks; and she spent a bad night in one of the ship's lifeboats, pulling at an oar with a nurse, and a worse day bobbing in the broiling sun, tending to the injured and wounded, until they were finally picked up by a British destroyer and landed at Oran. Mentioning Ike's name was enough to get her a telephone, and within minutes Ike was on the line, relieved and happy that she had survived, and promising that he would send a plane to pick her up and bring to Algiers. The next day she was reunited with Ike (as well as with her fiancé, newly promoted to lieutenant colonel), and with Telek at Villa dar el Douad, Ike's tree-shaded house with its spectacular view over the Mediterannean. She replenished her wardrobe in the shops of Algiers, paying particular attention, if her second memoir is to be believed, to her lingerie—Algeria was part of metropolitan France, rather than a colonial possession, and received its fashions from Paris; and the only underthings Kay could find in her size were "see-through" bras "cut out at the nipples" and lace panties in the shape of a fig leaf. She spent Christmas Eve with Ike at a dinner party given by General Smith—turkey, plum pudding, and champagne. The evening was only slightly spoiled by the big news of the day: Admiral Darlan, "the little guy," as Mark Clark referred to him contemptuously, had been assassinated.

This was something less than a tragedy to most of those sitting around the table—indeed, it must have seemed like a blessing in disguise. Darlan

* In fact, Rommel was opposed to the idea. He believed it made better sense to abandon North Africa to the Allies altogether and withdraw the German forces there to Italy or southern France before they were bottled up in Tunisia and destroyed; but even a visit to Hitler's headquarters in East Prussia to plead his case directly with the Führer accomplished nothing except to undermine Hitler's faith in Rommel, and vice versa.

had served his purpose by ordering the French forces to stop fighting the Allies, and was now an embarrassment who moreover had nothing further to offer them. He had been repudiated by the marshal, he was anathema to the Gaullists, and with the scuttling of what remained of the French fleet he had no further card to play. "He was a weak character," Ike wrote to Mamie, on the subject of Darlan (he did not mention the recent arrival of Kay in his letter), "at least nothing more than an opportunist, but so far as I could ever find out, he played square with us here."[27] Who ordered the admiral's death is a question that still stirs speculation, over sixty years later, and remains unsolved. His assassin, an idealistic young French aristocrat named Fernand Bonnier de la Chapelle, was a member of a royalist group, but there was no convincing reason why the death of Admiral Darlan would help in any way to bring the comte de Paris to the throne as king of France. Bonnier de la Chapelle had also been a member of one of the resistance cells organized by Robert Murphy; this fact prompted the suspicion that having used Darlan ("They squeezed me like a lemon," as Darlan had complained), the Americans, with or without British connivance (*l'Albion perfide* again), had arranged to get rid of him, but no proof exists. Bonnier de la Chapelle, believing that his execution would be faked and that he would eventually reemerge from hiding as a national hero, was given a hasty and perfunctory trial, found guilty, and, surely to his surprise, shot by a firing squad—his coffin had been prepared before the trial had begun. The answers to any questions on the subject of Darlan's murder went to the grave with him, as they were no doubt intended to. Darlan received a full state funeral, with Ike and his staff in the congregation at the cathedral in Algiers, as well as Harold Macmillan, recently arrived from London as minister of state and Winston Churchill's adviser to Ike. Macmillan, with brutal political realism, gave Darlan his ultimate epitaph: "Once bought, he stayed bought."[28]

These were among the kinder words spoken about Darlan after his death, but they were true enough. His assassination, however convenient for the Allies, left the French without an authority figure in North Africa. Apart from General Barré's 10,000 men in Tunisia, who for the moment were merely adding to General Anderson's logistic woes, the bulk of the

French forces in North Africa remained inactive while their generals pondered the advantages of joining the Allies or remaining loyal to Vichy (even though the latter was already a rapidly sinking ship), and tried to choose between the leadership of General Giraud and that of General De Gaulle. Despite Marshall's cabled advice to "leave international diplomatic problems to your subordinates and give your complete attention to the battle in Tunisia" (easy enough to suggest from Washington), an ever-increasing amount of Ike's energy was devoted to the tangled political problems of French colonial administration and military command now that Hitler, in a fit of anger or petulance, had effectively ditched Vichy by occupying the whole country, leaving French officials overseas to decide for themselves to whom or to what they owed loyalty.

In the meantime, a dangerous situation was rapidly developing. Montgomery, whose British Eighth Army in the desert was slowly driving Rommel's Afrika Korps back toward Tripoli in a series of carefully prepared set-piece battles, to Rommel's despair, was both out of touch and out of sympathy with the Allied forces in Tunisia. "The party in Tunisia is a complete dog's breakfast* and there is an absence of good chaps there,"[29] Montgomery wrote to General Brooke in his usual brisk, acerbic style, and he was increasingly anxious about the lack of any news of progress from Ike. It had always been Monty's opinion that the only sensible way to take the all-important port of Tripoli, and cut Rommel off from his supply line, was to do it from the west ("by Eisenhower's forces attacking through Tunisia").[30] Indeed, the main objective of Torch had been, from the very beginning, to take one by one the ports through which the German army

* A favorite phrase of Monty's, meaning a mess. As a young man the author met Field Marshal the Viscount Montgomery of Alamein (as he became after the war), when I was at a boarding school in Switzerland. Monty occasionally visited the school—he liked the company of young people—and sometimes asked the older British boys to have tea with him. One learned quickly to understand and appreciate his trademark phrases and his sharp-tongued, acid descriptions of his colleagues in the war ("hopeless" and "dim" were two of his favorite adjectives). His clipped, high-pitched voice was unforgettable, as were the piercing eyes that missed nothing, and his loud, mocking laugh. He was not an instantly likable person—in fact he rather resembled a bossy scoutmaster, with a savage sense of humor, a strangely inappropriate taste for smut and practical jokes, and occasional sentimental moments that as boys we found uncomfortably out of place in a war hero. One had the impression that he enjoyed these tea parties a lot more than we did.

in North Africa was supplied and squeeze Rommel between the western army (Ike) and the eastern army (Monty) like a nut between the jaws of a nutcracker. Monty was no admirer of General Anderson,[31] and he rightly interpreted the silence from Algiers as proof of Anderson's lack of progress in Tunisia. This was dismaying. Even after Monty had taken the Libyan port of Benghazi, which the Germans had largely destroyed and rendered useless, he was still 760 miles away from Tripoli, and would have to fight his way there with a supply line that stretched almost 1,500 miles back to Cairo. "We must find out what is going on in the West," he pleaded to General Sir Harold Alexander, British Commander-in-Chief in the Middle East, deeply concerned that "the Western Army" might be "seen off"* by the Germans, leaving Arnim and Rommel free to combine and attack the Eighth Army before it had a chance to build up its supplies after its long advance across Libya.

When one considers the importance that disagreements over strategy between Ike and Monty would have in prolonging the end of the war in Europe until May 1945 (entailing the loss of hundreds of thousands of lives), it is worth bearing in mind that the root of the trouble lay here, in the last weeks of 1942, when it first became apparent to Monty that the "Western Army" was stuck. Ike was not swooping down out of the Tunisian mountains to take Tripoli and cut off Rommel, and it began to seem to Monty that he himself would be obliged to take Tripoli and then advance into Tunisia to rescue Ike. Considering the number of men Ike had at his disposal, and Ike's direct access to virtually unlimited supplies from the United States that could be sent across the Atlantic to Moroccan ports, Monty not unnaturally resented this.

Even without this basic difference of opinion it is unlikely that Ike and Monty would ever have really warmed, and not just because Monty had told Ike to put out his cigarette the only time they had met so far. Ike was gregarious, seldom had anything bad to say about anyone, and, on the surface at least, was relaxed and good-natured, or at least liked to appear so. Monty was a loner, arrogant, vain, unforgiving, professionally brilliant, and

* A phrase of Monty's signifying "defeated."

utterly convinced that he was always right. Whereas Ike's good humor was genuine, unaffected, and affectionate, Monty's was cruel and mocking and always carried a sting. Matters would not be helped by the fact that Monty had all the combat experience Ike lacked—as a captain in the First Battalion, the Royal Warwickshire Regiment, Monty had been shot in the chest leading his platoon in an attack on the German trenches at Le Cateau in 1914; lay for four hours between the lines in no-man's-land under a pile of British dead; was shot a second time, in the knee by a sniper; and came home to be awarded the Distinguished Service Order (a decoration for gallantry in combat reserved for officers, and only one step below the Victoria Cross).[32] He recovered from his wounds, returned to the front, and served with distinction through some of the worst battles of World War I, ending the war as a lieutenant-colonel. His dislike of anybody's smoking around him was because of his lung wound, rather than a mere personal aberration or eccentricity, but it is typical of Monty not to have courted sympathy by explaining this—he never minded making himself unpopular; indeed he reveled in it.

Monty was a complicated and secretive character. People would remark on the fact—he meant them to—that he had a portrait of Rommel hung on the wall of his caravan (or trailer) at his tactical headquarters, above the desk, where he could look at it while he worked. But the painting he chose of Rommel (by the German artist Willrich) is a shocking one, not at all the bluff, energetic, square-faced, confident general one sees in most photographs and official portraits of Rommel. Below the trademark forage cap and sun goggles, Rommel's face is thin, lined, and careworn, with the skin drawn tightly over the bones, that of a man who has been pushed far beyond his emotional and physical limits, with haunted, empty eyes that look directly at the viewer, as if they had seen one battle too many. Because he had been promoted to the rank of field marshal by Hitler so young, his fellow German field marshals often referred to Rommel contemptuously as the *Bube-marschall* (the baby marshal), but in this painting he looks like an old man. In the background, far away on the horizon, thick black smoke rises from burning tanks. It is impossible to know if the tanks are Allied or German, but the expression on Rommel's face suggests that it no longer

The portrait of Rommel Monty kept above his desk.

matters to him much—either way, men are burning to death out there, because of his orders. Monty worked facing this picture of Rommel, as the press endlessly reported; but it was an introspective, defeated Rommel he was looking at, not the vigorous Rommel of legend. It was not enough for him to have beaten Rommel at El Alamein; it was as if he needed to possess Rommel's soul, and perhaps also felt a certain sympathy for his great opponent. There were hidden depths behind the brash, self-confident pose Monty presented to the world.

Both Ike and Monty fiercely craved adulation, professional success, and fame, Ike for the most part secretly, Monty unashamedly. Neither of them took criticism easily, and both of them were impatient of people with minds slower than their own. On the other hand, Ike had a remarkable ability to cope with even the most difficult of Britons (with the exception of Monty), while Monty remained, in a very insular way, critical of Americans in general, and unable to disguise the fact that he thought they were "soft," amateurish, poorly trained, and worse led, and he had neither curiosity about nor interest in America. Also, Monty's manner—his clipped speech, his air of superiority, and his eccentric clothes—might have been calculated (though it was not) to rub Americans the wrong way. Americans related more easily to "dressy" British generals like Alexander than to Monty, who was apt to survey the battlefield in baggy corduroys, suede chukka boots, an old khaki sweater, and a black beret, while carrying a flapping unfurled umbrella. Monty's persona was, in fact, both self-indulgent and a way of turning himself into a popular "character" for the benefit of British troops and the British public—he had no idea of the effect it had on Americans, and did not much care in any case what they thought. That the Allied war effort in Europe would depend, to a remarkable degree, on the working relationship between two such very different men was at once amazing and frightening, and it is to Ike's credit that he managed to contain his feelings on the subject of Monty (except in private) until long after the war was won.

Monty's skepticism about events in Tunisia was shared at even higher levels. Churchill grumbled that Ike's army was all tail and no teeth (and there was no denying that given the large numbers of men at his disposal, Ike had very few of them actually fighting in Tunisia) and that he had intended North Africa as a springboard, not a sofa. Roosevelt, still smarting about the criticism he had received over the Darlan deal, for once said no to a request from General Marshall and refused to promote Ike to the rank of general until he had produced some kind of victory; consequently, the Allied Commander-in-Chief continued to wear three stars while many of his British subordinates were entitled to four. In London, General Alan Brooke confided to his diary, "I am afraid that Eisenhower as a general is hopeless! He submerges himself in politics and neglects his military duties, partly, I am afraid, because he knows little, if anything about military matters. I don't like the situation in Tunisia at all!"[33]

To do him credit, Ike liked the situation in Tunisia no more than Brooke did, and perhaps less, since it was at least in part of his own making. The news that Churchill and Roosevelt were discussing what would soon be called a summit conference to be held in Casablanca cannot have cheered him up, since the failure to take Bizerte and Tunis rapidly would surely be on the agenda. (Stalin declined to attend the summit, saying that he could not leave the Soviet Union, an excuse which, for once, may have been true, since the Battle of Stalingrad was at its height.) Under the circumstances, not surprisingly, Ike ignored Marshall's advice to stick to his desk and went up to the front more often, hoping to gain control of the battle. What he saw there was deeply discouraging. The Allies' line ran nearly 400 miles from the Mediterranean in the north to Tébessa in the south, but several ambitious attempts to attack in strength in the north and push the Germans back were sharply defeated, and by Christmastime General Anderson had his hands full merely trying to hold on to the line to which his forces had withdrawn. Less as a result of planning than by accident, his left was held by British troops, the right by Americans, and the center by General Barré's French. This was not a convenient or a happy arrangement. Barré's troops were woefully short of supplies and ammunition and had no modern antitank guns or radio communication equipment; nor could Barré reconcile himself to taking orders from Anderson. General Marshall's old friend General Fredendall, in command of the American troops on the right, was already showing signs of erratic judgment and a marked reluctance to put his own life at risk that would soon undermine the confidence of his own troops.

It is easy to be wise in hindsight, of course, but one cannot help wondering how it was possible for Ike to ignore Fredendall's problems when they seem to have been obvious to everybody else, and why Ike did not replace him then and there with Patton, who was wasting his time in Morocco hunting wild boar, riding an Arab stallion provided by General Noguès, and waiting for the Rifs and Berbers to rise in colorful revolt, as in Sigmund Romberg's *The Desert Song*. The answer is probably that Ike had not yet learned to be ruthless—he was still trying to be a nice guy, at least toward West Pointers of his own generation. But great commanders are never nice guys—like a prime minister, a commander needs to be, in Gladstone's telling phrase, a "good butcher," willing to sacrifice even a best friend without hesitation when it becomes necessary. This willingness does not always come as quickly or easily to a military man as it does to a politician. Even the young Bonaparte had to learn to be as ruthless with the generals who served under him as he was toward the enemy. And reluctance to be ruthless was Grant's greatest—perhaps his only—weakness as a general, until he learned to rely on John A. Rawlins, his chief of staff, to swing the cleaver on his behalf. Ike would come to rely, in much the same way, on "Beetle" Smith, but the learning process was long and unhappy, and it needed a full-scale calamity to drive the lesson home.

This was not the only lesson he had to learn. When Anderson told Ike he wanted to give up Medjez-el-Bab, which he felt he did not have enough troops to hold against increasing pressure from the Germans, Ike took one look a the map and "forbade" it. Medjez-el-Bab might not look like much of a town on the ground, but if there was one thing Ike could do, it was read a map. Gettysburg hadn't been much of a town, either—the only thing it supposedly had to attract Lee's army in 1863 was a warehouse full of shoes—but a look at the map showed Lee that every major road in the area met at Gettysburg. And so it was with Medjez-el-Bab: this was a road center, a railway junction, and the connecting link with the French forces on Anderson's right. Giving it up might help Anderson to straighten his line, but this would isolate the already shaky French, and therefore weaken his center, and it would be necessary to retake Medjez-el-Bab in order to launch an offensive. Ike ordered it held at all cost and took "personal responsibility" for the fate of the garrison. This too was part of his learning experience in generalship. Up until now, he had given broad orders—men had died as a result of them, but only because all movements in war result in a certain unavoidable number of deaths. Now he was commanding a battle. Knowing Anderson as Ike did, he can have had no doubt that he was committing the men in Medjez-el-Bab to stand and die there, rather than retreat to better or safer ground, just as Grant had refused to order his army to fall back after the first bloody day in the Wilderness battle against Lee in 1864. Ike was determined that Medjez-el-Bab would be held; many men, British and American, would die to hold it; and that was that. With this decision, Ike the staff officer took the first, crucial step to becoming Ike the general. He had seen, in one glance, what had escaped the attention of even so seasoned a general as Anderson,* and then given an order for which he took full responsibility. Even critics of Ike's generalship, such as Monty and Brooke, had they been there, would have recognized that he had, at last, with that one order, joined the ranks of fighting generals. Fighting generals have blood on their hands—it is an inescapable part of the profession.[34]

He had not yet joined the ranks of successful generals, however. This would be made clear to him at Symbol, the Casablanca conference forthcoming in January, where even the presence of his mentor General Marshall was not sufficient to prevent President Roosevelt from showing impatience with progress in Tunisia. Fortunately for Ike, there were graver problems to occupy the president and the prime minister than the failure to capture Tunis and Bizerte quickly.

In today's age of jet travel, in big, pressurized, comfortable, quiet airliners fitted with all the "mods and cons" from seats that recline into beds to instant communication, it is difficult to describe the danger, discomfort, and sheer novelty of wartime long-distance air transportation in 1943. Moreover, Roosevelt and Churchill were both men in their sixties, one of them confined to a wheelchair, the other more robust but overweight and short of breath. Though nobody knew it but his doctors (who may not have shared the knowledge with their patient), Roosevelt was already suffering from the cardiovascular problems that would kill him in two years; and Churchill had already had a heart attack, in 1941 (while he was a guest in the White House), that had been kept secret from him—and from the rest of the world—by his doctor, Sir Charles Moran. Both Churchill and Roosevelt drank more than was good for them, were heavy smokers, and for obvious reasons were subject to enormous stress and anxiety, which, it must be said, they concealed with singular grace and good humor in public. Neither of them would probably have been permitted by doctors today to undertake a long, noisy, exhausting journey by plane, in the age before pressurization made flying comfortable, or safe for elderly chain-smokers.

* Experienced though Anderson was, Monty, with his usual cutting sarcasm, described him as "a good plain cook," in English terms a strong put-down.

Roosevelt flew to Casablanca in the strictest of secrecy, for obvious reasons beyond his natural glee at eluding the press and his fondness for springing surprises on people. He traveled by many stages—the journey took him five days, starting with a train from Washington, D.C., to Miami. He then went by flying boat to Brazil by way of Trinidad; then took a flight across the Atlantic to Gambia, and from there to Casablanca, in the first "presidential" airplane, an olive drab Douglas C-54 four-engine Skymaster transport (the military version of what would become the DC-4), which required a long wooden ramp so that he could be wheeled up to and down from the door.[35]

The prime minister flew from England in a converted RAF B-24 Liberator four-engine bomber in conditions of extreme cold and discomfort, and in some danger of attack by German aircraft. His aircraft, like Roosevelt's, had to keep well below 10,000 feet, since above that height oxygen masks would have to be worn, which meant that it couldn't "fly above the weather" as modern aircraft do, and traveled at speeds of less than a third those of today's jetliners. Churchill's physician, Sir Charles Moran, who traveled with him, reported that he and Churchill slept on two mattresses laid side by side in the tail of the bomber, and that when the primitive heating system failed, he woke to find Churchill trying to stuff blankets against the fuselage to stop the freezing drafts. "He was shivering," Moran wrote later; "we were flying at 7,000 feet in an unheated bomber in mid-winter. . . . The P.M. is at a disadvantage in this kind of travel, since he never wears anything at night but a silk vest. On his hands and knees, he cut a quaint figure with his big, bare, white bottom."[36] General MacArthur would remark of Churchill's journey that at the prime minister's age it ought to qualify him for the Victoria Cross, and while that is something of an exaggeration, typical of the general at his most effusive, there is no question that both leaders' journeys required courage, amazing powers of endurance, and a good reason.

The reason for the conference was simple but urgent. The United States and Britain had still not formulated a joint strategy for waging the war, and inevitably Stalin was growing impatient. The "emperor of the east" and the "emperor of the west," as Harold Macmillan referred jocularly to

Churchill and Roosevelt, were well aware of Stalin's displeasure and suspicion, and of the need to engage British and American forces on the European continent as quickly as possible, but there was no agreement as yet between them about when and where to begin the process. The president, carefully watched over by General Marshall to make sure he didn't weaken on the subject, still favored a landing in France as soon as possible (a variant of Sledgehammer), whereas the prime minister still urged a less ambitious and risky landing in the Mediterranean—Sicily, Italy, or the Balkans. There was bound to be contention on the subject, and Churchill was determined to have it out with the president—when he was advised to continue being tactful with American feelings after Pearl Harbor, he had replied, "Oh, that is the way we talked to her when we were wooing her, now that she is in the harem, we talk to her quite differently."[37] In the event, in the presence of Roosevelt, he did his best to restrain himself, but there were in fact profound differences between the two powers that could not be papered over.

The main issues to be settled were the crisis in the North Atlantic (the Allies' shipping losses to U-boats reached nearly 1 million tons in the month of November 1942 alone, a rate which, if continued, threatened to starve the United Kingdom and cripple the Allied armies); the commitment to an Allied "combined bomber offensive" against Germany; the degree to which American resources and troops should be split between the European and the Pacific wars; the solution to the controversy between Giraud and De Gaulle; and above all the decision about where Allied forces should attack the Germans next, after Tunisia had been won. Other areas of disagreement included Roosevelt's desire to have China treated as one of the great powers (an idea to which Churchill was instinctively opposed, dismissing the Chinese contemptuously as "five hundred million pigtails"); Churchill's hope of bringing Turkey into the war on the Allies' side; and the future of Tube Alloys, the code name for the atomic bomb program.

The British came better prepared—Churchill had sent ahead a ship, HMS *Bulolo*, converted into a floating information and communications center, and they were thus in a position to buttress their arguments by producing plans and information papers at a moment's notice, to the anxiety and envy of the Americans. One American remarked ruefully, paraphrasing Caesar, "We came, we listened, we were conquered."[38] And indeed, on the surface, the British seemed to have won most of the points they cared about: the war against Germany would continue to take priority over the war against Japan (to the disgust of Admiral King of the United States); Allied forces in North Africa would be landed in Sicily after the Germans had been defeated in Tunisia, and from there would attack Italy; the RAF Bomber Command would attack German cities by night and the USAAF Eighth Air Force would attack "strategic targets" by day; the production of escort vessels would be "maintained or increased" to defeat the U-boats in the North Atlantic; and the president and the prime minister would undertake to force the two feuding French generals to shake hands.

Sledgehammer—now in the form of seizing a "bridgehead" in Brittany—was kept on life support as a gesture toward the Americans' desire to land on the continent in 1943; but in fact everybody recognized that Husky, a major amphibious landing in Sicily, would almost certainly move Round-Up, the full-scale invasion of Europe, ahead to 1944, if for no other reason than the shortage of landing craft. Marshall swallowed his disappointment manfully, and made a mental note that in the future the United States would arrive at conferences with all the plans and information necessary to support the positions it intended to take.

Ike was asked to attend the conference for only one day, basically to explain why things were moving so slowly in Tunisia to a fairly critical audience of the Combined Chiefs of Staff. His B-17 lost two of its four engines flying over the Atlas Mountains, and he was obliged to spend part of the flight standing at the bomber's open doorway with his parachute on waiting to bail out. This was surely the reason why Roosevelt remarked, on seeing him, "Ike seems jittery."

Casablanca was not a happy conference for Ike. The Combined Chiefs of Staff must have looked to him very much like a hanging jury; and he also had a private heart-to-heart talk with General Alan Brooke, just the two of them, in which Brooke pointed out all the flaws in Ike's plans for defeating the German forces in Tunisia. ("As a result of our talk a better plan was drawn up," Brooke commented afterward in his diary.) Given Brooke's acerbic and impatient manner, this talk cannot have been a happy experience for Ike. He had a long talk with the president, and while he presents a very sunny picture of it in *Crusade in Europe,* in fact Roosevelt pressed him hard on the date by which he expected to have defeated the Axis forces in Tunisia, and finally pinned a reluctant Ike down to May 15. Reading between the lines, one can infer that the president had neither forgotten nor as yet forgiven the heat he had been obliged to take on Ike's behalf over the Darlan deal, and that he was still waiting impatiently for a victory to justify it in the eyes of his critics. Behind the reflexive affability and the good humor that came naturally to Roosevelt whenever he could get away from the White House and the demands of politics, he expected Ike to deliver.

That was not Ike's only challenge at Casablanca. Before he had even left the conference to return to Algiers, the British had produced a scheme for "unity of command" in the Middle East, which, while it sounded perfectly reasonable, was in fact a well-disguised end run to take the control of the fighting in Tunisia out of Ike's hands. Since Montgomery's Eighth Army would shortly be advancing into Tunisia from the south and making contact with Ike's forces in northern Tunisia, it was undeniably important to put the entire battle under a single commander before the two armies met, rather than continue to have an American commander in Algiers and a British commander, General Sir Harold Alexander, in Cairo. The British would naturally have preferred to see Alexander in that role—the prime minister liked and admired him; he was a "fighting soldier" with a brilliant record in both world wars; and his charm and good manners had an emollient effect on everybody who came into contact with him, even Monty. But they also recognized that any attempt to place a British general, however universally respected, in sole command would provoke a difficult and unpleasant struggle with the Americans, and one the British would surely lose. They therefore made the tactful suggestion that Ike himself should be given overall command, with General Alexander serving as his deputy and also commanding the ground forces, Admiral Sir

Andrew Cunningham continuing as his naval commander, and Air Chief Marshal Sir Arthur Tedder as his air commander.

The president, General Marshall, and Ike himself were at first pleased that the British had voluntarily suggested putting an American in the position of supreme commander. Not until the Americans had a chance to reflect on the arrangement did it dawn on them, too late, that they had been snookered. In effect, the British proposal was intended to kick Ike upstairs to become—to use the analogy of modern business terms—chairman of the board of directors, while the actual operations of the company were run by three British CEOs, each of whom outranked him. As Brooke wrote in his diary, "We were pushing Eisenhower upstairs into the . . . rarefied atmosphere of a supreme commander, where he would be free to devote his time to the political and inter-allied problems, while we inserted under him our own commanders . . . to restore the necessary drive."[39]

More was at stake here than British reservations about Ike's skill as a fighting general. The American military men were strict believers in "unity of command," and they suspected that the British wanted to replace it with what they described as "command by committee," a notion that was deeply unacceptable to Americans. Certainly it was true that no British commander had the kind of complete authority that General MacArthur enjoyed in the South Pacific (and none would have been allowed to exercise it in so high-handed a fashion) or that Ike would eventually have in Europe. The British approach to high military command was influenced in part by an instinctive national desire for consensus, and in part by the experience of World War I, in which Field Marshals Sir John French and Sir Douglas Haig, the successive commanders of the British Expeditionary Force in France, had insisted on and exercised almost unlimited authority and, in the view of many after the war, used it to squander hundreds of thousands of British lives in Flanders for no meaningful gain. The Battle of the Somme still haunted British politicians of all parties, and ensured that no British general would ever be given that long a leash again. Moreover, the personality of Winston Churchill, who was both prime minister and—by his own invention—minister of defense, prevented the rise of any authoritative British military figure. Churchill was himself a former soldier; every aspect of warfare interested him, from the smallest details of armaments to the broadest grand strategy; he was prodigiously well-informed, curious, determined to express and defend his opinions (and have them respected); he expected to be consulted about even quite minor matters in military affairs; and he insisted on his right to intervene (or, in the view of his advisers, interfere) in anything that caught his omnivorous attention. Whereas President Roosevelt seldom meddled in military affairs and was happy to delegate them to those he trusted, provided his general intentions were respected, Churchill was Britain's only "supremo." He would tolerate vigorous argument from those whom he respected (and who were brave enough to try arguing, like General Brooke), but he would not tolerate failure to carry out his wishes once he had made up his mind. Inadvertently, the British chiefs of staff, looking for a way to strip Ike of his military command, had made possible his promotion to a new role as the Anglo-American "supremo" both they and the prime minister had been determined to avoid. But of course at the time it did not seem so, to them or to Ike; indeed it looked more like a humiliation.

As the full meaning of his new role dawned on him, Ike seethed, but there was very little he could do about it for the moment. Perhaps fortunately, it was overshadowed by President Roosevelt's impromptu and unilateral announcement at the concluding press conference that the Allies would insist on the "unconditional surrender" of Germany and Japan—a statement that surprised and "deeply offended" Churchill, who thought that announcing it was unwise and premature—and by the awkward though carefully staged ceremony at which Generals De Gaulle and Giraud briefly, and with deep and undisguised mutual distrust and contempt, shook hands for the cameras of the press corps. "The bride was very shy," Harold Macmillan wrote, referring to De Gaulle, "and . . . I never thought really that we would get them both to the church."[40]

By that time Ike was back in Algiers, facing the familiar problems of a static front, bad weather, and a stubborn enemy. Marshall (accompanied by the American chiefs of staff) paid him a brief visit after the end of the Casablanca conference, bearing the news that like it or not Ike was now definitely the overall commander in the Mediterranean with Alexander as his deputy, and that he should be patient about his fourth star, which would be coming soon, partly no doubt to resolve the slight embarrassment of having a British full general serving as the deputy of an American lieutenant general. Ike had expressed himself as willing to serve under Alexander in the interests of unity of command and Anglo-American goodwill, but this was not acceptable to the president or Marshall, and at the same time would have thwarted the British attempt to put Alexander in charge of the fighting while Ike served as a kind of military figurehead. Ike was at last able to write to Mamie about what he had been doing. "By now the papers have told you about the 'big wig' party we had in this area. They had a grand meeting but I'm glad they're all gone."[41] He could also report to her that on Groundhog Day the sun was shining in Algiers, but that he expected (correctly) six more weeks of winter in Tunisia.

Ike had plenty to worry about. Rommel was now on the Tunisian border and was beginning to reinforce the "Mareth Line," which the French had built before the war to discourage an attack by the Italians from Libya, and with which Rommel now hoped to hold off Monty's British Eighth Army while the bulk of his own army moved behind it into southern Tunisia. From there, Rommel could either move north to join up with Arnim's Fifth Panzer Army and attack Anderson's British First Army, or attack west at the lightly held and widely spread-out American forces holding the passes at the southernmost end of the Eastern Dorsal Mountains, between Gafsa and Faïd. A successful attack here might take Rommel all the way to Tébessa, eighty miles behind the Allied line, forcing Anderson to retreat all the way back to Bône and the Algerian border. Indeed, except for the mountainous terrain and the omnipresent mud, the situation was not unlike that which had presented itself to Rommel and the other German panzer commanders as they broke through the Ardennes and wheeled north for the Channel ports in June 1940, rolling up the French front line as they raced for the sea. Ike could see what was coming clearly enough, and he urged Anderson to hold back the U.S. First Armored Division as a "strong, mobile reserve" so that it could be used against the Germans once they attacked—probably, Ike thought, against the relatively weak U.S. Second

Corps.[42] Anderson did not disagree, but he was deeply concerned with the weakness of the French troops at the center of his line, and was constantly obliged to reinforce them, and thus unable, or perhaps unwilling, to carry out Ike's orders.

On February 10 Ike learned that he had at last been promoted to four-star general—the same rank that Pershing had held in World War I, and that MacArthur had held as Army Chief of Staff and now held again. Whatever satisfaction this might have given Ike was somewhat marred by the fact that he received the news from Harry Butcher, who had heard it announced on the BBC, instead of in a formal telegram from the War Department—Ike was sufficiently a West Pointer and an old Army man to set high store by the traditional courtesies and protocol of Army life.[43] Also, Ike had a gnawing feeling in the pit of his stomach that Rommel was on the scene and about to attack him while the British Eighth Army was still in Libya, no doubt "regrouping" before assaulting the Mareth Line. The "Desert Fox"—Rommel—was ill, exhausted, and convinced that Tunisia could not (and should not) be held, but his name still retained its old magic, and nobody underrated his ability to catch an opponent off balance and attack with astonishing speed and strength just where he was least expected. Anybody familiar with Rommel's reputation could also guess that he would be unable to resist attacking one of the two armies against him before they had joined forces, and Ike apparently reached this conclusion before Anderson did. Henri de Jomini's classic textbook on Napoleon's tactics had been part of the curriculum at Rommel's officer cadet school, as it had been at West Point, and both Ike and Rommel were familiar with the emperor's belief in serial attacks—faced with two enemy columns or armies, he would always attack the weaker or the nearer of the two quickly, before the enemy could unite and consolidate forces. The idea of Torch had been to trap Rommel between two Allied armies, but the armies were still too widely separated to squeeze him, and it would be his instinct to destroy one of them as quickly as possible, while holding off the other.

Ike wrote to Mamie on February 11 from Algiers, hinting at his promotion. He still seems to have been defending himself against criticism from her about Kay—"There are also a couple of WAACS around the office, but I never use one unless Marshall [a male clerk] is so busy that I am forced to do so," he wrote, adding the comforting remark, "You're a slick gal—or sumpin'—and you suit me to a T."[44] Connoisseurs of Ike's letters home soon come to recognize (as doubtless Mamie herself recognized) that when he drops into dialect or baby talk, he is putting on the old charm to avoid a reply to a pointed question. A hint of what had provoked Ike's comment regarding the WAACS occurs in one of Mamie's own letters to her parents about a report she had received from Mickey Keogh on the subject of Ike's New Year's Eve party in Algiers: "11 men & 5 women—suppose the latter were WACs—durn 'em."[45] If Ike thought his promotion to full general was going to stanch questions from home on this topic, he was an optimist.

Ike wrote to Mamie at greater length on February 15, eloquently expressing the loneliness of command, the weight of his responsibilities, and his resentment at those who criticized him over the Darlan deal or because the battle in Tunisia was going so slowly. "Subordinates can advise, urge, help and pray," he wrote, "but only one man, in his own mind and heart, can decide 'Do we, or do we not?' The stakes are always highest, and the penalties are expressed in terms of loss of life or major and minor disasters to the nation. No man can always be right."[46]

It is amazing that Ike found the time to write a long, speculative, and self-revealing letter on this of all days. What he did not, and could not, tell Mamie was that on the previous day the German attack had begun. On February 13 Ike had driven forward to Faïd Pass to see for himself the positions held by American troops there. His "gut instinct" was that this was a likely direction for Rommel's attack, and he was right, although Army Intelligence (G-2) and General Anderson thought otherwise. He did not like what he saw. The troops were spread thin, their positions were poorly prepared, and although they had had forty-eight hours in which to plant minefields and string barbed wire, they had failed to do so—a clear indictment of their officers and NCOs. The U.S. First Armored Division, which Ike had ordered Anderson to keep concentrated as a mobile ready reserve, was in fact spread out in small detachments. To Ike's further dismay, the engineers of Second Corps, instead of being used to fortify the line General Fredendall's man were supposed to hold, were busily occupied with pneumatic drills and explosives, excavating a "safe quarters" for Fredendall himself deep in the rocky side of a ravine, where he would be protected from shelling and bombs, and from which he emerged only with great reluctance—inevitably affecting his men's morale. Alarmed at the "complacency" and "unconscionable delay in perfecting defensive positions in the passes" that he had observed, and disturbed by the evidence of Fredendall's incompetence—and worse, by what appeared to be sheer cowardice—Ike set off to confer individually with Fredendall's commanders on the front line, driving through the night over rough tracks and roads. At dawn he was delayed at Sbeïtla by bursts of gunfire; went forward, his .45 pistol in hand, accompanied by his aide Tex Lee, to see for himself what was happening; and deduced correctly that the German attack had begun and it was now too late to make any of the "changes in disposition" he had been about to urge on Fredendall.*[47]

In fact, Rommel had struck even harder than Ike might have anticipated. Although hampered by dual command and unprofessional schadenfreude between himself and Arnim, Rommel had persuaded Arnim to attack in the Allies' center and seize Sidi Bou Zid, then shift the Twenty-First Panzer Division quickly south to join the Afrika Korps in attacking Gafsa. After this, Rommel would drive straight for the Kasserine Pass to dislodge the American forces from southern Tunisia altogether and open up the road to Tébessa. From there he could roll up the Allied line—a one-two punch, beginning with Arnim's Fifth Panzer Army, then, when the Allies were reeling, the knockout blow in the south by the reinforced Afrika Korps. Ike had been far enough forward—possibly farther forward than a four-star general, however new, ought to go—to catch the opening moves of Arnim's attack on Sidi Bou Zid and his unexpectedly rapid advance on Sbeïtla, and to get a close-up look at the disorganization and panic that affected Fredendall's men as they came under fire. Two

* Those who think of Ike as a "deskbound" general should keep in mind that there was, so far as I know, no other British, American, or German four-star general in either world war who walked forward at dawn, pistol drawn, into what sounded like a firefight.

American counterattacks on February 15, with heavy losses in men and tanks, failed to slow either Rommel's or Arnim's attack; and by February 16 the former was in possession of Gafsa and the latter of Sbeïtla. By February 17 they were both driving hard from the north and the south toward Kasserine, where they would join up. Rommel's advance had in fact been so swift that Arnim decided not to bother sending him the Twenty-First Panzer Division, reasoning that he didn't need it, and that events would probably be decided one way or the other before it reached him. Not only was this a mistake, but it set off several days of contention between Rommel and Arnim, and appeals from Rommel to Kesselring in Rome. However, despite the confusion at the top, on the ground the combination of heavy tanks; seasoned, battle-hardened troops; and close cooperation between panzer units and the Luftwaffe was proving to be irresistible.

Although he was delayed when his driver ran the car into a ditch,* Ike returned overnight to his headquarters to assess the damage and try to organize reinforcements. He "scraped the barrel," to use his own words, although given the number of men he had in Algiers, Oran, and Casablanca it is hard to see how that can have been true, or what they were doing that was more important than getting to the front with a rifle and bayonet. Having set this in motion, he went back to the front again—he not only wanted to see what was happening but wanted the troops to see him. "The situation could not have been worse," Kay Summersby wrote, describing Ike's return to Tébessa, "there were bad casualties and talk of retreat. . . . They were all concerned about the General and . . . no one knew exactly where he was. The battle lines were fluid and it would have been very easy for his driver to mistake the enemy's patch of mud for our patch of mud."[48] It is debatable whether the best place for a four-star general during a battle was bouncing around on muddy dirt tracks near the front line in an open scout car while his staff wondered where he had gone,

* Kay Summersby was not the driver. She drove Ike in his car from Algiers to Constantine or Tébessa, and he then switched to a military scout car with a soldier as driver, to go up to the front. She insisted on driving him as far forward as he would let her, and wore "boots, slacks, a man's battle blouse and an old Air Force flying jacket," in addition to a steel helmet (recommended by Patton) in case they were strafed by German aircraft. Even allowing for a certain degree of exaggeration on Kay's part, driving as close to the front as Tébessa during a full-scale German attack was not without danger.

but Ike seemed to feel that this was necessary, and he may have been right. Perhaps he merely felt that it was up to him to set a better example to the troops than Fredendall. Somehow, in the middle of the chaos of battle, and despite his own exhaustion and concern, he managed to project a combination of calm, self-confidence, and optimism, and the troops responded to it.

Like Grant at the end of the first awful day of Shiloh, Ike was shaken but not dismayed. When Sherman found Grant that night, the general was sitting on the ground under a tree in the pouring rain, his black slouch hat pulled low over his eyes, calmly smoking a cigar and whittling a stick with his penknife. Sherman dismounted and led his horse over to where Grant was seated. "Well, Grant," Sherman said—he did not as yet know Grant all that well—"we've had the devil's own day, haven't we?" Grant, at the end of a single day of battle that had cost more casualties than those in all previous American wars combined, kept on whittling, and without looking up said, "Yes; lick 'em tomorrow, though." And so he did. Though Grant could not know it, the Confederate commander, A. S. Johnston, had been wounded in the leg and bled to death on the battlefield, his army was almost as exhausted and demoralized in victory as Grant's was in defeat, and with Johnston's death command had fallen on P. G. T. Beauregard, who had never wanted to attack at Shiloh in the first place.

In much the same calm, determined way as Grant, Ike steadied Fredendall as much as he could (while making a mental note to replace him with Patton at the first opportunity), ordered Anderson to move enough forces down to Sbiba and Thala to block Arnim's attack to the north, resigned himself to a humiliating and costly defeat in the Kassserine Pass, and ordered everybody to dig in on high ground and hold it. Rommel's attack, he was convinced, would wear itself out—despite its initial success Rommel had neither the troops nor the supplies to sustain it; and so it proved. Above all, Ike knew he had to hold Tébessa, and hold it he did, though by the skin of his teeth. By February 22 Arnim and Rommel had run out of steam; a depressed and weary Rommel was at last given complete command of all Axis forces in North Africa (now that it was too late); General Alexander, newly arrived from the desert, had taken up his duties as Ike's deputy and commander of ground forces; and the situation in Tunisia was stabilized. Ike was deprived of the chance for a serious counterattack against Rommel when Fredendall, true to form, took counsel of his fears and persuaded him that the Germans still had enough strength left for another big attack and that it would be safer to form a defensive line and wait for it. Needless to add, this attack never came. Instead Rommel, taking advantage of his new authority, turned south and east to attack Montgomery before the latter could bring up the bulk of his forces from Tripoli and Benghazi. Having failed to break one jaw of the nutcracker, Rommel now had no option but to break the other before they both closed on him.

Now that Alexander was in place, Ike's resentment about his appointment as commander of the ground forces seems to have simmered down. This was, in part, a reflection of Alexander's unassuming charm, good manners, and unflappable professionalism—he came, he saw, and he conquered Ike—and in part a dawning realization on Ike's part after the Battle of the Kasserine Pass that he needed a deputy and a ground forces commander, British or not. For his part, Alexander was horrified by what he found in Tunisia, particularly by the way American, British, and French units were scattered together and intermixed without rhyme or reason. He rightly thought this was a dangerous weakness. He quickly reshuffled the ground forces into distinct national sectors; created a real mobile reserve that could not be frittered away or broken up to fill in gaps in the front line; and tried to establish some form of coordination with Monty's Eighth Army and eliminate once and for all the problem of fighting the Tunisian campaign with two completely separate armies each operating on its own, which had created as many difficulties for the Allies in North Africa as it had for their opponents Rommel and Arnim.

In the meantime, Ike was at last free to do something about the lessons he had learned from the disaster. The numbers tell the story—in a week of battle the American divisions suffered losses of "192 killed, 2624 wounded, 2459 prisoners and missing."[49] Since, as Ike pointed out, almost as many men had been captured or surrendered as had been wounded, this was not a record to be proud of. Trying to look on the bright side, he cabled to Marshall, "A certain softness or complacent attitude that was characteristic of all units only a few days ago has disappeared," and "all our people from the very highest to the very lowest have learned that this is not a child's game." But Marshall was not cheered by this—indeed, he was surprised and distressed by the message, since it referred to "two of our best divisions" (the First Armored and the First Infantry), both of which had already had "battle experience under fire" and taken casualties during the landings—and he wrote to General Jacob Devers, chief of the armored forces, "You have a real problem on your hands."[50] Ike was determined that "no unit from the time it reaches this theater until this war is won will ever stop training," and borrowing from Monty's experience, at the suggestion of Alexander, he instituted "battle schools" in which officers and men refined their battle skills under conditions that were as near to real combat as possible, and trained with the weapons they would use in battle, instead of receiving these at the last minute. Perhaps most important, Ike finally replaced Fredendall as commander of the U.S. Second Corps with Patton, confident that Patton's energy, thirst for battle, and passion for training would reform matters quickly, as in fact they did. (Fredendall was given a relatively "soft landing," being sent home to the United States where he took command of an army to train and was promoted to lieutenant general. In the future, though, neither Ike nor Marshall would be so generous with those who failed to measure up in battle.)

As for Rommel, he had been made overconfident by his victory at Kasserine, and his attack against the Eighth Army at Medenine, in front of the Mareth Line, was a stunning failure. He had urged on his men the need for "speed and fierceness," and they delivered both; but the British outnumbered him in men, tanks, and artillery, and fought a defensive battle in well-prepared positions, against an uninspired frontal attack by three panzer divisions. Monty was so confident of victory that after watching the opening moves of Rommel's attack, he snapped, "The Marshal has made a balls of it,"[51] and went back to his caravan for the day to write letters. By eleven o'clock that night Rommel had withdrawn his crippled forces, having lost almost a third of his tanks; and three days later he flew home on sick leave, never to return to North Africa. He had failed to damage the other jaw of the nutcracker, and the total destruction of Axis forces in Tunisia, as he had predicted to Hitler, was now a foregone conclusion.

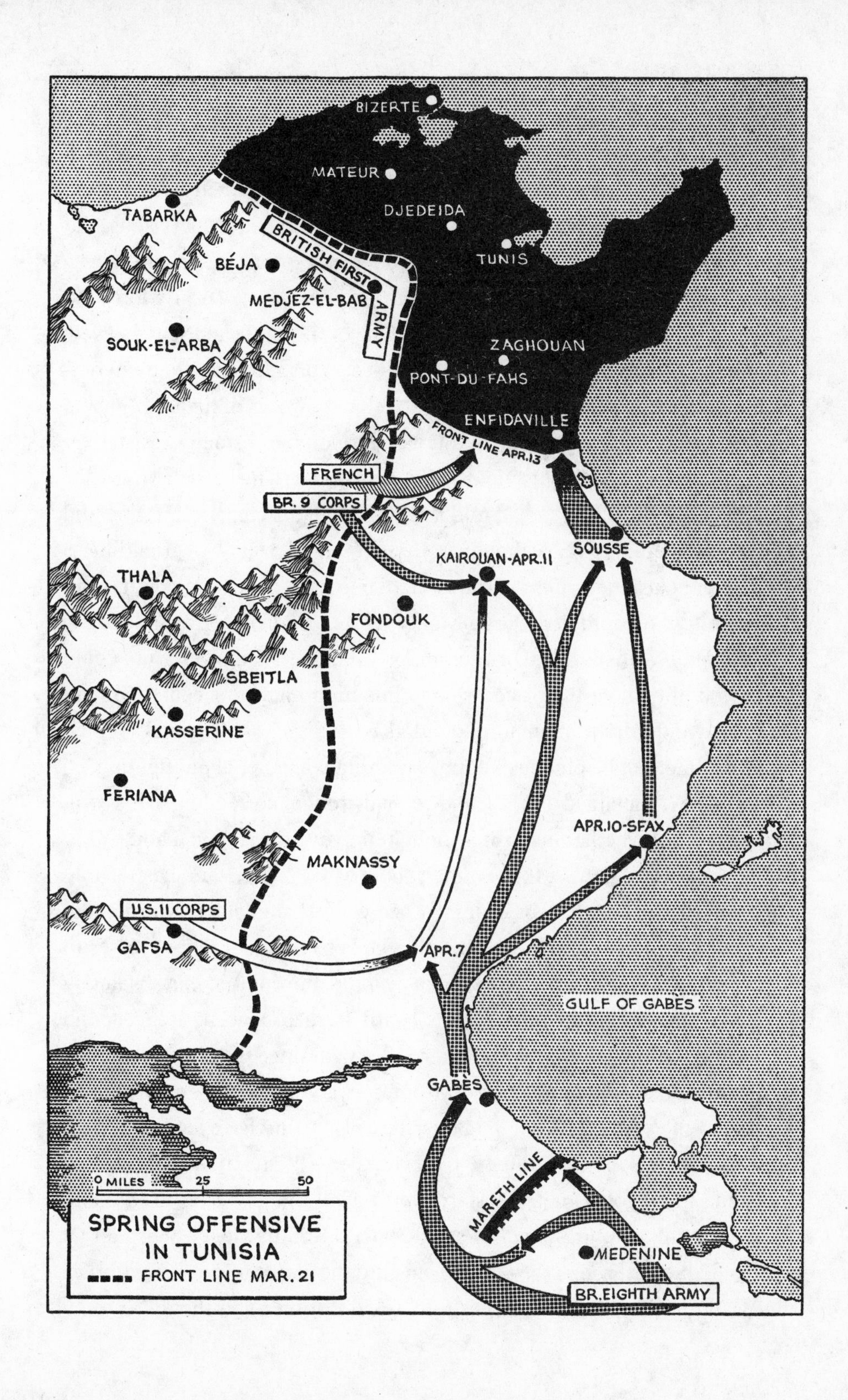
BIZERTE
MATEUR
TABARKA
DJEDEIDA
BRITISH FIRST ARMY
TUNIS
BÉJA
MEDJEZ-EL-BAB
SOUK-EL-ARBA
ZAGHOUAN
PONT-DU-FAHS
ENFIDAVILLE
FRONT LINE APR.13
FRENCH
BR. 9 CORPS
SOUSSE
KAIROUAN-APR.11
THALA
FONDOUK
SBEITLA
KASSERINE
FERIANA
APR.10-SFAX
MAKNASSY
U.S. II CORPS
GAFSA
APR.7
GULF OF GABES
GABES
MARETH LINE
0 MILES 25 50
SPRING OFFENSIVE IN TUNISIA
FRONT LINE MAR. 21
MEDENINE
BR. EIGHTH ARMY

. . .

On March 20 the Germans once against demonstrated their powers of recovery and resilience, even without Rommel in command, in a bitter week-long battle, the biggest in North Africa since Alamein, as Monty sought to break through the Mareth Line against stubborn resistance. This time it was Monty who suffered from overconfidence, and he was very nearly forced to retreat until he changed tactics in the middle of the battle and sent the New Zealander General Freyberg out into the desert on a long left hook to outflank the enemy, while supporting his attack with waves of fighter-bombers directed from the ground—then a novel development, stoutly resisted by the fighter pilots. After this victory, Ike flew down to meet with Monty for the first time since Monty had told him to put out his cigarette, in England in the summer of 1942. Although both men put on a show of friendliness for the photographers, their private reaction to each other was skeptical, to put it mildly. After their meeting Monty wrote to General Alexander, now Ike's deputy, with his usual brittle candor, "I liked Eisenhower. But I could not stand him about the place for long; his high-pitched accent, and loud talking, would drive me mad. I should say he was good probably on the political line; but he obviously knows nothing whatever about fighting."[52] Monty also complained that Eisenhower had brought no bedding of his own, and wondered whether his visitor thought they were living in a hotel.

To General Marshall, Ike confided about Monty, "He is unquestionably able, but very conceited. For your most secret and confidential information, I will give you my opinion which is that he is so proud of his successes to date that he will never willingly make a single move until he is absolutely certain of success—in other words, until he has concentrated enough resources so that anybody could practically guarantee the outcome."[53]

Neither man would change his opinion about the other during the course of the war, or after it, and this situation would have serious consequences for the Allies' strategy. Unfortunately, the relationship between Ike and Monty got off to a bad start over—of all things—a bet. Talking to Ike's normally cautious chief of staff General Smith, Monty boasted that he would break through the Gabès Gap onto the plains of southern Tunisia and capture the port of Sfax early in April. Smith, who thought Monty could never bring it off, uncharacteristically told Monty he could ask for anything he liked if he did that, and Monty said he wanted his own B-17 Flying Fortress, "complete with an American crew on their payroll," the whole thing to be "his personal property until the war ended." Smith, no doubt supposing that it would never happen, or that Monty would forget all about it, agreed, and Ike was dismayed to receive a message from Monty on April 10: "CAPTURED SFAX EARLY THIS MORNING. PLEASE SEND FORTRESS TO REPORT FOR DUTY TO WESTERN DESERT AIR FORCE AND THE CAPTAIN TO REPORT TO ME PERSONALLY."[54]

Whether Smith had told Ike about the bet or not, Ike had no choice but to pay up, although it involved him in endless and contentious difficulties with the USAAF and with the War Department, neither of which was amused that an American four-engine bomber and its crew had been given away to a British general for the duration of the war as his own personal property. And since Monty, understandably, could not prevent himself from telling people about the bet he had won, and the fact that Ike had been obliged to pay up, it quickly became common knowledge and part of the "Monty legend," to Ike's annoyance and embarrassment. Rather fortunately, the aircraft crashed in July 1943, and since Monty had not stipulated that it had to be replaced if it was destroyed, it was not. All the same, the incident came to typify the relationship between Ike and Monty, which had started off on the wrong foot and did not improve.

Even the scent of victory was not enough to change their minds about each other. By the time Monty's Flying Fortress had landed at Gabès, the Germans were in full retreat, although still fighting with obstinate courage and efficiency. On April 11, Monty's Eighth Army and Patton's U.S. Second Corps met up at Kairouan and, joined by the French, now commanded by General Juin, they pushed the Germans back into a small pocket in northern Tunisia. Since the mountainous terrain favored infantry attacks over tanks, Ike replaced Patton as commander of the Second Corps with his old West Point classmate and friend Major General Omar Bradley (referred to with genial contempt by Patton as "Omar the Tentmaker"), sending Patton back to Seventh Army Headquarters to start preparing the troops for Husky, the forthcoming invasion of Sicily.

Hard fighting was still to come, but now the result was no longer in doubt, and the presence of Alexander, despite Ike's initial misgivings, helped to make swift transfers of British and American units possible where they would formerly have been resisted or ignored. When Monty's advance from the south bogged down against fierce resistance and formidable terrain along the strongly held Enfida line, Alexander was able, at Ike's request, to switch several divisions from Monty's Eighth Army to the British First Army, and attack the German right flank with overwhelming power—something that Monty would have resisted strongly had it been requested by anybody other than Alexander. Even the French, rearmed by the Americans and led by Juin, a soldier's soldier if ever there was one, attacked aggressively, and on May 13, two days before Ike had promised Roosevelt he would win the war in Tunisia, the Germans surrendered. Ike's forces took more than 250,000 prisoners, including Colonel General von Arnim and General Hans Cramer, who had taken command of the Afrika Korps after Rommel's departure.

When Arnim was taken through Algiers on his way to a prisoner-of-war camp in Britain, somebody reported to Ike that the German wanted to call on him, a polite gesture from a defeated commanding general to a victorious one. Monty, as it happened, had gotten into hot water after his victory at El Alamein by inviting the captured General von Thoma to dine with him in his tent, to the great anger of the British press; but whether Ike knew that or not, he had already made up his mind to say no.

"The custom," he would later write, "had its origin in the fact that mercenary soldiers of old had no real enmity toward their opponents. . . . A captured commander of the eighteenth century was likely to be, for weeks or months, the honored guest of his captor . . . . For me World War II was far too personal a thing to entertain such feelings. . . . The war became for me a crusade. In this specific instance, I told my Intelligence officer, Brigadier Kenneth Strong, to get what he possibly could out of the captured generals but that, as far as I was concerned, I was interested only in those who were not yet captured. . . . Not until General Jodl signed the surrender terms at Reims in 1945 did I ever speak to a German general, and even then my only words were that he would be held personally and completely responsible for the carrying out of the surrender terms."[55]

It had been seven months since Ike had sent his armies off to land in French North Africa to carry out a plan of his own devising, and on a scale hitherto unknown in the history of warfare. He had emerged—and matured—at last as the supreme commander of the war in Europe, the man on whose shoulders the entire responsibility for the Anglo-American contribution to victory rested.

On May 26 he wrote to Mamie, preparing her for the news, "His Majesty, the King, has conferred on me the Grand Cross of the Most Honourable Order of the Bath. This is the highest military award the British can give and I must say I feel very humble—though I cannot avoid also a deep feeling of pride."[56]

It had been just over two years since he had been "misidentified in the newspapers as 'Lt. Col. D. D. Ersenbeing.' "

CHAPTER 12

Sicily

At Ike's villa in Algiers, 1943. *(Left to right, front row)* Anthony Eden, General Sir Alan Brooke, Winston Churchill, General George C. Marshall, Ike; *standing*, Air Marshal Sir Arthur Tedder, Admiral Sir Andrew Cunningham, General Sir Harold Alexander, Monty.

Anybody looking at a map of the Mediterranean will see at once the importance of certain islands, the largest and most important of which is Sicily. This was the case as long ago as ancient times, when it was a Greek colony for just that reason—he who controls Sicily dominates the passage through the Mediterranean, whether in the age of the oared ships of Grecian times or that of the modern bomber.

Other islands matter too. Malta was so important for the same reason that possession of it made the Knights of Malta rich and powerful for centuries. In modern times, in the hands of the British, Malta controlled the sea-lanes between ports in Italy or Sicily and those in Tunisia; as a result, the Luftwaffe bombed Malta mercilessly by day and by night from 1940 through 1942. At one point the island's air defense was reduced to three obsolete RAF Gloster Gladiator biplanes known as Faith, Hope, and Charity; and it was considered worth the sacrifice of a British and an American carrier to German U-boats to supply Malta with more modern Hurricanes, flown from the flight deck directly into combat over the beleaguered island. The small force of RAF bombers on Malta was enough to make the task of supplying Rommel by sea difficult, costly, and dangerous, and the Luftwaffe squadrons based in Sicily exacted a constant toll on Allied convoys traveling through the Mediterranean from Gibraltar to Alexandria.

These concerns had played a part in the decision made by the Combined Chiefs of Staff at Casablanca to invade Sicily—a decision that had produced considerable, and unusual, acrimony between the Allies all the way up to the level of the president and the prime minister. Nobody denied the importance of Sicily, but given the shortage of landing craft and trained men, invading it in 1943 would inevitably delay a major cross-Channel invasion until the summer of 1944—bad news for Stalin, and a bitter pill for General Marshall and President Roosevelt to swallow. Once Sicily was taken, moreover, it would be almost impossible to resist going on from there to the Italian mainland—the attack on Germany's "soft underbelly" which Churchill had been preaching for so long to deaf ears (deaf in Washington, at any rate). On the other hand, the Americans had a substantial number of troops sitting idle in Morocco (too many, in the opinion of some people); the landing craft used for Torch still remained for the most part in the Mediterranean (to the fury of Admiral King and General MacArthur, who wanted them sent to the Pacific); and taking Sicily, while it would not end the war, might at least knock the Italians out altogether, a substantial setback for Hitler. As a small bonus it would permit moving Allied bomber forces in the Mediterranean closer to their targets, particularly the vital oil fields and refineries of Romania, without which the Wehrmacht's fuel supply would be drastically reduced. There was no great enthusiasm for Husky among the Americans, and they regarded themselves, with some shame (and determination not to repeat the mistake) as having been talked into it by the wily British, but it came to seem like an inevitable next step, and in the absence of Sledgehammer, now defunct, or of a determined cross-Channel landing in 1943, a necessary one.

Even though he would command the invasion of Sicily, Ike was not a great enthusiast for the idea. Strategically, he would have preferred to take Sardinia and Corsica, since they extend down the length of the western coast of the Italian peninsula over 400 miles,[1] from Genoa to Naples, and possession of them would force the Germans to keep a large number of divisions in Italy to deal with an invasion at any point. By contrast, crossing the Strait of Messina, only just over two miles wide, from Sicily to the Italian mainland* would merely land Allied forces on the "toe" of the Italian boot; they would be obliged to fight their way over rugged, mountainous terrain, which the Germans could easily defend, even to reach Naples, itself something of a strategic backwater that led nowhere. Indeed, Italy itself seemed to many a strategic dead end—it would take forever and cost countless casualties for the Allied armies to fight their way north across the rugged mountain barriers of central Italy; and even if they succeeded, where would that success put them? At the Alps, which they could neither cross—Switzerland being in the way—nor go around.

In the planning that followed the Casablanca conference, it had long since been decided that General Alexander, in his role as Ike's deputy and ground commander, would be in control of the ground battle, with Monty and Patton respectively commanding the British and the American forces. Indeed, Patton had been replaced by Bradley during the climactic battle for Tunisia so that he could return to Morocco and take charge of preparing the American troops for their role in Husky. Monty was determined to take his Eighth Army troops with him as the nucleus of his force, in part because he knew, and had selected, their commanders, and could therefore rely on them; and in part out of simple vanity—to the British public Monty and the British Eighth Army were virtually synonymous. That there were obvious difficulties seems to have been ignored by everyone. The Eighth Army had made its reputation in the Libyan desert, and the role of its armored formations (and also their reputation for making swift, wide-flanking movements in open country) was ill suited to the task of advancing over rocky, mountainous country, densely filled with centuries-old villages and

* Today there are plans to build a bridge across it.

farms that had been built for the express purpose of resisting invaders; over steep, narrow roads with many bridges; and against an enemy that would fiercely defend every strongpoint. In Sicily, unlike North Africa, tanks could not operate off the roads—very often there would be a precipice on one side and a steep cliff on the other. At any place where the enemy could hold them up it would be necessary for the infantry to make a flanking movement in which every stone wall, rocky hill, vineyard, orchard, ancient chapel, or fortified village would have to be assaulted before the tanks and "soft-skinned" vehicles could move on. The Germans could be expected to meticulously demolish every bridge behind them as they retreated, so bridging equipment—always bulky, vulnerable to shelling and dive-bomber attacks, slow-moving, and difficult to transport on narrow roads—would be at a premium, and engineers would have to build their bridges while under fire from snipers and artillery.

Sicily would be a tough nut to crack. Quite apart from having a terrain that favored the defense, it was garrisoned by over 150,000 Italian and 20,000 German troops, with the possibility of ferrying at least one more German division across the Strait of Messina as soon as the invasion took place, followed rapidly by as many as three more. (By the time the invasion actually took place the number of Axis troops in Sicily would have risen to over 300,000, against the 478,000 that Ike proposed to land.) Although the reputation of the Italian army was poor, as was its level of equipment, the more elite Italian units (and the Italian air force) had fought better in North Africa than they were given credit for, and were in general held to be better in defense than in attack. Also, the Allies' planners considered it likely that the Italians might fight with more enthusiasm when defending their own country than they did when attempting to expand Mussolini's colonies by conquering Egypt under the command of a German general. As for the Germans, nobody doubted that they would throw in the best divisions they had in Italy to defend Sicily. They had delayed the capture of Bizerte and Tunis by six months against overwhelming odds, operating at the end of a lengthy and precarious sea supply line; in Sicily, they would be fighting in an area not much larger than New Hampshire, with a sea crossing of hardly more than two miles for supply and reinforcement, and with dozens of good airfields from which the Luftwaffe could operate—indeed, part of the rationale for invading Sicily in the first place was to take the airfields there for the use of the Allied air forces. From the perspective of Berlin—and of Field Marshal Kesselring's headquarters in Rome—there was every reason to hope that if the Allies invaded Sicily they would receive a bloody nose.

Of course the real problem that faced the Germans was whether the Allies intended to land in Sicily or elsewhere. Germany occupied—together with its Italian ally—the entire northern coast of the Mediterranean Sea, from the Franco-Spanish border in the west to the border between Greece and Turkey in the east, and all of the principal islands except Malta, a vast coastline to defend. Hitler was faced with the famous warning of his hero Frederick the Great: "He who attempts to defend everywhere, defends nowhere." That the Allies intended to invade somewhere in 1943 was obvious—why else were the harbors of Tunisia, Morocco, and Algeria packed with landing craft?—but the crucial question remained *where*. Sardinia and Greece both seemed likely places, and even the Balkans were considered a possibility.

Working parallel to the planners, the Allied secret intelligence agencies—the American OSS and Britain's SOE and MI-6—attempted to point the Germans in the wrong direction, a useful rehearsal for the invasion of France. Perhaps the most famous of these dirty tricks, as they were known in the trade, was "the man who never was."* A body was placed in the sea (from that busy British submarine HMS *Seraph*) where it would wash ashore in Spain. It was dressed in the uniform of a senior British officer and bore documents that named Sardinia as the objective of the Allied invasion. This ruse, along with many others, set off alarm bells in Berlin, and was successful enough to cause the Germans to reinforce Sardinia. Of course, the moment the invasion fleets appeared off the Sicilian coast the Germans would realize they had been fooled, but by then it might be too late.

It was unfortunate that Husky was a perfect example of planning by committee. During the initial phases of planning Ike's attention was still

* The episode became a book that was later made into a very successful movie called *The Man Who Never Was*, starring Clifton Webb.

focused on winning the battle in Tunisia, as was that of his deputy Alexander, whom even his most fervent admirers did not in any case credit with a Napoleonic gift for strategy. As a result, Husky was a product of staff planning in Algiers, Cairo, London, and Washington, and had all the disadvantages of a "by the book" military operation drawn up by deskbound officers. It should have been clear to anybody that the overriding objective was to get to the Strait of Messina as fast as possible at all costs and choke off the Germans' reinforcements and supplies, but both Allied air forces wanted the airfields seized quickly, since without these they could not provide fighter cover, while both navies wanted as many ports as possible captured in working order in order to supply the armies, since they did not believe they could meet the needs of the armies "over available beaches." The initial result was a plan intended to keep everybody happy, "safe" in the sense that there would be at least two important landings in southeastern Sicily, one led by Patton and one by Monty, with a naval diversion in the northwest of the island to distract the Germans, possibly to be expanded into a third landing near Palermo. Within the landing areas far too many of the troops were spread out in "penny packets," to use a favorite phrase of Monty's, rather than concentrated for a single big thrust inland. To paraphrase Frederick the Great, he who attacks everything attacks nothing.

Matters were not helped by the conflicting personalities of the ground commanders, both notoriously difficult prima donnas. Patton had been obliged to attend as an observer, much against his will, one of Monty's famous "battle schools" for senior officers in the desert, and remarked afterward, "I may be old, I may be deaf, and I may be stupid, but it just don't mean a thing to me."[2] He was bored by logistics and planning, and thought Monty was too slow and cautious, unwilling to move before everything was "just so," down to the last detail. Monty in turn had dismissed Patton as "an old man of about 60,"[3] and indeed in a photograph taken of him at the battle school Patton does look old, impatient, befuddled, and furiously hostile, with his trench coat buttoned up to the neck and with two oversize stars (confiscated by an aide from the uniform of a captured Italian general) sewn on his forage cap. Time would not improve the relationship between the two men, or cause either of them to revise his initial judgment of the other.

Ike—somewhat reluctantly—accepted the planners' logic with regard to landing a strong force in the southeastern corner of the island, since they hoped it would lead to the rapid capture of the port of Syracuse. This would be followed by a series of "assaults by echelon," in the south and perhaps in the north, so that each landing, if it proved successful, would be able to provide air cover for the next. The obvious danger of the plan was the risk of "defeat in detail" if the Germans reacted quickly enough, and the fact that it "was complicated and that is always a disadvantage." However, his hands were tied by the shortage of troops—the battle for Tunisia was still going on, and in any case General Marshall was determined that the invasion of Sicily should be done "on the cheap," without the Allies' "being drawn into a campaign that would continuously devour valuable resources" or "a commitment to indefinite strategic offensives in the area." In other words, there was to be no drain on the manpower being trained and assembled for the cross-Channel invasion, and no Churchillian attempt to parley victory in Sicily into a full-scale attempt to take Italy.

Patton *(standing at left)*, Ike, and President Roosevelt, 1943.

As it happened, Monty's first use of the B-17 he had won in his bet with Smith was to fly to Algiers in April, three months before the launching of the invasion of Sicily, and see for himself if the rumors he had picked up about the plan for Husky were true. He was consternated to find that in fact the plan contradicted everything he held dear—it lacked the virtue of simplicity, he complained; it failed to concentrate the available forces for a single, powerful attack; and it spread units all over the map in "penny packets," separated from each other by time and distance. Even before his journey to Algiers, Monty had been highly critical, describing the plan for Husky in a letter to General Brooke as "a hopeless mess," and commenting, not without reason, "Alex, is, I suppose, in charge of it. But he hasn't time to bother about it."*[4] About Ike, Monty's opinion was, and remained, succinct: "Nice chap, no soldier."[5]

At lunch in Algiers with Ike and Alex, Monty made his case for giving the Eighth Army the responsibility for a strong, concentrated attack in the southeast of Sicily, which the other Allied armies (i.e., Patton) would "support"; and for moving the Eighth Army's headquarters from Tunisia to Cairo immediately to get on with detailed planning, which he would oversee himself by flying back and forth "to ensure that both operations functioned smoothly." How he can have concluded that Ike and Alex had agreed to this proposal is hard to imagine, let alone that they had put him in command of Husky, but shortly afterward he flew to Cairo and started to redraft the plan for Husky in the light of what he took to be his new authority, puzzling Brooke, Ike, and Alex and positively infuriating Cunningham and Tedder, the naval and air commanders, respectively, who "felt that Monty was too 'junior' in the Husky hierarchy to be dictating planning policy." Whatever else Monty had done, he had started an inter-Allied, interservice row.

It is to Ike's credit that he accepted many of Monty's reservations

* Monty's attitude toward Alex, while mildly affectionate, brings to mind General Hoffman's toward Field Marshal von Hindenburg. When showing distinguished visitors over the battlefield of Tannenberg, site of Hindenburg's decisive victory over the Russians in 1914, Hoffman would say: "That is where the Field Marshal slept *before* the battle of Tannenberg. . . . And there is where he slept *after* the battle of Tannenberg. . . . And here is where he slept *during* the battle of Tannenberg."

about Husky despite his irritation with Monty (his blood pressure would have been sent soaring had he known what Monty was writing about him in letters to Brooke). Ike dropped the original "encircling plan," about which he had never been all that enthusiastic and in which beach-supported landings would take place to the west, south, and east of the island; instead he adopted a modified version of Monty's plan, in which the British forces would land on the eastern coast of Sicily with Syracuse as their first objective, while the Americans strike straight north for Palermo, to cut off the retreat of all Axis troops in the western part of the island. Partial acceptance of his ideas did not, of course, satisfy or silence Monty. Still under the delusion that he had been appointed chief planner by Ike and Alex, he proposed in mid-April to concentrate his attack, so as "to secure lodgement in a suitable area and then operate from a firm base," in part by eliminating the attacks on the airfields that Tedder was counting on. Given Monty's reputation for not moving until his base was completely secure, this was something like a shot across Patton's bows, as well as quite the opposite of what Ike wanted—Patton's preference for attack over preparation was already well known. Admiral Cunningham and Air Chief Marshal Tedder instantly relayed their "complete disagreement" with Monty's proposed plan to Ike, Cunningham remarking, with typically British understatement, "I am afraid Montgomery is a bit of a nuisance." Actually, he was worse than a nuisance; he was now threatening that Husky could not take place in July at all unless his plan was adopted, and demanding a face-to-face meeting with Ike and Alex to "thrash things out." Thus, by May, less than two months from the invasion—the date of which Ike was in no position to postpone, since he had already alarmed both Marshall and Churchill by moving the date from June to July—Husky was in disarray, with the planners in Cairo working day and night to complete Monty's plan while those in Algiers were working on the final details of Alex's. Monty, Ike complained to Kay Summersby with a sigh, was a thorn in his side.

Under the circumstances, it is hardly surprising that Ike's health suffered. He complained to Mamie that he was putting on weight, and not getting enough exercise, and that one of the enlisted men in his headquarters who had not seen him since he left London told Butch he had aged five years since then. Ike himself noticed that he was smoking more heavily than ever, and of course paying the price for it in the form of coughs, colds, and bronchitis. He suffered from time to time from "African trots," or, as the British called it "gippo tummy," debilitating diarrhea. He tried not to let the casualty lists depress him, and to ignore it when the war reached out and took somebody he knew. One of his young aides, whom he had allowed to go to the front to get a little experience, was reported missing, and Ike had the task of telling Kay Summersby that her fiancé, the dashing Colonel Richard Arnold, had been killed. The news, when it reached the United States, drew from Mamie, who was apparently still fretting over the presence of Kay Summersby and the WAACs at Ike's headquarters, a letter to which Ike replied on June 11 with unusual sternness:

> A very strange coincidence occurred this morning. I had two letters from you . . . and in one of them you mentioned my driver, and a story you'd heard about the former marital difficulties of her fiancé. You said it was a "not pretty" story. Your letter gave me my first intimation that there was any story whatsoever—I didn't know anything about it. In any event, whatever guilt attached to him has been paid in full. At the same moment your letter arrived I received a report that he was killed. [6]

This was, by Ike's standards, a fairly sharp rap on Mamie's knuckles—he had found it very difficult to break the news about Dick Arnold's death to Kay, who had been worried all day because Ike seemed withdrawn and silent, not at all his usual cheerful, chatty self, she thought. Back at the villa, he got out of the car "heavily like an old man," took her into the sitting room, and lit a cigarette. She asked him if he would like her to make him a drink. "That would be a good idea," he told her, his face grim. "Make one for yourself, too." Then he sat her down and said, "I'm going to give this to you straight,"[7] and told her that Dick Arnold had been killed a week ago, though the news of his death had come through to Algiers only that morning. He handed her a handkerchief when she started to cry, then as he handed her one fresh one after another, she realized that the thoughtful Mickey Keogh had put a whole dozen on the coffee table, knowing what was coming.

Ike sent her off to the country for a few days at a "rustic Algerian house" oddly named "Sailor's Delight" that had been provided for him at General Marshall's insistence, so he would have somewhere to relax. There, by herself, she "touched bottom," as she came to the conclusion that she scarcely even knew the man who had been her fiancé. "Ike knew more about me and had seen more of my family than Dick ever had. . . . The realization that I had never known [Dick] was as shocking as the knowledge that now I never would."[8]

Ike's wartime life, it is apparent, was like that of most supreme commanders, a tight, fiercely loyal and protective little world that revolved for the most part entirely around him. Whatever the exact nature of his relationship with Kay, it was hardly unusual among American senior officers in the European theater to have a female companion. Patton was accompanied by an attractive young woman dressed in a nurse's uniform and described as his "niece," whose existence he tried desperately to conceal from his wife, Beatrice; and even Ike's dour and unsmiling chief of staff General Smith had a "wartime romance," in the popular phrase of the time, with a Nurse Wilbur. Mamie's concerns were not altogether without foundation and were shared by many other Army wives.

By the time of Dick Arnold's death Kay's job as Ike's driver had expanded to include answering his mail—like any celebrity, he was deluged with requests for autographed photographs of himself, and heartfelt letters from total strangers, every one of which he insisted must be answered. Many were simply fan mail, others were pleas on behalf of a loved one in the Army, and a few were simply eccentric, like one (forwarded by General Marshall) from John A. Petroskey, of Box 157, Lyon Mountain, New York, who had just read in a newspaper that Ike drank cold water with his meals, and urged him to give up the habit for the sake of his digestion. "General Rommel," Mr. Petroskey pointed out, "would not have been the successful desert fox if he was caught with a cylinder or two missing in his brain by bad digestion." Even in this modest role, Kay attracted Mamie's wrath. Mamie began to suspect that Ike's letters to her, when they were typed rather than handwritten, had been dictated by him to Kay, or worse yet, even drafted by Kay for his signature. Mamie sharply requested that Ike not send her typed letters in the future.

Since Kay's luggage had gone down with the ship that brought her to Algeria, her Military Transport Corps uniform was shabby and worn by now, so Ike arranged to have a tailor make her two new uniforms, from the same cloth as his own. Though she was, in fact, a civilian, the new uniforms were slightly more military in appearance than her old ones, judging from photographs. In the short, tight-waisted tailored jacket that was already becoming known as an Eisenhower jacket, she looked (surely not by accident) more like a WAAC officer than a British MTC driver. In one snapshot of her taken at the time she is shown trying out a U.S. Army Harley-Davidson motorcycle in the North African desert. She is wearing her new uniform, and a forage cap is perched at a saucy angle on her head above piles of very unmilitary-looking, carefully coiffured hair. Ike's olive drab Packard can be seen in the background.

It would be a mistake to read too much into this—though brushing it under the rug or attempting to ignore it altogether would also be wrong. That there was a strong bond of mutual attraction between Ike and Kay seems clear enough—almost everybody around them from Winston Churchill down remarked on it—but Ike may simply have been reaching out to Kay for companionship and relief from the loneliness of high command, and she would hardly have been the first woman in the world to have read much more into their relationship than that. Even Kay makes it clear in her second memoir that they lived in a fishbowl, with little or no privacy. Harry Butcher, Tex Lee, the other aides, and Mickey Keogh were in constant attendance, and there were armed military guards outside and inside the house at all times—so however much Ike may have yearned for privacy, it seems unlikely that he would have found it, with or without Kay. Still, this is not to say that Mamie would have approved of Kay's place in Ike's household; indeed, even loyalists like Butcher occasionally questioned the wisdom of that.

Apart from the few moments of relaxation over a drink or a game of cards, however, the impression one carries away of Ike's life in Algiers is of constant, high-tension stress and almost unbearable responsibility, rather than romance or high jinks. By 1943 Ike had more than 2 million men under his command, an area of responsibility that ran from the Sahara to the northernmost tip of Scotland, the armies of three nations reporting to him, and an invasion to launch with nearly 500,000 men. The amount of paperwork that crossed his desk was probably the largest in the history of warfare; his subordinate commanders were nearly all difficult and demanding characters requiring patient treatment; and his superiors in London and Washington were impatient for victories. From time to time Ike may well have envied his old boss (now a rival four-star general) Douglas MacArthur. With the exception of a few Australians, MacArthur had no other Allied armies serving under him; no equivalent of the troubles caused to Ike by Generals De Gaulle, Giraud, Brooke, and Montgomery; no British admirals and air marshals to contend with; and no high-level visitors to deal with, since hardly anybody wanted to visit the South Pacific. North Africa, on the contrary, since it was within comparatively easy flying distance and the only major piece of territory occupied by Allied troops, attracted visitors including King George VI (who cut Kay dead when Ike introduced her to him)[9] and Winston Churchill, each of whom took up a large chunk of Ike's carefully rationed time, and tried his patience by asking questions and offering suggestions.

The dispute between Monty and almost everybody else about the plan for Husky was still not resolved. (Monty claimed that he had won Bedell Smith's support for his plan "in the lavatory" at Ike's headquarters, but it seems more likely that Smith was merely trying to get out of the toilet, where Monty had him pinned down.) Indeed, this dispute escalated as Monty claimed to have Patton under his command, and even tried to ensure that Alex and Ike would not move their headquarters to Sicily—"I must have there the Main HQ of all three Corps," he wrote to Alexander, adding tactlessly, "There will be no room for anyone else."[10]

To his diary, Monty confided, "The proper answer is to cut [Alexander] right out of Husky. I should run Husky. With the new plan it is a nice tidy command for one Army H.Q., and the Eighth Army should run the whole thing. I would then deal direct with Allied Force H.Q. [Eisenhower]."[11] Of course Monty was right that Husky had two many layers of competing commands, and that a single commander on the spot might have been more effective (and "tidy"), had the right person existed to play this role. But besides his natural tendency to overestimate his own generalship, Monty seems to have totally ignored the fact that putting American troops (and Patton) under the direct command of an English general for the first invasion of the European continent would have been politically unacceptable to most Americans at this stage of the war, and particularly to the president. It was one thing for Alex to be Ike's "deputy and ground commander"—he was clearly subordinate to Ike, and also very careful never to step on Ike's toes—but quite another for Monty to suggest adding a whole American army and its volatile commander to his own. Nor was everybody as convinced of Monty's genius or the superiority of the Eighth Army over other Allied armies as he was himself. Admittedly, he had defeated Rommel at El Alamein and then moved the Eighth Army nearly 2,000 miles from Egypt to Tunisia, but some people thought he could have moved a good deal faster, and not all of those who thought so were American.

In the end, however, under the pressure of time and the need to reach some sort of decision about the shape of Husky, Monty got a surprising amount of what he wanted, to Patton's disgust. Patton would support him on his left flank, while the Eighth Army made a strong attack north to seize Syracuse and gain the Navy a port, and to secure the airfields at Gela that were the absolute minimum Tedder demanded for the air forces, then advance rapidly to Messina and cut the Germans off. Although the rivalry between Patton and Monty and the "race for Messina" have since caught people's attention, the real quarrel was less between Monty and Patton than between Monty and his fellow British commanders. Monty's insubordination toward Alexander was concealed by the latter's instinctive courtesy and good manners, but no such smokescreen hid the real dislike and anger on the part of Air Chief Marshal Tedder and Admiral Cunningham, who, like many other senior officers in the British armed services, regarded Monty as "not quite a gentleman"* and as an unforgivable self-promoter with a tendency to dither on the battlefield and "regroup" after victory. As for Patton, he recognized and respected Monty's professionalism, though he disliked Monty as a man. Patton's real quarrel was with his old friend Ike, whom he regarded, with increasing bitterness, as having sold out to the British, and as having become more interested in running a smooth coalition than in giving American forces a chance to distinguish themselves in battle.

Ike was under no illusion, despite the optimism he habitually showed in public, that this was a happy team, but with his usual pragmatism, he knew it was the best one he had at his disposal. He kept a firm grip on Patton, and let Alex deal with Monty, very sensibly recognizing that whatever command arrangements were decided on beforehand, everything would ultimately be decided on the battlefield. Ike would give Monty a good deal of what he wanted, but once his troops were onshore Monty would have to deliver what he had promised.

This, as it turned out, would be Monty's Achilles' heel. Even his own admiring biographer would write of him later, "His conceitedness irked even those most devoted to him"; he had a tin ear for American sensitivities and an ill-concealed contempt for the fighting ability of American troops (while overrating that of his own); he disliked "coalition warfare" as much as Patton did; and for all his boastfulness and his need to play the starring role, his pursuit of Rommel after his victory at Alamein had been as disappointing as Meade's pursuit of Lee after Gettysburg. In terms of sheer professional and technical ability, Monty was perhaps the Allies' best general of World War II, as Patton himself recognized grudgingly, and he was one of the very few who could match the proficiency of such

* In his biography of Patton, Carlo D'Este attributes this remark correctly to General Lord Gort, Commander of the British Expeditionary Force in France from 1939 to 1940; but it had been said about Monty as long ago as before World War I, and many people continued to feel that way about him even after his elevation to the peerage as Viscount Montgomery. In part this was a reflection of the fatal British preference for dashing, attractive upper-class amateurs over "swots," and many applauded General Lumsden, the Eighth Army's flamboyant tank commander whom Monty sacked in 1942, when Lumsden explained the incident on his return home to London in the bar of White's by saying, "The desert wasn't big enough for *two* shits."

German commanders as Manstein and Guderian; but he invariably oversold what he intended to do, then tried to explain away what looked to other people like a stalemate or a partial success as a triumphant victory. Whatever Patton and Monty might think of him, however, it was Ike's job, and first concern, to be a "coalition commander"—however difficult it might be for an American officer to be totally impartial.

In the meantime, Ike, rather surprisingly, perhaps motivated by a need to prove that he could make war without Alex or Monty to help, launched an attack (oddly code-named Hobgoblin) on the small island of Pantelleria, almost halfway between Cape Bon, Tunisia, and the southern coast of Sicily. Pantelleria had been fortified before the war by Mussolini, and was well known for its sweet dessert wines, capers, raisins, dried figs, and volcanic thermal baths, though any attempt to portray it as an island paradise is compromised by the fact that the Romans used it as a place of exile for failed politicians. The idea of taking it was in part an effort to please or perhaps momentarily silence Winston Churchill, whose ample mind was furnished with an exhaustive atlas of Mediterranean islands that he believed could be taken at a small cost in casualties, of which Pantelleria—somewhat grandiosely described as "the Gibraltar of the central Mediterranean"—had been one for some time; and in part a concession to the RAF and USAAF, which not only wanted Pantelleria as a base for fighter planes, but also were anxious to prove that it could be taken by bombing the garrison into surrender, as a demonstration of airpower. At this stage of the war, the dominating reality of air operations was still the short range of existing American and British fighters. The P-40, the Hurricane, and the Spitfire still had no provision for auxiliary "drop tanks," and their useful combat radius was only 200 miles or so, exactly twice the distance from Pantelleria to the landing zones in southeastern Sicily. There was thus certainly a plausible case to be made for taking Pantelleria, which had a good airfield, even though it risked drawing the Germans' attention to Sicily as the Allies' next objective. The island itself was difficult to assault—it had a rocky coastline, with no beaches; nature might have formed it for the express purpose of being an inhospitable place on which to land paratroopers; and it had only one small port. But for a period of six days and nights the RAF and the USAAF bombed it relentlessly, dropping between them over 5,000 tons of high explosives on an island of thirty-two square miles. Not surprisingly, the Italian garrison of 11,000 men surrendered eagerly on June 11 before the assault troops even went ashore, giving Ike a secure fighter base for the Sicily landings in case Monty failed to deliver Gela on schedule. Ike himself went to Pantelleria to observe the operation in person from the bridge of a British cruiser, which at one point proceeded at flank speed through a mined area, Admiral Cunningham remarking casually like the old sea dog he was that at the speed they were going the bow waves would probably push aside any mines that might be in their way, and if they *did* hit one it would just be bloody bad luck. Ike had adroitly managed to prevent the prime minister, who was then in North Africa, from accompanying him, arguing that Churchill's life was too important to risk. Being excluded rankled Churchill, and he neither forgave nor forgot it—a year later it would fuel his determination to observe the D-day landings from close to the beaches.

Victory at Pantelleria seemed to buoy Ike up. It was small, but the operation had gone like clockwork, and cooperation between army forces, air forces, and naval forces had for once been flawless, though admittedly resistance had been minimal. Still, it seemed like a good augury for Husky, which would take place four weeks later. Ike reported to Mamie that he'd just had some of the busiest days of his life, and that he'd just taken "quite an interesting and somewhat exciting trip"[12]—no doubt a reference to the mines.

Churchill mellowed as the date for Husky approached and with the news that one of his pet Mediterranean islands had been captured. He had arrived before Hobgoblin, and his presence in Algiers did not make life any easier for Ike, who was obliged to keep the prime minister company until the small hours of the morning, drinking, smoking, and talking strategy until Churchill was ready for bed, which was seldom before two o'clock. Kay Summersby also reported that the prime minister, who reveled in being able to take a long bath with unlimited hot water—hot water was strictly rationed in Britain—would make Ike sit on the toilet of the PM's bathroom so they could talk while the PM lay in the tub under a layer of bubble bath smoking a cigar, until the room was steamy. Ike always had to "change into a fresh uniform afterwards, because the steam was murder on [his] trouser creases."[13] Churchill, at any rate, whatever his thoughts after the war, had no objection to Kay's presence—he enjoyed the company of pretty women at the dinner table, so long as they kept quiet, and whenever he invited himself to dinner at Ike's villa, he would always say, "Now, tell Kay to come. I want to see her." She in turn commented on his abominable table manners and his ruthless determination to hog the conversation. The two men got along famously, partly because Ike was genuinely fond of Churchill, and partly because he was always prepared to argue when he thought Churchill was wrong—something not everybody had the guts to do.

The purpose of the prime minister's visit so close to the date for the Sicilian landings was to reopen the discussion of what to do "once we had

Ike and Churchill, North Africa, 1943.

Sicily in our grasp," this time without the presence of Roosevelt. Churchill had come to Algiers directly from an exhausting and somewhat unsatisfactory two weeks of conferences with the president and the Combined Chiefs of Staff in Washington about the Allies' strategy, and went from there to North Africa by flying boat (the longest leg involved a seventeen-hour flight) via Newfoundland and Gibraltar, accompanied by General Sir Alan Brooke, the Chief of the Imperial General Staff, as well as by the prime minister's personal chief of staff and military adviser, the faithful and long-suffering General Sir Hastings Ismay. General Marshall and the American chiefs of staff flew from Washington separately, while Anthony Eden, Harold Macmillan, Generals Alexander and Montgomery, Kay's old boss Carl Spaatz, "Beetle" Smith, Air Chief Marshal Tedder, Admiral Cunningham, and Robert Murphy were summoned to Algiers or were already present to appear as needed. Waiting in the background were Generals Giraud and Juin in case the subject of the French forces in North Africa came up.

The British were there to support Churchill in expressing his continuing doubts about the wisdom of Round-Up (soon to be merged with Bolero, the buildup in the United Kingdom of the invasion forces, and renamed Overlord), the cross-Channel invasion, which remained the Holy Grail of the Allies' strategy but which the prime minister likened to putting all one's eggs in one basket; and to press for further operations in the Mediterranean once Sicily had fallen. The Americans were there to resist any operation that would delay or weaken Round-Up. To nobody's surprise, Ike reported, "Mr. Churchill was at his eloquent best in painting a rosy picture of the opportunities that he foresaw opening up to us with the capture of Sicily."[14] Many of those present had heard all this before several times in Washington, where the prime minister's eloquence had failed to change the mind of President Roosevelt and General Marshall, and those on the British side of the table were already so familiar with the prime minister's views on the Mediterranean that some of them were seen to nod off at the first mention of Turkey, the Greek islands, or the Balkans.

It is typical of Anglo-American relationships at the time, and of Ike's new role as the supreme mediator between the two nations, that as the "host," as Churchill described him, as if it were a country house party, Ike sat at the head of the table, with the other Americans seated down one side, and the British, with the prime minister at the center, down the other. It was not quite a confrontation, but it looked like one, and at times threatened to become one. Churchill did not hesitate to point out that the British had "three times as many troops, four times as many warships, and almost as many aeroplanes available for actual operations as the Americans." They had "since Alamein, not to speak of earlier years, lost in the Mediterranean eight times as many men and three times as many ships" as their ally. They had, in short, paid their dues; and they expected to see some return on their investment in the Mediterranean, rather than having operations there cease altogether with the capture of Sicily in order to shift men and landing craft to the United Kingdom for an invasion whose success they doubted. Even Ike, though he was firmly committed to the cross-Channel invasion—doubly so in the formidable presence of Marshall—was nevertheless tempted by the chance to seize the excellent modern airfields in southern Italy, glittering prizes for Tedder and Spaatz, and by the possibility that Italy might surrender, obliging the Germans to move substantial forces from France to Italy.

Ike described the plan for the landing on Sicily—without dwelling on the issues that still divided the commanders—and said that he had had a long talk with Brooke, who had emphasized that the Soviet Union had the only army that could meet the Germans on equal terms and produce "decisive results" in 1943—to put matters in perspective, 218 Soviet divisions were at that moment engaged against over 100 German divisions on the eastern front. Ike said he agreed with Brooke, and himself felt that if an attack were to be made on Italy, it should take place as soon as Sicily had been captured, "with all the means at our disposal." In short, he believed that the Allies' strength in the Mediterranean should not be squandered on taking more islands or on remote Balkan adventures, but should be used quickly to knock Italy out of the war, then shifted immediately to Britain for the invasion of France. He estimated that by May 1944 there would be twenty-nine British and American divisions in Britain, including seven from North Africa—enough for the invasion of

France, though hardly for a full-scale encounter there with perhaps as many as seventy-five German divisions. Clearly, diluting the Allies' strength by irrelevant attacks in the Mediterranean would be fatal.

Brooke—whom Ike had initially judged to be "adroit rather than deep, and shrewd rather than wise," but whom he had come to respect as a "brilliant soldier"[15]—had had a hard time of it over the past two weeks in the constant company of the prime minister, whose demands on others were unsparing and who was, as usual, difficult for his military advisers to keep on what they regarded as a steady course. At one point in Washington, Marshall had remarked to Brooke, in mild despair at the prime minister's eloquence on the subject of Italy and the Balkans, that he looked on Britain's Mediterranean strategy with a "jaundiced eye" and thought that a cross-Channel invasion and the liberation of France would "finish the war quicker."[16] Brooke had replied sharply, "Yes, probably, but not the way we hope to finish it," pretty much summing up the difference between the two allies.

By this time, Ike and Brooke—an unlikely duo, the one gregarious and good-humored, the other frosty and sharp-tongued—were on good terms, despite Brooke's reservations on the subject of the cross-Channel landing and his doubts about Marshall's ability as a strategist.* On Brooke's first morning in Algiers, Ike had surprised Brooke by walking into his bedroom wearing around his neck the cross of a grand commander of the French Legion of Honor, General Giraud having just bestowed his own on Ike at a ceremony early that morning. Giraud's right to confer the Legion of Honor on anybody was in some doubt—De Gaulle was furious when he heard of what he took to be lèse-majesté on his rival's part—and Ike had promised Giraud not to wear it again until they met in a liberated Metz, but he slipped it back on to wear while waking up the exhausted Brooke.

Now, at the meeting in Algiers, Brooke, rather surprisingly, suddenly brought up the question of what would happen if Italy collapsed, and asked

* As a result Brooke had no difficulty in recognizing at once just how angry Ike was at Monty's insistence on being given an American B-17 and its crew, and gave Monty a severe scolding on the subject, though Monty, blind as always to other people's feelings, continued to believe it was a good joke that Ike had enjoyed as much as he did. Ike nurtured a dislike for Monty that had serious repercussions in the invasion and the battle for western Europe, and it persisted to the end of his life.

if Ike had considered on what terms he would accept its surrender. Even Churchill thought that "this was getting on very fast," and the subject was dropped. That was a pity, since Ike would be facing the issue much sooner than anybody at the table thought. In the end, and no doubt with some relief, the participants decided to leave further "exploitation of the Sicilian operation" to Ike's judgment, while encouraging him "to take advantage of any favorable opportunity" to "rush into Italy" in order to seize the Foggia airfields. Possession of those airfields would also oblige him to take Naples, since it was the only major port through which the air forces—with their enormous needs for fuel, bombs, and heavy equipment—could be supplied.

Though nobody seems to have remarked on it at the time, this left a huge responsibility in Ike's hands—the decision about Italy was his to make. He was in sole command. No other Allied commander in the war had ever been or would ever be granted this kind of authority. It confirmed once again Ike's unique position, not only as commander in chief of British and American forces in the Mediterranean and commander of all American forces in the European theater, but as the judge of whether to invade the mainland of Italy, when to do so, and ultimately on what terms Italy's surrender might be accepted. For somebody who had only just reached the rank of colonel in 1941, and had never expected to rise any higher, this was an astonishing transformation.

This had been a high-level meeting in every sense, and not only because of the presence of the prime minister. Also present throughout were the three men who were candidates to command the cross-Channel invasion when it took place. It would be crass to speak of "ambition" with regard to George Marshall, who was utterly selfless, but it was certainly his dearest wish to command the invasion of which he was the chief enthusiast and architect. He would never plead his own case, of course, but his feelings on the matter were well known to President Roosevelt, who had higher esteem for him than for any other American military figure. General Sir Alan Brooke, Marshall's British counterpart, was equally selfless, but his desire to cap his distinguished career as a soldier by leading the Allied armies in the invasion of Europe was well understood by the prime minister, who, always more moved than Roosevelt by sentiment, was fond of "Brookie," the irascible and often argumentative Ulsterman, and had great respect for him as a soldier. As for Ike, he was the only one of the generals present to have led a combined Anglo-American army in battle and commanded a major invasion—and slow and fumbling as the fight for Tunisia had been, it was undeniably a victory.

Toward the end of the meetings in Algiers, Ike took Churchill and the prime minister's party to his advance camp, where they spent the night under canvas. The prime minister was in good spirits, having driven to Carthage in the morning to address a large number of British troops gathered there in a Roman amphitheater, where the "acoustics were so perfect that no loudspeakers were necessary." At dinner, under the stars, he mused to Ike that he had spoken "where the cries of Christian virgins rent the air whilst roaring lions devoured them, and yet I am no lion and am certainly not a virgin."[17] If Ike seemed to feel a slightly guilty fascination with the prime minister, whose views on any number of things were often at odds with his own and those of the United States, it is easy enough to understand. What other leader of a great nation could reach such flights of fancy in the midst of war, while still remaining a hardheaded negotiator for his country's interests? Even the presence of Churchill's bumptious and argumentative son Randolph could not spoil the evening.

A month before Husky was to begin, Ike took another serious risk, announcing to assembled international journalists the timing, the place, and the date of the landings on Sicily, and thus breaking the long-standing tradition of maintaining complete secrecy about military operations until the last minute. He attributed this decision to "an inborn hatred of unexplained censorship, and to "confidence . . . in the integrity of newsmen," which was as unusual among senior military officers then as now. His reasoning was flawless—it was impossible to hide from the press the preparations being made all over North Africa for a major invasion, and in the ordinary course of events the correspondents would naturally speculate in their stories on the purpose of all this activity. From these stories, the enemy might very well be able to conclude that Sicily was the target. By taking the correspondents into his confidence—he told them everything there was to know about the forthcoming landings—Ike placed on them the moral responsibility for any leaks or stories that might help the Germans, and their self-restraint was more effective than even the most rigorous censorship could have been. It is hard to say who was the more astonished at Ike's openness with the press corps—Ike's commanders or the correspondents themselves—but it worked. This underlined once again Ike's strong dislike of censorship, or of "doctored" official news, which had revealed itself before when he refused to block the publication of stories about race relationships between white and black American troops in England in 1942.

A few days before the invasion Ike, Alex, and Admiral Cunningham moved to Malta, only fifty miles from the southern beaches of Sicily, where the Royal Navy's "splendid naval communications," apparently unlike those at Gibraltar, made it the natural place for a forward headquarters. The day before the landings was tense—almost like a rehearsal of D-day. Strong winds from the west—once again the Mediterranean was proving treacherous, despite its benign appearance—increased steadily in force during the day, raising doubts about whether the troops could get ashore. The landings on the east coast were protected from the wind by the island itself, but those on the south coast were exposed to its full force. Ike spent a difficult day trying to decide whether or not to cancel the entire operation. As evening fell, he and Cunningham took a stroll outside to look at the sea, and at the wind indicators, which were registering a Force 5 gale, causing even the normally sanguine British admiral to shake his head in dismay. Ike received an urgent message from Marshall in Washington asking if "the attack was on or off." The truth was that Ike didn't know. The time was fast approaching when it would be too late to cancel the landings, and the wind was still rising, producing high seas and heavy surf on the beaches. Ike felt the full burden of his responsibility. It was his decision to make, and his alone, and although the wind velocity was still increasing, he finally decided that the landings would proceed at dawn as scheduled—"those on the east," he thought, "would surely get ashore and we would have less confusion and disadvantage than would result from any attempt to stop the whole armada."[18]

The landings began at two-thirty in the morning on July 10. Patton's Seventh Army was on the left and Montgomery's Eighth Army on the right, and despite the bad weather, they were initially successful—indeed, the weather caught the Germans off guard, since it looked too severe for an invasion to be possible. In the tradition of Lord Nelson, Admiral Cunningham, unable to resist being near the action, commandeered a Royal Navy destroyer and went off in it to see for himself what was happening. He reported back to Ike that the landings in bad weather "constituted one of the finest exhibitions of seamanship it had been his pleasure to witness in forty-five years of sailoring." From Syracuse on the east coast to Licata on the south, along a coastline of almost 150 twisting miles, the Allied forces encountered minimal resistance. Indeed, on the extreme left the U.S. Third Division made history by using the new LSTs, LCIs, and LCTs, which allowed them to go directly from embarkation port to landing beach without transferring at sea to smaller Higgins boats, and to arrive on the beach with all their tanks, vehicles, and artillery—a formidable step forward in amphibious

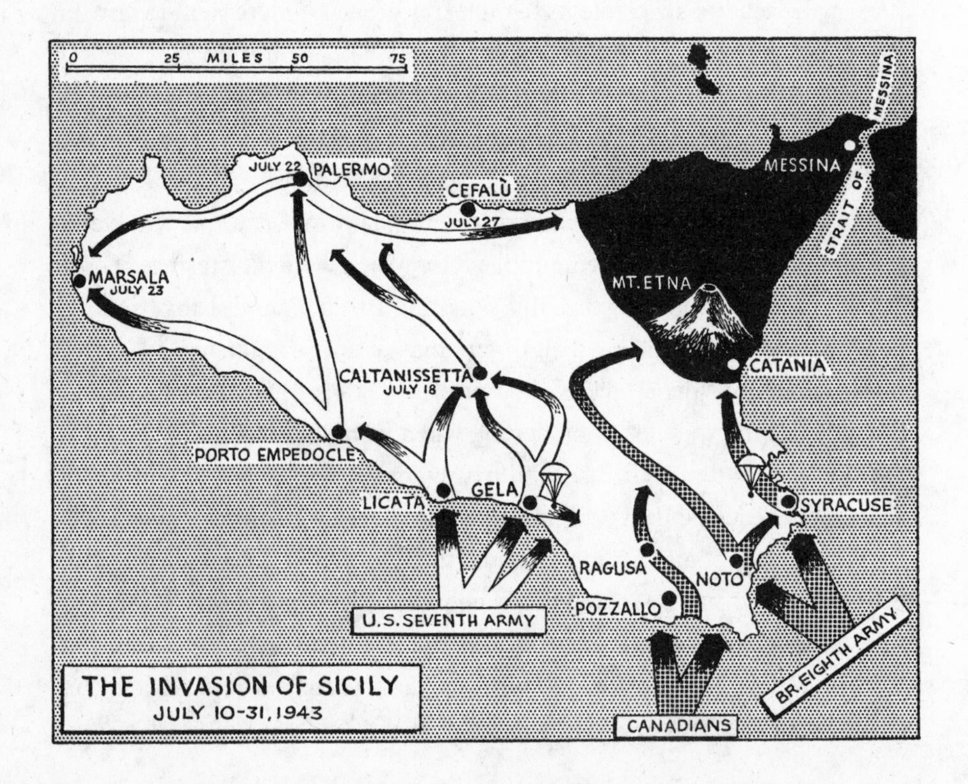

warfare. Another new weapon that made a critical difference in the landings was the DUKW, a wheeled, load-carrying four-wheel-drive truck shaped like a small barge that could "swim" to the beach under its own power, then continue on land. Ike would later comment on the fact that the "weapons" that made the most significant contribution to the Allied victory were all unarmed—the DUKW, the Jeep, the bulldozer, the 2.5-ton four-wheel-drive truck (known as the deuce-and-a-half), and the Douglas C-47 transport plane.

Husky was a triumphant success—the only disappointment was that the airborne landings once again demonstrated their extreme vulnerability. American paratroopers were blown off course by the high winds. British airborne troops, landing in gliders, were not only scattered all over the countryside, but fired on by the ships of the invasion fleet—a problem that would be repeated several times over the next three days as antiaircraft gunners all over the fleet opened fire at the first sight of the C-47s and asked questions later. Even more tragically, some of the gliders were cast loose from the aircraft towing them too far out by nervous pilots, and landed in the sea. The problem of getting transport pilots who were neither trained nor equipped for combat to fly a straight, steady course at low speeds and heights through antiaircraft fire until they reached the designated drop zone would recur on D-day.

On July 11 Ike came from Malta to see for himself what was happening, just in time to watch a full-scale German armored attack on Bradley's divisions at Gela, which was eventually repulsed, despite heavy casualties. He landed onshore at Cape Passero to give the Canadian division there "a welcome to the Allied command." He himself was awed at what he saw: "Up to that moment no amphibious attack in history had approached this one in size. Along miles of coast line there were hundreds of vessels and small boats afloat and antlike files of advancing troops ashore. Overhead were flights of protecting fighters."[19]

By July 15, Allied forces had reached a line more than thirty miles inland, captured the ports of Syracuse and Augusta, and taken several airfields. The Germans had been duped into believing that the Allies would land in western Sicily, nearer to the ports of Algeria and Tunisia, and had moved the bulk of the available forces there, leaving the southeastern coastline thinly defended. They hastily ordered the withdrawal of Italian forces in western Sicily, but the effect was an almost total collapse of Italian resistance. Ordered to stand and fight, the Italians might have done so, but in retreat all except the elite Italian divisions simply dissolved. By July 22, just over ten days after landing in Sicily, and after a speedy end run that would have done credit to Stonewall Jackson, Patton was in Palermo and almost three-quarters of the island was in the Allies' hands.

At this point, however, things started to go wrong. Part of the problem was due to simple geography—the famous volcanic Mount Etna lay between Monty's forces and Messina, its eastern slopes descending directly to the sea. Montgomery would have to go around it to move north toward Messina, either by clinging to the narrow coast road or by attacking to the west in the direction of Catania and then moving around Mount Etna over the rugged terrain on the other side, where there were very few decent roads. He chose to try the latter, but by now the Italian forces that had collapsed before the Allies' rapid advance were being replaced with German divisions brought across the Strait of Messina from the Italian mainland. An "outflanking attack," made by the First Canadian Division and the British Fifty-First (Highland) Division, failed to dislodge the Germans.

Montgomery had continued to hold the notion that Patton's role was to "support" his advance toward Messina on his left, and he still had the illusion that he himself was in command. As usual, his meticulous planning had paid rewards—the vital port of Syracuse fell to the Eighth Army on July 11, and except for some rather sharp actions against the unfortunate paratroopers, casualties were low. Monty wrongly assumed, for some reason, possibly wishful thinking, that Patton's Seventh Army "was not making very great progress," but he anticipated that this would change as he pushed forward, "loosening resistance in front of the Americans." After that, the Americans, he thought, could "draw enemy attacks on them," permitting him to "swing hard with my right." This was a singularly incorrect and even insulting view of the situation, but also irrelevant, since Patton had no intention of drawing enemy attacks on the Seventh Army for Monty's benefit. Patton had already quarreled bitterly with Omar Bradley, accusing him of not being "aggressive enough" because Bradley wanted "to consolidate his bridgehead" before advancing inland, and complaining to Ike about him. The mild-mannered Bradley did not forget this when circumstances would shortly reverse their positions and place him in command of Patton.* Now, however, nothing was going to stop Patton from moving forward as fast as possible, even if, tactically, he was going in the wrong direction. Monty had not been altogether wrong—a single, strong commander with an objective, wide-angle view of the battle and a firm control over it was exactly what was needed at this point, though he was not the right man. Unfortunately, neither was Alexander, who was too easygoing to restrain either Patton or Monty, and acted like a rather remote CEO of the battle, with Ike playing the role of an even more distant chairman of the board. One large main attack might have succeeded in taking Sicily, if everything had been poured into it, but instead the battle deteriorated into two almost unconnected attacks in opposite directions. On the night of July 12 "the sky above the Catania plain filled with unexpected German transport planes," as the Germans, recognizing that their Italian ally was not going to make a fight of it, reacted at last in force. Two regiments of the German First Parachute Division, the Twenty-Ninth Panzergrenadier Division, and elements of the elite Hermann Göring Division were on the scene. In the north, where the coast road from Palermo to Messina was the only possible route, the Germans skillfully slowed down Patton's advance, and in the south they managed to block Monty from taking Catania or breaking through to the west of Mount Etna, in some of the bitterest fighting the Eighth Army had ever experienced. Any chance of a quick coup de main, if there ever was one, had evaporated. In Ike's words, "along the great saw-toothed ridge of which the center was Mt. Etna the German garrison was fighting skillfully and savagely. Panzer and paratroop elements here were among the best we encountered in the war, and each position won was gained only through the complete destruction of the defending elements."[20]

* Bradley was the wrong man for Patton to pick on: first, because despite his quiet demeanor he would not allow himself to be bullied; and second, because he had been a classmate of Ike's at West Point. His opinion of Patton was succinct: "I would have relieved him instantly. . . . He was colorful but he was impetuous, full of temper, bluster, inclined to treat the troops and subordinates as morons. . . . He was primarily a showman. The show always seemed to come first."

It did not help matters that the Catania plain was a "pesthole" of malaria. Ike had brought experts on malaria in from the United States, but the troops still got sick in larger numbers than in any other area of the Mediterranean. Difficult as Patton might be—and he was about to plunge into the hottest water of his career—and disloyal as he was capable of being (he was now in the habit of referring to Ike in his diary as "D.D." for "the Divine Destiny" because of Ike's determination to remain impartial between the American and the British armies under his command), Ike nevertheless had more confidence in him than in Monty. Patton, in his view, understood the importance of speed in warfare, and speed mattered more than anything else. "Speed," Ike would write, "must be redoubled—relentless and speedy pursuit is the most profitable action in war. . . . To insure rapidity of action all commanders, and troops, must recognize opportunities and be imbued with the burning determination to make the most of them."[21]

Since Monty had conducted perhaps the most leisurely pursuit in the history of modern warfare in advancing from the Egyptian border to Tunisia after his victory over Rommel at Alamein, and was in the process of "digging in" south of Catania, this comment about Patton may be taken as criticism of Monty—about as sharp and pointed as Ike would permit himself to be in writing about his own commanders once the war was over. And indeed, although Monty was in one of his sniffiest moods, Patton's dash, whatever his motives, paid dividends. Patton took Palermo, even though Monty thought it didn't need taking, then drove his troops hard along the coast road toward Messina, which Ike described as of the " 'shelf' variety, a mere niche in the cliffs interrupted by numerous bridges and culverts that the enemy invariably destroyed as he drew back fighting." Attempting to outflank the Germans the only way he could, Patton carried out a series of amphibious operations with the help of the U.S. Navy, landing behind the Germans as they retreated down the coast road and attempting to cut off their retreat. Bradley dismissed these actions as useless showmanship (and wasteful of lives); and it was true that the troops landing from the sea often had to be rescued by the troops advancing up the road, fighting their way forward against fierce resistance to meet them, which meant that casualties were mounting to no purpose. But in the meantime, the Germans were undeniably falling back until they were within a couple of days' march from Messina. Patton was always in motion. He sent off another major attack inland, directly to the east, north of Mount Etna, intending to make a kind of "march to the sea," in Sherman's tradition, which would meet with the advancing British Army at Taormina. This began with a brutal, bitter battle at Troina, which Ike described as "one of the most fiercely fought small actions of the war," and in which the (U.S.) First Division was subjected to no fewer than twenty-four separate counterattacks that left the rocky, arid terrain littered with corpses, by no means all German.

In fact, neither Patton nor Monty seriously upset the Germans' schedule for methodically withdrawing to Messina, but the collapse of the Italian army, the Germans' evident decision to carry out a fighting retreat and evacuate their forces to the mainland, and the impending loss of Sicily to the Allies produced a significant and unexpected political result in Rome. On July 25, after a stormy meeting of the Fascist Grand Council, King Victor Emmanuel III dismissed Mussolini from office, and replaced him

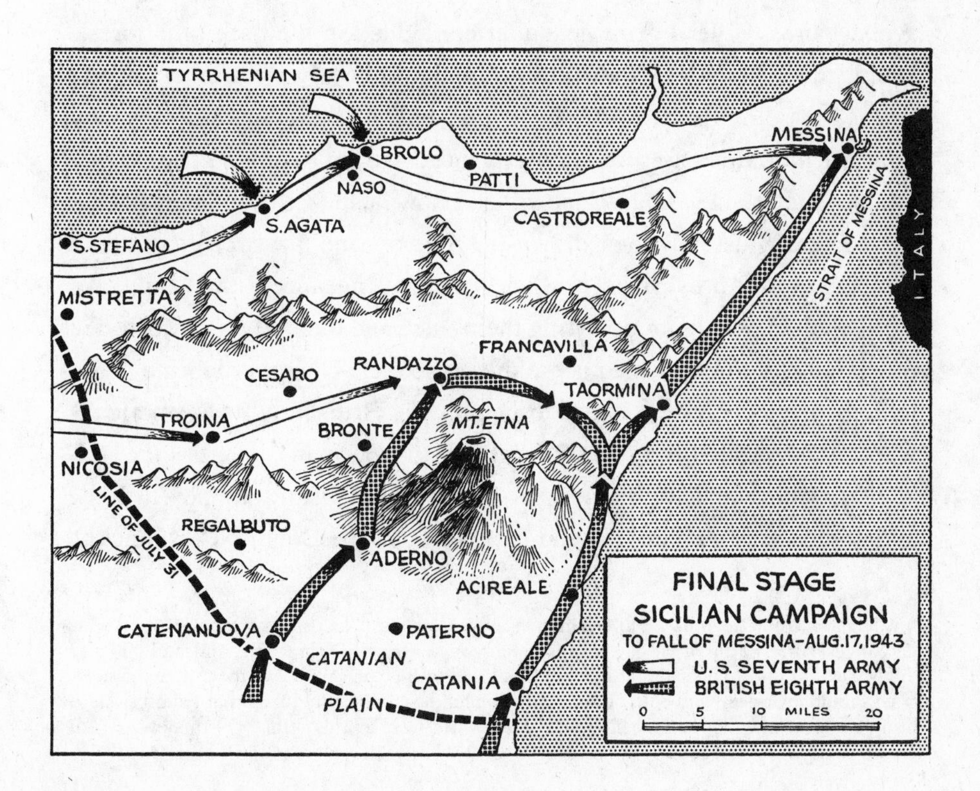

with the elderly Marshal Pietro Badoglio. The Sicilian campaign had not been flawless in conception or execution, and it was not by any means over yet, but already it had paid a huge dividend. Hitler's partner, friend, and onetime political mentor was gone at last, shorn of his power and under arrest. Mussolini's popularity had been in sharp decline since 1942—his attempt to carve out the jackal's share of Hitler's conquests had cost the Italians their entire African empire and inflicted on them an endless, mournful string of defeats everywhere from Abyssinia to Russia. But in one heart, at any rate, Mussolini was still admired; the Führer had long ago pledged loyalty to him, and nothing, not even defeat or disgrace, would cancel that baleful friendship, or release Mussolini from its obligations.*

The Italians might rid themselves of the Duce, but they would not rid themselves that easily of the Germans. Before the fighting was even over in Sicily, the new Italian government was already sending out "peace feelers" to Ike, desperate to take Italy out of the war before the Germans overran it and the SS executed everybody involved in overthrowing Mussolini. At Wolfschanze, Hitler's grim headquarters hidden in the dark pine forests behind the eastern front, the Führer was already deeply involved in plans to rescue Mussolini from his captors, punish Italy for betraying its leader, and fight off the Allies as soon as they landed on the mainland, as nobody doubted they soon would. Whatever else now happened—and hardly anybody foresaw just how long-drawn-out, bloody, and disastrous the Italian campaign would prove to be, or what a terrible vengeance Hitler would take on the Italian people for deserting their Duce and their ally—it was to Ike that Badoglio now appealed, as soldier to soldier. The most delicate negotiations of the war were about to begin. The British and the Americans wanted Italy to join the Allies in fighting the Germans, whereas the Italians, very conscious of the number of German divisions already in Italy, and of Germany's ability to send more through the Brenner Pass quickly,

* When Mussolini acquiesced in Germany's *Anschluss* with Austria on March 11, 1938, Hitler called Prince Philip of Hesse, his special envoy to Mussolini, in Rome, and said, "Please tell Mussolini I will never forget him for this. . . . I will never forget, whatever may happen. If he should ever need any help or be in any danger, he can be convinced that I shall stick to him, whatever may happen, even if the whole world were against him." On this promise at least, Hitler never reneged.

"wanted frantically to surrender . . . but only with the assurance that such a powerful Allied force would land on the mainland simultaneously with their surrender that the government itself and their cities would enjoy complete protection from the German forces."[22] Needless to add, the Italians were in for a bitter disappointment. Thus, less than a month after Churchill and the Combined Chiefs of Staff at the meeting in Algiers had dismissed as premature General Brooke's question about the terms on which Ike would accept Italy's surrender, Ike was in fact facing just that question.

In the meantime, neither Patton's reckless dash toward Messina nor Monty's slower, hard-fought progress up the eastern coast road from Catania, which he did not take until August 2, succeeded in trapping the German forces in Sicily. On August 15 Patton was still twenty miles from Messina, and Montgomery more than thirty miles. Messina did not fall until August 17, by which time, "despite Allied air and naval superiority," as *The West Point Atlas of American Wars* notes disapprovingly, the Germans had "evacuated some 100,000 troops, 9,800 vehicles and 47 tanks" to the mainland, including three intact German divisions.

To Monty's annoyance, Patton, as he had planned from the moment he reached Palermo, entered Messina before the British Eighth Army—a triumph for Patton which Monty affected to ignore, but which nevertheless rankled. Punishment for hubris was in the wings, however. On August 3, his own nerves no doubt frayed by the stubborn resistance of the Germans, Patton had flown into a rage at a shell-shocked soldier in a base hospital, calling him a coward and personally throwing him out of the medical tent. Six days later he encountered another shell-shocked soldier. This time, Patton lost any semblance of self-restraint; he cursed at the soldier, threatened to put him up against a wall and shoot him, drew his pistol, waved it in front of the terrified young man's face, then slapped him in full view of the doctors and nurses. It was not easy for the medical personnel to restrain the commanding general, but they eventually managed to intervene to protect the soldier, while Patton, now completely out of control, ranted on about cowards who "should not be allowed in the same hospital as brave wounded men."

Attempts to suppress the story failed, and rumors about it spread swiftly throughout the Seventh Army. Patton was so oblivious of the consequences of his own behavior that he boasted about the incident to Bradley. Despite his own disapproval, Bradley did his best to prevent the report of the incident by the surgeon in charge of the Ninety-Third Evacuation Hospital from making its way up the chain of command to Ike. The story had already reached correspondents, however, and in the long run there was no way to stop it from appearing in newspapers at home. When Ike learned about it, he was infuriated. He chose to attribute it to Patton's "highly emotional state," and to see in it a reflection of "the emotional tenseness and . . . impulsiveness [which] were the very qualities that made him, in open situations, such a remarkable leader of an army."[23] But it seems more likely that Patton, who constantly worried about his own courage and had to struggle—albeit always successfully—to suppress his fears when under fire, was reacting to exactly the weakness he sensed in himself. Bradley, in his calm, commonsensical way, was surely correct in telling Patton that "every man has a breaking point, some are low, some are high," though the truth is also, as he surely knew, that some men can screw up their courage to perform remarkable deeds of bravery, at great cost to their psyche, while others are naturally brave, and that Patton by his own admission was in the former category, and suffered before and after every moment of danger. In any case, Bradley was wasting his breath. Patton was incorrigible in his inability to restrain this kind of outburst, and remained so to the end of his life.

Ike realized better than anyone else that Patton had incurred a court-martial by striking an enlisted man, and understood immediately that the incident could cost Patton his career. But once he had calmed down, he concluded that he could not afford to lose Patton, and with his usual brisk efficiency set about putting out the fire that Patton had lit—a decision for which Patton should have been, but was not, everlastingly grateful. Instead of informing Marshall, Ike kept the incident "in the family" for as long as possible. He met with the correspondents who had the story, firmly refused their demand that he fire Patton in exchange for their suppressing it, and told them frankly and in detail that he intended to keep Patton and how he intended to punish him. He did not attempt to apply censorship—"They were flatly told to use their own judgment"—and when the story did eventually reach home it was in the form of gossip from servicemen. At this point it was picked up by Drew Pearson and created a scandal, though the effect was softened by the fact that it was already old news.

As for Patton, Ike wrote him a severe letter or reprimand; promised to relieve him of his command if such an incident ever happened again; and ordered him to apologize to the two young soldiers, the medical personnel who had been present, and "representative groups" of officers and enlisted men from each of the divisions he commanded. So far as Ike was concerned, that closed the incident, though Patton, with his natural instinct for showmanship, went so far overboard that his apologies bordered on insubordination. He told one group of soldiers assembled to hear his apology, "I thought I'd stand here and let you fellows see if I am as big a son-of-a-bitch as you think I am," and received an ovation from them. Patton understood, as many of the correspondents and the politicians at home did not, that most soldiers had as little sympathy as he had for a man who broke down under fire—breakdowns endangered their lives, and if tolerated encouraged a fatal decline in morale. Writing to Ike, however, Patton took a humbler tone: "I am at a loss to find words with which to express my chagrin and grief at having given you, a man to whom I owe everything and for whom I would gladly lay down my life, cause to be displeased with me."[24] It would not, unfortunately, be the last time Patton "displeased" Ike.

Interestingly, Ike's private view of this incident was more restrained. It was not what Patton had done but where he did it that offended Ike most. "In forward areas," he wrote later, with plenty of time to put the matter in perspective, "it is frequently necessary . . . to use stern measures to assure prompt performance of duty by every man." Hesitation or "shirking" must be "sternly repressed."[25] What Patton had done might not in fact be an offense at all in a platoon or a company under fire, but it was unforgivable within the confines of a hospital, away from or after actual combat. In short, Ike recognized, as every professional soldier must recognize, the absolute need to enforce discipline and punish shirking in combat by any means, and the impossibility of condoning cowardice—no group under fire, he thought, can hold together if one member refuses to do his duty, or if his superiors allow that soldier to get away with it. What Patton had done would be, in battle, merely "an incident," Ike wrote, which would have passed without notice—it was the time, the place, and the great disparity of rank between Patton and the soldiers he assaulted that made the slapping incident so grave.

Although it took awhile for the depth of his disgrace to sink in, the immediate result was that Patton would be stuck in a backwater commanding a diminishing Seventh Army as its troops were siphoned off to be shipped to the United Kingdom for the cross-Channel invasion, in which there was so far no role for Patton himself. Meanwhile his junior in years and seniority Omar Bradley was named to command the American forces designated for the invasion—a role Patton had wanted and expected to play.

There was no role for him either in the forthcoming invasion of Italy—perhaps unfortunately, because speed and concentration of forces were exactly what was missing from the plan. The great difficulty was the Machiavellian diplomacy of the new Italian government, which was negotiating with the Germans to remain in the war on their side in the hope of preventing Germany from taking over Italy, while at the same time sending agents to neutral Lisbon to negotiate secretly with the Allies on the terms for Italy's surrender. Ike sent his Chief of Staff Bedell Smith and his British intelligence officer Major-General Kenneth Strong to Lisbon as his emissaries, and they were soon involved in complex negotiations that seemed to have had as their main purpose the Italians' need to spin things out for long enough to allow the Allied armies time to invade Italy and save the new government from the Germans. An apparently endless stream of official and unofficial Italian agents arrived in Lisbon, most of them bearing messages that contradicted everybody else's. There were also representatives of the Vatican who were desperate to prevent Rome (and Vatican City) from being bombed or becoming a battleground.

At one point, Ike contemplated an airborne assault by the Eighty-Second Airborne Division to seize Rome and its airfields, code-named Giant 2, a daring coup de main that was canceled at the last moment. Brigadier General Maxwell D. Taylor, as well as several American OSS agents (two of them disguised as Irish seminarians)* had been sent secretly into Rome

* One of the OSS agents disguised as seminarians was Michael Burke, later president of the New York Yankees, and for a time president of Madison Square Garden.

behind the enemy lines. There—at great risk of torture and death at the hands of the Gestapo if they were captured—they negotiated separately, and often in conflict, with various competing Italian elements. The pattern Robert Murphy had set in Algiers before Torch was, unfortunately, repeated in Rome—General Taylor negotiated with Italian generals, former fascists, royalists, right-wing political figures, and the Vatican, while the OSS agents met with labor unionists, leftist political figures, and intellectuals. Ike was plunged into a world of operatic melodrama and conspiracy that involved every shade of Italian political opinion, from communists on the left to repentant fascists on the right, as well as the Vatican. All of this came to nothing in the end. Badoglio was simply too timid, too convinced of German retaliation, or too fearful of igniting an Italian civil war between left and right to commit himself fully to the assault on Rome; and whenever an agreement seemed possible a new German division was "discovered" close to Rome, making the airborne drop impossible. Part of the problem was that the Italians wanted at least fifteen Allied divisions in Italy to protect them from the Germans, whereas the best Ike could do was put two divisions under Monty across the Strait of Messina (operation Baytown) with a further four divisions to land at Salerno and take Naples later on (Avalanche)—he had neither the troops nor the landing craft to do more. Letting the Italians know how few divisions Ike had to spare for Italy would encourage them to break off the negotiations (not to speak of the danger that they might leak the information to the Germans); but the Italians were realists, and that was exactly what they wanted to know before they made a move.

In these circumstances, it seemed to Ike that the best thing to do was to force their hand by letting Montgomery go, so on September 3 two British divisions crossed the Strait of Messina and landed on the "toe" of Italy—the first landing by the Allies on the continent of Europe since the failed raid on Dieppe in 1942. Ike had wanted them to go ten days earlier, but as usual Monty was slow to move until everything was organized to his satisfaction. Nine years later, when Ike wrote *Crusade in Europe*, he would still be unable to hide his irritation with Monty's delay at Messina in 1943. Monty moved north up the toe even more cautiously, very conscious that ahead of him were three German divisions, which had escaped from Sicily. Monty was not wrong to move cautiously, however—Calabria, the entire southern "foot" of Italy, is rugged, rocky, mountainous, inhospitable country, with poor roads in those days, strewn with villages that had been built to resist the invader as long ago as the times when he might be Phoenician, Athenian, Carthaginian, or Arab. It was ideally suited for a well-prepared rearguard defense by the Germans who were already there.

Monty was further slowed down by the failure to decide what he was to do once he was on the mainland. The Italians, as they seesawed on their way to surrender, would have been even more alarmed had they known that the Allies' next big move was Avalanche, which would inevitably leave the Germans in firm control of Rome, hundreds of miles to the north, as well as of central and northern Italy. From the point of view of the Italian government an Allied amphibious landing at Salerno was like a lifebuoy thrown into the sea several feet too far from a drowning swimmer for him to reach—by the time the Allied armies had taken Naples, Marshal Badoglio and his cabinet expected to be swinging by the neck in Rome from a piano-wire noose fastened to a butcher hook, the Gestapo's newest and most painful form of execution. On the other hand, Ike was trapped by a different set of demands. Without long-range fighters, Salerno was as far north as the Allied air forces could promise him air cover—at the same time, there was no point in securing important air bases like those at Foggia without also securing a significant port through which to supply them, and this meant that Naples had to be taken. What is more, Ike was operating within a very narrow time frame; not only units but above all landing ships urgently needed to be sent back from the Mediterranean theater to Great Britain if the cross-Channel invasion was to be mounted in the summer of 1944. Winston Churchill had exploded in exasperation because "the fate of two great nations seems to be determined by some god-damned things called LSTs and LCIs," but it was true.* The shortage of landing

* He later tidied up this remark, writing, in a long message to General Marshall, "How is it the plans of two great empires like Britain and the United States should be much hamstrung and limited by a hundred or two of these particular vessels will never be understood by history."

craft, to which Roosevelt had given a low priority early in the war in the interest of rebuilding the fleet after Pearl Harbor, was the Achilles' heel of the Allies' strategy. There were never enough landing craft to go around, the needs of the Pacific and the Atlantic could never be balanced, and every operation, great or small, depended on their availability.

The Salerno landing has gone down in history as a near-disaster, though the truth is that both the American and the British armed forces were still learning about large-scale amphibious operations by doing them, and that without this painful learning process the D-day landings would have been impossible. In any case, the difficulties of Avalanche were formidable. To begin with, command of the invasion was given to Mark Clark, who despite his unquestioned bravery in landing secretly by night in North Africa from a submarine before Torch had never commanded a large formation in combat, let alone a combined Anglo-American force of four divisions (plus U.S. Rangers, British Commandos, and a "floating reserve" of two regimental combat teams) carrying out an opposed landing on the mainland of Europe. The Fifth Army (or, as it would henceforth be rigorously known in press releases and news stories, "Lieutenant General Mark W. Clark's Fifth Army," for Clark was as much of a military prima donna as Patton and Monty) looked strong enough on paper, but too many of the American divisions were new and inexperienced, and the best British divisions had been left in Monty's hands. Many of the Liberty cargo ships used at Salerno were still manned by civilian crews, who insisted on putting in only a normal eight-hour workday. Also, most of the American troops hit the beach with no prior bombardment of the German positions facing them. The reason was a hope of catching the enemy by surprise—but this decision was a serious error, as it turned out. The situation in Italy remained confusing and unsatisfactory; the Italian government, which had signed a secret armistice on September 3 under tremendous pressure from Ike, had reluctantly agreed to announce Italy's surrender at the same time that he did, but then reneged on this agreement once the airborne landing near Rome had been canceled. Angered and out of patience, Ike announced the Italian surrender unilaterally on September 8, the day before Salerno. By the time the troops were landing on September 9, fighting had broken out between Italian and German forces in Rome, and the Italian fleet had sailed from its home port at Taranto to surrender to the Allies at Malta, pursued and attacked all the way by the Luftwaffe.

Whatever expectations Ike may have had that the Italian army would switch sides or that the German army in Italy would collapse did not materialize. The Führer had been persuaded with difficulty to permit the withdrawal of German forces to northern Italy if the Italians joined the Allies, but had also decided that if the Allies landed at Salerno—a fairly obvious possibility—it must be held long enough to allow the Geman divisions to the south to disengage and "extricate themselves." This task fell to the Sixteenth Panzer Division, which would provide a model defense, employing everything from huge Tiger tanks that actually fought gun duels with the Allies' destroyers offshore to such new weapons as the Goliath radio-controlled miniature tank, which could be guided into position and then exploded, and two versions of a radio-controlled glider bomb, which was guided to its target from the airplane that launched it. Quite apart from these innovations, the Sixteenth Panzers hastily organized "a mobile defense . . . mine fields along the shore; strong points sited to block exits from the beach; supporting mortars and artillery emplaced to cover the whole area; and tank units in reserve for counterattacks." In short, they were ready for Mark Clark, and not at all surprised when the Fifth Army arrived.

Steeped as we are today in the successful Allied invasions that began with Torch and ended with Overlord, the Normandy invasion, people are inclined to suppose that success was always more or less guaranteed. The film footage always shows vast fleets of ships and aircraft, landing craft hitting the beach, vehicles and tanks streaming onshore, and the enemy surrendering en masse—but this was never the case. Landing an army on a beach against an entrenched and experienced enemy is never an easy proposition, and Salerno would turn out to be (as Wellington described Waterloo) "a damned close run thing." American military historians have compared Salerno to Gettysburg—a desperate battle with heavy losses, in which the victor was at first driven back by a series of furious attacks, and after being almost overwhelmed, hung on in the end by his fingernails. At Salerno, the

Germans came so close to driving Mark Clark's forces into the sea that withdrawing the Allied forces from the beaches was considered. Indeed, Ike, informed of the situation (which Clark's chief of staff described as "near disaster"), intervened to prevent Clark from "evacuating Fifth Army Headquarters," saying that "headquarters should be moved last of all and the Commanding General should stay with his men to give them confidence. He should show the spirit of a naval captain, and, if necessary, go down with his ship."[26] These were harsh words, but they were merited.

The German counterattack, when it came, was furious, and besides the Sixteenth Panzer had elements of five additional German divisions, including the powerful Hermann Göring Division, which had been evacuated from Sicily. It was touch and go whether Clark could hold the beaches on September 12 and 13, and he was ultimately saved by the amazing courage of his artillerymen, who, literally, stuck to their guns and kept firing under direct attack by German tanks; by British and American light cruisers and destroyers, which came close inshore in shallow, mined waters and fired until they ran out of ammunition or, in the case of one American destroyer, until the tubes of the guns were white-hot; and by the timely arrival of General Matthew Ridgeway's Eighty-Second Airborne Division, which had been returned to Clark's command—just in time—after the cancellation of the assault on Rome, and made two consecutive night drops onto the Salerno beachhead, going straight into action after landing.[27]

The fact that Salerno came so near to disaster became another issue that embittered the relationship between Montgomery and his American counterparts. Monty had not been impressed with the idea of trying to take Rome by an airborne assault, and was doubtful about the value of the Italian army if, in fact, it decided to change sides. "I wouldn't trust them a yard," he wrote, "and in any case they are quite useless when it comes to fighting."[28] He had met Mark Clark briefly before crossing the Strait of Messina, and it seemed to both men that they had reached a kind of understanding and had formed what amounted to a mutual admiration society. Clark, who was no slouch at flattering officers senior to himself, deferred to Monty; and Monty, who was not only senior but older, and with more combat experience, would decide, mistakenly but typically, "My own observations lead me to the conclusion that Clark would be only too delighted to be given quiet advice as to how to fight his army."[29]

There was of course no doubt in Monty's mind about whose advice Clark would most benefit from. At their very first meeting Monty not only told "Mark"—as he immediately began addressing Clark, who continued to address him as "sir"—to feel free to call him at any time of the day or night, but also advised him to ignore any instructions from Alexander, "their common commander," that he thought were not "right"—a suggestion that shocked the normally unshockable Clark, who, whatever his other faults as a commander, had an instinctive respect for rank.

As the fighting to hold on to the beaches at Salerno raged, the American commanders, including Clark, began to suspect that Monty was moving too slowly, or was more interested in pursuing his own attack up the Adriatic coast of the Italian peninsula to take the port of Bari and the airfields at Foggia than in supporting the Fifth Army. This idea did not, of course, occur to Monty, who was convinced that by engaging the German forces south of Salerno, he had "saved their [i.e., Mark Clark's] bacon." Ike's naval aide Harry Butcher noted in his diary that Monty had missed "a great chance to be a hero," and that "Ike wondered what the result would have been in the toe of Italy if Patton had been the commander instead of Montgomery." For his part, Monty thought Clark's decision to take Naples after the "near debacle" of Salerno was incomprehensibly bad strategy—he wanted the Fifth Army to stay in the "Salerno area" and serve as a "firm pivot" while the Eighth Army struck inland, supplied and reinforced from the south through the port of Taranto. Taranto had been taken by the British First Airborne Division, landed from naval ships commanded by Admiral Cunningham in one of his many efforts to emulate the "Nelson touch."

Ike himself arrived at Salerno on September 16 to see for himself what had happened and to allot the blame. Perhaps inevitably, blame was quickly shifted onto the shoulders of Major General Ernest Dawley, the American Corps Commander, though not without a good deal of effort on Clark's part to place it there, and some soul-searching on Ike's, who would later remark that "inept leadership must be quickly detected and instantly removed." There was no denying that Dawley's Thirty-Sixth Division and elements of the Forty-Fifth had broken at the height of the German counter-attack; indeed the situation on September 12 had verged "on panic," as described in the relatively unemotional and unjudgmental prose of the editors of *The West Point Atlas of American Wars*. Nor can it be denied that the German thrust had almost reached the "rear areas along the beaches," and that their reaching these areas would not only have made a withdrawal necessary but almost certainly have turned it into a rout. But the real problem lay not in the fact that Salerno had been a near-failure, but rather in the continuing lack of any sensible decision about how far to commit the Allied forces in Italy, and to what purpose. By September 16 it was already apparent that the Italians, having surrendered, had no great enthusiasm for fighting their former ally, and that a long campaign up the mountainous central "spine" of Italy to liberate Rome would offer little reward and would be a drain on the buildup of Allied forces for the cross-Channel landing in 1944. Of course, a campaign up the spine would tie up a certain number of German divisions—there were already sixteen in Italy, and with the arrival on the scene of Rommel, there would be more—but that advantage would be balanced by the number of Allied divisions fighting them, which would not be available for the invasion when the time came. In short, the Allied leaders had allowed General Marshall's worst fears to come true—having placed themselves in Italy reluctantly, a bit at a time, they could not now withdraw, but they did not want to reinforce.

Monty summed up the problem astutely, in a letter to Brooke: "I presume you want the Rome airfields. Do you want Rome for political reasons, and to be able to put the king back on his throne? Do you want to establish the air forces in the Po Valley? Do you want to drive the Germans from Italy? Are you prepared to have heavy losses to get any, or all of the above?"[30] There were no clear answers to any of these questions, except for the one about the king, whom everybody wanted to get rid of except Churchill, who always had a wistful soft spot for monarchs in distress and an unfashionable faith in the principle of hereditary monarchy, and was at least lukewarm about the future of the House of Savoy. In the end, the German army would remain in Italy until May 1945, holding an ever-decreasing area; Rome would not fall to the Allies until just before D-day, and the Italian campaign became a kind of strategic backwater, the fighting there every bit as slow-moving and bloody as it had been during World War I. Neither Ike nor Monty would be there to deal with it, however—it would be left to Alexander and Clark to fight the Italian campaign to the bitter end.

The fate of Italy was one of the many reasons for the forthcoming meeting of Roosevelt, Churchill, and Stalin, which had been the subject of protracted and somewhat ill-tempered diplomatic negotiation and correspondence between the three principal figures and their foreign ministers. Now that "the tide had been turned," the Russians, who had been secretive, uncooperative, and demanding enough when they were losing, were becoming even more difficult with their western allies. A host of problems which had been placed on the back burner in the interests of common survival so long as the Germans were still winning victories were now demanding urgent attention. One of these problems was the future of Poland—always a bitterly contentious subject because the British had, after all, gone to war on behalf of Poland and recognized a Polish government in exile in London which was fiercely determined to prevent the Soviet Union from simply replacing the German occupation with its own. A second problem was the terms on which the Soviet Union would accept Finland's defection from the German side and its surrender. A third was the future borders of Europe. A fourth was the place of China among the great powers—another contentious subject, given that the president's sentimental friendship with Generalissimo Chiang Kai-shek and Madame Chiang Kai-shek was not shared by Churchill or Stalin. Of course, the biggest and most abrasive issue between the western allies and Stalin was the timing and size of the Anglo-American invasion of Europe. Stalin had jokingly remarked to the foreign secretary, Anthony Eden, when Eden visited Moscow that while he thought Churchill was as keen as he was on hurting Hitler, Churchill "had tendency to take the easy road" for himself, "and leave the difficult jobs to the Russians." Eden, though his reply was diplomatic, was not amused—Stalin's humor almost always had a sharp edge, and neither Eden nor anyone else was in any doubt that his patience on the subject was wearing thin. Stalin had been deeply angered by the decision to delay the cross-Channel invasion from 1943 to 1944—so much so that in August 1942 he had subjected Churchill and Averell Harriman (who accompanied Churchill on this Anglo-American equivalent of the journey to Canossa,* which the prime minister likened to "carrying a large lump of ice to the North Pole") to a withering blast of invective and rage. At the end of it, he calmed down and said, "History will be the judge of this." Churchill replied, "Yes, and I intend to be the one who writes the history," a brilliant Churchillian retort which did not bring a smile to Stalin's face.

One reason for Stalin's doubts about the cross-Channel invasion, beyond his normal suspicion about the plans and motives of his allies, was that nobody had been named to command it. This question had been sputtering back and forth between the president and the prime minister for some time, and had even reached the press in the United States. Roosevelt's preference for General Marshall was well enough known to be a subject of editorials: the anti-Roosevelt Hearst and McCormick papers took the line that Roosevelt was proposing to marginalize Marshall, who was widely admired, and get him out of Washington by appointing him to a combat command, which would be, in effect, a demotion, replacing him with a more pliant figure. In the Republican view, it was only Marshall's fearsome rectitude that kept the armed forces from the meddlesome hands of the president's liberal New Deal friends. This was not a view of himself or his position that Marshall shared—he was unyielding in his respect for the president—but it embarrassed Roosevelt and perturbed Churchill. Much as Churchill wanted a decision, Roosevelt was determined not to look as if he had been pushed into making it by hostile newspapers, so the matter was shelved. Churchill assumed that Marshall would be the choice, and although he had at one time virtually promised the job to Brooke, he assured the president that Marshall would be "entirely satisfactory," on the condition that Marshall's deputy was British. He urged making the announcement as soon as possible, since "Rumor runs riot . . . [and] the impression of mystery and of something to be concealed is given. This is a

* In 1077 Henry IV, the Holy Roman Emperor, was obliged to make a winter journey on foot and wearing only a hairshirt to apologize to Pope Gregory VII, and was kept waiting three days outside the walls of Canossa in the freezing cold until he was admitted.

fine field for malicious people."[31] But the president was not to be moved, and in any case the question was swamped by a new sharp disagreement between the British and the Americans about whether the "supreme commander," as the post was beginning to be called, should have command of British and American forces in the Mediterranean as well as Europe. The British dug in their heels at this suggestion—since they had agreed to have an American in "supreme command" of the invasion of Europe, they wanted to make sure that the supreme commander in the Mediterranean at any rate was British. "You should leave Admiral Leahy in no doubt," the prime minister cabled to Field Marshal Dill in Washington, "that we should never be able to agree to the proposals of putting the 'Overlord' and Mediterranean commands under an American Commander-in-Chief."[32]

Thus the question of who was to command Overlord was still unanswered when Roosevelt left for Cairo—for the meetings between the Allied heads of state had finally been planned to take place in three stages, in an attempt to satisfy everybody. Roosevelt and Churchill with their staffs would meet in Cairo for military discussions before meeting with Stalin in Tehran, which was as far as Stalin would travel outside the Soviet Union; then they would meet again in Cairo on the way home, this time with Chiang Kai-shek present, to discuss the war in Asia and the Pacific. (This last set of talks was held because Stalin would not accept Chiang as an addition to what was becoming known as the "Big Three"—nor was Churchill willing to accept that—and because the western Allies did not want representatives of the Soviet Union sitting in on discussions about Asian and Pacific strategy, since the Soviet Union was not at war with Japan.)

The prime minister, typically, "preceded the President into our area," and stopped off in Malta for a tête-à-tête with Ike. He was still trying to convert Ike to his Mediterranean strategy, about which he waxed enthusiastic, "while the project of invasion across the English Channel left him cold." Ike was by then practiced in the art of listening politely without agreeing. The prime minister told him, over and over again, "We must take care that the tides do not run red with the blood of American and British youth, or the beaches be choked with their bodies."[33] This vivid language did not shake Ike's firm adherence to Overlord, however, and he refused to be influenced by lengthy arguments in favor of a landing in the Balkans, or of concentrating more manpower in Italy.

Ike then flew to Oran to meet the president, who was arriving on a warship, and flew on with him to Tunisia, where a villa, oddly enough named the "White House," had been found for him and ringed with troops. Kay Summersby was present to meet Roosevelt and Ike at the airfield outside Tunis, and was irritated when the head of the president's Secret Service detail refused to let her drive Roosevelt. "No woman will ever drive the President," he told her, banging his fist on the top of the car. "No woman ever has—or ever will as long as I'm boss here."

However, when she arrived at the villa, she was called into the library, where the president, who was sitting in front of a big picture window with Ike at his side, said, "Why didn't you drive me? I heard you were going to be my driver and I was looking forward to it."[34]

Roosevelt roared with laughter when he heard that the Secret Service had forbidden Kay to drive him, and asked her to drive him for the rest of his stay. The next day she drove the president on a tour of the Tunisian battlefields—Telek, the Scottie Ike and Kay shared, leaped into the car as they were leaving and immediately made himself at home on the president's lap. Nothing could have been better calculated to put the president in a relaxed and sunny mood; he and Ike sat in the back with Telek, like a replica of Fala, on the presidential lap and a pretty young woman up front driving. Later on, when it came time for lunch, the president chose a grove of trees for a picnic spot—Kay described it as resembling an oasis—and when she opened the picnic basket Ike said, "No let me do that, Kay, I'm very good at passing sandwiches around."

Ike got out of the car, went to the front passenger seat, and selected a chicken sandwich for the president; as a Kansas farm boy, he said, he was a pretty good judge of a chicken sandwich from all the Sunday-school picnics he had gone to as a child. The president bit into it and agreed he was a good judge of chicken sandwiches, then patted the seat beside him, and said, "Won't you come back here, child?" and shared his sandwich with Kay. He told stories that had them both laughing, and she told him about London in the blitz, and about the WAACs with whom she shared a villa in Algiers.

"It seems to me you might like to join the WAACs," the president said.

Kay said that nothing would please her more, but she wasn't an American citizen, so the WAACs wouldn't have her.

"Well, who knows?" the president said, grinning. "Stranger things have happened."* [35]

Part of the plausibility of this idyllic scene is that it was a desert version of exactly what the president liked to do best when he was at home in Hyde Park, New York. Whenever he could escape from the Secret Service and from Eleanor, he loved to go off for long picnics with "Daisy" Suckley, his cousin, neighbor in Rhinebeck, and constant companion, in his little convertible, specially fitted with hand controls. He even built a small "picnic house," all on one floor because of his wheelchair, with a view over the Hudson, as a kind of retreat. The company of pretty and attentive women, whether in uniform or not, always brought out the best in the president. Flirting was one of his great pleasures, as Mrs. Roosevelt had long since come to accept with resignation.

After the battlefield tour, with Kay continuing as his driver, Roosevelt left by plane for Cairo, and from there he went to Tehran to meet Stalin and Churchill. He spent a night at Ike's "cottage" near Carthage, where he revealed that he still had not made up his mind about his choice for supreme commander. He praised Marshall, but then, with his usual ability to convey conflicting messages, he told Ike that it was always a mistake to change a winning team.

In any event, Roosevelt kept his cards close to his chest at Tehran, at least on this subject, leaving Churchill under the impression that his intention was to give Marshall the job of supreme commander and bring Ike home to replace Marshall. Stalin, on the other hand, addressed the matter with his usual brutal realism. No operation as big and hazardous as this one was realistic until somebody had been appointed to command it. That choice was for the American and British governments to make, of course, and no concern

* The president did not forget. Although a British subject, Kay Summersby was commissioned as a second lieutenant in the Women's Army Corps in October 1944. The Women's Auxiliary Army Corps (WAAC) had by then been renamed the Women's Army Corps (WAC) by Congress, late in 1943. It will therefore be referred to as such henceforth in this book, except in quotations from letters in which the writer continues to refer to WACs as WAACs.

of his, but he doubted that preparations for the invasion could "be carried on successfully" unless there was a supreme commander, and demanded that the announcement be made not later than a week after the conference.

While the president and the prime minister were in Tehran, Ike, obeying General Marshall's order to get a rest and a change of scene, took Kay on a visit to Cairo, Luxor, and the Valley of the Kings, and then to Jerusalem, where they stayed at the King David Hotel and visited the garden of Gesthemane. There, he gave her the postcard on which he had written, "Good night—there are lots of things I could say—you know them. Good night,"[36] and a four-leaf clover, which was to remain one of her most treasured possessions. It is hard to know exactly what to make of this. On the one hand, it was not, as some have alleged, a romantic trip à deux—Ike was accompanied by aides, and Kay by two other women who worked in Ike's headquarters in Algiers (Marshall would hardly have countenanced the excursion otherwise), although the presence of the others would probably not have been of any comfort to Mamie, had she known. The likelihood is that the trip was a good deal less intimate and heavy-breathing than Kay Summersby makes it sound in her second memoir, but a lot more sentimental and close than Mamie would have wanted.

On his return, Ike—again accompanied by Kay—went to see for himself what was happening in Italy. He was in the process of transferring his headquarters from Tunis to Naples, to be closer to the fighting. What he found there was disappointing—"wretched weather," he wrote, "had overtaken us," and both the Fifth Army and the Eighth Army were stalemated along a muddy, mountainous front stretching from Minturno, on the Mediterranean coast, to Vasto, on the Adriatic. The Germans were dug in as firmly as they had been in World War I on the western front—more firmly, in fact, since here they were often dug into solid rock. From Monte Camino, at the center of Mark Clark's front, it was only 100 miles as the crow flies to Rome, but it might as well have been 1,000. "Men and vehicles sank in the mud. . . . Heavy rains fell and the streams were habitually torrents." In both armies, the armored bulldozer and the mule became more important weapons than the tank. Monty fought a bitter, bloody battle to cross the Sangro River, which became enshrined in British army legend; but despite his dream of breaking through the German line to make a "left hook" and take Rome, he was forced to conclude that it would be March 1944 before this could be done, and then only if good weather dried up the ground first and he received reinforcements.

When Monty invited a young Australian pilot who had been shot down near his mobile headquarters to lunch and asked what he thought should be done, the young pilot startled the general by recommending, "Stop friggin' about."[37] This struck Monty not as insubordinate but as a perfect description of the mess the Allied forces were in. They were "friggin' about" in Italy, in miserable country with two armies—fourteen American, British, Commonwealth, and French divisions dispersed along a front 100 miles long, facing an estimated thirteen and a half German divisions—and with not enough strength to permit a single, concentrated thrust that would break the German line and open up a war of movement in which Allied tactical airpower and armor would matter. So awful was the weather that Monty sent to London for "a Mackintosh suit" from Cording's, the kind of thing Englishmen wear to shoot pheasants in the rain, and settled in impatiently to wait for spring—or, he hoped, the call to a more important command.[38]

The question of who would command Overlord—and who would be the deputy commander and the ground forces commander—continued to keep every general in the Mediterranean theater on his toes. Brooke was still considered a possibility, at least by the British generals, who did not know that Churchill had already conceded supreme command to an American. Alexander was also considered a possible candidate for supreme commander or, failing that, for deputy and ground forces commander, though Monty nurtured hopes of getting that himself. There was even some discussion that General Sir Henry Maitland Wilson, fondly known as Jumbo throughout the British army, might get the job of ground commander, though Monty doubted it. Speculation was so lively that Monty arranged a special code with the War Office so that he could be informed instantly who the choice was. The code names were flowers, and his own was Daffodil.

In the meantime the president and the prime minister returned from

Tehran. Churchill took to his bed, stricken with the flu; Roosevelt faced up at last to the choice of supreme commander. Roosevelt broke the news of his decision to Marshall personally, explaining, with great tact, that he would not be able to sleep at night without the presence of Marshall in Washington, and dictating to him a brief message for Stalin, which he signed "From the President to Marshal Stalin—The immediate appointment of General Eisenhower to command of Overlord operation has been decided upon—Roosevelt, Cairo, Dec. 7. 43."[39]

A few days later, on his way home, the president landed in Tunis to meet Ike, who was "scarcely seated in the automobile when he cleared up the matter with one short sentence.

"He said, 'Well, Ike, you are going to command Overlord.'"[40]

From the President to Marshal Stalin

The immediate appointment of General Eisenhower to command of Overlord operation has been decided upon

Roosevelt

Cairo, Dec. 7. 43

Dear Eisenhower, I thought you might like to have this as a memento. It was written very hurriedly by me as the final meeting broke up yesterday, the President signing it immediately.

G.C.M.

PART III

Victory

CHAPTER 13

"Supreme Commander, Allied Expeditionary Forces"—1944

Ike and Monty. (Tedder, with pipe, is on the left.)

In retrospect, the choice seems inevitable, but of course it was not. On the other hand, it is now easy to understand why the job fell to Ike, once the president had made it clear that supreme command must go to an American general. Marshall was a great man, the indispensable organizer of victory, almost inhumanly objective and unselfish, but he had no experience of commanding large forces in battle. Ike himself lacked combat experience, of course—this was one bone of contention between himself and Monty (not to speak of Brooke, Patton, and even his old boss MacArthur)—but by the winter of 1943 nobody could deny that he had successfully carried out three important assaults, and done so with a green army and in partnership with an ally which did not at first accept an American in command easily or uncritically. Torch had been the largest and most daring amphibious operation in history, and although mistakes had been made, the French had created severe military and political difficulties, and the Germans had managed to hold out in Tunisia for six months, it indisputably ended in victory. The landings in Sicily had been carried out successfully, and although it took much longer than anybody had anticipated to capture the island and three German divisions had escaped intact across the Strait of Messina, this too was a victory, and one that had led to the downfall of Hitler's partner Mussolini and the surrender of Italy. Under Ike's command, American and British forces had landed at last on the mainland of Hitler's *Festung Europa* ("fortress Europe"), and were now within 100 miles of Rome. The Allies had generals with, perhaps, a sharper strategic and political vision than Ike—Marshall and Brooke, for example—though he would quickly catch up with them, for he was a fast learner. There were also generals who were more experienced at "fighting a battle," like Patton, Monty, and Alexander, although in that respect too Ike was learning on the job. But there was nobody who had anything like Ike's record of leading an alliance—always the most difficult feat in warfare—or of commanding military operations on a huge, daring, and unprecedented scale. What is more, Ike somehow inspired people: civilians and ordinary soldiers of both nations, even cynical political figures and the always troublesome French. Something about his big grin; his long-limbed, loose American way of walking (the Kansas farm boy grown to a man); his easy, familiar way of speaking to everybody from King George VI down to privates in both armies; his lack of pretension; his evident sincerity; and his willingness to accept unimaginably heavy responsibility made people like Ike. They were willing to be led by him. They were willing to have him command their sons and husbands in battle. They trusted him. They were willing to die for him. It is hard to imagine Alan Brooke or George Marshall winning people's confidence, affection, and trust the way Ike did, apparently without effort or design, and it was typically astute of that supreme master of politics Franklin Roosevelt to see that quality in Ike at once, and to recognize, however much he admired Marshall, what a formidable weapon it was. And why not? It was one he and Ike shared.

On Christmas Eve President Roosevelt announced the news to the nation and the world, as well as revealing Ike's new and unexpected title: Supreme Commander, Allied Expeditionary Forces. Already the pressure on Ike to choose those who would serve under him was building up. Churchill had relinquished supreme command to an American reluctantly—after a pained conversation with Brooke, to whom he had promised it—but he was determined to get some return for bowing to the inevitable. He had accepted Roosevelt's decision, which the president told him about in Cairo, in what is now the Mena House Hotel, overlooking the Pyramids; and on Roosevelt's last day Churchill insisted on taking him to see the Sphinx. "Roosevelt and I gazed at her for some minutes in silence," he wrote later, "as the evening shadows fell. She told us nothing and maintained her inscrutable smile."[1]

Roosevelt had nothing to learn from the Sphinx about inscrutable smiles. Even once he had chosen Ike, he was still thinking about what the title should be, and he added the word "Supreme" at the last moment in order to emphasize "the importance of the undertaking." It was always present, at the back of the president's mind, that the British did not share his enthusiasm for Overlord, and he therefore wanted to signify, by the commander's title, America's unyielding commitment to this "new venture," which was itself to be "supreme," above and beyond all others. No Mediterranean strategies, no tempting raids on Greek islands, none of Churchill's hopes for a powerful and ambitious thrust in Italy toward Vienna or through the Balkans to stab Germany's "soft underbelly," not the prime minister's plan to land in Norway, not even his cherished desire to bring Turkey into the war on the Allies' side—none of these things was to interfere with or delay the supreme operation of World War II: the invasion of Europe and the conquest of Germany. Ike's resounding new title was intended to convey that message to the world; its meaning was meant to be heard as clearly in London as in Berlin.

Roosevelt may not have had Churchill's sweeping command of history, but he knew enough history not to repeat the past. In 1918 the Allies had defeated the German army in the field, but they did not invade Germany—the Germans got rid of the kaiser by themselves; negotiated an armistice; marched their troops home in relatively good order; and adopted, with disastrously mixed feelings, a new, democratic form of government. Their failure to come to terms with their own defeat, or to make that new government work, led inexorably to World War II. Marshal Foch, a pessimist (though not the only one) among the many optimists at the Versailles peace conference, had memorably dismissed the Peace Treaty of 1919 as *une armistice de vingt ans* (a twenty-year armistice), and unfortunately he had been dead right. It was exactly twenty years later that World War II began, as a continuation of the previous war, and, from the German point of view, an attempt to reverse the outcome. Roosevelt's great wisdom was to see clearly that this time the German army would have to be totally crushed and destroyed, and Germany itself conquered, invaded, and, if necessary, laid waste, so that a new Europe could be built on the ashes of the old—a Europe in which Germany could be contained and bound like Prometheus, rather than continually renewing itself to dominate its neighbors.

For this reason Roosevelt had declared at Casablanca, to Churchill's annoyance and surprise, that the Allies would accept nothing less than unconditional surrender from Germany. These were not, as Churchill supposed them to be, empty words or mere rhetoric—Roosevelt had chosen them carefully, and for once they meant just what they said. Churchill supposed, although he thought it was improbable, that if the Germans overthrew Hitler they might be able to come up with a government with which it would be possible to negotiate a peace. But Roosevelt had no "mental reservations" on the subject of unconditional surrender. That is why he displayed so little interest in the German anti-Nazi resistance, such as it was, or in the attempt on Hitler's life on July 20, 1944, by senior Army officers. He knew there would be no easy way, or clever strategy, or stroke of good fortune to bring about what he wanted; the German army would have to be fought out of France and back into Germany, and defeated on its own soil. He wanted the Allied armies commanded by a man who understood this and would do it, and he was not open to compromise on the matter.

Since 1945, almost everybody has had a say about the supposed mistakes that were made in the last year of the war—especially the presumed failure on the part of the western Allies to take Berlin and the failure to confront the Soviet Union over the borders and the independence of the eastern European countries. Many if not most of these have been blamed on Roosevelt, but it should always be borne in mind that the president did not live to write his own memoirs, or to criticize those of others. Ike, when he came to write his, was careful not to join in the postwar criticism of Roosevelt. Ike himself had shown no interest in wasting the lives of American soldiers to get to Berlin, and several times he offended and even angered Churchill by going over the heads of the prime minister and the president to deal directly with Stalin, as if he were himself a head of state, to ensure that there would be no accidental clashes between Allied and Soviet troops as their front lines began to touch.

Ike had understood better than anyone else what his mission was—it was simply to defeat Germany, once and for all. The exact wording of his orders from the Combined Chiefs of Staff left no doubt about that: "You will enter the continent of Europe and . . . undertake operations aimed at the heart of Germany and the destruction of her armed forces." It was not the first time in American history that a president had promoted a general to high command for just such a purpose. In March 1864, when Lincoln called Grant to Washington to promote him to lieutenant general (then the highest rank in the U.S. Army) and put him in command of all the Union armies, Grant began at once by abandoning the strategy that had, until then, been the common objective of every commander of the Army of the Potomac—to advance on Richmond and take it. For over three bloody years a succession of generals had fought from every direction to take Richmond—some of them got close enough to see the spires of its churches without the help of binoculars—but Richmond was still the capital of the Confederacy. Grant, with Lincoln's blessing, chose another strategy altogether. He would ignore Richmond—he had no desire to capture it, and still less to see it once it had fallen, if it did fall. Instead, he would seek out and destroy Lee's army. Day by day, from April 1864 to April 1865, Grant fought Lee's Army of Northern Virginia head-on in some of the bloodiest and most continuous fighting of the Civil War. He would not retreat. As he wrote to Washington from the James River after the dreadful three-week bloodletting of the battles of the Wilderness and of Spotsylvania, "I intend to fight it out on this line if it takes all summer." Grant would not take the pressure off Lee for an instant, no matter what the cost in casualties. He did not attempt to outmaneuver or outwit Lee; he was confident that he could replace casualties more easily then Lee could, and that Lee would therefore eventually run out of men. It was a simple strategy, brutal but profound, as well as mathematically correct, and it worked. Grant would bleed Lee's army to death, not in one big battle as McClellan, Burnside, and Hooker had all tried to do and failed, but by wearing the Confederates down, day by day, in an endless series of battles big and small until Lee had no option left but complete and unconditional surrender.

Whatever other mistakes the president may have made, he achieved the objective that was most important to him: the concentration of the full power of the United States, the United Kingdom, and France in the hands of one man, who would use it for the sole purpose of defeating Germany, and not allow it to be frittered away in campaigns of secondary importance. This time the Germans would have to be defeated on their own ground, and defeated so thoroughly that they would be unable to deny or mythologize their defeat. And since they still had the most powerful and most professional army in the world, it would have to be defeated in battle remorselessly, ground down by superior numbers and sheer, overwhelming firepower until it could no longer replace its losses, or supply those soldiers who remained with food, ammunition, fuel, and equipment. Symbolically (though no doubt unintentionally so), the president had stayed a night in Ike's cottage in Tunisia, overlooking the ruins of Carthage—and what he intended was a "Carthaginian peace" like the one Rome had inflicted on Carthage when Scipio Aemilianus finally razed the city in 146 BC after a century of war. Roosevelt chose Ike for the task that mattered most to him, as Lincoln chose Grant. He could not have chosen better.

A bad case of the flu which turned into pneumonia—"the old man's friend," as Churchill's doctor Sir Charles Moran described it jovially to his patient, explaining that it carried the elderly off without pain or fuss—did not prevent the prime minister from striking a hard bargain. If Ike was to be supreme commander, then his deputy and ground commander must be British, as Ike himself acknowledged. Ike's choice was Alexander, who had served in this role in Tunisia. Ike had "developed for him an admiration and a friendship," and regarded "Alex" as "Britain's outstanding soldier," who was, "moreover, a friendly and agreeable type—Americans instinctively liked him."[2] But Churchill liked Alex too, and was unwilling to relinquish him. Lying in bed in Ike's villa in Tunis, under the firm care of Sir Charles and a platoon of nurses, and amusing himself as best he could by having *Pride and Prejudice* read aloud to him by his daughter Sarah ("What calm lives they had, these people," he remarked without envy), Churchill dug in his heels. Unlike Ike, he still had high hopes for the "Italian operation" and wanted Alex to command it. (No doubt he also thought, at the back of his mind, that a great victory over the Germans in Italy might make Overlord unnecessary and open the way to other ventures in the Balkans.) Ike would have to take Monty instead of Alex. Even eight years later, when Ike came to write his memoirs, his lack of enthusiasm for this decision shines through the veil of tactfulness that he adopted when writing about people he didn't like. Monty, he makes clear, was not chosen by him but "assigned" to him by the prime minister. He praises Monty's ability to elicit "intense devotion and admiration" among "British enlisted men," carefully leaving unsaid what American enlisted men thought of Monty; and damns him with faint praise by describing him as "careful, meticulous, and certain"[3]—not exactly the qualities that Ike admired in Patton, or that made Rommel such a formidable foe.

If he was obliged to take Monty as commander of ground forces (as well as commander of the British forces), Ike was at any rate not prepared to have him as deputy supreme commander, and managed to get Air Chief Marshal Sir Arthur Tedder for that post instead. Tedder, an altogether more pliant and engaging Britisher than Monty, was well liked by his fellow airmen, British and American (this unanimity was rare, since they were a singularly contentious and envious group). Also, although Tedder had a certain reputation for intrigue, interference in the plans of others, and waspishness at the expense of people less intelligent than himself, he was at least a firm believer in the "allied" principle so dear to Ike's heart, and could be counted on to rise above national prejudice. Ike and Churchill agreed without difficulty on Admiral "Bert" Ramsay of the Royal Navy as naval commander-in-chief, since Admiral Cunningham was returning to Britain to become first sea lord. Ramsay's energy, sharp intelligence, and willingness to try the new and unexpected (rare in an admiral) were invaluable—he had almost single-handedly put together overnight the fleet of "little ships" that made the evacuation of the BEF from Dunkirk possible—and he got along well with Americans, even with most American admirals (for the rivalry between the Royal Navy and the U.S. Navy could never be totally suppressed). Most important of all, Churchill liked him. Ike insisted on retaining Bedell Smith as his chief of staff, and, more important, also managed to get the prime minister to agree that Monty would command the ground forces of the invasion only until Ike himself moved his headquarters to France to "assume direct operational control,"[4] at which point Monty would relinquish command over the American forces. This crucial limitation on his appointment, it almost goes without saying, Monty almost instantly put altogether out of his mind. "Ike has told me that he wants me to be his head soldier and to take complete charge of the land battle,"[5] he wrote to Brooke, without any mention of the fact that the moment Ike set foot on French soil overall command would revert to him, if he chose to exercise it.

When Churchill was at last well enough to get away to Marrakech to recuperate, he had reason to be content. Ike's deputy would be a British air marshal, his ground commander a British general, his naval commander a British admiral, and his air commander another British air marshal: Air Chief Marshal Sir Trafford Leigh-Mallory. Leigh-Mallory was an odd choice, and perhaps the least successful one of the lot—he had been a good leader of fighter pilots, but he had no real knowledge of air support of ground forces or parachute operations, both of which would be crucial to Overlord; and he was famous in the Royal Air Force chiefly for a successful cabal he had mounted against Air Chief Marshal Sir Hugh Dowding, the deeply respected Air Officer Commanding-in-Chief of Fighter Command during the Battle of Britain.* Leigh-Mallory was disputatious, opinionated, and

* The immediate cause of the conflict between Leigh-Mallory and Dowding was the former's conviction that the Germans should be attacked while they were still over the Channel by large numbers of British fighters, rather than waiting until they were over England and attacked by individual squadrons. This became known as the "big wing" theory, and its most outspoken advocate was Wing Commander Douglas Bader, the remarkable and obstinate legless RAF fighter ace. Dowding thought the big wings would merely increase the RAF's losses, and he may have been right, but he lost the public relations war.

out of his depth; he was also a controversial figure within the Royal Air Force; and he would be the source of much of the friction between Ike and the "bomber barons"—Harris in the RAF and Spaatz in the USAAF—friction which another airman might have smoothed out. Moreover, he was one of the few senior commanders to lose their nerve on the eve of D-day. It is difficult to avoid the impression that Air Chief Marshal Sir Charles Portal and the Air Council were getting rid of a square peg in a round hole when they recommended Leigh-Mallory to Ike, in the tradition, common to the armed forces of all nations, of recommending a man everybody dislikes for a promotion that makes him somebody else's problem.

Ike was more sure of himself when it came to the Americans he wanted serving under him. He had already made it clear that Bedell Smith would remain his chief of staff; and he had been sufficiently impressed by Bradley's performance in Tunisia and Sicily that he now, in a long message to Marshall, earmarked Bradley as a potential United States "Army Group Commander." In the same message, he noted that Patton should probably serve as one of Bradley's "Army Commanders"—in short that Bradley should be in command of all the American ground forces for Overlord, and that Patton should serve under Bradley. Ike praised Bradley as "a sound, painstaking, and broadly educated soldier," and as "a keen judge of men [and] absolutely fair and just in his dealings with them," high praise indeed from a man who prided himself on just such qualities. He also remarked that Bradley was "emotionally stable"—no doubt implying a certain caution about Patton, for whom the same could certainly not be said.

Ike had planned to arrive in England by January 10 to take up his new command, but he received a "Christmas telegram" from Marshall, urging him to come to Washington immediately, for "short conferences" with him and with the president, and for a "brief breather." Although Ike, in his memoirs, uses the word "urge," in fact Marshall went well beyond "urging" Ike—he couched the message in the form of an order. "I am not interested," he wrote, "in the usual rejoinder that you can take it. It is of vast importance that you be fresh mentally and you certainly will not be if you go straight from one great problem to another. Now come on home and see your wife and trust somebody else for twenty minutes in England."[6]

For reasons of security, Ike was unable to tell Mamie that he was returning home, but she was informed in confidence by Marshall, who was not above playing Cupid in the private life of his most valued generals. Ike worried—correctly, as it turned out—that he would not have enough time to devote to Mamie once he got home, that too many people would compete for his attention and time, and that it would not be easy to resume married life with her after more than a year apart. Perhaps to make the trip a little easier on himself, he decided to take Harry Butcher back with him, while Kay Summersby and the rest of his staff moved on to London. He may or may not have handed Kay a note reading, "Think of me—you know what I will be thinking,"[7] as they drank a last cup of coffee together in his villa in Algiers before he left for home and wife for "two weeks" ("Twelve days," he corrected her precisely). We have only her word for this, but certainly he must have guessed that his homecoming would not be easy. On his way back to Washington, he landed at Marrakech to visit Churchill.

Although Churchill was supposed to be recuperating, he had not been idle. While General Marshall and the Combined Chiefs of Staff had been determined to return to the United Kingdom from the Mediterranean as many units and, above all, landing craft as they could in preparation for Overlord, the prime minister had been fighting an equally determined rearguard action to make use of them there before they were moved.

Churchill's vast correspondence (undiminished by illness or distance from London) throughout December is full of minutely detailed references to landing craft. Others might be content with a broad, general view of the subject—they were urgently needed for Overlord and in the Pacific, and that was that—but the prime minister wanted a worldwide "audit" of just how many landing craft there were of every type in all theaters of the war, and what condition they were in. This was the kind of thing at which he excelled—nobody could match his energy, persistence, memory, thirst for details, and ability to chivvy the facts out of reluctant naval commanders. He counted landing craft like a miser, unearthing them in unexpected places, determined to put them to use in the Mediterranean before they were relinquished to Ike.

On Christmas eve, from his sickbed, he reached the conclusion that there were 104 LSTs in the Mediterranean, but that by mid-January this number would have been reduced to "only thirty-six." He had located another fifteen in the Indian Ocean, and was determined to get his hands on them, but then he discovered to his dismay that no less than eighty-eight were needed for two divisions.* On Christmas day he was busy signaling the British chiefs of staff and the first sea lord on the subject of these fifteen LSTs in the Bay of Bengal, which, he thought, "would not arrive in time for Anzio, but would play their part in repayment of 'Overlord' a little later."[8] Since the Allies were already committed to landings in Burma, Normandy, and the south of France (the much disputed Anvil), this was a little like robbing Peter to pay Paul, but that did not deter the prime minister. He plunged into the question of how much time would be needed for training (or retraining) the LST crews and the troops who would be carried by them, and with the help of a naval aide came up with the optimistic conclusion that only eleven days were required before the operation—"three days for initial loading, six days for rehearsal, two days for reloading." He worried about how long it would take the crews to adapt to the "problems of tide," but assumed that if they were "good seaman," brief instruction should suffice.[9]

By December 28, he had involved President Roosevelt in the details of the controversy over the LSTs. The president reluctantly fell in with the prime minister's plans, though his message to Churchill has the ring of somebody who is throwing his hands up in the air, and hopes never to hear another word again on the subject. He specifically refused to agree to anything that risked delaying Overlord, which had been promised to Stalin at Tehran, rather casually and prematurely, for mid-May—a rash promise

* This was to be the beginning of a long dispute between Churchill and the ground commanders about the size of the "tail" required by American and British divisions, in contrast to fighting troops. The dispute went on until it reached a crescendo on the eve of D-day. It was intensified when he discovered that his precious LSTs had landed nearly 18,000 vehicles at Anzio for a total force of 70,000 men, and that only 380 of these vehicles were tanks. He rather petulantly suggested that the LSTs (landing ships, tanks) should instead be called LSVs (landing ships, vehicles), and complained that he had hoped to hurl "a wild cat on shore, but all we got was a stranded whale."

which now lay like a pall over all plans for future operations and also specifically ruled out any attempt to take Rhodes or other islands in the Aegean. This was a blow to Churchill, whose fondest hope had been to take Rhodes and whatever other Greek islands he could, both as the first step to opening up a new front against the Germans in the Balkans and as a way to prod Turkey into the war on the Allies' side; but resistance to the idea in Washington was too entrenched for him to overcome, and even Brooke and the British chiefs of staff were unconvinced that it would lead to anything except a substantial loss of naval ships.

The prime minister had prepared a fallback position, however: an amphibious landing of at least two divisions, plus airborne drops, near Rome, possibly at Anzio (code-named Shingle), as a shortcut to avoid the apparently endless task of driving the Germans back across the mountainous central "spine" and innumerable rivers of Italy from Naples to Rome, and thus bringing the Allied forces within reach of Germany.

At the back of his mind was the possibility of reaching not Rome, but Vienna, thus causing the Third Reich to crumble and placing British and American troops as conquerors in central Europe before the Red Army got there. Arguing with Winston Churchill on military matters was a full-time job—involving an endless barrage of cables; detailed reports; and lengthy, carefully drafted messages—so even those most opposed to the landings at Anzio, like Marshall, began to give ground on the subject, perhaps because Anzio seemed at any rate a more realistic and useful goal than Rhodes. Shingle therefore gradually acquired, despite deep reluctance on both sides of the Atlantic, the aura of a settled fact.

In Marrakech Ike was subjected to the full force of Churchill's charm and powers of persuasion. Churchill, supported by the presence of Mrs. Churchill, Sarah Churchill, and Lord Beaverbrook, was in good spirits—the president had sent him a Christmas tree that had been felled on the Roosevelt tree farm in Hyde Park; and he had received congratulations on his recovery from two ill-matched well-wishers, the communist Marshal Tito of Yugoslavia and the fascist Generalissimo Franco of Spain. All the same, Ike was not entirely won over. He pointed out, correctly, that "the landing of two partially skeletonized divisions at Anzio, a hundred miles beyond the front lines as then situated, would not only be a risky affair but that the attack would not by itself compel the withdrawal of the German front. Military strategy may bear some similarity to the chessboard, but it is dangerous to carry the analogy too far. A threatened king in chess must be protected; in war he may instead choose to fight!"[10]

This is, of course, exactly what would shortly take place at Anzio, but Ike was very conscious of the fact that the Combined Chiefs of Staff had already agreed to the operation. It was scheduled to take place toward the end of January simultaneously with a renewed attack on the German fortified position at Cassino in the hope that Mark Clark's British and American forces could break through the "Gustav Line" and meet up with Allied forces as they broke out of the Anzio beachhead. The shipping and landing craft were being obtained for Shingle, largely owing to Churchill's persistence and refusal to take no (or even maybe) for an answer; and the Russians had been informed that the operation was going to happen, so it was too late for pessimism or any significant change in the plan. In any case, the responsibility was no longer Ike's; it was that of his replacement as Commander-in-Chief, Mediterranean, General Sir Henry Maitland Wilson (Jumbo), on whose toes Ike was not anxious to tread. Ike heard the prime minister out politely, and as tactfully as he could, "continued to voice doubts of such an optimistic outcome."[11]

It was not only on the subject of Shingle that Ike was more cautious than Churchill. Rounding up the LSTs necessary for the Anzio landing was one of the factors that had made it necessary to move Overlord from mid-May to late May, and also made it doubtful that Anvil, the landing in the south of France, could take place at the same time—there were simply not enough LSTs to go around. Churchill had never liked the idea of Anvil, so that did not concern him. But Ike, with the precision of mind that was his specialty in military affairs, had already recognized that with or without Anvil, the last week in May did not present the right conditions for the Normandy landings, nor could he be certain that he would have all the LSTs he needed by then. He already knew that he had to approach the French coast by moonlight, to give cover to the invasion fleet and the airborne troops, but he also wanted exactly the right amount of daylight for the bombardment and the landings. Too short an interval would curtail the bombardment, which was intended to stun the defenders; too long an interval would give them a chance to recover. At the same time, it had been determined that the landings should take place at exactly three hours before high tide. If the water was too high, the mined obstacles placed in such profusion on the beaches by Rommel would be invisible to the helmsmen of the landing craft; if it was too low, the troops would have to cover too much open beach under enemy fire. In addition, the tide varied from beach to beach, so the landings would require very complex synchronization.*

From late May on, the first lunar period that would satisfy all these conditions was between June 5 and June 7, and whatever promises might or might not have been made to Stalin at Tehran, June 5 was beginning to look to Ike like the best date for D-day, assuming the weather cooperated. If the weather was bad on all three days, the invasion might have to be postponed for a full month to ensure the right combination of moon and tide. Under the circumstances, Ike may be forgiven for not having argued as strongly against Shingle as he wanted to—it was already a foregone conclusion, and he had other things on his mind.

On his way back to Washington, Ike had been fired on by trigger-happy neutral Portuguese antiaircraft gunners as his plane passed too close to the Azores. Every effort had been made to keep his return to the United States secret, with the result that he and Butch had to take the freight elevator at the Wardman Tower up to the floor on which the apartments of Mamie and Ruth Butcher were situated. Ike surprised Mamie by walking in on her while she was clearing up after her regular bridge party, which she had thought it better not to cancel.[12] Their son John noted later that she was struck by the fact that Ike had put on weight, no doubt because of lack of exercise, and that his face showed clearly the strain he had been living under for a year and a half. He looked older than she remembered; he

* The best, most concise, and most commonsense explanation of the necessary combination of conditions for D-day is by Winston Churchill in *The Second World War*, Vol. 5, *Closing the Ring*, 523. I am indebted to him for making sense of it for the reader.

had changed in other ways too—he was "preoccupied and impatient," his manner was more abrupt, and when visiting with Mamie the few people who were in on the secret of his presence in the United States he simply stood up and announced when it was time to leave without consulting her.[13]

There is no question that the relationship between the two of them was strained. This was hardly surprising—Ike had gone away a recently promoted and almost unknown major general and returned a four-star general who was already one of the most famous, admired, and most instantaneously recognizable people in the world; who moved as an equal among presidents, prime ministers, and kings; and on whose shoulders rested responsibility for the liberation of Europe and victory over Hitler's Germany. It may not have been true that Ike occasionally called Mamie "Kay"—that may have been wishful thinking on Kay's part at the end of her life—but even so, it cannot have been an easy time for either of them. They made a secret visit by private railroad car to West Point to see John, flew to Kansas to see Ike's mother and some of his brothers, and spent a couple of days together at White Sulphur Springs in West Virginia. Still, the tone of their reunion was set when Ike was invited to the White House to meet the president and also spend some time with Eleanor Roosevelt, but no invitation was extended to Mamie, who was humiliated and furious. According to her granddaughter Susan, Mamie was still "riled" by this years later, and she forgave neither the Roosevelts nor Ike.

Whatever they talked about while they were in Sulphur Springs, Mamie's letters to Ike after his return to London reveal only too clearly what was still on her mind—even her granddaughter admits that she put Ike "on the defensive." In one reply to her he tried to make light of her concerns—"I know that people at home always think of an army in the field as living a life of night clubs, gayety and loose morals"—and emphasized that she shouldn't believe what she heard from "gossips"; but clearly he was under attack on the home front and resented it. Perhaps he also felt that he was at least in part responsible; after all, Kay had picnicked with the president, while Mamie had not been invited to the White House. If so, this can only have made him all the more uncomfortable and resentful.

One senses also that Ike's choice of Harry Butcher to accompany him on his trip home was a mistake. Ruth Butcher was Mamie's next-door neighbor and closest friend, and the fact that Harry was having an affair with a woman Red Cross worker in North Africa was by this time common knowledge in military circles—he would eventually divorce Ruth to marry her. Harry Butcher was big, jovial, and generous, the life of the party wherever he went, but Mamie may well have thought that he was not setting a good example for Ike, or looking after her best interests; and of course he was cheating on her best friend.

In any event, Ike arrived back in the United Kingdom to assume his command on January 15, 1944, his aircraft having been forced to land in Scotland again because of the fog. His private railway car, code-named Bayonet,* was sent to bring him to London, and Kay Summersby picked him up at the station near midnight in the middle of an impenetrable London "pea-souper"—the dense fog, thickened by the soot from millions of coal fires, for which London was then still famous—so thick that a man had to walk in front of the car carrying a flashlight. Ike had brought Kay—ironically, in view of Mamie's having been excluded from the White House—a photograph of Roosevelt, personally inscribed to her by the president, who had not forgotten the picnic in Tunis. When Kay finally got back to 20 Grosvenor Square she and the guards had to form a human chain to guide Ike across the pavement from the car to the door in the fog.

The staff had found him a "lovely house" just off Berkeley Square, London's most fashionable area; but, predictably, Ike took one look at it and decided to live in Telegraph Cottage again, though in the end he would keep the Berkeley Square house for nights he had to spend in London. Poor Telek, unfortunately, was quarantined for six months—Britain's strict regulations on quarantine for dogs and cats entering or reentering the country could not be lifted even for a four-star general and supreme commander (or even for members of the Royal Family)—but otherwise Ike's "military family" was reconstituted just as it had been before his

* Grand as this was, Monty traveled in greater state. Attached to his private railway carriage, code-named Rapier, was a boxcar for his Rolls-Royce.

departure for Gibraltar on the eve of Torch. One of the attractions of Telegraph Cottage was that it was close to Richmond Park so that he and Kay could go riding together in the mornings (this was the only exercise he got). Another attraction was that it was even closer to Bushey Park, near Kingston, since he intended to move Supreme Headquarters, Allied European Forces (SHAEF) there immediately, lock, stock, and barrel, from generals to privates, removing it in one stroke from Grosvenor Square and the temptations of London.

One cannot help feeling that the tension between Ike and Mamie may have been less about whether or not he was sleeping with Kay Summersby than about the huge difference between Mamie's life at the time and his. She was living in a modest apartment in Washington, while he had at his disposal in Britain a four-engine aircraft and crew, a private railroad car, a house in Berkeley Square, a country house, a large staff, and a car and driver. Kay not only was a part of that, as his driver, secretary, and constant companion, but also shared in a glamorous wartime lifestyle, instead of Mamie. Even if the relationship between Ike and Kay was entirely chaste, as it may well have been, it would still be enough to make Mamie grind her teeth in jealousy. Ike was experiencing the greatest moments of his life, however heavy his burdens might be; and another woman—young, pretty, and adoring—was sharing them with him.

The burdens were, in fact, heavier than ever. Heaviest of all was the plan for Overlord. Ike had seen it earlier, at the end of October 1943 in Algiers, before he had been selected to command it, and had not been impressed. Indeed, Harry Butcher had confided to his diary, "Ike said the planning seemed such a mess that he really doubted if he wanted a command there. Not enough wallop in the initial attack."[14]

Ike's reaction was temperate compared with Monty's. While Churchill was recuperating in bed in Marrakech late in December, he had asked Monty to read the latest version of Overlord, and Monty took the plans upstairs to his own bedroom for the night and read them with his usual care. It was, he told the prime minister the next morning, "impracticable," and it contained the same errors that had almost doomed the Salerno landing, and would very shortly render the Anzio landing more of a problem for the Allies than the Germans.

This was not just a question of "not enough wallop in the initial attack,"[15] though Ike had correctly seen that as the major flaw and thought that five or even six divisions would be needed to land on D-day, instead of the three that the plan called for. Monty agreed with that, but also raised other objections. The beachhead was too narrow, he thought—the beaches would be "fatally congested." He wanted "the broadest possible front," so that the British and American armies would have their own separate beaches, and so that the individual corps would have well-marked beaches of their own "from which to develop their operations." Given the differences between the two armies in command structure, weapons, and ammunition, Monty was certainly correct that the British and American landings "should be kept separate" to avoid confusion, and also that trying to pass one corps through a beachhead assigned to another that had already landed was a recipe for chaos. These errors, he thought, were symptomatic—the major flaw was that this was an operation that had been designed by planners. It was an intellectual exercise that was being handed over as a finished product to those who would command it, instead of a plan drawn up by men who had experience in "fighting a battle" and would be in command of this one.

Churchill was impressed by this (the words "fighting a battle" could be relied on to get his attention), and he urged Monty to "grip the show"—and the next day Monty claimed to have repeated it all to Ike during the latter's brief visit to Churchill's bedside at Marrakech. Ike's reaction to the plan in its present form had been identical to Monty's, so he cannot have been surprised by Monty's opinion, although whether he in fact asked Monty to reexamine the question in detail in England, "and gave him the necessary authority to do so,"[16] is more doubtful. Monty certainly acted on that assumption, but Ike does not even mention this meeting in *Crusade in Europe*.

In any case, Ike went off to Washington, and Monty flew to England more than ever convinced that "the original plan was thoroughly bad," and immediately called the Overlord planners to a meeting at his headquarters in St. Paul's School. There, rather like the headmaster of St. Paul's, Monty heard the planners out silently, then, after a twenty-minute break, stood up, demolished their plan, and sent them back to their headquarters at Norfolk House to start all over again. There was to be a broader—much broader—beachhead; they were to find a way to add at least two more divisions; and they should make every effort to plan for seizing a port. Since the date then scheduled for D-day was twelve weeks away (it would eventually be postponed by a further four weeks), the consternation of the planners, who had been working on Overlord day and night for nearly a year and a half, may be imagined.

Ike did not disagree with any of this, though he would surely have managed to communicate it with a good deal more tact than Monty. But when he arrived in London and came to consider it, he insisted on adding the airborne landings, a British airborne division on the left and two American airborne divisions on the extreme right, which would be dropped by parachute and gliders at night so as to "seal off" the beaches from flank attacks by the Germans while the troops were coming onshore, inevitably the most critical moment of the invasion.

The man chiefly responsible for the planning of Overlord was Lieutenant-General Sir Frederick Morgan. His title, Chief of Staff to the Supreme Allied Commander (COSSAC), had been given to him at the Casblanca conference in January 1943, long before the supreme commander had been chosen, and he had been involved in drawing up the invasion plans since 1942. Morgan took over the task from earlier planners who, at Churchill's direction, had been hard at work since 1941 on what had at first seemed the rather chimerical problem of how to put a British—and after December 7, 1941, an Anglo-American—army back on the continent of Europe, liberate France, and defeat the Germany army in the West. Morgan was widely respected in the British army, as well as by Churchill and General Marshall, and the dispute between himself and Monty over the plan for D-day—which began on January 3, 1944, at St. Paul's School and sputtered on throughout the rest of the war and long after it in works of history, television documentaries, memoirs, autobiographies, published diaries, and biographies—is hard to place in its proper context. Because Ike and Monty insisted on expanding the beachhead and adding two divisions to the initial assault, plus the airborne attacks, they have generally been credited with "saving" the plan for D-day—indeed, Monty took credit for that for the rest of his life. In his memoirs, however, Ike praised "Freddie" Morgan (as he was known throughout the British army) warmly, although Monty had no use for him. The truth is that by January 1944 the most important decisions about D-day had long since been made by Morgan, and were, as the saying goes, "carved in stone"—it was far too late to change them, nor was it necessary, since Morgan got most of them right. Morgan and his staff had decided on landing in Normandy instead of Brittany or the Pas de Calais; they had dropped the idea of seizing a port after the disastrous episode of the Canadians at Dieppe in 1942; they analyzed the available beaches and chose the ones that would be used on June 6; they prepared the myriad of special weapons and installations that were required to land an army—two armies, in fact—on open beaches without capturing a port,

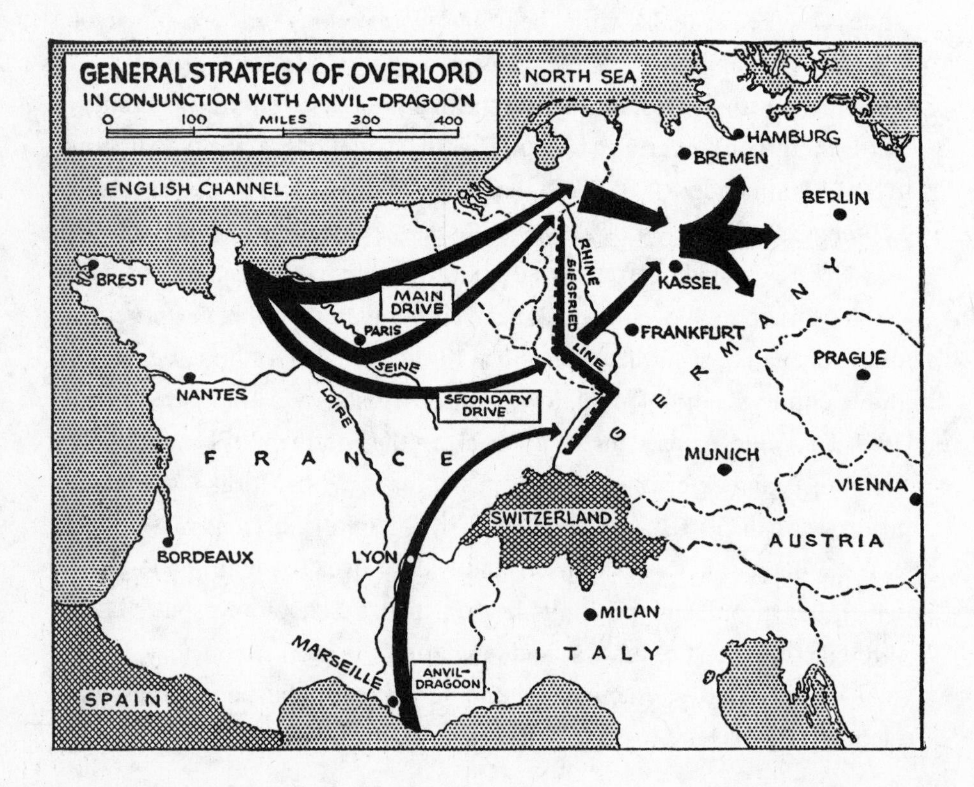

from the artificial harbors and breakwaters, and the pipeline under the sea, to the "swimming" tanks and General Hobart's "funnies"; they took three years of painstaking research into the smallest topographical, military, and maritime details and turned it into a practical military plan that covered, among other things, "transport arrangements, organizational planning, diversionary campaigns [and] security precautions"; and they even made a stab at determining occupational zones to be set up once Germany was defeated. Morgan's plan was inventive, audacious, immensely detailed, well-prepared, and correct in most of its basic assumptions—he picked the right place, the right beaches, and the right time for the landings, and devised the means of getting the men there. Overlord would never have taken place without him.

Morgan's limitations were those that had been forced on him. Just as the Normandy beaches had been picked long before Morgan himself entered the scene, in part on the basis of what was in 1941 the maximum combat radius of British fighters operating from southern England, in 1942 and 1943 Morgan could plan ahead only on the basis of the number of landing craft that the naval commanders predicted would be available to him in the late spring of 1944, and that had dictated the three-division assault. It was not for nothing that General Alan Brooke, telling Morgan of his appointment as COSSAC, had sent him away with the words, "There you are, it won't work, but you must bloody well make it."[17] Morgan had done just that, but he got scant credit for it.

Morgan was in part a victim of Ike's postwar biographers, who wanted to establish that Ike was responsible for the planning of D-day; and in part a victim of Monty's desire not only to claim credit for the last-minute changes to the COSSAC plan, but to draw a veil over the protracted battle he fought to capture Caen, on the extreme left of the beachhead. Monty had been overoptimistic about Caen before the landings, and in the view of the Americans he was overcautious afterward. On the need to capture Caen quickly, as in many other respects, the plan for Overlord was firmly fixed by January 1944, and nobody could have changed it.

A perfect example was the position of American forces on the right of the invasion. This was not a choice of Ike's, or Monty's—it had been dictated by the simple fact that when American troops were first shipped to the United Kingdom early in 1942, they were sent, at the request of the prime minister, to Northern Ireland to assume garrison duties and free the British troops there for service elsewhere. As American troops began to pour across the Atlantic in ever-increasing numbers, it made sense to keep them as close together as possible, since their command structure and their logistic and supply requirements differed radically from those of the British, so they were shipped directly from ports on the northeastern coast of the United States to ports on the west coast of the United Kingdom (Bristol, Liverpool, and Glasgow). This also made sense in terms of transportation—it was a straight line from the great embarkation center in New York Harbor to Liverpool, and from there only a short distance to training camps in the west of England. Thus the American ground forces in the United Kingdom were from the very beginning concentrated in the west (and in Northern Ireland), where they could most easily be supplied directly from America (since the British could not supply their needs), and where there was, in any case, more room for them to train than in the more heavily urbanized and industrialized southeast of England. Since logic is the basis of "logistics," this meant that when the time for the invasion came, American ground forces would necessarily have to embark from the ports closest to them, in southwestern and western England, and that would inevitably place them on the right when they landed in France.

Morgan had foreseen all this—hence it was the responsibility of the British forces on the left to take Caen, and it would be that of the American forces on the right to take the port of Cherbourg. Hence too, once a breakout had been achieved, and the Allied forces turned on the all-important "hinge" of Caen, the Americans would stay to the right, pivot to the southeast, and make for the Seine, above or below Paris, while the British would attack in a straight line eastward toward Belgium, Holland, and eventually the Ruhr. Inevitably, that would make the focus of the American attack the southern part of Germany, in the direction of Munich, Vienna, and Prague, while the British attacked in the north of Germany, taking the German ports one by one. This would also mean—Morgan was nothing if not farsighted—that the British occupation zone of Germany would be in the north, and the American zone in the south (the need for a French zone of occupation was not, at the time, foreseen). The path of the armies—and, more important, the path of the separate supply lines they trailed behind them—would determine the occupation zones, and so it proved. Even though it was President Roosevelt's ardent desire and determined intention that America should have northern Germany, with the great ports, as its occupation zone—he felt that Americans would want to have access to the ports in case it was ever necessary to bring their troops home quickly*—this could not be done without hopelessly entangling the American and British supply lines. So the United States ended up with the zone General Morgan had selected, centered on the cities of Frankfurt, Stuttgart, and Munich, while the United Kingdom got the ports and the north of Germany, centered on the cities of Cologne, Münster, and Hamburg. Even the wishes of the president dissolved when confronted by the consequences of the implacably logical mind of Freddie Morgan.

Thus the addition at the last minute of two more divisions and the airborne units has to be placed in its proper perspective—it was undoubtedly important to the success of D-day, but it was by no stretch of the imagination the kind of total revision that Monty demanded, or later described. The basic plan was sound, and even if it was not, Ike and Monty would have to live with it.

Even the limited changes they insisted on came at a price. Adding the extra divisions meant that more landing craft were needed, and Churchill's insistence on Shingle (and Marshall's equally stubborn insistence on Anvil) meant that there were none to spare. It was largely due to this that Overlord had to be moved from early May to mid-May, then to the end of May, and finally to June 5, 6, or 7, despite the concern of Churchill and Roosevelt that Stalin would regard any postponement past the middle of May as breaking the promise they made to him at Tehran.

* For much the same reason, Churchill wanted the north of Germany as the United Kingdom's zone of occupation. In case of trouble (and he was by then thinking of the Russians, not the Germans), he wanted to be able to reinforce or evacuate British forces directly across the North Sea, not have them stuck in the far-off mountains of Bavaria.

Like so many promises, this one produced a good deal of soul-searching and wishful thinking as it came due. Churchill, stretching the truth slightly, wrote to the president that he did not himself remember giving "Uncle Joe" a particular date, and that he did not feel "a June date . . . would in any way have broken faith with him."[18] Roosevelt wrote back that it was his "understanding that in Teheran Uncle J. was given a promise that 'Overlord' would be launched during May," urged that "no communication should be sent to Uncle J. on this subject" for the time being, and pointed out that "the psychology of bringing this thing up at this time would be very bad."[19] In short, the prime minister couldn't remember whether they had made a specific promise of a date or not; and the president wasn't sure either but thought they might have, and guessed that the best thing to do was not to bring the subject up with Uncle Joe in the hope that he would forget about it. In the event, Stalin for once took the postponement with relatively good grace when he finally learned of it, and lived up to his promise of launching a major attack at the same time as Overlord to keep the Germans busy.

Measured in terms of the space devoted to it in *Crusade in Europe*, the most critical problem facing Ike was the obduracy and refusal to cooperate on the part of the "bomber barons." Ike had no doubt expected nothing else from Air Chief Marshal Harris—Tedder would have warned him of the counterattack that was sure to come from High Wycombe, headquarters of RAF Bomber Command—but he must have expected better from General Spaatz, who was an old friend and whom he had brought back from North Africa to take command of the USAAF Strategic Air Force in Europe, which included the U.S. Eighth Air Force in the United Kingdom. One of the many reasons for postponing the invasion had been to ensure better weather—and therefore clearer visibility and more accurate bombing—for the "preparatory bomber offensive." The destruction of critical points on the railroad lines, as well as bridges and main roads, was deemed vital to the success of Overlord. At all costs, even substantial (and inevitable) French civilian casualties, the Germans had to be prevented or at least hindered from bringing their reserves into action. For this purpose, the "heavies"—the four-engine Lancasters, Halifaxes, and Stirlings of Bomber Command; and the B-17s and B-24s of the U.S. Eighth Air Force—were essential. The difficulty was that both Harris and Spaatz were deeply engaged in waging their own private air war against Germany. Harris was involved with Pointblank, the bombing campaign intended to destroy the German aviation industry, and the saturation bombing of German cities by night (his intention was to methodically destroy the sixty major German cities, at the rate of two and a half per month). Spaatz was involved with the "Oil Plan," consisting of massive daylight raids on every refinery his aircraft could reach. Neither of them wished to divert his forces to Ike and Tedder's "Transportation Plan," which was in essence the bombing of the French railroad system. They were tough nuts to crack too. Harris had the ear of the prime minister, and Spaatz had the ear (and the support) of Generals Arnold and Marshall.

Although they were natural rivals, and never hesitated to discredit each other's theories about bombing, Harris and Spaatz were willing enough to cooperate with each other when it was a question of resisting Ike's demand to control the heavy bombers during the period before D-day, since it was anathema to both of them. Harris's resistance was fortified by doubt that his crews could actually hit targets like railroad junctions and bridges by daylight, since they had been trained to bomb the center of large cities by night—Harris's specialty was the "thousand-bomber" raid of a German city in an attempt to start a firestorm that would cause terrific destruction; in short, his preferred weapon was the cudgel, not the rapier. Spaatz's crews had only a marginally better reputation than RAF Bomber Command's for "pinpoint bombing," as it was called—even though the Eighth Air Force flew by day, its planes had been known to attack the wrong city, including, in one regrettable incident, Schaffhausen, in neutral Switzerland. In neither case was there any enthusiasm for trying to hit a specific target like a railroad viaduct with four-engine bombers from 10,000 or 20,000 feet.

The argument about how bombers should be used had been going on, with various degrees of acrimony, since the last two years of World War I, and even today it raises the hackles of otherwise placid historians. Harris, it is sufficient to say, was a true believer in the bomber, indeed *the* true believer. He had been appointed to the post of Air Officer Commanding, Bomber Command, in 1942, at the point when Bomber Command was sunk in the slough of despond, unable to do any serious damage to the Germans or to defend itself from its critics at home and its rivals in the RAF and in the other services. The aircraft it had at its disposal were slow, were inadequately armed, and carried a minimal bomb load; the crews had demonstrated that in daylight they were sitting ducks for German fighters and antiaircraft guns, and at night they were unable to find their target. Cast in the shadow by the luminous achievements of Fighter Command in the Battle of Britain, Bomber Command was an embarrassment. Harris, with his formidable energy, his attention to detail, and his fiery temper, changed all this overnight—he was a twentieth-century aerial version of Sherman. He brought in the big four-engine "heavies"; he introduced a new and more sophisticated level of intelligence and applied science; he got rid of the deadwood and promoted tough, hard thrusters like himself. An aggressive personality, he turned Bomber Command into the perfect instrument of aggression. He was, among other things, a gifted bureaucrat as well as a born commander of men—tirelessly, he dictated long, detailed, formidably well-informed and logical memos, minutes, and reports, bristling with hostility, and rebutting in detail every criticism or doubt from any quarter. He was not afraid to be blunt, abrasive, and rude, even to the prime minister (though in Harris's case this approach was balanced by his elephantine but effective gift for flattery).

Harris, even more than Spaatz, was almost a law unto himself. In theory he received his orders from the British Combined Chiefs of Staff, via the Chief of the Air Staff, Air Chief Marshal Sir Charles Portal; but in practice he was relatively free to pursue his own campaign against Germany and pick his own targets, and he was not averse to going above Portal's head to deal directly with the prime minister. Spaatz's boss was General H. H. "Hap" Arnold, Commanding General of the United States Army Air Forces, who reported directly to General Marshall, and when necessary to the president. Spaatz too enjoyed broad and independent authority, so it was hardly surprising that neither he nor Harris wanted to be placed under Ike's command, even temporarily.

This was a fight that had been going on for some time. As early as 1942, when Harris took over Bomber Command, he had believed that if he was given enough bombers he could win the war. In a lengthy paper to Churchill he argued that any attempt to invade France and fight the German army would be wasteful and involve "enormous losses," whereas a bombing campaign, if it was fully supported,* would reduce the invasion to a mere "mopping-up operation." Until Monty's victory at El Alamein, Bomber Command was Britain's only instrument for punishing Germany, and so long as that was the case, Churchill was willing to humor Harris. Churchill replied to this particular paper with no more than what was, for him, a delicate rap on the knuckles: "You must be careful not to spoil a good cause by overstating it. . . . I do not however think that Air bombing is going to bring the war to an end by itself."[20]

Perhaps needless to add, this caution was like water off a duck's back. Harris remained committed to the idea that "area bombing" would bring the Germans to their knees, apparently oblivious of the fact that the Luftwaffe's bombing of British cities had merely made the British more determined to fight. He needed to demonstrate what he could do, so he wrung out of Churchill permission to assemble a hitherto unthinkable raid of over 1,000 RAF bombers against a single German city in one night (code-named, by Harris, Millennium), and destroyed 600 acres of Cologne, including its historic mediaeval center, with a loss of only forty aircraft.† The smoke

* What Harris and Spaatz had in mind by "full support" can best be illustrated by the USAAF's bombing campaign against Japan in 1944 and 1945. Developing the Boeing B-29 was the single largest and most expensive industrial achievement of the war, and cost $1 billion (in 1944 money) more than the atomic bombs it would eventually deliver ($2 billion for the Manhattan Project, $3 billion for the B-29). It also required keeping China in the war, so that the planes could be based there, and when the Japanese reacted with fury and overran many of those airfields, made necessary the amphibious conquest of the islands leading toward Japan to serve as air bases for the B-29 force, at the cost of some of the bloodiest fighting of the war. What Harris and Spaatz wanted was nothing less than the total commitment of Anglo-American industrial, scientific, manpower, and raw material resources to the bombing campaign.

† The word "only" perhaps needs to be put into context. This represents 3.8 percent of the bombers dispatched, well below what was considered the "acceptable" loss of 5 percent; but losing forty RAF heavy bombers meant the loss of 280 to 320 trained men in one night. (American heavy bombers, much more heavily armed because they flew by day, usually carried a crew of ten. British heavies carried a crew of seven or eight, since they did without a copilot and carried only three gunners.)

and flames from the burning city rose to over 15,000 feet, 250 factories were destroyed; and huge numbers of people were made homeless. Harris made Churchill an instant if still skeptical convert to "area bombing" and at last convinced almost everybody, even his enemies on the Air Staff and in the Ministry of Economic Warfare, that Bomber Command was a serious weapon. Even the Germans were convinced at one stroke. Göring, when he was told by telephone the news of the raid on Cologne, shouted that it was impossible, "a stinking lie," that nobody could drop that many bombs in one night, and demanded of the hapless Cologne police commissioner, "How can you dare to report such fantasies to the *Führer*?"[21] But to the clearer north German minds of Speer and Goebbels the writing was on the wall. Cologne was "a real catastrophe," the first of many, and the unmistakable sign that Armageddon was about to descend on Germany from the air.

From the moment of Ike's return to Britain he was plunged into meetings on the question of if, when, and how he would command and direct the heavy bombers in support of the invasion. This was not a matter on which he felt it possible to compromise—the bombers were an integral part of his plan, the equivalent of a massive preparatory artillery bombardment, and disrupting the Germans' ability to move their reserves and tank divisions toward Normandy quickly to seal off the Allies' beachhead was indispensable to the success of the invasion. Knowing his adversaries, he was not prepared to settle for "directing" Harris and Spaatz; he wanted to "command" them. Harris was a formidable and experienced infighter, however, who did not feel himself obliged to "fight fair" in defense of his unfettered control of Bomber Command. He and Spaatz took turns alarming the prime minister on the subject of French civilian casualties, which they estimated at about 80,000, causing in Churchill and the British Cabinet "a grave and on the whole an adverse view of the proposal to bomb so many French railway centers."[22]

Ike was, as it would turn out, correctly skeptical of these figures, and believed them to have be "grossly exaggerated"—he had his own statistician, in the person of a distinguished zoologist and newly hatched expert on the effect of bombing on people and buildings, Sir Solly Zuckerman. But the figures stuck in the prime minister's mind, and he eventually felt obliged to write a long, detailed message to Roosevelt about "this slaughter . . . among a friendly people who have committed no crimes against us."[23] The president took a considerably more philosophical view of French civilian casualties, and replied: "However regrettable the loss of civilian life is, I am not prepared to impose from this distance any restriction on military action by the responsible commanders that in their opinion might militate against the success of 'Overlord' or cause additional loss of life to our Allied forces of invasion."[24] This opinion would be reinforced when the French themselves were finally consulted on the matter. De Gaulle indicated that he saw no problem; and when asked his opinion, "Major General Pierre Koenig, the commander of Free French Forces in the United Kingdom, cold-bloodedly replied, 'This is war and it is to be expected that people will be killed. . . . We would take twice the anticipated loss to be rid of the Germans.' "[25]

Checkmated on the issue of humanitarianism, Harris and Spaatz retreated to the old argument that the best way of helping the invasion to succeed was by keeping the pressure up on the German homeland, or that, in Harris's words, "the only efficient support which Bomber Command can give to Overlord is the intensification of attacks on suitable industrial centres in Germany."[26] When that too failed, shot down by Portal in a series of long, brilliant, acidly polite minutes, they fell back on the argument that they would not take orders from Leigh-Mallory, since as a former commander of fighter pilots he had no experience of bombers. Ike, by now infuriated with these "prima donnas," as he called them, threatened to resign his command and go home; and at this, a compromise was worked out whereby his deputy Air Chief Marshal Tedder, instead of Leigh-Mallory, would communicate SHAEF's needs to the bomber commanders. As things happened, Harris and Spaatz did exactly what Ike wanted them to do, and did it very effectively. By D-day the British and American "heavies" had attacked railroad targets all over France, dropping no less than 66,517 tons of bombs; paralyzing most of the rail network; and destroying innumerable vital targets like the Saumur railroad tunnel, which was blocked by nineteen 12,000-pound bombs dropped by RAF Bomber Command, and which the Germans never succeeded in reopening. A measure of how faithfully Harris and Spaatz did their job in the end, once Ike had forced them to do it, is that in the two months before D-day the Allied air forces lost nearly 12,000 men and 2,000 aircraft over France. French civilian casualties were under 10,000.[27]

Although there are innumerable examples of Ike's temper, he usually set a remarkable standard for patience under pressure. He got what he wanted, but without the theatrics of Harris or Monty, and without the long papers or minutes which were intended to be the bureaucratic equivalent of the 12,000-pound bomb, and also serve to build up a record for the future. In some respects, Churchill was the worst offender at this—he seldom wrote anything of consequence without considering how it would be read later on, after the war, and even after his own lifetime, and because he was a gifted polemicist, as well as an indefatigable writer, his view of affairs has often prevailed. Indeed, as David Reynolds has argued in his book, *In Command of History: Churchill Fighting and Writing the Second World War*, even more than half a century after Churchill completed the six volumes of *The Second World War*, his version of events still dominates the history of the war, at any rate in what he called the "English-speaking world." Even today, we still tend to see the war unconsciously through Churchill's eyes, if only because he had a gift for turning even the driest of diplomatic exchanges into a great story with himself in the foreground, and because he zealously guarded his notes, papers, and letters for his own future use, so that he often gives himself the last word. It helped, of course, that Roosevelt and Stalin—and of course Hitler—never wrote their war memoirs to contradict him. Churchill alone was left to tell the tale. This was emphatically not Ike's way. One never has the sense, reading his books, letters, or papers, that he is writing for posterity, or building a case for his side of the story. It would be hard to imagine a more fair-minded memoir than his, in fact. Even on the subject of the recalcitrant bomber barons, who sorely tried his patience, he is noticeably fair-minded.

By the time Ike officially took up his command on January 15, the preparations for D-day were well under way—men (almost 3 million of them), equipment, and supplies were being slotted into place all over Britain, so that they could all be moved to the ports and loaded at the right time and in the right order. The most immediate question facing him was, of course, the "Transportation Plan" and the need to have the bomber forces placed under his command temporarily. But scarcely less important was the need for more landing craft, discussions of which always returned to whether Anvil, the landing in the south of France, should be canceled or not. The British, led by the prime minister, had always been in favor of a Mediterranean strategy and skeptical about D-day, but had now changed sides with the Americans. The British now wanted Anvil scrapped to release more men and landing craft for Overlord, while the Americans, who had hitherto been the enthusiasts for Overlord and opposed Mediterranean adventures, insisted on Anvil. Churchill, Brooke, and Monty were strongly opposed to Anvil, while Marshall was a fervent supporter, as was Ike, who believed that it was vital to confront the Germans with a "secondary attack" and above all to secure the port of Marseille for the Allies' use before the Germans could destroy it.

The major problem confronting Ike was not, however, the invasion itself—that would either succeed or fail, and very little more could be done at this point to ensure its success—but what to do once the Allied armies were firmly established on the Normandy beachhead. Obviously, they would have to break out of the beachhead as quickly as possible; but once they had done so, in which direction should they attack, and with what as their goal? On this subject there was intense disagreement, particularly between Ike and Monty, his ground force commander.

Monty had been on his best behavior during Ike's absence in the United States, but in fact he was simmering, since no public announcement had been made that he was going to command both the British and the American forces on D-day. Indeed, the general assumption was that Bradley would command the American forces and Monty the British; that they would be, in effect, equal partners, under the supreme commander. Ike did not get around to announcing that Monty would be the overall ground forces commander until January 21—when Monty mentioned the fact himself while addressing American troops, it was censored from the news reports—and the delay caused Monty a good deal of embarrassment. He had spent much of the time since his return to the United Kingdom addressing American troops, and from January 13 to January 20 he had succeeded in speaking to no fewer than six American divisions, "at three parades of about 3,000 to 4,000 each in each Div. area"—something no British (or American) general had ever attempted before. The Americans found him a little strange and exotic, with his black tanker's beret, RAF fleece-lined leather flying jacket, and baggy corduroy trousers, not to mention his schoolmaster's manner and his sharp English accent; but they seemed impressed by his determination that they were going to win, and for the most part listened politely. He understood how to talk to soldiers of any nationality. He would stand on the hood of a Jeep, one hand in his trousers pocket, point at one of them, and ask, "You, what's your most valuable possession?" The soldier would give the reply standard in every army: "My rifle, sir." Then Monty would say, "No it isn't, it's your life, and I'm going to save it for you." Not surprisingly, this always got him cheers and applause from infantrymen.[28]

Naturally, Monty took it for granted that he had been a great hit with the Americans, and this confidence fueled his determination to get rid of every trace of General Morgan's plan and replace it with his own. He had already sent the COSSAC planners back to the drawing board to expand the beachhead and add at least two more divisions, but he also had strong opinions about the strategy that should be pursued once the initial landings were secure, and these were to become a bone of contention between the British and the American commanders in the months to come, as well as between Monty and Ike for the rest of their lives.

These differences, and the acrimony that still accompanies them today, centered on the historic, ancient town of Caen, where William the Conqueror, duke of Normandy, who invaded England in 1066 and assumed the throne, is buried; and where Robert, first earl of Gloucester, the illegitimate son of King Henry I of England, was born in 1090. Caen has been very much a part of Anglo-French history for over 1,000 years—the famous Bayeux tapestry depicting William's victory over King Harold at the Battle of Hastings is displayed only a few miles from Caen—and the town was besieged, taken, sacked, and rebuilt innumerable times through centuries of warfare between the English and the French.

There is a reason for this. Like the town of Gettysburg, Caen is a road center, linking the main ports of Normandy, Cherbourg in the west and Le Havre in the east, with each other and with Paris. In medieval times it was also a center of river commerce, and in modern times it became a railroad center. It is close to the sea—no more than thirteen miles away—and its importance as a road and rail center meant that without it the Germans would have great difficulty in bringing their reserves up to the Normandy beachhead. Far more important, without it they would have no chance at all of bringing into action the elite armored divisions that the Führer was planning to hold in the Pas de Calais during the first twenty-four hours of the invasion. It was with them in mind that Rommel had told his trusted aide Captain Hellmuth Lang, "Believe me, Lang, the first twenty-four hours of the invasion will be decisive. . . . The fate of Germany depends on the outcome. . . . For the Allies, as well as Germany, it will be the longest day."[29]

The importance of Caen was recognized by everyone. On the German side, if the Allies landed in Normandy (the preponderance of opinion among Germans from Hitler down, including Rommel, still favored the Pas de Calais), Caen would have to be held at all costs in order to carry out the counterattack on which everything depended. On the Allies' side, Caen had always been thought of as the "hinge" around which the British and American armies would swing when they broke out of the beachhead. General Morgan had seen this clearly from the beginning, but he did not think Caen could be captured with an initial invasion force of only three divisions; and even Monty, who had, some people felt, gone out of his way to humiliate Morgan publicly during his criticism of the COSSAC plan for Overlord, could not disagree. With the expansion of the initial landings to five divisions, plus the airborne divisions, and the enlargement of the beachhead to give the troops "more elbow room," the possibility of taking Caen quickly was increased. Even so, it was an iffy proposition. If Generals Morgan, Eisenhower, and Montgomery could see the importance of Caen, so would Field Marshals von Rundstedt and Rommel the moment the Allies landed on the Normandy beaches. Caen was certain to be defended vigorously, unless it could be taken by a coup de main before Rommel could reinforce it. However, Monty, in redrafting Morgan's plan, included the taking of Caen as one of his objectives by D+5 (five days after D-day), though this startled the American generals less when they saw his "phase lines" for the battle than the fact that he did not plan to reach the Seine or the Loire before D+90. Worse still, shortly before D-day, he "rashly committed himself to getting into Caen by the afternoon of the 6th," i.e., the day after the invasion.*[30]

This may have been part of a growing tendency toward *folie de grandeur*, which led to some curious actions on Monty's part. Monty had always been "insufferably" vain, arrogant, and boastful; but in his new position as ground forces commander for Overlord, and perhaps also because he was tasting for the first time, now that he was home, the full adulation and hero worship of the British public, he repeatedly went too far. Ike's fame never went to his head; Monty's fame went straight to his. On being told that the king disliked his habit of wearing a black beret, Monty lectured his majesty at an audience on the morale and propaganda value of his trademark headgear, which he estimated was worth an additional corps to him; he argued with Secretary of War Sir James Grigg; he commissioned a portrait of himself from Augustus John, the great portraitist, then rejected it when he saw it, and refused to pay John for it; he wrote a long, detailed letter to the Archbishop of Canterbury proposing a special service to be held in Westminster Abbey "in coronation regalia" to "consecrate the nation's strength," and laid out the exact service with himself as the central attraction. (The archbishop replied with baffled politeness and passed the problem along to the king and the prime minister, who promptly squashed it.) A joke making the rounds at the time had "Churchill saying to the King, 'I'm very worried, I think Monty is after my job,' and the King replying, 'Gosh! I'm very relieved to hear that; I thought he was after *mine*.' "[31]

Monty's furious energy kept him always in motion, and while much of it was fortunately channeled in the right direction, some of it overflowed in

* D-day was at the time still scheduled for June 5.

ways that would come to haunt him in the future. Nothing had a more serious effect than the claim that he would be in Caen twenty-four hours after D-day, when in fact he would not reach it until July 9, and then only after an attack by the RAF Bomber Command that reduced whatever remained of the town to a mass of rubble blocking the streets. In the meantime, Ike was focused on Cherbourg, not Caen. Cherbourg was the prize—an important deepwater port that could be reached directly across the North Atlantic from ports on the East Coast of the United States.

To sustain the kind of attack Ike had in mind, supplies would be, as always, the critical factor—he needed to take Cherbourg, if possible before the Germans destroyed the port facilities; and once his armies moved beyond the Seine, he would need to take Antwerp, for the same reason. If the British could take Caen quickly, and advance to the east, they would be in a position to attack toward Antwerp, while the American armies, having secured Cherbourg, wheeled in a southeastern arc on the hinge of Caen to drive the German forces back toward the German border. It was Bradley's job to break out of the beachhead, secure Cherbourg, then attack toward the Seine. It would be Patton's job—he was being kept in England commanding a phantom army to deceive the Germans, under strict orders from Ike to keep his mouth shut and stay out of trouble or be sent home—to carry out the all-important end run, wheeling the U.S. First Army from the Normandy beaches to the southeast in a great arc, which would eventually place him to the south of the Ruhr. Meanwhile Monty, with his British and Canadian armies would advance to the north of it, giving them an opportunity to cut Germany off from its own industrial heartland, or forcing the German army into a battle it could not win to defend the Ruhr.

Ike was not a general who planned everything the way Monty—the master of the "set-piece battle"—did, but he had a very clear idea of what he wanted to accomplish, and like Grant he believed in using American resources on the largest possible scale. Ike intended to use his forces not in a single thrust, but across a wide front extending from Switzerland to the sea, attacking the enemy constantly until superior numbers inevitably ground the Germans down. It did not matter, in the long run, that the Germans had better tanks or antitank guns; they could not replace these, let alone their losses in troops, whereas Ike could—and in any case the Allies' tactical air strength would more than outweigh any superiority of German armor.

All the same, Ike was well aware that once he was firmly established in France he faced two serious problems. The first was the possibility that the Germans might fall back on their own and adopt a strong defensive line, confining the Allied armies in Normandy and perhaps Brittany, and bringing about a return to the static warfare of the western front in 1914–1918. Fortunately, Hitler's own experiences as a soldier in the trenches during World War I, his strong instinctive belief in the value of attacking, and his determination never to give up ground for any reason prevented that from happening; nor were Rundstedt and Rommel the kind of generals to dig in, even had the Führer allowed them to. Ike's solution was one that came naturally to him, particularly after the experience of the drawn-out, static campaign in Tunisia, and that was to keep moving. Churchill complained constantly about the number of "non-fighting" vehicles, mostly trucks and Jeeps, that were to be landed in France, estimating that there was one for every 4.84 soldiers (this also meant that something like one out of five soldiers was driving a truck instead of fighting), and wondering indignantly what had happened to the idea of an army marching on its feet; but in fact the vast number of vehicles was central to Ike's concept of war. Not for nothing had he helped to carry out the U.S. Army's first transcontinental truck convoy in 1919—a war of movement on a vast scale was what he had in mind, one which would keep the Germans off balance and unable to form or hold a defensive line. But that, of course, intensified Ike's second problem, which was supplying four Allied armies on the move. (From north to south they were: Canadian First Army, British Second Army, U.S. First Army, and U.S. Third Army; plus two more, U.S. Seventh Army and French First Army in southern France after Anvil, renamed Dragoon, and now scheduled for August.)

The logistic problem was awesome and unprecedented, and would ultimately be dominated by the need for fuel, and this in turn meant that the great ports of Cherbourg, Antwerp, and Marseille had to be captured—intact and as soon as possible. Otherwise, the armies would be limited to the amount of fuel that trickled across the Channel from the United Kingdom through PLUTO (the pipeline under the ocean), plus whatever could be carried by truck and train 500 miles to the northeast from Marseille after mid-August. By that time the Allied forces in France would consist of at least thirty-seven divisions (twenty American, twelve British, three Canadian, one French, one Polish), each of which would require a minimum of 600 to 700 hundred tons a day. All of it would be shipped from the beachhead or the ports to the front line in trucks, which would then turn around immediately and go back for more, adding enormously to the consumption of what the British army calls POL (petrol, oil, lubricants). Thousands of tons more a day would be required for the construction of advanced airfields (and for fueling the aircraft that operated from them), as well as heavy equipment for the repair of bridges, roads, and railway lines.

Given the immense difficulties facing him, Ike may not at first have noticed that Montgomery had cannily slipped an "escape clause" into his statement that he would take Caen quickly. Whether he took Caen or not, Monty argued, the purpose of attacking it was to pin down the German armored forces on the left flank so that Bradley could break out on the right and take Cherbourg—the British and Canadian armies would be, in effect, supporting the Americans. Monty's slogan would be, "Caen is the key to Cherbourg"; but this was not a role for the British and the Canadians that the prime minister would have accepted for a moment, and Monty's own plan had included not only the capture of Caen but, even more important, the ground to the east of it. The fallout from this misunderstanding, stretching over the four weeks it took Monty to capture Caen and the three weeks it took the U.S. First Army to reach Cherbourg (by then the Germans had succeeded in largely destroying the port facilities and planting huge quantities of mines in the harbor), was not only a serious delay but the beginning of a deepening rift between the British and the American commanders—a rift that Ike had worked hard for over two years to prevent. Above all it would adversely affect the relationship between Ike and Monty, and set the pattern for more serious problems in the autumn and winter of 1944.

Ike's intention was clear-cut: "to bring all our strength against [the enemy], all of it mobile, and all of it contributing directly to the complete annihilation of his field forces."[32] He intended "to give priority to the left," since the land north of the Ruhr was more open, and since there were already alarming rumors about German "secret weapons" aimed at England that were being prepared in the Pas de Calais and along the North Sea—the V-1 buzz bomb, a jet-propelled flying bomb; and the V-2 ballistic rocket, which would very shortly be brought into action. The attack from the left through Belgium and Holland into north Germany could be supplied through the big, modern port of Antwerp, which would be close to the front lines, instead of trying to supply it from Cherbourg and the invasion beaches, over 250 miles behind. This was a sensible plan, but it depended on Monty's taking Caen, then advancing quickly to the Seine; and as the fight for Caen dragged on, even those American commanders who had not yet made up their minds about Monty would begin to feel that he was letting them down badly.

As the winter ebbed and the invasion date grew closer, Ike seemed to many people to be numbed by the sheer volume of work and decisions. Kay Summersby, who was by then acting as his appointments secretary as well as his driver, notes down a typical day:

10:30 General Betts
11:30 Conference of the Commanders in Chief
12:30 General Prentiss
1:30 Lunch with the Prime Minister
8:00 Dinner with Admiral Stark
10:00 General Spaatz[33]

The period between dawn and ten-thirty in the morning was spent going through the night's messages from Washington and the commanders in the field, and answering them. Lunch with the prime minister might drag on for several hours, but it was better than dinner with the prime minister at Chequers, his official country house, which often did not end until two in the morning or later. Although it added immeasurably to his workload, Ike insisted on traveling all over England and Scotland inspecting the units that would be going to France, as well as key factories and the hospitals that would be caring for the wounded who were evacuated from the beaches. Everywhere he greeted people intensely, personally, smiling broadly, shaking hands, talking until his voice was reduced to a hoarse growl. Wherever he went, he wanted to see everything—the men's quarters, their mess, their training area—and to ask as many men as he could where they came from. Kansans and Texans always got a special smile from the "boy from Abilene." But if a soldier was from Minnesota, the supreme commander would talk to him about fishing; or if he was from the South, about southern cooking, a subject on which Ike was something of an expert. He did not give soldiers a lecture like Montgomery; or a sonorous, carefully rehearsed neo-Churchillian speech like Douglas MacArthur; or a rousing, profane pep talk de haut en bas like George Patton. He simply treated them like fellow citizens, and with his plain, unadorned uniform; his earnest, worried face; and his unmistakable sincerity, if it hadn't been for his age and his four stars, he might have been one of them. Soldiers, both American and British, factory workers, and ordinary citizens took to saying "Good old Ike!" when they saw him, mobbing his car, pumping his hand, and cheering him—to his evident embarrassment, but he understood that this was part of his job. Sometimes, when he got back in the car, Kay Summersby saw him dabbing at his eyes with his handkerchief.

He suffered from a sore throat, from an endless succession of colds, from "a cough that lasted for months"—conditions not improved by his cold, cheerless, drafty office; his huge consumption of cigarettes; his nervous tension; or his constant late nights. His eyes were red-rimmed with exhaustion and reading, and his temper was badly frayed. His only relaxations were an occasional game of bridge, cooking (southern fried chicken or chili), and planting a vegetable garden (a return to his boyhood in Abilene). He dealt patiently with the demands of many "prima donnas," as he called them—Patton, who disobeyed Ike's orders by opening his mouth in front of reporters and promptly putting his foot in it again; De Gaulle, who was making difficulties about his role in the invasion that would not be solved until the very last moment; Churchill, who deluged Ike with ideas and demands for information, most often in the form of telephone calls beginning at midnight; Monty, who seemed determined "to do everything his own way and in his own time" (prompting Patton to observe, "Why the American army has to go with Monty I do not see, except to save the face of the little monkey"),[34] to mention only a few. He kept up his end of the correspondence between himself and Mamie, writing to her on Valentine's Day 1944, "I never forget that 28 years ago I brought out the West Point class ring to 1216 McCullough, proud as a peacock!!"[35] In another letter he recounted at length a typical Anglo-American misunderstanding ("a comedy of irritating errors," Ike called it), in which he gave a tailor some material he liked to make up into a bathrobe for her and the tailor made it into a topcoat instead. In a later letter, in April, he reminded her that it would soon be time to pay their income tax. Whereas many of his subordinates, above all Monty and Patton, sought the limelight, Ike tried—but failed—to avoid it. He resented intrusions into his privacy, disliked giving interviews, and sympathized with Mamie's anger at being hounded by the press. Few people can have hated fame when it finally embraced him as much as Ike did, though in later life, as president, he seems to have come to terms with it at last.

Ike was less aware than Mamie of his fame at home. He was surprised to learn that his goings and comings were so well reported in American newspapers, particularly since he tried to slip away quietly at the last moment, after having been "bothered" on one trip by reporters. He did his best to stay out of the papers, and was very careful to avoid the kind of embarrassment that seemed to dog his old boss MacArthur. For example, a Republican congressman "made public" an imprudent exchange of letters in which MacArthur, whose prose style was perhaps deliberately opaque, seemed to be agreeing that he might accept a "draft" for the Republican nomination to run for the presidency against Roosevelt. "Thank the Lord no misguided friend has been able to embarrass me seriously in this manner,"[36] Ike commented piously, restraining what must have been a certain satisfaction at MacArthur's faux pas—he himself had been outraged and angered by the publication of some of his letters to his mother, although they were harmless enough.

One of the events he looked forward to with sharp anticipation was his son John's graduation from West Point. In what was, for Ike, a rare use of his rank and position, he asked General Marshall to authorize John to spend his leave with him, before going into advanced training. Marshall was enthusiastic about the idea, and pulled whatever strings were necessary to fly a new second lieutenant fresh out of West Point to the United Kingdom to join his father for a month, recognizing what a boost it would be to Ike's morale—and in any case, it had been a general's prerogative, if he cared to exercise it, to have his son on his staff ever since the Civil War, when the practice was commonplace. Few things, including his high rank, pleased Ike more than the fact that his son was following in his footsteps. Ike had seen very little of his boy for the past three years, but their correspondence reveals a kind of instinctive and good-natured closeness—at one point John made fun of his father for wearing his cap at a nonregulation angle, and Ike apologized for not living up to the West Point dress code. Ike had kept himself informed of John's progress at West Point without ever making the mistake of interfering, even when John seemed unable to choose between the infantry and the field artillery after graduation, an uncertainty that created a certain degree of fatherly impatience in Ike, though he himself had faced much the same problem in his last year at West Point, when he was offered the coast artillery and insisted on the infantry. John would arrive a week after D-day, and be at his father's side through some of the most tense moments of the war.

The preparations for D-day rapidly turned the entire south of England into a giant armed camp. It seems amazing that the Abwehr failed to notice this—but the Abwehr was hampered, of course, by the fact that MI5 and Scotland Yard's Special Branch had long since caught all its agents in the United Kingdom and given them a choice of being hanged or being "turned"*—that is, reporting false information to Berlin on their radio transmitters. In eastern England the British had built a secret training ground where every possible kind of obstacle that might be waiting for the

* Some were hanged, including one unlucky spy who landed by parachute within sight of the DeHavilland Aircraft factory, where the Mosquito bomber was produced.

Allied troops when they landed was meticulously duplicated—minefields, pillboxes, underwater obstacles, antitank defenses, everything the troops might find in their way was there for them to become familiar with. Beaches that resembled those in Normandy were located in odd places all over the United Kingdom, and troops practiced landing on them and fighting their way inland. Special units—camouflage units, medical units, communications units, frontline engineer units with their armored bulldozers and mine-flail tanks, German-speaking prisoner interrogation units of the Field Security Police, RAF forward air controllers to accompany frontline combat units and direct air strikes, demolition experts, and an endless array of specialists, all of them under Ike's command—moved into place, each to be transported to the right ship, with their equipment, and landed at the right place on the right beach in exactly the right sequence. As the weeks went by, all nonessential civilian road and railway traffic between the south of England and the rest of the United Kingdom was forbidden; travel to and from the United Kingdom and neutral Eire was stopped, "since enemy spies abounded" there; coastal shipping and fishing were halted; and finally, in a step that infuriated foreign countries and ought to have alerted the Germans to what was about to happen, the government "even took the unprecedented step"[37] of stopping all diplomatic communication of any kind to and from the United Kingdom. The south of England was sealed off, "tight as a drum," with more than 2 million troops encamped from Bristol to Hastings, nervously cleaning their weapons and waiting for the moment to embark.

Their supreme commander had visited as many of them as he could—he had already "visited twenty-six divisions, twenty-four airfields, five ships of war, and numerous depots, shops, hospitals, and other important installations"[38]—but inevitably it was the remaining big issues about the invasion plan that took up his time. The shape of the invasion, the beaches themselves, the units that would be landed, where they would be landed, and the timetable for landing them—about all these things, there was little or nothing Ike could do except to hope that the right decisions had been made. But some other things still remained to be fought out. Perhaps the most contentious of these was the decision to use the airborne troops. In

1944 the jury was still out on the subject of airborne troops (to some degree it still is today). Infantry commanders objected to having their best and fittest officers, NCOs, and soldiers siphoned off into more glamorous "elite" units with higher pay and special badges. Also, the inescapable fact remained that when you dropped airborne soldiers what you got, assuming they landed in the right place, was light infantry with no way of fighting against regular infantry supported by tanks and artillery. On the occasions when airborne troops had been used so far in World War II, their performance had mostly varied between disappointing and catastrophic, except at Salerno, where a night drop of two battalions of the U.S. Eighty-Second Airborne Division had saved the day. The number of aircraft needed for dropping a whole division was enormous, and not all the transport pilots were prepared for going into combat as "sitting ducks."

All the same, as Ike looked at the map, he was determined to use them. The most vulnerable point of the landings, in his opinion, was on the right. He had insisted on a landing on the east coast of the Cotentin Peninsula, without which it would be impossible to capture Cherbourg, but nobody liked the look of "Utah beach," the only one that was available. Ike himself described it as "miserable"—on the landward side was a shallow, wide lagoon, crossed by narrow causeways, on which the Germans would certainly direct their artillery. If the Germans could hold on to the exits from these causeways, the troops would be trapped on the beach and "slaughtered" there.

Ike's solution was to drop two airborne divisions by night to seize the "vital" exits from the causeways. Quite apart from the "normal" difficulties of a night drop, almost everything about the area "was highly unsuited to airborne operations." Troops would drown, dragged down by the weight of their equipment, if they landed in the lagoon. The country behind it was *bocage*—small fields divided by high hedgerows and thick stone walls that would split the paratroopers up into small isolated "pockets" unable, in the dark, to make contact with one another and unable to form larger units. The coastline over which the transport planes and gliders would have to fly was studded with antiaircraft batteries. Ike could see no choice. If he couldn't get his troops off the beach at Utah in the first twenty-four hours of the assault, the Germans could mount a counterattack starting from there that would roll up the flanks of the landing areas one by one from west to east while the men were still precariously packed into narrow beachheads, trapped between the sea and the higher ground where the beaches ended.

Bradley agreed that the airborne assault was necessary, and Major General Matthew B. Ridgway, the senior airborne commander, felt it was feasible. Still, it remained the most controversial and risky element in the invasion plan, though not the only such element. Just as risky was the plan to drop British airborne forces to the east of Caen, between the Orne and the Dives rivers, to block the German approach to Caen and the beaches. Leigh-Mallory vigorously opposed this, but was overruled by Ike and Monty, since capturing Caen was the only way to secure the left flank of the invasion. Seizing the two drawbridges* over the Caen canal near the town of Ouistreham was considered especially vital, since they were the only way across the canal in the immediate vicinity of the invasion beach. They would also be the most direct way to get British tanks off the beach and inland, so it was essential to capture the bridges intact and hold them, instead of merely destroying them.

Some sense of Ike's caution as D-day approached can be obtained from his reply to a message from Marshall regarding his query as to whether the president should make an announcement about Overlord on D-day, and emphasizing how important it was for him to read it before it was given. The president declined to commit himself one way or the other as to whether he would do it or not, but sent a draft of what he might say to Ike via Marshall. Ike firmly cabled back that "based on military considerations alone," he did not think that either the president or the prime minister should make any announcement "until the success of the Allied landings was clearly demonstrated."[39] Ike was determined to not count his chickens before they hatched, and equally determined that nobody else should do so. Unlike Monty, who was if anything brashly overconfident,

* These bridges were later renamed Pegasus Bridge, after the flying horse sleeve patch worn by the British Sixth Airborne Division; and Horsa Bridge, after the Horsa gliders that British airborne troops used to land within less than 200 feet of the second bridge on June 6. The British actor Richard Todd was among the parachutists who landed at Pegasus Bridge, and later played in the movie *The Longest Day*, in which the attack was re-created.

Ike was realistic—nobody knew better than he did that, powerful as the forces at his disposal were, the Germans, and even the weather, were capable of thwarting them.

As the winter dragged on into the spring—a period marked by a resumption of the blitz on London, though on a smaller scale—Ike got busier and busier. He noted that he had his best ideas early in the day, when he was fresh, and that it was therefore a pity that Churchill very often saw him late at night. He wrestled with the question of whether or not to send Patton home after Patton made an indiscreet political remark ("Frankly," Ike wrote to Marshall, "I am exceedingly weary of his habit of getting everybody into hot water through the immature character of his public actions and statements"), but decided to keep him on, after another severe warning. He fended off General Giraud's request to be called to London as a "technical adviser," since his presence could only offend De Gaulle. (The place and date of the invasion were kept rigorously secret from the French, to De Gaulle's fury, but both he and his opponents were well aware that it was coming soon, and were determined to play a role.)

Everybody, including Hitler, knew the invasion was imminent, and as the spring days grew warmer and longer, it was as if the whole world were holding its breath. Mamie made careful plans to vanish from sight when the news came (she would fail, since D-day was earlier than she had thought it would be). Rommel redoubled his visits to the beaches to inspect the fortifications and the minefields. All over England people watched the columns of trucks going south night and day, carrying men, supplies, and tanks toward the staging areas. From the cliffs and beaches of the eastern seacoast, people could hardly miss the passage of huge mysterious platforms towed by oceangoing tugs, and the endless lines of landing craft of every type and size. From Telegraph Cottage Ike wrote home to Mamie to apologize for having forgotten to write to her on Mother's Day—it was not a holiday observed in Britain, so nobody had bothered to remind him of it—and added, rarely for him, "God knows I am busy."[40]

As the end of May approached, he prepared to move to his "advance headquarters" outside Portsmouth, with all his staff, including Kay, who remarked on the fact that he was smoking more than ever.

In Amsterdam, in the "hidden annexe" behind Otto Frank's small office building on Prinsengracht, Anne Frank, like so many others all over occupied Europe, waited impatiently and anxiously for the news of the invasion—for those Jews who still remained alive, the question was whether or not it would come in time to save them.

CHAPTER 14

Triumph—D-Day, June 6, 1944

It was symbolically appropriate that when Ike finally bought a house of his own—until then he and Mamie had always lived in rented houses and apartments, or in government quarters from Fort Sam Houston to the White House—the location would be Gettysburg.

D-day and Gettysburg had a good deal in common. Both were climactic battles, in the sense that although neither ended a war, each ensured that the losing side would never win. After Lee failed at Gettysburg, there was never any realistic chance that the Confederacy could win the war it had begun; that dream, whether noble or not, ended effectively with the defeat of Pickett's attack on Cemetery Ridge at two-thirty in the afternoon of July 3, 1864. Similarly, once the German army had failed to destroy the Allied forces on the Normandy beaches within twenty-four hours after the landings on June 6, 1944, it was clear, even to the majority of Germans, that Germany had lost the war.

Like all great battles since the beginning of organized warfare, both Gettysburg and D-day were struggles to gain the high ground. At Gettysburg, Lee fought for three days to drive Meade off the high ground, which Meade just barely held. On the Normandy beaches, the struggle was to get off the beach and onto the higher ground where the beach ended. In most places this was not a great elevation, and in some places it was only a few feet; but in warfare all height is precious, and has to be paid for in blood. The three days of fighting at Gettysburg cost over 53,000 Union and Confederate casualties; the first twenty-four hours of the invasion of Normandy cost about 17,000 American, British, Canadian, and German casualties. In both cases the battle took on, almost at once, such mythic qualities and gave birth to such a quantity of memoirs and historical works that almost every foot of ground has a story attached to it, so that both battles seem to

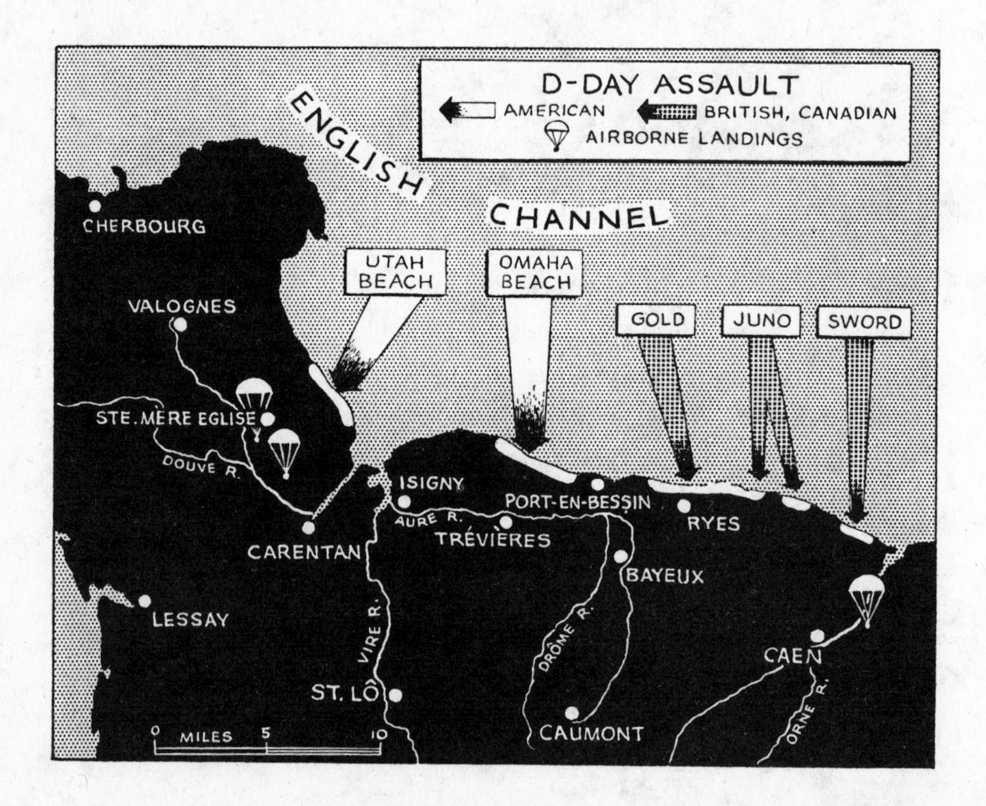

many people as if they had been fought just yesterday, rather than, respectively, 142 and 63 years ago. D-day, like Gettysburg, was fought in a relatively confined place, with the result that characters stood out in it visibly during the battle, and still stand out for us today. Brigadier General Lewis Armistead stuck his hat on the point of his sword and held it high in the air to rally his men to him as he ran toward the Union guns on Cemetery Ridge ("the high water mark of the Confederacy," as the stone commemorating his death there still reminds us). On D-day Private William Millin of Lord Lovat's (British) First Special Service Brigade Commandos marched back and forth ceremoniously among the dead and the dying on Sword beach in a storm of shells and bullets, playing, at Lord Lovat's request, "Highland Laddie" and "Over the Isles" on his bagpipes to greet the men as they struggled onshore. (This prompted one of them to shout, "Get down and shut up, you mad bugger!")[1] A grandson of President John Tyler took part in Pickett's charge as a private; and a son of President Theodore Roosevelt, Brigadier General Theodore Roosevelt, cane in hand and walking up and down Utah beach "as if he were looking over some real estate," was the only one of the Allies' generals to land with the first-wave of troops on D-day (and surely the only general whose son also landed in the first wave: Captain Quentin Roosevelt came ashore on Omaha beach). Then too, though soldiers are not ideologists or armchair patriots, these were battles fought by men who believed, at some level, in their cause—states' rights versus the preservation of the Union at Gettysburg; democracy versus dictatorship and German hegemony in Normandy. And finally these were both battles fought not by professionals or mercenaries but by citizen soldiers who would have preferred to be doing something else.

The airborne troops—to some of whom Ike* had come to wish good luck on the night of June 5, barely holding back his tears—began to drop onto the continent of Europe at around one in the morning on June 6. The U.S. Eighty-Second Airborne was to secure a vital bridgehead across the Merderet River for the capture of Cherbourg; the U.S. 101st Airborne was

* In a day when four-star generals may have 150,000 or 200,000 troops under their command, it is worth reminding ourselves that on D-day Ike commanded more than 2.8 million men.

to seize and hold the exits and causeways from Utah beach, on the extreme right of the invasion; the British Sixth Airborne Division was to seize the crossings over the Orne River on the extreme left of the invasion, about five miles northeast of Caen. Scattered far and wide in the night as they came down, often far from where they expected to be and with no visible landmarks, and separated from one another and often from their NCOs and officers, the paratroopers urgently clicked little tin noisemakers in the shape of a toy cricket that they had been issued to identify themselves to one another in the dark. At both ends of the invasion area, they made the transition from the sheer terror of hurling themselves out of an airplane into total darkness to the confusion of firefights with unseen enemies at close range.

Nearly 18,000 paratroopers landed during the night, and at dawn many of them would be found dead, drowned by the weight of their equipment in rivers and ponds, or hanging lifeless from the limbs of trees, where they had dangled helplessly until the Germans shot them. One was hanging from the steeple of the old church in the village of Sainte-Mère-Eglise (he was wounded, and had managed to feign death to avoid being shot where he hung while a fierce firefight went on below him).[2] All the same, the losses were nothing like the 70 percent that Air Chief Marshal Leigh-Mallory had predicted, when he had protested to Ike against the "futile slaughter" of the airborne divisions; and by the time Ike woke early in the morning of June 6, first reports made it clear that the airborne assaults were a qualified success.

On the left, the British Sixth Airborne had seized the two vital bridges Monty needed with a glider assault in pitch darkness and were holding out until British commandos could fight their way from the beach to join them. On the right, despite fierce fighting, the Eighty-Second Airborne and the 101st Airborne had achieved most of their immediate objectives.

At six-thirty in the morning (British double summer time) underwater demolition engineers ("frogmen") began to demolish the mined obstacles that protected the beaches to produce and mark channels fifty yards wide for the landing craft, and the first "swimming tanks" were already going onshore to support the infantry. By seven in the morning, the assault troops in the first wave were beginning to wade through waist-high water onto the beaches, while overhead a fearsome naval bombardment—the heaviest in history—began to pound the German defenses from Cabourg on the left to Quinéville on the right. A German officer reported back to his headquarters that he was looking at "ten thousand ships,"[3] and was warned not to exaggerate; but gradually, as the full extent of what was happening made its way up the German communications system, doubt was extinguished except at the very top.

Field Marshal von Rundstedt, breakfasting at his headquarters at OB West outside Paris, still thought that the landings were a diversion, and that the real invasion would take place in the Pas de Calais, though he took the precaution of alerting two of the four Panzer divisions over which Hitler retained control: the Twelfth SS Panzer Division and the Panzer Lehr.* Hitler was asleep at his house in Berchtesgaden, and his naval aide decided not to wake him until more information was available. Field Marshal Rommel was at his home in Herrlingen, and in the confusion nobody thought to call him with the news until ten-fifteen. He alone understood at once that this was not a diversion, but the real thing—the invasion he had been preparing for. "How stupid of me!" he repeated over and over again as his aide gave him the details over the phone. By one in the afternoon, he was in his car and on the way back to France, urging his driver on insistently—"*Tempo! Tempo! Tempo!*"[4] (Faster, faster, faster!) he called out, as the big Horch sped down the road, staring ahead through the windshield and tapping his "informal" field marshal's baton, a short black stick with a tasseled silver top, against his knee impatiently.

Cautiously, the most Ike would say was that the landings were going "fairly well"; but as the messages began to come in from the beaches, he saved for history the scrap of paper on which he had written what he would tell the press if the invasion failed, and at nine-thirty-three in the morning he at last ordered his press aide Colonel Ernest Dupuy to announce: "Under the command of General Eisenhower, Allied naval forces, supported by strong air forces, began landing Allied armies this morning on the northern coast of France."[5]

* Panzer Lehr was built around a nucleus of officer and NCO instructors from the German army's armor training schools.

Characteristically, Ike did not make the announcement himself, but this made no difference—around the world people stopped what they were doing to listen to the radio. In Britain work ceased and men and women stood and sang "God Save the King" spontaneously. In Philadelphia the Liberty Bell was rung, and church bells rang out in jubilation all across the United States, the United Kingdom, and Canada. In America so many people called friends and families to tell them the news that "telephone switchboards were jammed" across the nation. In the House of Commons Winston Churchill, artfully keeping the members in suspense as he told them of the liberation of Rome, said, at last, "I have also to announce to the House that during the night and the early hours of this morning the first of a series of landings in force upon the European continent has taken place—in this case the liberating assault fell upon the coast of France." He went on to say, "There is a brotherhood in arms between us and our friends of the United States. . . . There is complete confidence in the Supreme Commander, General Eisenhower." From all over the world—except, as Churchill would say, "in the abodes of the wicked"—congratulations poured in. Even Stalin cabled, "It brings joy to us all." The normally staid *London Times* commented, "At last the tension has broken."

All over occupied Europe the news was heard, broadcast by Radio Berlin (which had, in fact, "scooped" Ike's announcement), as well as picked up by those who risked their lives by listening to the BBC clandestinely. In POW camps the word spread rapidly, as it also did in the concentration camps, where it caused a collective—and, as it would turn out, sadly premature—burst of hope among those who were destined for extermination at the hands of the SS. In her cramped quarters in the Frank family's hiding place in Amsterdam, fifteen-year-old Anne Frank wrote in her diary: "English broadcast in German, Dutch, French and other languages at ten o'clock: 'The invasion has begun!'—that means the 'real' invasion. English broadcast in German at eleven o'clock, speech by the Supreme Commander Dwight Eisenhower. . . . I have the feeling that friends are approaching. We have been oppressed by these terrible Germans for so long, they have had their knives so at our throats, that the thought of friends and delivery fill us with such confidence! I may yet be able to go back to school in September or October."[6] In this, Anne Frank spoke hopefully for millions.*

Ironically, Lieutenant General Sir Frederick Morgan, the architect of Overlord, who had been elbowed aside by Monty, heard the news at the side of a road, when "the BBC warn[ed] listeners to stand by for a special announcement . . . and [he] told his driver to stop the car for a moment. He turned up the volume on his radio—and then the author of the original invasion plan heard the news of the attack."[7]

In West Point, New York, where Mamie had arrived on June 5 to attend John's graduation the next day, her plan for keeping out of sight collapsed when she was woken at seven in the morning by a reporter asking her what she thought of the invasion. When she came downstairs at the Thayer Hotel, the lobby was full of reporters and photographers waiting for her, and when "John finally went up to the podium to accept his diploma, the name Eisenhower elicited a rapturous ovation."[8]

By nightfall on June 6, 156,000 troops were on shore along the Normandy beaches. In some places the beachhead was still paper-thin; in others, a precarious foothold had been won inland. But with each hour that passed, the likelihood that the Germans would be able to drive Allied forces back into the sea receded. As Rommel had predicted correctly, the first twenty-four hours were decisive—after that, the Allies' buildup would proceed rapidly; and although the Germans might still hope to prevent the Allied forces from breaking out, the combination of Rommel's absence on June 6 and Hitler's delay in releasing the panzer divisions under his control would prove fatal. As night fell over the beaches, Ike, who had fretted all day because Monty, rather than he himself, was in command of the ground forces, surveyed a mixed but hopeful picture.

From Varreville and Sainte-Mère-Eglise on the extreme right to Caen on the extreme left was a distance of some forty miles, as the crow flies.

* Tragically, the Allies' advance was too slow to save her. The Franks were betrayed by a Dutch informer and seized by the Gestapo on August 4, 1944. They were sent to Auschwitz, where Mrs. Frank died. Anne and her sister were sent from there to Belsen, where they both died in March 1945.

Though the boundary lines between the corps touched neatly on the map, in fact the invasion beachheads as they developed during the course of D-day were not contiguous. The two American beachheads, Utah and Omaha, were separated by the Carentan estuary and the three rivers that ran into it, a distance of some twelve miles. A further gap of seven miles separated Omaha from Gold, Juno, and Sword, the British-Canadian beaches. Nowhere had the troops reached Monty's "planned objectives"; but by the end of the day, a good deal of ground had been gained at Utah, in part because of the efforts of the airborne forces, and from Arromanches to Lion sur Mer, on the right, the British and Canadian forces had pushed well inland, and the British Sixth Airborne had seized substantial ground on the east bank of the Orne River as well as the vital bridges over the Orne and the Caen canal. (The initial success of the British-Canadian landings was due in part to the extensive use of General Hobart's "funnies," the special-purpose tanks that the Americans had rejected. Among these were the Crocodile, a flamethrowing tank, which although slow-moving proved to be a surefire destroyer of pillboxes; and the even more useful flail tank, which cleared a path through minefields with revolving chains mounted on a drum in front of it.)

Everywhere, the profusion of obstacles, mistakes by the helmsmen of the landing craft, and the difficulty of recognizing landmarks on the beaches in the heat of action and under the smoke from the bombardment, from a small boat bobbing in heavy seas, conspired to place the troops some distance from where they had expected to be. Also, these factors often split units up, so that they arrived on the beach fragmented and in some cases leaderless. On Utah, troops landed more than a mile east of where they were supposed to be, and this was not untypical. Clausewitz's famous dictum about the "fog of war" was never truer than on D-day; but in most cases officers and NCOs somehow managed to find and reunite their men, gather up their own equipment (or someone else's), and move the troops forward under heavy fire to assault the exits that would get them off the beach. It helped that the situation was obvious to even the lowliest (and most frightened) of privates—the beaches were killing grounds, and there was no way to retreat from them, so the only way to survive was to fight your way off them onto the higher ground held by the Germans.

The two major concerns for Ike were Omaha beach and the situation developing around Caen. Omaha was the least satisfactory site of the day. Owing to "strong lateral currents and poor navigation" troops were landed far from their planned objectives and were confusingly intermingled. Not only did they face the unexpected presence of an extra German division, a first-line unit moved there at the last moment and undetected by the Allies' intelligence, but much of their heavy equipment was disembarked too far out to sea and sank straight to the bottom, together with the crews—a demoralizing sight for infantrymen who had been told that DD, or "swimming," tanks would blast a way through the German defenses on the seawall for them. In the circumstances, it is hardly surprising that, to quote the calm, restrained prose of *The West Point Atlas of American Wars*, many of "the survivors of the assault wave huddled behind the shingle—demoralized, confused, and, in many cases, without leadership." Eventually, "courageous leaders rose to the occasion,"[9] and by nightfall elements of the Fifth Corps had made their way across 200 yards of open beach and up the cliffs overlooking the beach to points between 1,000 and 2,000 yards inland, and taken the principal exit road off the beach. Still, it was touch and go. Only a small fraction of the supplies the troops needed had been landed; casualties had been heavy; and only the lack of a clear-sighted, commanding presence on the German side prevented the enemy from making a counterattack that might have driven the Americans back into the sea. Once again, the absence of Rommel was a critical factor—he was the one man who might have recognized the opportunity and acted on it immediately, but he did not arrive back at his headquarters at La Roche-Guyon to take control of the battle until ten that night, by which time it was too late.*[10]

Ike's second area of concern was Caen. At Caen, unlike Omaha, the British and the Canadians had come very close to reaching Monty's first

* "Do you know, Lang," Rommel told his aide on the drive back to La Roche-Guyon from Germany, "if I was commander of the Allied forces right now, I could finish off the war in fourteen days."

objectives, though not quite close enough. The paratroopers and glider-borne infantry of the Sixth Airborne had seized the bridges over the Orne and the Caen canal at dawn and held them for thirteen hours, despite intense fighting, until, amazed and incredulous, they heard the noise of bagpipes approaching them and saw Lord Lovat's commandos coming down the road toward them, with Private Millin in his kilt calmly marching at their head playing "Blue Bonnets Over the Border" to the accompaniment of German machine-gun fire and artillery.[11] All the same, Caen remained in German hands. Although the Führer finally released the panzer divisions he had held back, only the Twenty-First Panzer had been close enough to attack on June 6. The attack had been plagued by misfortune—the Allies' bombing of Caen had blocked the streets with rubble, forcing the Twenty-First Panzer to lose time making its way around the town. As a result, the tanks failed to make contact with the infantry that was supposed to support them, and were stopped dead by British antitank guns already dug in on high ground. In Rommel's absence, the panzer attack, which was intended to roll up the easternmost flank of the invasion beaches, fizzled out instead.

Still, when Ike looked at the map what he saw was that Monty, who had unwisely claimed that he would take Caen on the first day, had failed to do so.

This disturbed and angered Ike at the time and for the rest of his life. In the sixty-three years since D-day this issue has spilled more ink than any other on the subject of the invasion; and by now it is firmly fixed in the minds of most American military historians that Monty dithered on D-day (and would continue to dither for three more weeks while he tried to take Caen), whereas British military historians take the "You're another!" view that the Americans' attack on Cherbourg was just as slow (against significantly weaker German forces), and much less well planned.

That something went wrong at both ends of the invasion front is undeniably true. The failure to take Caen quickly infuriated the air marshals, particularly Air Chief Marshal Leigh-Mallory, Ike's Air Commander, who strongly disliked Monty (Monty reciprocated by referring to Leigh-Mallory as a "gutless bugger"),[12] and Air Chief Marshal Tedder, Ike's deputy, who "once described Monty as 'a little fellow of average ability who has had such a build-up that he thinks of himself as Napoleon—he is not!' "[13] Their promises of air support had been based in part on rapidly setting up advance airfields on the flat, dry ground east of the Orne; and acting on Monty's optimism about taking Caen in the first day or two after the invasion, they had sent RAF ground personnel onshore in the early stages of the assault, despite the fact that those men were neither trained nor equipped to defend themselves as infantry.

The setback at Caen infuriated Winston Churchill, momentarily shaking his confidence not only in Monty but in the fighting ability of British soldiers. Churchill wondered if British troops had lost the fighting spirit they showed in World War I, and even the usually levelheaded Brooke confided to his diary that the senior officers of the British army were not much good because the best ones had been killed off in the previous war. Much of this breast-beating was unedifying and untrue, and it was more than matched by the Americans' concern that their ordinary infantry divisions, unlike such elite units as the First Infantry Division ("Big Red One") and the paratroopers, were woefully lacking in fighting spirit, field craft, and marksmanship.

The truth is that in both armies, green troops with no battle experience were slow to close with the enemy, relied too much on artillery and air support instead of their rifles, worried too much about their flanks, and tended to come to an abrupt halt if their officers were killed. (The Germans, whose NCOs were trained to take over when there were no officers, and who were taught "When in doubt, attack," rather than halt and "regroup" or dig in, were better off in this respect; and recognizing this weakness in the Allied troops, they made something of a specialty of picking off British and American officers by sniper fire.) All the same, there was a learning curve (as it is now called) for the infantry, just as there was for the more sophisticated tasks of fighter pilots or naval crews. Green units needed time to learn how to cope with the harsh realities of battle, and, inevitably, to find and weed out NCOs and officers who were incompetent or lacked fighting spirit. Neither British regimental spirit and tradition nor American optimism and lavish supplies could adequately substitute, in the harsh environment of combat, for good leadership, battle experience, and the right equipment.

Following Churchill's lead, British military historians have also portrayed the average German soldier as something of a superman, in part to explain the failings of British troops against the Germans; but this too is an exaggeration. The best German units were certainly formidable, but they were in short supply, and even the Twenty-First Panzer Division, despite its superior tanks and battle experience, did not accomplish much on June 6. Also, many of the German units defending Cherbourg were second-rate, poorly equipped, and unenthusiastic about a last-ditch defense when they were already cut off and unable to retreat.

The British—including Montgomery himself—indulged in self-flagellation and finger-pointing over the alleged battle-weariness and cautiousness of experienced divisions, even Monty's beloved Seventh Armoured (the "Desert Rats"), that had already fought in Libya, Tunisia, and Sicily. But more often than not the problem was the acknowledged superiority of German tanks and antitank guns and the fact that divisional commanders were changed so frequently that the men found themselves fighting under unfamiliar senior officers. There was in any case no practical way for the citizen soldiers of a democracy to develop the particular combination of military professionalism, political fear (and, in some cases, like the SS divisions, zealotry), and a sense of fighting with their backs to the wall that motivated the elite German units, and would continue to motivate them to the very end.

And yet, putting to one side the differing national views of military historians, the fact remains that the British did eventually take Caen, despite the best efforts of the Panzer Lehr and the Twelfth SS Panzer to stop them; the Americans did take Cherbourg, although the Germans had by then rendered the port almost unusable; and the combined Allied armies did finally break out and, at the Battle of the Falaise Gap, hand the Germans one of their most severe defeats of World War II. Anybody reading most of the more recent books about D-day might easily suppose that the Allies had lost the war, instead of winning it.

Ike's own view was more balanced than that of most of the people writing about him, and certainly more balanced than that of Churchill. Like Grant after his historic victory at Vicksburg, Ike understood that there was a lot more fighting to be done, and that there was still plenty of fight left in the enemy. He also understood that for Churchill, like many of his countrymen, the D-day landings, after five years of war, represented a bittersweet triumph—as the Soviet forces advanced toward Germany, and the buildup of American forces continued, Britain faced a severe manpower shortage. Its forces would soon be dwarfed by those of its two principal allies, and when that happened, its power to influence events would diminish proportionally. At the same time, quite apart from war-weariness, men who had fought for four or five years had a natural reluctance to be killed at the end of the war, when peace, if not yet in sight, at least seemed a rational prospect.

The crux of the problem, however, was that Monty, with his peculiar combination of boastful self-confidence and unshakable self-esteem, had promised more than he could deliver, and absolutely refused to admit it. He had raised people's hopes too high, and having disappointed them, he stubbornly insisted that everything was going exactly as he had planned it—that in fact the longer it took him to capture Caen, the better, since day by day he was grinding up the German panzer divisions, and thus freeing the Americans to take Cherbourg and, eventually, to break out of the beach head. QED, he might have said, had he been a schoolteacher, but it seemed to everybody except himself (and his chosen few, the young aides who surrounded him and tried to ensure the almost complete isolation he required to think) that the British army was stuck.

Certainly it seemed that way to Ike, and this may be one reason why in photographs taken of him in the days after June 6 his expression is grim and careworn, quite unlike his usually ebullient appearance. Late on June 7, Ike, determined not to be sidelined away from the action as he had been at Gibraltar during the first days of Torch, arrived off the British beachheads in HMS *Apollo* to meet with Monty, who had sailed from Portsmouth the night before on HMS *Faulknor* and would land the next day. Bradley had come aboard Monty's ship from his own headquarters on USS *Augusta* early in the morning on June 7, and the two had a grim discussion about what was happening onshore. Monty was worried that if the British and American corps didn't "join up" rapidly, Rommel might be able to attack them "in detail" one by one. Bradley's concern was whether the troops at Omaha could hang on—the "situation on the Omaha beach is critical,"[14] Bradley's aide recorded during the conversation between the two men. Monty's solution was the only possible one: in his opinion, the first priority was to "join up—before pushing west or north." This meant that the British would have to attack inland and take the medieval town of Bayeux, home of the famous tapestry, to relieve the pressure on Omaha; and that the U.S. Seventh Corps on Utah beach would have to attack eastward rather than westward toward Cherbourg. Monty then sought out Ike's ship, went aboard, and reported these decisions to Ike. He later remarked that he thought Ike was suffering from "acute anxiety,"[15] and not showing the sangfroid and complete lack of self-doubt which were Monty's own trademark as a commander—and which would shortly get him into hot water. It seems more likely that Ike was already regretting having accepted Monty as his ground forces commander instead of insisting on Alexander—and surely that he would rather have been dealing directly with Bradley himself, rather than through Monty. He even expressed the opinion that he wished he had George Patton over there; given the fact that Patton was still in disgrace for once again opening his mouth in front of reporters, this makes it clear how dismayed Ike was.* Whatever Monty's opinion of the situation was, the supreme commander knew that in London the prime minister was anxious to hear that Caen had been captured, and in Washington the president and General Marshall wanted the port of Cherbourg in the Allies' hands as soon as possible. Clearly, neither of these things was about to happen. On the way home, Ike's ship ran aground at full speed, and he had to be transferred to another. (The next morning

* Patton was being kept in England in command of a phantom army group for the moment, while waiting to take the U.S. Third Army into action, as part of a scheme to persuade the Germans that the main invasion was still to come in the Pas de Calais. He felt strongly that even Monty's five-division plan for Overlord was too small, and he wanted a ninety-mile invasion front, as opposed to the fifty-mile invasion front Monty had demanded. He referred to Monty as "Chief Big Wind," complained that Ike had sold out to the British, and confided to his diary that Monty's attack was "on too narrow a front and may well result in another Anzio, especially if I am not there."

Monty's destroyer ran aground too—an indication that warships were operating close inshore, in very shallow waters.)

Almost everybody noticed that Ike was ill at ease when he finally gave a press conference to war correspondents at Southwick House on June 8. Even loyal Harry Butcher thought Ike seemed "tired" and "listless," not words which are generally used about Ike (and few people knew him better than Butcher). Ike cautioned the reporters to remember that the situation was still "critical" and to avoid giving a "false picture of optimism"—not exactly a rallying cry. Kay Summersby noted, "There was a terrible let-down after D-Day. Everyone felt it. Ike was tired, as if he had run out of steam. And he was very much depressed."[16]

Part of the reason was uncertainty. The successful invasion was a triumph on a scale unequaled in military history; but the battle showed signs of breaking down into a series of small, sharp, local attacks and counterattacks across a fifty-mile front, and this was the kind of fight that Rommel might hope to win, or at least sustain for a very long time. The Allied forces not only needed to expand their beachhead (and solve their rapidly growing supply problem) but also needed to strike a knockout blow, and there were no signs of such a blow. The air forces were hampered by bad weather, poor cooperation (and ill will at the top) between air and ground, and the continuing argument about what, exactly, they should be attacking. At the same time, the weather seriously hampered the efforts of the navies to supply the Allied armies over the beaches.

It grated on Ike's nerves to be in Portsmouth or, worse still, in London sorting out the supply problems with his old friend General Lee (who everybody except Ike thought was autocratic, uncooperative, and not up to his job). Ike, who more than anyone else wanted to be at the front, was obliged to deal with an endless succession of what were then beginning to be called VIPs demanding to be allowed there, the most determined of them being the prime minister, and by far the most difficult General De Gaulle. This particular task was not made easier by Monty, who didn't want to be interrupted, now that his tactical headquarters was onshore, by Churchill, De Gaulle, Ike, or even the king. On June 12, with the arrival from Washington of General Marshall, General Arnold, and Admiral King, who had come to have a look at what was happening, Ike finally got to land in France, at Omaha beach. His spirits were lifted by being, at last, in the combat zone and by the fact that everywhere he went the soldiers cheered him. Since Churchill also landed in France that day, accompanied by General Brooke and Field Marshal Smuts of South Africa, Monty's privacy was severely breached. His optimism was unshaken, however—everything, he told visitors, was going exactly as he had planned it.

That same day, a major event of a different sort took place across the Channel. The first of Hitler's *Vergeltungswaffen* (vengeance weapons), the V-1 pilotless jet bomb, manufactured by slave labor in deep underground caverns and soon to become known to the British as a doodlebug or buzz bomb because of the noise it made, began to land in England. Between June and September 1944, the Germans launched over 8,000 V-1s, damaging over 750,000 houses, "twenty-three thousand of them beyond repair";[17] killing over 6,000 civilians; and seriously injuring nearly 18,000. The V-1 was followed very shortly by the even more terrifying V-2, a ballistic missile that carried a warhead of 1 ton. Over 1,300 V-2s were fired against London, killing nearly 3,000 civilians and seriously injuring over 6,000. The V-1 flew at low altitudes until its engine ran out of fuel, at which point it fell to the ground and exploded—the sudden silence served as a chilling warning that the bomb was falling. The V-2 reached a height of over fifty miles, then fell to earth at over 4,000 miles per hour, so that it gave no warning at all of its coming—the high-pitched wail of its passage through the air came after the enormous noise of the explosion.

Hitler had high hopes for these weapons and had invested huge industrial efforts in them, and in others to come, which were less successful. But they did not have the decisive effect he expected. The British were certainly shaken by this new blitz, and large numbers of people were evacuated from London; but Britain's industrial output was scarcely touched, nor was the determination of the British to get on with the war and finish Hitler off. The same amount of money, labor, and raw materials invested in tank production, or, more terrible to contemplate, in nuclear physics, might have won Hitler the war, so it was perhaps a blessing that he chose to renew the bombing of London instead. In any event, the presence of the V-1 launching sites in France would add to the pressure on Ike, since the only sure way to stop the attacks was to overrun these sites.

Ike himself was reluctantly persuaded to use the bomb shelter Churchill had built for him at Telegraph Cottage for the first time. Despite the many advantages of living outside of London, Ike was in the direct path of the flying bombs, and at night he sometimes shared his cramped shelter with Kay, Harry Butcher, three enlisted men, and, for nearly three weeks, his son John. Newly graduated from West Point, Second Lieutenant John Eisenhower had arrived on June 13—his mother had discovered only on graduation day that he would be going to Europe instead of spending his leave with her. Since the arrangements for this had been made between Ike and Marshall, Mamie had not been informed, and she was understandably upset to have both her husband and her only living child at risk at the same time, but like a good Army wife, she accepted it without complaint. "John will be in soon," Ike wrote to her on June 13—he had sent his private train, Bayonet, to bring John from Scotland to London—"maybe in one and a half hours. I'm really as excited as a bride—but luckily I have so much to do I haven't time to get nervous!"[18]

This contrasts with Kay Summersby's observation, "Ike was not a particularly doting father. . . . Sometimes I thought he was supercritical."[19] Of course, as any parent knows, love and criticism are not incompatible when it comes to one's own children; and for that matter, it cannot have been easy for John, as the most junior of junior officers, to be in the constant presence of a four-star general and supreme commander—let alone in the presence of Kay Summersby, whose position in Ike's military "family" seemed to most people, on the surface, to say the least, ambiguous.

If Kay is to be believed, Ike thought his son "should toughen up" and that John's bridge game needed "sharpening up" as well. Ike took John on several inspection trips to the invasion area, which quite sensibly he didn't mention to Mamie, for fear, no doubt, of worrying her. His letters to Mamie during John's visit are occasionally, and perhaps deliberately, quite opaque. On June 18 he wrote to tell her that John was "having a good time," and mentioned offhandedly that his "driver, you know, is British, and has been taking him [John] here and there."[20] If by "driver" Ike was referring to Kay Summersby, he apparently thought it prudent not to mention her name to John's mother. Four days later, Ike wrote a slightly overbearing letter to warn Mamie on the subject of visiting John when he started his training at Fort Benning, advising her that "it would be unwise to stay too long." This was good advice—no second lieutenant in training would want his mother hanging around, after all—but whether it was something Mamie wanted, or needed to be told, as an Army wife for twenty-eight years, is unknown. On the other hand, on June 29, at the end of John's leave, Ike wrote her a letter "to be delivered by our son," which must surely have surprised her, to say that he was sending John home in his own B-17, accompanied by Colonel Lee, Ike's aide; Captain Pinette, a WAC officer at SHAEF; Sergeant Farr, Ike's steward; and "my secretary Mrs. Summersby," who, he wrote, "is going to try to find Mrs. Arnold, mother of her late fiancé."[21]

What Ike did not say in his letter was that Kay's flight to the United States, as well as Tex Lee's, was in part connected to President Roosevelt's promise to her in North Africa about a commission in the WACs. Kay herself, in her own account, does not mention Mrs. Arnold at all, but merely describes the trip as "a holiday." If that was the case, it was not an altogether easy vacation. Nobody had prepared Kay for the heat of Washington in the summer, or for the fact that Mamie would be at the airport when she arrived in Ike's plane. She claims to have become aware for the first time of the rumors about herself and Ike while in Washington, though this seems unlikely, since rumors had been widespread in North Africa and London. Certainly she was uncomfortably out of place among the Army wives who were Mamie's friends, if only because she was in uniform and they were not. Invited to a party by Mamie, Kay later complained that she was being "scrutinized," though part of the trouble may merely have been that she was young, beautiful, British, and a lot more assertive than may have seemed appropriate, in somebody who was merely a chauffeur and a secretary, to the rank-conscious wives of senior officers. Matters were not improved between Kay and Mamie when John insisted on taking Kay to New York City, where "he showed her the town"[22] and took her to see *Oklahoma*—a very "hot ticket" indeed at the time.*

Kay returned to Britain on Ike's B-17 without Tex Lee, who had stayed behind to pull the necessary strings at the War Department to get Kay her commission in the WACs. Even with the president's help, the commission did not come through until October, when Ike would pin her gold second lieutenant's bars on her new uniform himself.

It is difficult to know what to make of all this. Certainly, it seems unlikely that Mamie would have received Kay or that John would have taken her on a trip to New York City if either of them had really believed she was Ike's mistress, or indeed that Ike would have wanted his son to escort her had that been the case. On the other hand, Kay was an enlisted, uniformed member of the British Motor Transport Corps—and sending her on a "holiday" to America in wartime on his personal B-17 was not exactly standard operating procedure between a general and his driver, or even his secretary. Nothing, one would have thought, would have been more likely to ruffle Mamie's feathers than sending Kay to Washington with John, innocently or not, but no sign of this appears in Ike's letters to Mamie, so it is possible that Mamie simply does not get enough credit for classy behavior and self-restraint—and perhaps a realistic understanding that making a fuss would backfire on her. She must surely have remembered how angry Ike had been when she confronted him over his golfing companion Marion Huff in the Philippines; confident in the strength of their marriage, she was smart enough not to repeat the same mistake.

In Kay's absence, Ike had taken every opportunity to get to the front and see for himself what was happening. His most dangerous excursion took place when he ordered Major General Elwood Queseda, the commander of the U.S. Ninth Air Force fighters and fighter bombers, to squeeze him into the tiny backseat of a converted two-seat P-51 Mustang and fly him over the front lines. This was a daredevil stunt for a four-star general. Because of

* The author was taken to see *Oklahoma* in 1944, and can vouch for the fact that it was a huge hit and that tickets were hard to get, and very expensive—being taken to see it was a very big deal indeed.

the Luftwaffe, German antiaircraft fire, and the normal accident rate of high-performance fighter planes flown from improvised and often muddy grass strips and serviced by ground personnel who had just graduated from a quick course in aircraft maintenance, a flight over the front lines in a P-51 would certainly not have met with General Marshall's approval, had Marshall known about it. Ike may have been influenced by the fact that Winston Churchill had persuaded Air Chief Marshal Leigh-Mallory to fly *him* over the front lines in a captured two-seat Fiesler Storch, the German equivalent of a Piper Cub—how the prime minister managed to squeeze himself into the backseat remains one of the enduring (and endearing) mysteries of World War II. Anyway, for Ike, who had wanted to be a military aviator until Mr. Doud made him choose between that and marrying Mamie, and who had taken flying lessons in the Philippines in her absence, the opportunity of flying in a P-51, even as a passenger, must have been well-nigh irresistible.

Ike managed to get to the front by car too, near enough to the fighting to be on the receiving end of an artillery barrage—and despite the danger he remained perhaps the only senior officer in the United States Army in Europe who almost never wore a helmet.* What Ike saw continued to displease him. He had very little patience with Monty's belief in a tidy "battle plan" and would later comment that "rigidity inevitably defeats itself." In his view, this was exactly what was taking place in the battle for Caen. Though Ike, like Grant, was not an admirer of Napoleon, he would have agreed with the emperor's curt explanation of how to fight battles: *On s'engage, et puis on voit* (One engages the enemy, then waits to see what happens). Monty preferred to work to a fixed plan, and in addition, seemed to be suffering from what an exasperated Lincoln, referring to General McClellan, called, "a case of the slows."

* I can find only a very few photographs of Ike wearing a helmet during World War II, and he looks mighty uncomfortable and grumpy in them. This may have been a prejudice he picked up from his years with General Douglas MacArthur, who never wore a helmet even when leading troops in combat during World War I. British senior officers were always amazed by the fact that American generals like Patton and Bradley wore a helmet and carried a sidearm. British generals, on the contrary, made a point of going unarmed and unhelmeted, nonchalantly wearing the red-banded, gold braid-badged cap of a general officer and armed only with a walking cane or a swagger stick, even when under fire; so Ike may also have picked up this habit from the British generals who served under him. He also felt that it was insulting to the troops for a general to dress up like a combat soldier.

Whether or not as a result of a misunderstanding between Ike and his commander of ground forces, Ike clearly expected that Monty would not only take Caen but break out across the Orne River with his armored forces toward the southeast, threatening to cut the lines of communication of the German forces in Normandy and Brittany. The ground on the east bank of the Orne was flat and relatively dry, ideal country for tanks, unlike the *bocage*—the innumerable small fields separated by deep ditches, stone walls, and dense high hedges—that faced the Allied forces as they moved directly inland from the invasion beaches, or advanced westward toward Cherbourg. Unlike Monty, Ike had a dread of time passing, which could lead to a stalemate. He understood what Monty said his strategy was, and did not disagree with it, but he did not suppose that Field Marshals von Rundstedt and Rommel would continue to tilt at windmills around Caen forever with their Panzer divisions simply because Monty wanted them to. More German armor was coming up from the south, including the formidable elite Second SS Panzer Division "Das Reich," which paused on its way from Toulouse to Saint-Lô to slaughter 642 men, women, and children at the village of Oradour-sur-Glane, herding them into the church and burning it in retaliation for sabotage by the resistance of the railroad line on which their tanks were being moved to the front—the kind of incident that should serve as an enduring reminder of how different the war was for those who had the misfortune to be occupied by the Germans.[23]

Ike was more conscious of this than the British generals, or his own government, for that matter—perhaps because of the time he had spent in France touring the war memorials of World War I. His visit to Verdun seems, among other things, to have inoculated him against the rampant Francophobia that complicated British and American attitudes toward De Gaulle, and still colors Anglo-Saxon foreign policy toward France today. However difficult and stiff-necked De Gaulle could be—and few could match him in that regard—he and Ike respected each other as professional soldiers. When De Gaulle had threatened not to broadcast to France on the eve of D-day if Ike broadcast before him, and had rejected among other things the idea of using American-printed occupation currency in France, Ike chose calm and compromise instead of the hard line that the president and the State Department wanted to take against De Gaulle. As a result, De Gaulle arrived in France on June 20 (despite a deeply resented last-minute attempt by Churchill to stop him) for the first time since 1940 aboard the Free French destroyer *La Combattante*—it would have been unbearable to him to have made the historic return journey across the Channel on a British or American warship—and paid a brief courtesy visit to Montgomery, who managed to infuriate him by trying to curtail his visit to Bayeux, which had just been liberated. (Monty noted in his diary afterward, with typical blindness to the greatness of other men, that De Gaulle was "a poor fish and gives no inspiration.")[24]

De Gaulle, on his part, expressed his confidence in Montgomery's plans, but noted, with a precise, acid touch for the uncomfortable truth, that he was certain matters would proceed just as Montgomery had predicted, though "with neither haste nor temerity."[25] Though seething, De Gaulle went straight on to Bayeux; walked on foot through the town, surrounded by cheering crowds, to the *sous-préfecture* (where the portrait of Marshal Pétain was just being removed); and received a visit from the bishop of Bayeux. Then, in his own words, "for the first time in four terrible years, a French crowd heard a French leader say that the enemy was the enemy, that their duty was to fight him, that France would be victorious."[26] He went on to Isigny, "cruelly destroyed, where the corpses were still being carried out of the debris,"[27] and where the town "paid me the honor of its ruins": and from there to Grandcamp and Courcelles. Everywhere De Gaulle went, crowds cheered him, and everywhere he immediately placed in authority his own men. By the time *La Combattante* was back in Portsmouth that night, any lingering notion in London or Washington that De Gaulle did not represent *la République française* or that his government would not rule liberated France had all but evaporated.

Nobody had recognized better than De Gaulle Ike's abilities as a diplomat as well as a warrior, or his capacity to learn quickly and to take on himself momentous decisions. De Gaulle noted in him—perhaps because in that respect they were alike, both professional military men moved by events into the highest level of politics—"an attraction towards the wider horizons that history opened before his career,"[28] and praised him for a quality for which he has received too little credit over the years: "audacity." By June 1944 Ike had commanded the two largest amphibious operations in history—nothing in warfare had ever been more audacious than Torch and Overlord—and had kept together by his prudence, tact, and fundamental fairness an alliance of three nations whose interests often diverged sharply.

As De Gaulle himself recognized, Ike was largely responsible for the swift, smooth transition De Gaulle made from controversial exile to the unquestioned leadership of France. Not for the last time, Ike calmly ignored his government's policy and relied instead on his own common sense, his judgment about people, and his view of what would best serve the interests of his forces. It would have gravely impeded the campaign against Germany if France, during the liberation, had collapsed into political strife or even civil war; and De Gaulle, however much the president might dislike him, was the best guarantee against that. Ike has received precious little credit for this (except from De Gaulle), but it is a perfect example of what a quick learner he was. He had suffered through the chaos and danger brought about by American policies in North Africa, when he had paid too much attention to Robert Murphy and the wishes of the president, and he was not about to let it happen again. No other American general, except perhaps Douglas MacArthur in his treatment of Japan at the end of the war, took on such heavy responsibilities or made such far-reaching political decisions on his own initiative. Ike was not just supreme commander—he was also a unique kind of American proconsul, bridging the worlds of diplomacy, politics, and warfare like Pompey or Caesar, but without the dangerous ambition and tendency toward authoritarianism that came to tarnish the proconsul's role in ancient Rome.

In the meantime, Ike, like De Gaulle, had noted to his dismay a lack of "haste" and "temerity" in Monty's fight for Caen. If he had had any doubts about the wisdom of turning George Patton—for whom "haste" and "temerity" were guiding lights—loose with the U.S. Third Army at the earliest possible opportunity, the deliberation with which Monty was trying to take Caen would have made up his mind. Ike's old friend Patton was in many respects a loose cannon, as dangerous to friend as foe, but if there was one thing he could do, it was to move with speed while ignoring risks that would have stopped anyone else. Ike also made up his mind to move his headquarters to France as soon as possible, mindful, even if Monty was not, that the moment he did so he would be able to terminate at will Monty's command of the "Twenty-First Army Group" (the two American and British armies at D-day). At that point Monty would command only the British forces, and Bradley the American; Ike would be his own commander of ground forces "in the field" while continuing to serve as supreme commander. Monty had been scathing about the consequences of this dual role, and continued to be—he would later argue that Ike "wasn't a great soldier" and "didn't know how to command in the field," and was still saying so two decades later, much to Ike's annoyance—but the Americans, particularly Bradley and Patton, chafed to get away from being commanded by a Britisher, particularly this one.

It was Ike's good fortune that Hitler had forbidden Rundstedt and Rommel to do the one thing that made sense, which was to retreat while their armies were intact; form a firm, well-entrenched line at or east of the Seine; and defend it tenaciously. The Führer saw in that a repetition of the awful stalemate of 1914–1918, in which he had served with distinction as a lance corporal; and besides, he was determined on principle—it was an ineradicable part of his personality—never to give up ground. His generals had wanted to retreat to a line they could defend during the dreadful winter of 1941–1942 outside Moscow, and he had forbidden them to—as he saw it, only his own willpower had prevented the retreat of the German army then, and his willpower alone would ensure victory now. Rundstedt and Rommel were therefore obliged to attack, at Caen, in the Cotentin Peninsula, and frontally against the Allies' positions. But their attacks, because of the Allies' air superiority and the collapse of the French railroad system under bombing and sabotage by the resistance, were never strong enough to succeed. Rommel, being who he was, never gave up attacking—whether it was with a company, a division, or a corps, Rommel demanded continuous attacks that never gave the enemy a chance to dig in or regroup—but he had neither the strength nor the supply chain for the one big, concentrated attack, the "knockout blow," which alone could stop the Allies from enlarging their positions. Although Rommel was holding on to Caen, he held it at the remorseless daily cost of experienced troops and armored vehicles he could not replace—indeed, the longer he made Monty fight to take Caen, the weaker his own forces were, by the simple arithmetic of attrition.

By June 18, even though the area of the Allies' beachhead was not significantly larger and Caen seemed as unreachable as ever, Bradley's Seventh Corps, commanded by Major General J. Lawton ("Lightning Joe") Collins, reached the Atlantic coast of the Cotentin Peninsula, effectively isolating Cherbourg. The importance of Cherbourg was emphasized the next day, when one of the worst storms in 100 years crippled one of the Mulberry artificial harbors and halted the unloading of supplies on the beaches for several days. Ike surveyed the beaches from the air the day after the storm and counted more than 300 boats and ships wrecked or driven ashore by the storm.

Collins's rapid advance, despite the infamous *bocage* country he was fighting over, was partly helped by the first use of Rhino. This act of American ingenuity fascinated Ike, the tank enthusiast and gadget-lover, who had taken such an interest in his mother-in-law's electric brougham in Denver and had taken an early tank to pieces "down to its last nut and bolt" with George Patton at Camp Meade in 1919 and not only put it back together again but made it run better. If there was one thing Ike understood, it was the American genius for invention. The dense hedges of the Normandy fields endangered even the tanks that managed to get across them, since the lightly armored underside of a tank was exposed to gunfire as the tank labored up the near side of the obstacle. A sergeant named Culin fashioned two scythe-like blades out of the steel German obstacles Rommel had spread across the Normandy beaches in such prodigious numbers; welded to the front of a tank, they allowed it to tear right through a hedgerow, carrying a huge chunk of earth and brush ahead of it as further protection against antitank shells—exactly the kind of thing which fascinated Ike but which Bradley had rejected when it was proposed to him as part of "Hobart's funnies." Very few people bothered to give Sergeant Culin the credit he deserved, but Ike did so, when it came time to write his memoirs.[29]

Perhaps because of the sheer volume of war memoirs and diaries—including those of Monty, Alan Brooke, and (posthumously) Patton—in which Ike is criticized for not being a "great soldier," he does not get the credit he deserves for generalship. It sometimes seems, in reading Brooke's view of events, let alone Monty's, that Overlord succeeded despite Ike, or with Ike along as a kind of four-star passenger. This is incorrect. Those who underrate Ike forget—as George Marshall never did—that Ike was by far the best-trained general in the U.S. Army (he had, after all, been mentored by generals Pershing, Conner, MacArthur, and Marshall), as well as the best card player. Ike had made the basic decisions on which the success of Overlord depended, and his view of the battlefield was broad and sure. He understood the importance of taking Caen—hence his irritation with Monty at the delay in taking it—but he never lost sight of the fact that the "right hook" southeast from Normandy toward the Seine would win the battle, and that it would be Patton's job to carry this action out. He wanted the right hook to be supported by a left hook from Caen to the south and the east, and saw no sign that this would happen, but with or without it, he knew that the right hook would deliver the main blow.

In the meantime, the Germans were providing Ike with enough good luck—the quality that Napoleon most esteemed in generals—to outweigh the weather and his difficulties with Monty. Cherbourg, Ike's first major objective, was surrounded by June 22. Three German divisions were dug in to defend it, but they were not first-rate troops, and their spirits were hardly raised by the fact that they had been cut off from any possibility of retreat or resupply—or by the fact that they were exposed to constant shellfire from British and American warships. Hitler's demand for an armored attack to relieve them set off a major quarrel between himself and the elderly, autocratic Field Marshal von Rundstedt, who pointed out that even if such an attack could be mounted at all, which he doubted, it could not begin before July 5, since the Second SS Panzer Division "Das Reich" was still on its way from Toulouse.

This produced a storm of fury in Berchtesgaden, to which Rundstedt and Rommel were summoned for a meeting with the Führer. Hitler stipulated that they were not to travel there by airplane, in case they were shot down; or by train, for fear of sabotage—the very problems that were prevent-

Generalfeldmarschall
Gerd von Rundstedt,
Commander in Chief West

ing the German armored divisions, including Das Reich from getting to the front in Normandy in the first place. The relationship between Hitler and the two field marshals was made all the more tense by his determination, over their strong objections, to hold General Eugen Dollmann, the commander of the Seventh Army, personally responsible with his life for the fate of Cherbourg—a difference of opinion that was resolved shortly after they arrived at Hitler's home by the news that Dollmann had suffered a fatal heart attack.[30] Still, the damage had been done. Rommel's faith in the Führer, already wavering in Tunisia, was now severely shaken, Rundstedt was outraged; and Cherbourg, despite Hitler's demand that the garrison fight to the last round, was surrendered by its commander, General von Schlieben, on June 27.

Dollmann's successor as commander of the Seventh Army demanded that he should be permitted to withdraw his forces "out of range of naval gunfire," but when Rundstedt authorized him to do so, the order was angrily countermanded from Berchtesgaten. Rundstedt then called Field Marshal Keitel in Berchtesgaten to protest, and after an acrimonious conversation, Keitel asked the *Herr Generalfeldmarschall* what he thought they should do, and Rundstedt replied, "Do? Why make peace, you fools, what else can you do?"[31] That was it. Although Hitler felt a certain guarded affection for

Rundstedt—at moments of crisis, he was fond of saying, "So long as the old *Marschall* grumbles, everything is all right!"—on June 30, Germany's senior and greatest general was relieved of his command by the Führer, and replaced as OB West by Field Marshal Günther von Kluge.

The dissension and unrealistic thinking in high places in Germany very fortunately outweighed by far any problems the Allies were facing, and was sharply increased by the news that Stalin had at last launched the major spring offensive on the eastern front that he had promised Roosevelt and Churchill. This was operation Bagration, named after the great nineteenth-century czarist commander who was killed at the Battle of Borodino before Moscow in 1812, and is one of the heroic patriotic figures in Tolstoy's *War and Peace.* It tore a 250-mile hole in the German front, "overrunning twenty-five German divisions" almost at once, and would soon drive the German army back to the gates of Warsaw and even into East Prussia—the Reich itself.

Under the circumstances, Hitler and the Oberkommando der Wehrmacht (OKW) had more immediate problems than what was happening in Normandy. As the eastern front crumbled before the enormous attack by the Red Army, involving nearly 2 million men, thousands of tanks, and over 400 Soviet guns for each mile of the front (for the first time in the war, German and Russian casualties were about equal, numbering approximately 800,000 killed, wounded, and missing on both sides), Germany's chances of moving more men and tanks into the rapidly changing battle in Normandy had diminished to zero, even had the French railroad system, still under attack from the RAF, the USAAF, and the resistance, been able to handle them.

From Ike's point of view, of course, the Germans were still very far from being fatally weakened. The capture of Cherbourg was a hollow triumph; General von Schlieben, despite Hitler's order not to surrender, had prudently taken time to destroy the port with typically Teutonic attention to detail; thus Cherbourg would be largely out of business until September, and only a trickle of supplies would come through it even then. And Caen was still in German hands. Ike's life "was one of almost incessant travel," as he put it, much of it uncomfortable and dangerous, "conferring

(Left to right): General Blumentritt, General Speidel, Field Marshal Rommel, and Field Marshal von Rundstedt, at Rommel's headquarters, Château La Roche-Guyon, June 1944.

frequently with General Bradley and General Montgomery," Ike records. "Such visits with Bradley," he went on,"were always enjoyable."[32]

It is interesting to note that Ike does not comment on whether his visits to Monty were enjoyable. Monty was as confident, chipper, and impossible as ever, but even some of his supporters had begun to feel that his head was on the chopping block. Churchill was dismayed and furious; the air marshals, all of whom despised Monty for having failed to deliver the airfields he had promised them, spread with glee every anecdote about his not very endearing eccentricities at his spartan tactical headquarters in Normandy; and Patton fumed—only Ike, despite his growing personal dislike of Monty, continued to back him.

. . .

Had Ike sought it, the German side offered a perfect example of the dangers of changing horses in midstream. Kluge, Rundstedt's successor, a toughened veteran of the eastern front, had been personally briefed by Hitler, and warned to take a firm hold of Rommel, whom Kluge in any case regarded as an upstart, overawed by the Führer and a product of Dr. Goebbels's Nazi propaganda machine. No sooner had Kluge arrived in France and alienated Rommel, however, than he realized that Rommel was right. Hitler had fired Rundstedt only to have Kluge come to exactly the same conclusion as Rundstedt the moment he had seen for himself the reality in Normandy, and the perilous position in which the German forces found themselves. Worse still, Rundstedt, although he held Hitler in bitter contempt (he continued to wear the old tunic of the Kaiser's army instead of the new pattern), was unshakably loyal to him. Like every German soldier he had sworn an oath of loyalty to the Führer, and he would not, could not, break it. Kluge was less old-fashioned. He not only had doubts but was willing to listen to the doubts of other senior officers, and this ambivalence would very shortly cost him his life.

Although when we now look back on the period from June 15 to the end of July as the beginning of the end for German rule in France, it did not seem so at the time—either to Ike, presiding over a costly stalemate; or to Kluge, who was massing his forces in Normandy and continuing to hold Caen. Seen on a map of France, the area held by the Allied forces was tiny: less than seventy miles wide and, at its narrowest point at Carentan, only ten miles deep. The capture of Cherbourg had done nothing to relieve the supply situation, which was aggravated by Ike's success in bringing ashore over 1 million men and more than 170,000 vehicles. Hitler's belief that a single hard blow in the right place might split the Allies' beachhead and give Rommel an opportunity to break through from Caen and roll up the British line was not by any means out of the question, though it overlooked two factors: the Allies commanded the air above the battlefield, and the Germans' tank strength was a wasting asset.

Ike "repeatedly urged Montgomery to speed up and intensify his efforts to the limit," but in the end he accepted Bradley's view that there was no point in waiting for Monty to take Caen—the breakout would have to be concentrated at Saint-Lô, instead of farther to the south, even though "this placed upon American forces a more onerous and irksome task than had at first been anticipated."[33] German forces were building up at the base of the Contentin Peninsula. Now that Kluge had adopted Rundstedt's view of the situation, Hitler was willing to allow him to do what Rundstedt had not been allowed to: form a strong defensive line in depth on good terrain until enough armor could be assembled for a major attack. Time was not, therefore, necessarily on Ike's side, and he knew it. Another big storm, a major setback at Caen, or a combination of the two, might be enough to turn the Allied forces in Normandy into a beleaguered (though large) garrison trapped on the beaches and supported only by air strikes and naval gunfire, both of which were dependent on good weather. What caught Ike's eye when he looked at the map was the town of Saint-Lô.

He was not the only one looking in that direction. The small, quiet Norman market town of Saint-Lô, like Gettsyburg and Caen, was important largely because all the local roads met there. From Saint-Lô the roads went north to Cherbourg; west to Coutances, on the Gulf of Saint-Malo; northeast to Bayeux; and south toward Avranches, Vire, and Argentan. From Argentan it was only seventy-five miles to Paris. The roads were not modern autoroutes, which did not yet exist in France (although Germany had autobahns), but they were good, straight, well-paved French roads, with neat rows of poplars on each side. On such a road, a tank could move at twenty miles an hour. Even a civilian, looking at the map, can see why Bradley needed to capture Saint-Lô, as well as why field marshals von Kluge and Rommel, and SS Obergruppenführer Paul Hausser (who had taken over the Seventh Army after Dollmann's death) needed to hold on to it at all costs. On June 30 Bradley gave the order to attack toward Saint-Lô, and set off what Ike, who was not given to exaggeration, called "some of the fiercest and most sanguinary fighting of the war." This was "hedgerow fighting," in *bocage* country, where the fields were the size of a "building lot," and where every hedgerow, "affording almost the ultimate in battlefield protection and natural camouflage," hid

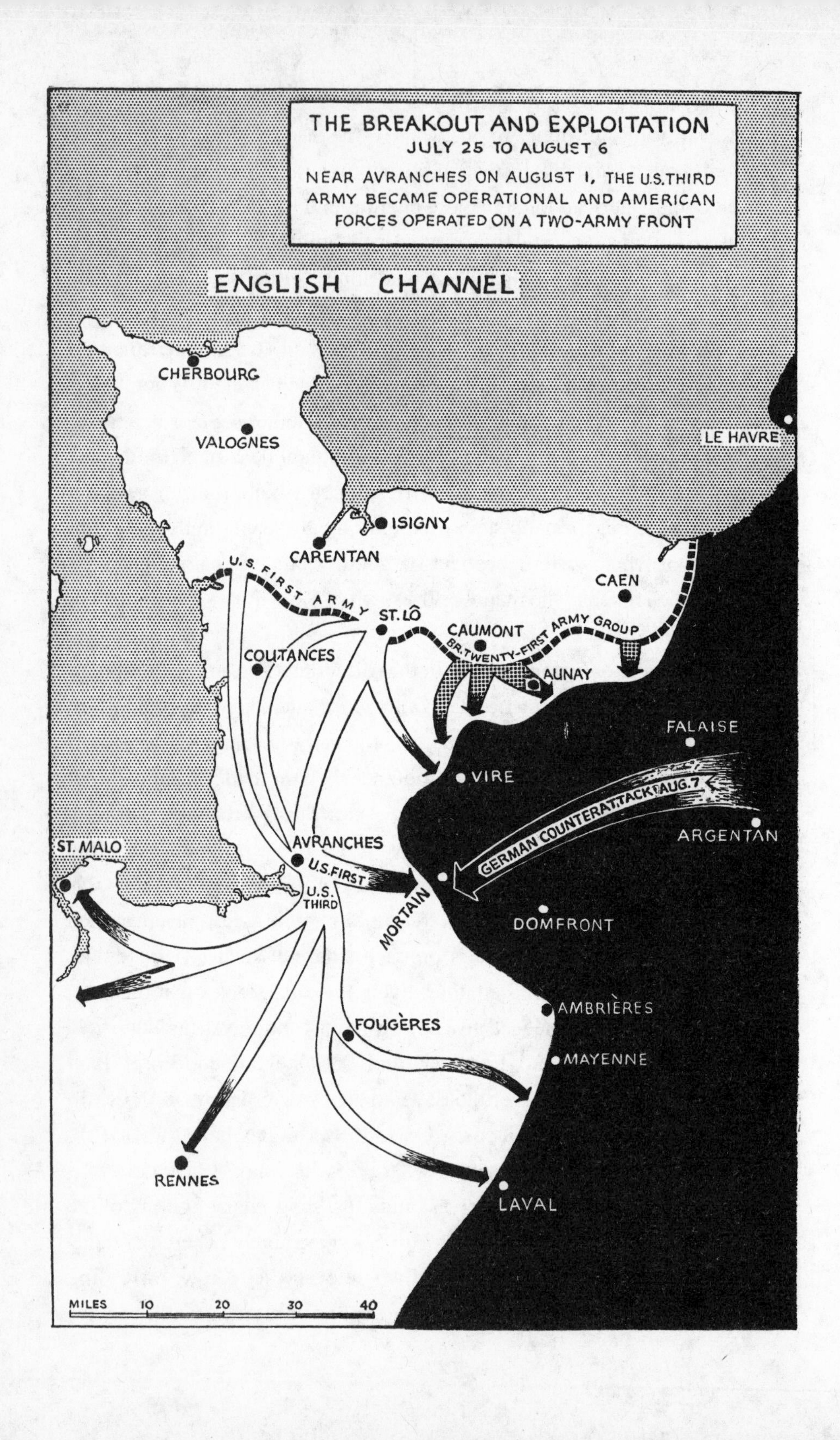

THE BREAKOUT AND EXPLOITATION
JULY 25 TO AUGUST 6
NEAR AVRANCHES ON AUGUST 1, THE U.S. THIRD ARMY BECAME OPERATIONAL AND AMERICAN FORCES OPERATED ON A TWO-ARMY FRONT
ENGLISH CHANNEL
CHERBOURG
VALOGNES
LE HAVRE
ISIGNY
CARENTAN
CAEN
U.S. FIRST ARMY
ST. LÔ
CAUMONT
BR. TWENTY-FIRST ARMY GROUP
COUTANCES
AUNAY
FALAISE
VIRE
GERMAN COUNTERATTACK AUG. 7
ARGENTAN
ST. MALO
AVRANCHES
U.S. FIRST
U.S. THIRD
MORTAIN
DOMFRONT
AMBRIÈRES
FOUGÈRES
MAYENNE
RENNES
LAVAL
MILES 10 20 30 40

a machine gun, a dug-in tank, or a mortar. For the attackers it was a nightmare battle of doggedly crawling toward these obstacles, assaulting them, and then within a hundred yards beginning the whole process all over again. Here Sergeant Culin's Rhino tanks, which were now in good supply, came into their own, as did Ike's idea of placing "an air liaison detachment in a tank belonging to the attacking unit," to direct the fighter bombers to their target. Ike had floated his idea, to nobody's great interest, all the way back in his days at Camp Meade with Patton. Now it was transformed into a "cab rank" system, whereby the fighter-bombers, U.S. P-47s or British Typhoons, circled high overhead waiting their turn to be called by someone on the ground to attack in support of the infantry. "All divisions," *The West Point Atlas of American Wars,* no more given to exaggeration than Ike was, remarks, "took heavy, demoralizing casualties." The five American divisions attacking Saint-Lô took over 11,000 casualties in twelve days, before the town finally fell to them on July 18. There are men still alive today who cannot forget the sight of German soldiers emerging from the smoke and debris of Saint-Lô waving white cloths or handkerchiefs, or the sight—and smell—of burned-out American Sherman and German Panther tanks at the side of the roads. Bradley didn't stop. He fought an equally hard battle to take the small village of La Haye-du-Puits, five miles west of Saint-Lô, in which five U.S. divisions were confronted by five German infantry divisions plus the Seventeenth SS Panzer Division, determined to capture the ground and the roads that would permit the big breakout, Ike's right hook—the knockout blow toward the Seine.

Prodded by Ike, Monty ordered attack after attack around Caen to support Bradley, culminating in a huge daylight attack by the RAF Bomber Command, which dropped over 5,000 tons of bombs on the villages to the east of the town in less than three hours. What little was left of Caen would finally be taken by the Canadians on July 20, but an attempt thereafter at a breakout across the Orne River to the southeast ground to a halt two days later with the loss of 400 British tanks.

In the meantime, more momentous events had been taking place elsewhere than the German loss of Caen. On the afternoon of July 17, Rommel, who had been visiting SS Obergruppenführer Sepp Dietrich's First SS Panzer Corps outside Caen (Dietrich had been Hitler's onetime chauffeur-bodyguard in the early days of the Nazi Party in Munich, and had risen to become one of the toughest armored commanders in the SS) was forced to leave the comparative safety of the minor roads briefly and take to the main road on the way back to his headquarters. There, one of the Allies' fighter-bombers attacked his open Horch staff car. The car skidded into a ditch, and Rommel was hurled onto the pavement, unconscious and with a serious fracture of the skull. Germany's most energetic general was sent home, apparently out of the war indefinitely; and Kluge, who already had his hands full as OB West, now also took over command of Rommel's Army Group B.

Three days later, at Hitler's daily briefing in the map room at the Wolf's Lair, his gloomy headquarters in East Prussia, Colonel Claus von Stauffenberg set off a bomb that very nearly killed the Führer—and also mistakenly set off an attempt on the part of certain Army leaders to depose the Nazi leadership. In Berlin the putsch was swiftly, and of course brutally, put down when it was discovered that Hitler had not been killed, and in Paris an attempt to arrest the SS and the Gestapo by some of the senior Army officers led to equally ruthless measures to restore the status quo. For those of any rank who had known of the plot, or suspected of it, or who had had discussions with the leaders, or even expressed sympathy with their aims, a terrible fate was in store, which would, as in ancient times, include in many cases their families. Nothing better illustrates the gangster quality of German political life under the Nazis than the kind of widespread revenge Hitler set in motion after July 20, a large part of it directed at the leadership of the Army, which was the only institution left with enough prestige and power to oppose him. The effects very soon spread to France. Strangely enough, they in no way prevented the average German soldier from fighting as hard as ever, but they had a dismaying effect on the senior commanders, for nobody could be sure what other officers who had already been arrested might confess under torture by the Gestapo, or who was under suspicion.

As for Kluge, his attention was fixed on Caen—and perhaps on his own precarious position, since many of those associated with the July 20 plot against Hitler's life were close to him. So when Bradley ordered Collins to attack to the west of Saint-Lô on July 25, leading off with a "bomb carpet" that saturated the ground in front of the American troops with over 4,000 tons of bombs (and that, owing to a last-minute screwup on the part of Leigh-Mallory, inflicted heavy casualties on the Thirtieth U.S. Infantry Division), the U.S. Seventh Corps was opposed by fewer than 5,000 German troops, mostly from the Panzer Lehr Division. By June 27 the German line had crumbled and in some places the Americans had advanced more than fifteen miles in two days of heavy fighting. The U.S. Eighth Corps took the town of Coutances after more heavy fighting, threatening to cut German forces off from the Atlantic coast of Normandy and from the garrisons in its ports.

At this point, Ike played what would now seem to have been his decisive card—George Patton, who had been kept in England champing at the bit in command of a fictitious army group to make the Germans believe that there was still another invasion coming, arrived in France and took command of Bradley's westernmost forces. These were soon to become, together with the forces Ike had landed on the beach with this in mind, the U.S. Third Army.

Ike now, at last, had what he wanted: a battle of movement. Difficult and disloyal as Patton might be at times, this was the opportunity he had been waiting for since 1918, and he was not about to miss his chance for glory. During the last days of July, the Third Army broke out of Normandy and stormed across Brittany, isolating one after another the important Atlantic ports of Brest, L'Orient, and Saint-Nazaire (all of which, unfortunately, the Germans methodically destroyed) and reaching Nantes and Angers by the second week of August—in the heart of France, at last, and no longer on its seacoast. More important, the Third Army swept east and then turned north, taking first Le Mans, then Alençon, in Ike's cherished "right hook," and threatening to trap the German Seventh Army in the "bag," the deep, narrowing "pocket" between two pincers: the British Second Army and the Canadian First Army in the north, and Patton's U.S. Third Army in the south, while Hodges's U.S. First Army pressed in from the southwest. It was, in fact, a classic pincer movement, just as Ike had planned it; and it succeeded, despite a fierce, desperate German armored

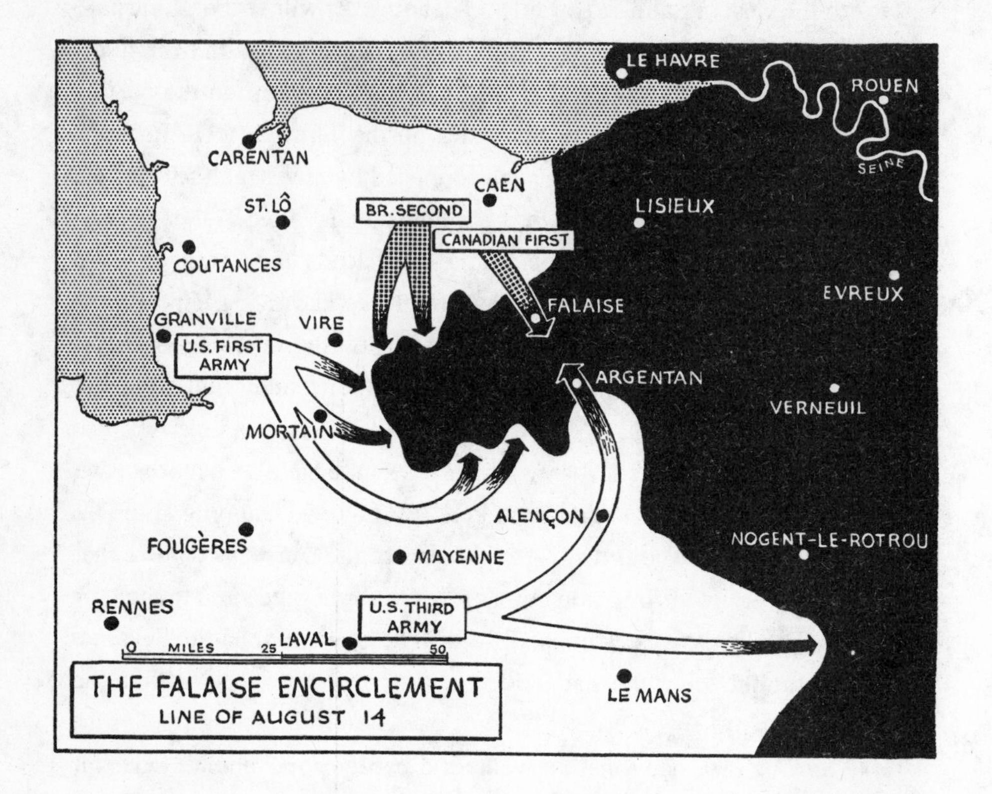

attack at Mortain, ordered by Hitler and intended to break out of the "bottom of the bag" all the way to Avranches, splitting the Allied beachhead in two parts. The gap was closed, very fittingly, when the tanks of the Second French Armored Division, serving in Patton's Third Army and commanded by General Philippe Leclerc,* who had marched from Lake Chad in central Africa to Algiers to join De Gaulle in 1942, made contact with those of the First Polish Armored division, serving with First Canadian Army, at Chambois, cutting off the Germans left in the pocket.

There were slipups, of course. Neither Ike nor Patton could forgive Monty for not having broken out toward the Seine; and Patton would continue to hold a fierce grudge against Bradley, who had halted Patton's

* Leclerc's real name was Vicomte Philippe de Hautecloque, but like many of De Gaulle's officers he had changed it to protect his family in occupied France from reprisal by the Gestapo.

advance to the north to sew up the "neck of the bag" for fear of American and Canadian forces running into each other. Still, by August 19 the "bag" had been firmly closed, and British and American fighter-bombers, flying in perfect, clear, hot weather, had a field day, undisturbed by the Luftwaffe. Fifty thousand German prisoners were taken; over 10,000 dead German soldiers lay rotting in the sun, their corpses choking the narrow farm roads between the Orne River and Chambois. There were also uncounted dead horses, for by now the German army was reduced to moving its equipment and guns with horses; so it was almost impossible to walk without stepping on the dead bodies of humans and horses. Those who still remember it describe it as the most terrible sight of the war, and photographs of it have much the same quality of horror as those taken after a large attack on the western front in the war of 1914–1918.

The failure of the German attack on Mortain and the closing of the Falaise gap shortly claimed another victim. On August 15 Field Marshal von Kluge's car was attacked by the Allies' fighter-bombers, and the radio car that accompanied him was destroyed. Kluge was trapped in a ditch and remained out of contact for a whole day. In the feverish atmosphere of the Wolf's Lair in East Prussia Hitler jumped to the conclusion that Kluge was conspiring with Allied generals to bring about a German surrender in Normandy; and on the August 17 he replaced Kluge with Field Marshal Walther Model, a fanatically loyal Nazi who had eventually managed to halt the Soviet advance after Operation Bagration, and whose faith in the Führer would remain unshaken to the very end. The unfortunate Kluge was ordered to return at once to Germany, and knowing what the Gestapo no doubt had in store for him (both because of the failure of the attack on Mortain and because of his contacts with some of those responsible for the July 20 assassination attempt against Hitler), he killed himself on the way home.

For the German army, the specter of defeat and the consequences of eleven years of acquiescence to rule by political terror and fanaticism were coming home to roost at last. Rundstedt had been replaced by Kluge, who had committed suicide. Rommel would shortly be given the choice between suicide and a trial for treason because of his suspected foreknowledge of the plot against Hitler, and in order to save his family took poison. Throughout

Germany and occupied Europe many other distinguished generals faced the same choice as Rommel, and those who were wise took his way out rather than suffer disgrace, torture, the humiliation of a rigged trial by a People's Court, execution, and the persecution of their family. In the circumstances, even Field Marshal Model did not find it easy to carry out the only thing which could preserve what remained of the German forces in France—an orderly retreat. His attempt to form a line west of the Seine came to nothing—on August 15, the long-awaited and still deeply controversial American landing in the south of France took place: originally Anvil, it was now called Dragoon, and it was still vehemently objected to by Churchill, Brooke, and Monty. Dragoon succeeded almost by default—the Germans had very little with which to oppose it—and the U.S. Seventh Army, commanded by Lieutenant General Alexander Patch, and closely followed by France's Armée B (soon to change its name to First French Army, commanded by the general and future marshal of France Jean-Marie de Lattre de Tassigny), was soon rapidly advancing north. Ike now had what he had always insisted on—a second blow that would not only reach the German border but deliver to the Allies the port of Marseille, intact—and with this, the position of the German army in France was rendered hopeless. On August 18 one of Patton's units reached an undamaged bridge over the Seine at Mantes-Gassicourt, west of Paris; and five days later he secured another bridgehead over the Seine, southeast of Paris, at Melun. Even Model could no longer hope to hold a position on either side of the Seine now, and he withdrew to the Somme and the Marne, his retreat beginning to turn into a panicked rout as German soldiers started to abandon their weapons and their units and make for home with as much loot from France as they could carry. The roads were jammed with soldiers, momentarily disorganized, and stunned by the magnitude of their defeat.

Their morale, already sinking, was further compromised on August 25 by the liberation of Paris—perhaps the single iconic event that marked the rapid victory of Ike's armies in the ten weeks since D-day, and which all over the world, even in Germany, made it clear that Hitler's defeat was, at last, only a matter of time. No event had astonished the world more than the fall of Paris in 1940—Paris was, after all, not just the capital of France, but one of the world's most beloved and admired cities. For many Americans (including Ike) and even a few Englishmen it was a second capital of the heart, symbolizing in a way that no other city does romance, artistic and intellectual achievement, historical grandeur, fashion and the particular genius of the French for life. As a result no other moment in the war seemed to symbolize so perfectly its approaching end as the liberation, in circumstances of high drama, of *la ville lumière*.

Ike had not wanted to take Paris—quite the contrary, in fact. A huge city the size of Paris could simply absorb an army, particularly if the Germans chose to fight for it street by street, and speed was his greatest concern now. Equally serious was the fact that by liberating Paris he might become responsible for supplying the city with fuel and food, and his own supply lines were now another serious concern. The ports he had liberated were incapable of working to anything like their full capacity, and fuel in particular was reaching the frontline units in a trickle—many of the tank divisions were as short of fuel as their enemy counterparts, in fact. Adding Paris to his burdens—with the risk of becoming embroiled in street fighting, as well as civil and political disorder—was what he most wanted to avoid.

On August 20, however, General De Gaulle, who had been incensed by attempts to stop him from flying from Casablanca to France on the specious reasoning that his aircraft was unarmed, arrived at Ike's headquarters at Maupertuis, to express his concern that nobody was marching on Paris. He quickly perceived that the issue was not so much Ike's military concerns as President Roosevelt's desire to prevent De Gaulle's entering Paris and presenting the world with a fait accompli of Gaullist government—in Washington, De Gaulle was still seen as a potential military dictator. De Gaulle pointed out that the fate of Paris was "of fundamental concern to the French government,"[34] and that if Ike did not send General Leclerc's French Second Armored Division into Paris, as head of the French government he would give the orders to do so himself. De Gaulle did not address Ike in his usual icy manner—the two men seem to have had genuine rapport. De Gaulle was too tactful to bring up the subject of

President Roosevelt's wishes, and Ike was too embarrassed to mention them. Instead, De Gaulle pointed out, rather respectfully for him, that since he had lent General Leclerc's division to the supreme commander, Ike could surely not deny his right to borrow it back at a moment of "national importance."

De Gaulle noted later that he found "this apparent stubbornness of Washington's policy quite depressing,"[35] but he did not blame Ike for it, and within the next forty-eight hours all the attempts that had been made to hold Leclerc back from Paris evaporated one by one, as Ike removed them on his own authority. By August 24, Pierre Laval, the French Quisling, had given up his effort to make a deal with the United States to hand over Paris intact and fled to rejoin Marshal Pétain, who was now in German hands. In the streets, the resistance was already fighting the German garrison of Paris, and the leftists in it maneuvered to take political power before De Gaulle arrived.

Hitler had ordered that Paris was to be held to the last man and, contradictorily, that all the city's public buildings, monuments, and museums were to be mined and blown up before the German forces left. Paris was to be "set ablaze," he commanded, and from time to time he called impatiently to ask the commander of the Paris area, General Dietrich von Choltitz, if the work of destruction had begun yet. Choltitz, who was having second thoughts about going down in history as the man who destroyed Paris, made contact with the Swedish consul general, and agreed to a cease-fire. On August 24, Leclerc's tanks rolled through Paris to the Hôtel de Ville, the Louvre, and the Hôtel Meurice (General Choltitz's headquarters); and on August 25 De Gaulle, having received a telegram of congratulations from King George VI, but not from Roosevelt, entered Paris at last. All over the world, newspaper headlines celebrated the event. The *New York Times,* in a banner headline, interpreted the news just as Ike and De Gaulle would have wanted, and as the White House would not interpret it—"DE GAULLE REPORTED LEADING SMASH INTO PARIS." The *Times* of London reported, more succinctly, "GEN. DE GAULLE ENTERS PARIS."

Entering the office of the prefect of the Seine, De Gaulle was asked to "proclaim the Republic," and with that poise which never deserted him, declined to do so: "The Republic has never ceased. . . . Vichy always was and remains null and void. I am myself the President of the government of the Republic. Why should I proclaim it now?"[36] With that he went to a window, and greeted "the crowd that filled the square and proved by its cheers that it demanded nothing more."[37] The next day he went to the Arc de Triomphe and relit the flame of France's unknown warrior; then, despite snipers firing from the rooftops, he made his way on foot slowly down the Champs-Elysées, before a crowd of millions, or, as he himself put it, "Rather a sea! A tremendous crowd was jammed together on both sides of the roadway. . . . As far as the eye could see there was nothing but this living tide of humanity in the sunshine, beneath the tricolor."[38] Once again, without much hesitation and on his own authority, Ike had made policy.

He delayed entering Paris himself until August 27, not wishing to take the limelight from De Gaulle. Kay Summersby drove him into the city amid wild scenes of celebration that often stopped the car. Ike's invitation to Monty to join him for the occasion had been declined, she noted, with "a short, snippy message."[39] With a soldier's eye for the significant military detail, Ike noted that at the Palais de l'Elysée, where De Gaulle was already installed as head of the French government, the approach to his office was guarded by the traditional Garde Républicaine in their ornate Napoleonic uniforms. Ike did not pause to celebrate in Paris, however. He was at this point a man in a hurry. It seemed to many people—but not to Ike—that the war could be over in a matter of weeks, perhaps of days. Even Model's furious energy and professional ability could not halt the retreat of the German army; and not even the best efforts (drumhead courts-martial and improvised roadside executions) by the Geheime Feldpolizei (the German army's uniformed version of the Gestapo), the Feldgendarmerie (the military police), and the troops of the Waffen SS could get the soldiers to rejoin their units and dig in. Ike's difficulty was that the Allied forces were running low on fuel and could no longer keep pace with the Germans' retreat.

Monty's snit about the entry into Paris was not without cause, however unforgivable it may have seemed to Kay. The long-delayed other shoe had dropped with a thump—on August 20 Ike had announced at his headquarters in Normandy that effective September 1 he would, at last, take over personal control as ground commander of the British and American armies, in addition to continuing as Supreme Allied Commander. This was the escape clause on which he had insisted when Churchill persuaded him to accept Monty instead of Alexander as his ground commander in Tunis. Ike had been entitled to do this ever since he moved his advance headquarters to Normandy, and both Bradley and Patton (who seldom agreed on anything) had urged him to. He had not felt it was appropriate to do so while the battle was in doubt, but now that the Allied armies were advancing everywhere, he felt that the time had come, and that major policy decisions—above all where and how to invade Germany to bring the war to a rapid end—would make it necessary for him to assume direct control of the armies before the battle moved toward the German border. As of September 1, therefore, Bradley would command the American Twelfth Army Group and Montgomery would command the British and Canadian Twenty-First Army Group; the two men would have "equal status" and report directly to Ike. There was of course no way for Monty not to see this change as a demotion—after all, he had been ground commander of both armies since before D-day, and had just won a huge and historic victory. The fact that it was always in the cards from the moment American forces in Europe began to outnumber those of the British and Canadians did not soften the blow—for that matter, Churchill, Brooke, and the British public were outraged too, when the story was prematurely leaked. Brooke acerbically summed up the British position when he wrote in his diary, "This plan [of Ike's] is likely to add another 3 to 6 months on to the war!"[40]

Monty argued against it vehemently, and on principle—nobody, he believed, could serve as supreme commander and also be his own commander of ground forces; the battle was too vast in scale for that. Indeed, Monty felt so strongly about this that he even urged Ike to replace him with Bradley, if it was a question of choosing an American instead of himself, and agreed to serve under Bradley. But this was to misunderstand Ike, who was not moved by personal ambition, or by his dislike of Monty. Ike simply sensed the need for one person to control the battle, and believed that it was his responsibility. Monty assumed wrongly that it was at least in part a question of Ike's ego, but it was nothing of the kind. It was a repetition of Grant's decision to leave Washington and retain personal control in the field over Meade's Army of the Potomac in 1864, as well as commanding all the Union armies. It was not that Grant did not respect the victor of Gettysburg as a soldier (although he did not much like Meade, from whom he felt no personal warmth, his feelings about him very much resembling those of Ike about Monty); it was simply that he had his own strategy for ending the war, and it was not Meade's—Grant intended to wear down the Army of Northern Virginia until Lee had no option but surrender.

When Ike looked at the map of western Europe, he saw what Grant had seen when he looked south toward Richmond from Washington. Ike's conclusion was instinctive, and was deeply rooted in American military tradition; and it differed radically from Monty's—it would come to be called the broad-front approach as opposed to the narrow-front approach to the defeat of Germany. That is, Monty (and Churchill) wanted to attack in the north and seize or isolate the Ruhr, cutting the German army off from its industrial base, whereas Ike, relying on sheer numbers and air strength, wanted to advance against the Germans from the North Sea to Switzerland, inflicting on them day by day losses they could not replace, until they collapsed. He did not think a single, clever stroke would do it, however carefully planned or vigorously executed, thus rejecting the views of both Monty (who wanted to attack in the north) and Patton (who wanted to attack in the south), and he did not suppose that the Germans would be defeated by capturing any one city, or even the Ruhr—they would stop when their Army was, at last, "fought out." This was Grant's philosophy of victory, updated to include tanks, trucks, and aircraft; Ike's indifference toward taking Berlin mirrors Grant's lack of interest in taking Richmond.

Although Monty's preference for a "pencil-thin attack"[41] has often been taken by British military historians as proof of his superior professional skill—the carefully planned rapier thrust rather than Ike's bludgeon—it in fact reflects the position in which the British now found themselves. Shortage of manpower had become a critical problem. Even by "cannibalizing" units—that is, breaking them up to use the men as replacements elsewhere, which is always a last resort in an Army with as much regimental tradition and pride as that of Great Britain—there was no way the United Kingdom could match the manpower resources of the United States, let alone the Soviet Union. By the summer of 1944, the cupboard was bare, and Churchill was anxious to bring the war to an end before this fact became obvious to everyone. A long, slow war of attrition against the Germans from the North Sea to the Swiss border was the last thing he wanted to see, and this appeared to be exactly what Ike had in mind.

This, as much as wounded vanity, explains the vehemence with which Monty pursued his case, even after it must have been clear to him that Ike was not going to change his mind.

CHAPTER 15

Stalemate

It is not "grace under pressure" that distinguishes great generals so much as courage and perseverance under pressure—and the willingness to take full responsibility when things, as they inevitably must, go wrong. Grant's greatness shone far brighter at Shiloh, when after a bloody day of confusion and defeat he turned what looked like a Confederate victory into a great victory for the Union, than it had at Fort Donelson, where victory was hardly in doubt. Napoleon's legend is far more appreciated in France for those moments of near-catastrophe that he overcame by the sheer force of his personality, like the retreat from Moscow, than for the glittering victories of his youth.

In the autumn of 1944, after two brilliant if knife-edge victories, Ike faced near-defeat and stalemate, and calmly, confidently overcame them;

this is perhaps the best measure of his greatness as a general. A more volatile and unstable man, like Patton, might have broken under the strain, or ranted at others for his own mistakes. A more arrogant man, like Monty, would have tried to deny that anything had gone wrong at all—Caen had not taught Monty humility; it had merely increased his vanity and further convinced him that he was always right, that in fact anybody who disagreed with him at all was a fool.

From the success of Torch and Overlord, however, Ike had learned first and foremost how to bear the heavy weight of personal responsibility and "the vicissitudes of fortune,"* which are always intensified by war. He was the supreme commander—he did not look for scapegoats when things went wrong or deny that anything had gone wrong at all, as Monty habitually did; he took the blame himself and tried to learn from the experience how not to repeat it. People around him confided to others, or let off steam somehow. Monty carried out a lengthy, constant—and frankly insubordinate—private correspondence with his mentor, CIGS Alan Brooke, often writing to him in sharply critical detail about Ike, as well as keeping a diary in which he was, if anything, even more outspoken. Patton kept an equally insubordinate (but more eccentric) diary, and also often let his hair down on the subject of Ike among his Third Army cronies over a drink in the evenings. Brooke himself kept perhaps the most stinging and unforgiving of diaries by any general in World War II (particularly where Ike was concerned), as an escape valve for the pressure of dealing with Winston Churchill day and night. By contrast, Ike's correspondence was either with Mamie, which was self-censored and included little or nothing about his work; or with General Marshall, which was always businesslike and carefully considered in tone. Ike let off an occasional burst of irritation about Monty in his moments of privacy with Kay Summersby, but most of the time he kept silence, no doubt at great damage to his cardiovascular system, as if he were always conscious that a supreme commander must above all present to the world a picture of confidence, self-control,

* "Vicissitudes of fortune, which spares neither man nor the proudest of his works, which buries empires and cities in a common grave."—Edward Gibbon, *Decline and Fall of the Roman Empire*, ch. 71.

and courage—"the three-o'-clock in the morning courage," in the words of Thoreau, "which Bonaparte thought was the rarest."*

With the failure of Operation Market Garden, Monty's daring airborne attempt to seize a bridge over the Rhine, and the slow fight to free the approach to Antwerp, his attention was fully occupied. In keeping with Ike's strategy of keeping up the attack on the Germans at all times, the torch was passed to Patton and Hodges. Hodges was to attack toward Cologne, which was now less than forty miles away from the front line of the American First Army, while Patton attacked toward Strasbourg and Mulhouse with the American Third Army. Neither attack was a complete success—bitter fighting, increasingly bad weather, and the shortage of supplies were combining to slow the Allied armies down. Despite the construction of double pipelines from Cherbourg to Paris and from Marseille to Lyon—both unheralded but miraculous achievements of the U.S. Army Engineer Corps—there was still not enough fuel getting to the front to sustain the broad advance Ike wanted, or two simultaneous major attacks in the north and the south that Patton and Monty had in mind. (Naturally, Patton wanted the main attack to be his own, with Monty supporting him, whereas Monty, with Churchill's backing, wanted the main thrust to be through the Ruhr then directly toward Berlin, commanded, of course, by himself, and supported by Patton.) The average daily tonnage coming in through the ports the Germans had damaged so badly was still well below what was needed, and the Germans, though cut off, surrounded, and under siege, were still holding out stubbornly in Brittany, at the ports of Lorient, Saint-Nazaire, and La Rochelle, and continued to control the mouth of the Gironde River, rendering Bordeaux useless as a port. At one point Patton unexpectedly captured a German supply dump containing 100,000 gallons of gasoline—and although there was an understanding that this kind of lucky find should be shared between the armies, Patton kept it for himself. It speaks volumes on the

* In fact, while Thoreau got the idea right, he got the quotation slightly wrong—Napoleon praised "*two* in the morning courage," by which he meant the courage to deal with a critical, unexpected surprise or crisis arising in the middle of the night, calling for immediate action. An example might be Lee's sudden decision late at night to halt his army's advance toward Harrisburg, Pennsylvania, on learning that Meade's army was close behind him, and concentrate all his forces toward Gettysburg instead—or Ike's decision on the night of June 4 to give the order for the invasion to go, despite the weather forecast.

subject of the supply shortage that to continue advancing, an American army should have been dependent on fuel captured from a fuel-starved enemy.

In Napoleon's words, "The most dangerous moment comes with victory," and this was borne out by the position in which the Allied forces found themselves in late September 1944. To some extent Ike was the victim of his own success. In the north, his forces were more than 200 miles farther east of the Normandy beaches than the plan for Overlord had predicted on D+120—the planners had assumed that by then, Allied forces would be holding a half-circle of territory from Nantes, on the Bay of Biscay, to just east of Dieppe on the English Channel, altogether excluding Paris, which they had intended to leave to "wither on the vine" as a salient. Instead, on D+120 Ike's line had reached Antwerp on his left and included Metz and Dijon at his center and most of the rest of France on his right—an astonishing victory, but one for which nobody had been prepared. Before the invasion Churchill had told Ike, "Liberate Paris by Christmas and none of us can ask for more."[1] But Paris had fallen to Ike in August, and by the autumn the Germans had been pushed back to—and in a few places beyond—their own border, with losses that would have certainly broken the resistance of any nation, army, or leader not sustained by sheer fanaticism.

All over France the main roads were converted to one-lane traffic (they were known as "red ball highways," referring to the warning signs put up by the military police), on which a vast army of trucks carried supplies and fuel from the ports to the front lines. No other traffic was allowed to interfere with the constant stream of loaded trucks: each truck ran at least twenty hours a day, stopping only for servicing and refueling. But still the demand for ammunition and fuel at the front far exceeded the supply—for, after all, the huge number of trucks consumed fuel too; and still more fuel was required for the engineers building airstrips (and for the planes that would use them as soon as the bulldozers left), as well as for the engineers who were working day and night with heavy equipment to repair the French railroad system, which the combined British and American air forces and the French resistance had been at such pains to destroy. Ike had set in motion a vast, complex organization to support the armies (the "tail" of nonfighting men, about the size of which Winston Churchill had already been complaining constantly before Torch), that included everything from trucking experts, bridge builders, and men who built and ran oil pipelines in Texas to specialists in locomotive and rolling stock repair. Many of these men had only recently been pulled from civilian jobs to put their expertise to work for the armies in France—within weeks of the liberation of Paris Ike's friend General Lee had over 8,000 officers and 21,000 men in and around the city. It was a spectacular accomplishment (though it did not change the fighting generals' opinion of Lee, whose haughtiness and privileged, luxurious lifestyle offended them all), but the unexpected speed of the advance and the thoroughness with which the Germans had demolished the port facilities still dominated the Allies' strategy. And according to Clausewitz's iron rules of warfare, desperate as the plight of the German army was—with the loss of the Romanian oil fields and the renewal, now that the bomber chiefs were freed once more from Ike's direct command, of the air assault against Germany's oil refineries and all-important synthetic oil plants—the Germans now benefited from interior lines of communications. That is, as the territory they commanded diminished back to the borders of prewar Germany, the length of their own supply lines shrank accordingly.

One secondary purpose of Market Garden had been to test whether the German army could still respond to a large attack. Montgomery had believed, optimistically, that it could not; but Ike had been more cautious, and the Germans' reaction at Nijmegen and Arnhem made it very clear to him that any notion Monty (and Churchill) might have had of a quick, painless attack from the Rhine to Berlin was illusory. Despite the Allies' control of the air, the Germans had found no difficulty in bringing up an impressive quantity of tanks: Tigers, Panthers, and even the new, monster sixty-eight-ton "King Tiger" tanks (fitted with turrets designed by Dr. Porsche). These tanks had played a decisive role in halting the attempt by the Thirtieth Corps to reach the Rhine Bridge five miles short of its goal—the Germans' reaction had been savage, well led, well supplied, and immediate. It was clear enough to Ike, even if not to Monty, that this would be no less the case once the Allied armies were across the Rhine and tried to encircle the Ruhr.

Although Ike tried hard to avoid the pomp and circumstance of his friend General Lee, his headquarters, code-named Shellburst, were becoming larger and less Spartan. He had moved from Normandy, where he had his office in a tent with a folding parquet floor; to Granville, a pretty seaside village in which he had a house overlooking Mont-Saint-Michel, one of the loveliest views in France; and then to Versailles, where he took over the luxurious Hôtel Trianon, set in eight acres of beautiful gardens on the edge of Versailles Palace, for his offices and the landmark eighteenth-century mansion Field Marshal von Rundstedt had used for his own living quarters. Kay Summersby got an office of her own, separated from his by a blanket, and he was joined by Telek the Scottie. This was not exactly roughing it, and perhaps explains why Ike spent so much of his time on the road, or flying up to the front in an L-5 spotter plane.

In October Kay was at last commissioned as a second lieutenant in the WACS, thanks to President Roosevelt; she also, thanks to Winston Churchill, received the Medal of the British Empire. Ike himself pinned the gold bars on her new uniform. Her job did not change much with her new status. She still drove Ike, looked after his schedule and his mail (the mail was still a sore subject for Mamie), and breakfasted every day with him (this would doubtless have upset Mamie even more, had she known). "The proximity to Paris,"[2] Kay noted, made Ike nervous, as did the chandeliers and museum-quality furniture of Field Marshal von Rundstedt's former residence, and he would shortly move his headquarters forward to Reims, to be nearer the front (and farther from Paris). Ike's military family had been reduced—he had moved Harry Butcher over to SHAEF's Public Relations Division temporarily. Reading between the lines, it seems possible that Butcher had briefly overstepped the line in his relationship with Ike, a common problem for those playing the delicate role of court jester or courtier; or perhaps the proximity of Paris was simply too much of a temptation for Butcher. Ike notes carefully, writing to Mamie, that he has "been in Paris only twice, each time for an hour," but it seems unlikely that Butcher followed his Spartan example. In any case, Butcher was soon forgiven, and he returned to Ike's headquarters, promoted to the four gold rings and gold-braided cap visor of a naval captain.

Ike's knee, which he had hurt in the forced landing of his plane, was still troubling him, and he mentions it several times in his letters home. "I have to be so d—— careful," he complained, and he was subjected to one and a half hours of treatment ("baking and rubbing") a day, about which he was something less than patient. As he himself noted, he was fifty-three, and his knee was in any case not going to heal as quickly as it did in his youth; and with the coming of winter the cold made it ache more. He responded to Mamie's worries about their relationship with patience and common sense, pointing out that no two people who had been separated for two years, except for one brief visit, could avoid being changed, but that no "problem" separated them except "distance."[3]

Be that as it may, other problems loomed large in Ike's life. Progress was slowing down all across the front, as German resistance stiffened and as Rundstedt's stern, old-fashioned professionalism took hold from top to bottom of the German armies in the west. Since Monty was fully occupied clearing the approach to Antwerp, Ike gave Bradley responsibility for attacking with his northern armies (Simpson's Ninth Army and Hodges's, First Army) in the direction of Cologne, supported by attacks farther south from Patton's Third Army and Devers's Sixth Army Group. They got within less than thirty miles of Cologne, but were stopped by bitter fighting in the thick, dense Hürtgen Forest, which offered General Zangen's Fourteenth Army an almost textbook-perfect area for defense, and for the kind of close-in fighting at which the German army excelled. Farther south, Patton and Devers succeeded in taking a big bite out of the German line, advancing in some places over seventy miles, taking Metz and Strasbourg and crossing the German border at some points. Strasbourg was taken by the Second French Armored Division; and Mulhouse, less than ten miles from the Rhine, by the French First Army of General De Lattre de Tassigny—both are examples of Ike's firm determination to have French cities liberated by French troops whenever possible.

By Thanksgiving of 1944, Ike's front line ran over 450 miles from the German-Swiss border at Basel to just east of Antwerp on the North Sea. It had three significant bulges: one in the north, in Holland, around Nijmegen and Venlo; one in the center, reaching out toward Cologne; and one in the south reaching out toward Karlsruhe. Nowhere were his forces over the Rhine, though at Nijmegen in the north and on the Swiss border in the south they were in sight of it.

In Italy—which the British complained had been drained of troops, air units, and armor by the landings in the south of France, just as they had predicted—Field Marshal Alexander's forces were nevertheless 200 miles north of Rome, and had taken the city of Florence. However, there seemed no likelihood that they could, in the near future, break through Kesselring's stubborn defense and fight their way through the hastily constructed but formidable Gothic Line.

In the east, of course, a war on an even larger scale was taking place—the front line ran from Memel on the Baltic to the Danube in Bulgaria, with a huge, threatening bulge developing in the south, where the army groups of Marshals Malinovsky and Tolbukhin were advancing swiftly to take Belgrade, surround Budapest, and approach within fifty miles of the German-Hungarian border. In the north, fanatical German resistance had stalled the Russians in East Prussia; but in the center the Russians at last held Warsaw. The Russians had paused for two months in sight of Warsaw to allow the Germans to savagely put down an uprising led by the Polish government in exile—which was based in London—so that Stalin could replace the so-called London Poles with his own communist Poles: one of the darkest and most cynical chapters in a war of stygian darkness.

From the view point of Berlin—from which Marshal Rokossovsky's First White Russian Army was now only 300 miles away—Germany's impending, and catastrophic defeat was obvious to any reasonable person; but of course the Führer was not by any definition a reasonable person. Despite his prolonged silence and his refusal to appear in public or tour the bombed-out cities of Germany, he still held the German people firmly in the web of the illusions he had spun for them over the past eleven years. They no longer required him to make dramatic speeches or public appearances—he had become an invisible tyrant. It was enough that he lived, to animate them with his own fanaticism, his irrational belief in victory, and, contrary though it might seem, his determination to go down fighting. The immense losses in lives, in conquered territory, in German cities and industries reduced to blackened rubble by the Allies' bombers, still had not shaken the German people's will to resist, least of all that of the Führer himself, for whom there was, in any case, no imaginable alternative. Berlin, like almost every German city, had been largely destroyed; by the autumn of 1944 whole areas, particularly the government center, resembled a lunar landscape; indeed there was constant debate within the RAF Bomber Command and the U.S. Eighth Air Force about whether it was worthwhile continuing to bomb the city. Nevertheless, from there, or from Berchtesgaden, or from his headquarters in the east, which the Russians were fast approaching, Hitler stubbornly clung to the belief that he could still win the war. Looking at the map—as he did every day for hours during the endless daily military conferences, in which he alternatively raged at his generals and interfered with military movements down to the battalion level—he constantly sought for the weak spot where Germany might still attack and change the course of the war. There was, as he well knew, only one chance to put together such an attack: Albert Speer, with his genius for keeping German industry functioning, could produce just enough new tanks, jet fighters, artillery, and fuel for one great attack, almost certainly the last. This attack would obviously be wasted in the east, where distance and the Soviet Union's overwhelming superiority of numbers would tend to make any action, on however ambitious a scale, vanish like a trickle of water in the sand. Besides, there was no decisive point toward which to attack in the east, no place the loss of which would bring the enemy's forces down like a house of cards. The more Hitler thought about it, the more his eye fixed on a place that was also the focus of Ike's attention: Antwerp.

Take Antwerp and the British and American armies in the north would be cut off from their most important and direct supply route, and separated, as the British and French armies had been in 1940. Once Antwerp was taken—and it was less than 100 miles from the German front line—the Allied armies would have to retreat, perhaps as far back as the Seine; and if they could be held there long enough, German troops in the west could be moved back eastward as rapidly as possible to deliver a stunning blow against the Soviet forces. It was a replay of General von Schlieffen's famous plan for the opening moves, in 1914, of what became World War I, and of Field Marshal von Rundstedt's attack on the French and British armies in May 1940—a knockout blow to the northwest arcing through Belgium to the Channel ports. Germany, most fortunately for the Allies, had no nuclear option—Hitler distrusted physics as a "Jewish science," and had been more impressed by the rocket engineers like Wernher von Braun. So any hope of ending the war on anything resembling his own terms depended on Speer, the German army, and the production of more "V-weapons" manufactured by slave labor. It was a slender hope, but it was the only card the Führer had left to play.

In the winter of 1944 German resourcefulness reached its peak. With his taste for the gigantic, Hitler was demanding the immediate production of Dr. Porsche's behemoth 188-ton tank, the "Maus," approximately the weight of four American Sherman tanks, and mounting a monstrous 128-millimeter gun. He also demanded increased production of the already huge sixty-eight-ton "King Tiger," which "eliminated its opponents with ease, on both Eastern and Western fronts," but which was so heavy, expensive to manufacture, and fuel-thirsty that there were never enough of them to make a significant tactical difference.

The German war economy was by then a curious blend of visionary weapons and medieval methods. The futuristic ME-163 "Komet" rocket fighter—so fast that none of the Allies' fighters could catch up with it and none of their bombers could shoot it down—was being assembled in underground caverns by starving, emaciated slave laborers in rags culled from the concentration camps; and the shortage of fuel for tractors sometimes made it necessary to pull ME-262 jet fighters out of their bombproof hangers onto the runway with teams of oxen. Speer's genius, together with his glacial objectivity and phenomenal energy (he was a workaholic's workaholic, before that word was in use), made him the perfect technocrat, and was vital to the German war effort; but his responsibilities were too vast and widespread for any one man to fulfill, even if that man was Speer. From the revolutionary Type 21 diesel-electric "Snorkel" submarine for the Navy to the mass-production of the new M.P. 43 automatic infantry rifle for the Army—made out of cheap stamped parts without time-consuming machining requiring trained craftsmen, it was the forerunner of the hugely successful Russian Kalashnikov rifle—Speer labored day and night to produce new weapons, new fortifications (it sometimes seemed to him that half of Germany was going underground) and new methods of production. He was constantly harassed by Hitler's grandiose demands, the armed forces' conservatism and old-fashioned perfectionism, and the rapidly decreasing supply of critical raw materials. He lived at a level of constant crisis, and yet he always managed to convey an almost surreal calm and competence. Still, like so many others in the Third Reich, he protected himself by a determination not to know what he did not want to know about, to the puzzlement of his judges at the Trial of the Major War Criminals before the International Military Tribunal at Nuremberg after the war and to future biographers. All of them wondered how it was possible for a man as intelligent as Speer, and as close to Adolf Hitler (who, in a burst of hyperbole, once called Speer "the greatest genius of all time"), not to know, for example, what was happening to the Jews. Of course Speer was not the only intelligent man to have suffered from this particular lapse—people as different as Field Marshal von Rundstedt and Dr. Ferdinand Porsche made the same claim after the war, as did millions of other less famous Germans.

By the end of November, Ike's forces were bogged down in bitter fighting all the way from the Swiss border to the sea. He described it as "the dirtiest kind of infantry slugging. Advances were slow and laborious. Gains were ordinarily measured in terms of yards rather than miles. Operations became mainly a matter of artillery and ammunition and, on the part of the infantry, endurance, stamina, and courage."[4] Losses were becoming high in rifle companies, and were increasingly hard to replace—partly because of the need for fresh infantrymen in the Pacific, and partly because of the low priority the War Department put on infantry manpower. Neither Ike nor Bradley ignored the possibility of a large German attack, but Ike insisted that the pressure on the Germans had to be kept up—that it would be fatal to go on the defensive for fear of a German attack, and risk a protracted stalemate like the situation in 1914–1918. Like Bradley, he suspected that the attack might come in the Ardennes, but was confident that it could be "pinched off."

The usual disagreements about strategy between the Allies led to a kind of military summit meeting at Maastrict, in Holland, in the first week of December. At this meeting, Monty tried once again to commit Ike to a large attack in the north toward the Ruhr, with himself in command. It did not help that Bradley was ill, with a lingering cold that had put him to bed for days; that Ike was tired, in pain from his bad knee, and impatient; or that Monty was at his worst, haughty, pedantic, and still harping on his idée fixe—being returned to the position of Ike's commander of ground forces, with Bradley (and the American forces) serving under him.

Ike made it clear that he would agree to nothing of the sort—he wanted Monty to attack to the north of the Ruhr and Bradley to attack to the south of it, and was unwilling to shift the whole effort to Monty's attack. That evening Monty confided in a letter to Brooke, "I personally regard the whole thing as dreadful. . . . I think now that if we want the war to end within any reasonable period you will have to get Eisenhower's hand taken off the land battle. . . . In my opinion he just doesn't know what he is doing."[5] The normally even-tempered Bradley wrote, "Montgomery made a very poor impression in that after putting forth his views and hearing everybody else's views, he refused to admit there was any merit in anybody else's views except his own."[6] Bradley also remarked that he would have to be relieved of his own command if Ike put the Twelfth Army Group under Monty, since he could not serve under Monty.

This was not exactly the cooperative spirit that Ike had always sought to engender among his commanders, and a photograph taken after the conference shows Monty off to one side angrily scowling, while Ike, Bradley, and Tedder are smiling at one another. Matters cannot have been helped by the fact that these three are all in neatly pressed uniforms, and wearing ties, while Monty, one hand thrust deep into his pocket, wears a sheepskin-lined RAF Irwin flight jacket, baggy corduroy trousers, and a patterned civilian scarf loosely knotted around his neck.

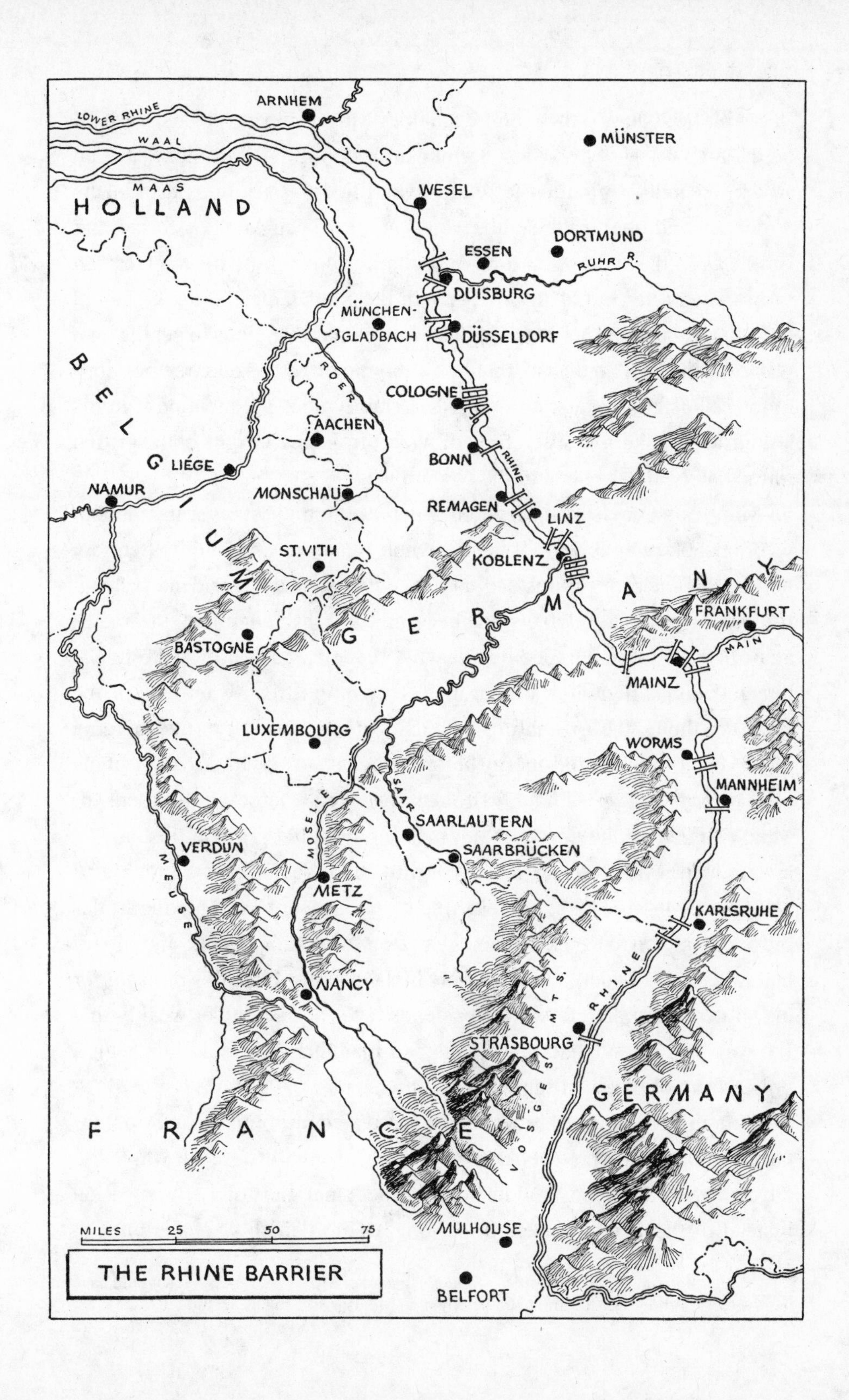

THE RHINE BARRIER

At no time during the conference does Monty appear to have dwelled on the possibility of a serious German attack. But in the event of such an attack, Ike's determination to retain two Allied thrusts, one north of the Ardennes and one to the south, would at least keep Monty's forces and Bradley's in the right place to offer a chance of pinching off the German thrust when it ran out of steam—as both Ike and Bradley thought it would before it reached the Meuse. Perhaps Monty was so anxious to get his own way over the "single thrust" that he simply failed to consider the possibility of Rundstedt's attacking first. He certainly does not mention it in his letters to Brooke and Secretary of War Sir James Grigg, both written immediately after the end of the conference.

Monty's angry letters triggered a flap at the highest level in London, where Brooke was about to bring the whole matter to Churchill's attention, hoping that the prime minister would take it up with the president. Instead, Ike, whose political instincts had become sure, reacted quickly; immediately flew to London, accompanied by Tedder; and beat Brooke to the punch. He was invited to dinner at 10 Downing Street, where the prime minister, though still rumbling with discontent, caved in to the supreme commander's charm and energy before Brooke had a chance to present his (and Monty's) case. Although Brooke presented a forceful argument for Monty's strategy the next day at a meeting with the prime minister, Ike, and Tedder, it was too late, and it got him nowhere. Ike refused to budge, and Churchill—no doubt not relishing the prospect of handing this strategic hot potato on to Roosevelt; and realizing something Brooke and Monty had failed to consider, that Ike surely had General Marshall's blessing for his strategy, and therefore the president's too—argued rather weakly that Ike was "his guest," and that it was bad manners to present him with a united front of British disagreement with his plans.*

Churchill may have also been more motivated by politics than by politeness as a host. He knew better than anyone else that Roosevelt wanted to have northern Germany and its ports as the American zone of occupation instead of the British zone, but so long as the British were the ones to

* Air Chief Marshal Tedder excepted—Tedder's dislike of Monty was such that it would have been almost unthinkable for him to come down on Monty's side in any argument.

attack toward the north of Germany, the logistics of the situation were likely to give them that area as their zone of occupation instead of the Americans. Given this reality, and the fact that it had been swept under the rug, he may have preferred to let sleeping dogs lie when it came to involving President Roosevelt in a strategic dispute between Ike and Monty. In any case, the outcome of the meeting was such that Brooke, according to his own account, "seriously thought of resigning, since "Winston did not seem to attach any importance to my views,"[7] though in the end he decided not to, if indeed he had ever really seriously considered this. As for Monty, he sulked at his TAC headquarters in Holland, surrounded by his handsome young aides-de-camp and his menagerie of farm animals, dogs, and canaries, and "absolutely and definitely" refused to send any of his staff to SHAEF at Versailles to help coordinate the planned attacks, on the grounds that he refused "to be involved in an unsound procedure."

It cannot be said that the flap over strategy had ended—in fact, it would continue unabated until almost the end of the war; it went on, even more shrilly, long after the war in books by Churchill, Monty, and Brooke; and it still remains a dividing line between British and American military historians. But at the time, this issue was subsumed by other events. Just two days later, in the late afternoon of December 15, Field Marshal von Rundstedt attacked at last, across a sixty-mile front from Trier to Monschau with two fresh panzer armies—the Fifth Panzer Army under Hitler's former chauffeur Sepp Dietrich, now promoted to SS Oberstgruppenführer, or full general; and the Sixth Panzer Army under the supremely competent, aristocratic General Hasso von Manteuffel—flanked by two fresh, reequipped German armies, a total of twenty-eight divisions, nearly 1,500 tanks, and over 2,000 artillery pieces.

Ironically, it was the same day that Ike learned his name had been submitted to Congress by the president for promotion to General of the Army, a five-star rank equivalent to that of field marshal in the British and German armies, marshal of France, or marshal of the Soviet Union. "From Lt. Colonel to 5-star General of the Army in 3 years, 3 months, and 16 days!" Butcher noted proudly in his diary, but by that time Ike had bigger things on his mind than his fifth star.

. . .

Neither Field Marshal von Rundstedt, OB West again; nor Field Marshal Model, commanding Army Group B, was entirely enthusiastic about the attack, which had been postponed twice since November. Both had argued forcefully with Hitler for a more limited offensive to pinch off the Allies' salient around Aachen, but while this was sound, orthodox military thinking, it would hardly have produced the apocalyptic effect which Hitler was anticipating and on which, indeed, the future of the Third Reich, if it had one, now depended.

Apart from gathering together every man and armored vehicle that could be found, the German plan for the offensive also included some unusual and ingenious elements. Some German tanks were disguised with sheet metal and painted olive drab with white stars to make them resemble American "tank destroyers"; and captured Sherman tanks were used to fool Allied soldiers into thinking that these tanks were their own. The colorful, swaggering SS Obersturmbannführer Otto Skorzeny* was handpicked by the Führer to put together a special unit of "commando" volunteers. Skorzeny, whose face was slashed by dueling scars, was a uniquely German combination of thug and Nazi Lawrence of Arabia; he had gained Hitler's confidence by carrying out the daring airborne rescue of Mussolini (as well as attempting to assassinate Tito in a raid on his own headquarters, and kidnapping the son of the regent of Hungary, Admiral Horthy). Some of his commandos were English-speaking and were equipped with Allied vehicles and dressed in American uniforms. Their object was to take the bridges over the Meuse in a *ruse de guerre* and in general to sow alarm and confusion behind the Allied lines. Skorzeny—whom the Allies' intelligence services would shortly describe as "the most dangerous man in Europe" (Dr. Goebbels himself could hardly have done better for him)—succeeded brilliantly at the latter task, so much so that Ike would be held in virtual isolation under armed guard during a good

* After the war, and his exoneration by the War Crimes Tribunal, Skorzeny took refuge in Franco's Spain; became an "adviser" to Colonel Nasser in Egypt and to Colonel Perón in Argentina; helped to set up and run "Odessa," the escape network for Nazi war criminals (this part of his story was turned into a novel, *The Odessa File*, by Frederick Forsyth); and became a wealthy man.

part of what came to be called the Battle of the Bulge, since Skorzeny had put about the rumor that he was planning a raid on Paris to seize or assassinate the supreme commander. Skorzeny's activities certainly led to panic and confusion far beyond what a significant German offensive might in any case have caused—including the immediate shooting of Germans captured wearing American uniforms* (and a number of captured Waffen SS) by Allied troops, and the massacre of American prisoners by the Waffen SS at Malmédy. But none of this can have been taken altogether seriously by such supreme military realists as Rundstedt and Model.

Rundstedt had insisted on waiting for bad weather—the worse, the better, from his point of view—which would keep the Allies' fighter-bombers on the ground, and he certainly got it. Everybody who fought on either side in the Battle of the Bulge remembers more clearly than anything else the snow, the fog, the bitter cold, and the frozen mud and slush. Seldom have the miseries of war been worse, and they were made more acute for many by the widespread belief that the war was nearing its end and that the Germans were all but beaten. Given this point of view, the sudden attack by twenty-eight divisions naturally came as a severe and demoralizing shock.

It did not, however, shock Ike, who had been expecting it, or something like it, although for once he had been badly let down by the Allies' intelligence, which failed to read the Germans' intentions accurately; and by the weather, which prevented air reconnaissance from detecting the movement of the Sixth Panzer Army, despite its puzzling disappearance from the map. He was neither taken by surprise nor uncertain what to do—he recognized at once that the Germans must be counting on capturing important Allied supply bases, the biggest of which was at Liège, and that the failure to do so would eventually stop them, since there was no way they could bring up supplies over the "miserable," narrow, frozen roads of the Ardennes, which were in any case blocked by their own troops advancing.

* Some measure of how calmly this kind of thing was taken is the laconic comment, quoted in the Imperial War Museum's *Victory in Europe*, from the diary of a British army chaplain: "A German SS officer captured in next village wearing a US uniform and cutting telephone wires—shot."

In that sense, the attack was improvisational, with nothing to back it up; but this did not, of course, make it less ambitious or less dangerous. It took very little imagination to see that if Rundstedt could drive a wedge deep enough between the British and the American armies, and then pivot north to cut the British off and threaten Antwerp and Brussels, the Allied front, spread out thinly for 450 miles, might collapse.

Indeed, one of Ike's first moves was to order General Lee, commanding the Services of Supply from his four-star hotel in Paris, to move his storekeepers and engineers forward to defend the Meuse crossings if necessary, and to prepare the bridges across the Meuse for demolition if worse came to worst. For the first couple of days it seemed as if Rundstedt might repeat his victory of 1940, on the same ground—the German advance was astonishingly rapid, and predictably ruthless, initially overwhelming one American unit after another. Retaining his calm, despite the confusion on the battlefield and the rumors about Skorzeny's commando teams, Ike did just what Grant would have done in similar circumstances. First of all, he decided that however badly the battle might seem to be going and however much the commanders on the spot clamored for reinforcements, he would not feed his reserves into the battle piecemeal,* since they would simply be destroyed one by one. What mattered was to build up forces for a powerful counterattack, even if this meant letting the Germans advance farther than would look comfortable on the map to armchair strategists in Washington and London, or to Monty. Second, he decided that under no circumstances would he retreat farther than the Meuse—the Allied armies would hold this line at whatever cost. Third, he ordered Monty to break off the attack in the north, and the even more unwilling and vehement Patton to do the same in the south, and begin moving their forces to seal off the flanks of the German attack. Also, he released the Eighty-Second and 101st Airborne Divisions, plus two newly arrived airborne and armored divisions, to Bradley, to hold the key towns of Bastogne to the south and Stavelot to the north. Bastogne, which the 101st got, like Gettysburg and Saint-Lô, was one of those otherwise unremarkable small market towns

* Ike remarked astutely, "This habit was a weakness of Rommel's," showing that Monty was not the only one to have studied Rommel's tactics closely.

whose importance was being the hub of numerous roads; and it was as clear to Ike and Bradley that Bastogne had to be held as it was to Rundstedt and Model that they had to take it. The importance of Stavelot, which the Eighty-Second got, was that the Germans would have to take it if they were going to reach Liège, with its precious supply dumps of fuel*—in fact the whole area from Stavelot to Liège was a mass of fuel and supply dumps and workshops.

The situation was not stabilized, and the German advance had not yet been checked, but Ike at least now had the means to check it. On December 19 he and Tedder went forward to Verdun, where Generals Bradley, Patton, and Devers had been ordered to meet him. "The present situation is to be regarded as one of opportunity for us and not of disaster," Ike told them. "There will only be cheerful faces at this conference table." Patton, who grasped Ike's strategy intuitively, smiled broadly and said, "Hell, let's have the guts to let the sons of bitches go all the way to Paris, then we'll really cut 'em off and chew 'em up!"[8]

This was farther than Ike wanted to let "the sons of bitches" go, but it was certainly his basic idea. Rundstedt was like Rommel at El Alamein: the farther and the faster he advanced, the more his already tenuous supply line would be stretched, and the more complete his eventual defeat would be. In addition, so long as Bastogne held out, Rundstedt's line of communications was fatally compromised.

Ike instinctively rejected the idea of building a defensive line to contain Rundstedt—he wanted to attack him and destroy his forces, now that the German army was at last out in the open from behind the defenses of the Siegfried Line. Here too, Ike made a quick decision. Although the weight of the German attack was toward the north, he decided merely to "plug the holes in the north" and if necessary even "give up some ground in order to shorten our lines"[9]—in other words, to go on the defensive there—while in the south "beginning a northward advance at the earliest possible

* Stavelot itself, though Ike did not know it, was the site of a fuel dump of almost 1 million gallons. He was enraged when he learned that Bradley had placed it there despite Ike's order to keep the fuel dumps on the west side of the Meuse. As matters turned out, the Germans ignored the presence of the fuel dump in Stavelot, but this was one of the many reasons why Ike thought Bradley's performance was flat-footed in the first days of the Ardennes attack.

moment." In practice this meant that the U.S. First and Ninth Armies would have to hold the Germans on the line formed by Stavelot, Malmédy, and Monschau, while Patton's U.S. Third Army would attack northward at all speed toward Bastogne—and that in the meantime Bastogne must be held at all costs.

The defense of besieged Bastogne became one of the legendary episodes of American military history, and deservedly so, because the entire battle hinged on holding the town against overwhelming odds and fanatical assaults. By December 20 Bastogne was completely encircled—the 101st Airborne Division was holding off three German divisions in furious hand-to-hand fighting—while to the north two SS divisions had almost surrounded Stavelot, fighting desperately against the U.S. Thirtieth Division to take the heights behind the town and cut it off. In the meantime, Rundstedt's panzer divisions were still rushing forward between those

THE ARDENNES
MAXIMUM GERMAN PENETRATION

two points, in an attack which would carry them another thirty-five miles farther west by Christmas day, and bring them less than three miles from the Meuse River.

In 1940, the crossing of the Meuse—and the sacrifice of the entire RAF "Battle" bomber force in France in futile attempts to destroy the bridges over it—had convinced the French that the war was lost, and had prompted Premier Paul Reynaud's fatal early-morning telephone call to Churchill to say, "We have been defeated—the way to Paris is open." Even so fiercely objective a professional soldier (and confirmed pessimist) as Field Marshal von Rundstedt must have had to repress a tiny sliver of hope at the news that the Meuse was in sight for his most forward tank units, but already any chance of repeating that triumph were fading fast. Patton performed exactly the kind of miracle that Ike knew he was capable of at his best, and switched his troops from their eastward attack to move north in miserable weather over frozen, rutted, backwoods roads that wound through hilly, heavily wooded country, to attack the southern flank of the German salient and punch open a corridor to Bastogne.

His biographer Martin Blumenson very rightly calls the execution of this "90-degree turn" by an entire army in the middle of a battle "the sublime moment" of Patton's career, "an operation only a master could think of executing." At the very tip of the German salient, the U.S. Second Armored Division attacked at Celles, to stop the Germans from making that final advance over those last few miles to the Meuse. Some of the bitterest fighting of the war took place in the week before Christmas 1944—indeed, the worst was yet to come. Hitler ordered Sepp Dietrich, the Munich brown shirt barroom brawler turned SS tank general, to attack with his two remaining uncommitted and intact SS panzer divisions, in weather that reduced visibility to near-zero, so that Allied troops, their nerves on edge because of Skorzeny's antics, often found it difficult to tell whose tanks were approaching them in the snow and fog, until it was too late. The impact and speed of the German attack shook many commanders at every level, and shattered many units. Ike's "steady, optimistic personal example" set the tone for commanders at every level, but it took some time for it to work its way down, and many units were initially ineffective, particularly when the leadership had broken down—Hodges's U.S. First Army was a good example of this kind of breakdown.

With a sixty-mile gap between the southern and northern sides of the salient, Ike concluded that it was impossible for Bradley to simultaneously command both the southern attack and the northern defense, and that he had "scraped the bottom of the barrel" to put every unit he could into the battle. Clearly, Bradley could no longer divide his attention between two widely separated fronts, and somebody else was needed to command the defensive battle in the north of the salient, where the German attack was now concentrated. Despite a certain reluctance, given his own feelings and those of Bradley and Patton, Ike did the only thing that made sense—he put Monty temporarily in command of the defensive battle in the north; asked him to move the British Thirtieth Corps, the only available reserve remaining, into the line; and then informed Churchill of the decision.* Churchill was supportive and enthusiastic, and Monty, who up until now had been carping from the sidelines at the progress of the Ardennes battle, seemed eager to accept. Apart from repeating, in various forms, "I told them so," Monty had been strongly in favor of moving Patton and the U.S. Third Army to the north, in support of his own attack—exactly the opposite of what Ike wanted to do. This was clearly another attempt on Monty's part to reopen the argument about the British plan for "one thrust" toward the Ruhr under his command, which Ike had rejected at the Maastricht conference and in London, but which would reemerge again and again until the Germans surrender made it academic.

Monty had complained to Brooke and Churchill of "a definitive lack of grip"[10] in the situation, and on December 20 took the rather more troubling step of sending a message to Brooke in the middle of the night predicting "a major disaster"[11] if he was not put in command of the battle. Brooke very sensibly ignored this message, which he did not receive until the morning, since his staff decided not to wake him up—there was in any

* Bradley quarreled bitterly with Ike over this decision, and also complained about it in his posthumous book, *A General's Life*, published forty years later. He threatened to resign during "a stormy [telephone] conversation" and told Ike he could not "be responsible to the American people if you do this." Ike replied, with amazing self-control, "Brad, I—not you—am responsible to the American people." Then Ike added, "Well, Brad, those are my orders."

case no way he could give orders to Ike, and he did not think it was his place to go behind Ike's back to Marshall. But news of the message made its way back to Versailles, where Ike was barricaded against the possibility of a raid by Skorzeny (he did not take the threat seriously, but his staff did); and to Luxembourg, where Bradley's headquarters was situated. Neither Ike nor Bradley was well pleased, but on the other hand they both trusted Monty's professional ability, and there was no other senior officer of his rank and prestige who could take command in the north at such short notice.

In the light of the furor between the Allies which Monty's subsequent comments caused—and for which Ike never forgave Monty—it must be said at once that in his usual precise way he instantly took command and control over two fairly reluctant American army commanders, Simpson and Hodges, and immediately began to "sort things out" and "tidy up the mess."[12] Even Monty's worst enemies—and he was about to make a lot more of them—agreed that he had a phenomenal gift for making order out of chaos, and while he did not exactly inspire enthusiasm in Bradley and Patton, neither they nor anyone else denied his military competence. What Simpson and Hodges needed was somebody on the spot with a firm plan who would tell them what to do, and while they would doubtless have preferred Bradley, they accepted Monty in the role, even though he had chosen to "show the flag" (and rub salt in their wounds) by turning up in his staff car flying the biggest Union Jack his aides could find, and accompanied by no fewer than eight Royal Military Police motorcycle outriders. Like MacArthur—indeed, like Patton—Monty knew how to put on a show, but he also knew how to establish his authority, and it had seldom been more urgently needed.

However, being Monty, he had already sown the seeds of bad feeling. To begin with, when Ike had called Monty at his TAC headquarters and asked him to take over command of the battle in the north, Monty merely shouted, "'I can't hear you properly—I shall take command straight away,' and put down the phone." It seemed to Ike then and later that Monty had hung up on him. Worse still, although Ike and Bradley were calmly certain that the German attack would be contained, and were already in the process of containing it, Monty managed to convey to Brooke and Churchill a much more pessimistic and dramatic view, reporting "great confusion and signs of a full-scale withdrawal,"[13] and even suggesting, in a joke that Brooke wisely cut from Monty's message before showing it to the prime minister, "We cannot come out by Dunkirk this time as the Germans hold that place!!!"[14]

The reference to the British army's evacuation from Dunkirk in 1940 was typical of Monty's sense of humor, but he cannot have failed to realize that it would increase the alarm—and the schadenfreude—already being felt in London. It is impossible not to conclude that Monty was deliberately making things sound much worse than they were, with the aim of painting himself as the savior of the American armies, the man who, to use the homely English phrase, "pulled the chestnuts out of the fire" for Ike. He bolstered this reasoning by assessing the importance of the British Thirtieth Corps far higher than Ike or Bradley did, since Ike had merely wanted them to cover the Meuse bridges in the north in case of a disaster. Monty was also outspokenly critical of Ike's assessment of the intelligence before the battle, although this was truly a case of "the pot calling the kettle black," since Monty had stubbornly ignored the intelligence reports from the Dutch underground, confirmed by RAF low-flying photoreconnaissance flights, of the presence of two SS panzer divisions in Arnhem before Operation Market Garden.

There were plenty of horror stories to go around, as Ike knew well—units that had broken and run; senior officers who had hastily abandoned their headquarters, leaving secret documents and maps behind them. But despite the claims by some of Ike's British critics that the battle in the Ardennes was a second Kasserine Pass, an American disaster, and proof of Ike's incompetence as his own ground forces commander, the truth is that most of the American troops fought well, in circumstances that could have caused a complete collapse and even though the bad weather deprived them of the air support on which Allied armies had come to rely. As for Ike, he recognized before anybody else the sheer magnitude of the German attack, and immediately took steps to confront it. He ordered the Seventh and Tenth Armored Divisions to the German breakthrough (despite

Bradley's initial hesitation to take the Tenth away from Patton, for fear of having to deal with an outburst from Patton, which Ike brusquely overruled); and he moved the two airborne divisions, the Eighty-Second and the 101st, which were still recovering from the part they had played in Monty's disastrous Market Garden, to the key towns of Saint-Vith and Bastogne, respectively. Without this decision, and the sense of urgency Ike conveyed to the commanders involved, the German attack might indeed have crossed the Meuse. Even more important was his quick recognition that all his armies must stop attacking and go on the defensive immediately, except in the Ardennes—a decision that Monty, who was preparing the attack in the north, resented. In later years, Ike responded in his usual measured way to the criticism he had received, taking full responsibility for what had happened. "We remained on the offensive," he wrote, "and weakened ourselves where necessary to maintain those offensives. This gave the German an opportunity to launch his attack against a weak portion of our lines. If giving him that chance is to be condemned by historians, their condemnation should be directed at me alone."[15]

By and large, historians have been kinder to him than the British were. Maintaining the offensive steadily from the Swiss border to the North Sea carried with it certain risks, the chief one being that the Germans would sooner or later attack what they took to be the weakest point in the line. But Ike considered this risk smaller than that of concentrating everything on the proposed British "full thrust" to the north through the Ruhr and toward Berlin, the southern flank of which the Germans would certainly have attacked with everything they had.

The battle went on until January 16, and indeed would reach a new crisis on January 3, when Ike's proposal to risk abandoning the recently liberated French city of Strasbourg in order to provide reinforcements for the Ardennes battle provoked a rift among the Allies. A furious General De Gaulle came to Ike's headquarters, threatening to move all the French forces to the defense of Strasbourg and lecturing him on the national importance of the city in French history.

Ike, who didn't want to be faced with a political crisis in France that might endanger his lines of communications, backed down, with

Churchill's blessing (the prime minister told him, "I think you've done the wise and proper thing") and modified his orders about Strasbourg. By January 3, very fortunately, the issue was academic—the Germans were retreating, and though the fighting remained intense, their attempt at a breakout across the Meuse had peaked and failed. It had cost the enemy 120,000 casualties and almost half the tanks and guns they had committed to the battle. The Allies' casualties, predominantly American, reached 78,500, of which an extraordinary 21,000 were captured or missing—indicating not only the ferocity of the battle, but also, unfortunately, the collapse of morale in the first few days of fighting. Still, in warfare only one thing matters, and that is winning; and by any standards Ike had won a great victory. The men and tanks Hitler had squandered on the Battle of the Bulge would be missing when the Russians made their final attack on Berlin, and when the American, British, and French armies crossed the Rhine—even the genius of Speer could not replace what had been lost.

Throughout the battle, Ike had complained that Monty was ignoring his orders to attack in the north, and even Monty's usual careful preparation and meticulous planning were not enough to explain the delay. This was not a small matter—Monty had about two-thirds of the U.S. forces committed to the battle under his command, including "Lightning Joe" Collins's U.S. Seventh Corps. This was a source of searing irritation to Bradley. Collins was, as usual, eager to attack; but Monty, while rather patronizingly praising Collins's "Irish vigour," held him firmly back, convinced that the Germans had another twenty or thirty divisions to throw into the battle, which Ike and Bradley were ignoring, and determined "to get the show tidied up" before passing to the offensive. Ike had envisaged an attack on the German armies from both the south and the north, crushing them in the salient—an objective made all the more plausible as the weather cleared at last, allowing British and American fighter-bombers to start flying again from their muddy, water-logged fields—but Monty's infuriating delay meant that only one of the "jaws" was doing the crushing. It seemed likely to Ike then (and continued to seem likely to him thereafter) that Monty, having stabilized the northern side of the salient, was more interested in preserving his forces for the full-scale attack he had been preparing than in finishing off the German forces in the salient.

Monty's slowness to attack in fact had brought Ike to see him on December 28, a seemingly endless journey by train in bad weather that Ike deeply resented having to make. The meeting would bring the relationship between the two men to an unhappy climax. Neither of them was in a conciliatory mood when Ike's heavily guarded train halted at the railway station in Hasselt, Belgium, Monty having refused to meet Ike in Brussels because the roads were too icy. Ike had come to tell Monty to get a move on and attack, but Monty cut him off with a whole list of grievances, returning to his cherished belief in a single large thrust to the north commanded by himself, with Patton "in support." Ike once again turned this down—it was not the strategy he wanted to follow, and, even more important, it was not the strategy that General Marshall and the president wanted him to follow. Despite this, Monty somehow came away from the meeting (which was between the two of them, at Monty's request, with neither of their chiefs of staff present) under the impression that Ike agreed with him. "He was very pleasant," Monty cabled to Brooke afterward "and the meeting was most friendly but he was definitely in a somewhat humble frame of mind and clearly realizes that the present trouble would not have occurred if he had accepted British advice and not that of American generals."[16] To paraphrase the duke of Wellington, if Monty presumed that, he would presume anything. Considering that Ike was seething with anger before, during, and after the meeting, and completely disagreed with everything Monty said, only somebody as monumentally self-satisfied as Monty, and with a tin ear for other people's feelings, could have reached this remarkable conclusion.

If it was possible to make matters worse, Monty promptly did so, by sending Ike a written confirmation of his views, reasoning that Ike had often changed his mind in the past after agreeing with Monty in a conversation—or, as Monty put it more bluntly to his chief of staff Major-General Francis ("Freddie") de Guingand, that Ike had welshed on him. This, predictably, Ike took as a slur on his integrity, quite apart from his amazement that Monty had left the meeting assuming Ike was in agreement with him. Monty's letter was hand-carried to Ike by the long-suffering and supremely tactful Guingand, who did not hand it over at once, since his first disagreeable task was to tell Ike that Monty would not attack on January 1, as Ike thought he had promised—that he needed several days more to prepare a careful fire plan and collect his reserves, and could not commit himself to an arbitrary date.

Since Bradley had already begun his own attack on the southern flank of the salient and was expecting Monty to attack the northern flank at the same time, this unwelcome piece of news caused consternation in Ike and Bedell Smith, both of whom were enraged. Monty's letter, therefore, could hardly have been opened and read at a less propitious moment, and it prompted Ike to prepare a "him or me" cable to Marshall, making it clear that he could no longer work with Monty. Already, the notion of replacing Monty with Field Marshal Alexander seems to have occurred to Ike—an astute one, since however big a hero Monty was to the British public, nobody, including the prime minister, could complain about replacing him with a figure so universally admired as Alexander. Ike had just received a message from Marshall about news stories in the British press calling for the appointment again of a British ground forces commander—curiously, they echoed Monty's demand—and urging Ike to "under no circumstances make any concessions of any kind whatsoever."[17] Coming from General Marshall, these were strong words indeed, and indicated to Ike that if he chose to press the matter, Monty would have to go. This same thought seems to have occurred to Guingand (but not to Monty). Guingand rushed back to Monty's headquarters and persuaded his boss "to eat humble pie" and write Ike a letter of apology. Monty seems to have been genuinely amazed that his letter should have provoked an Allied crisis, though what really got his attention was Freddie Guignand's mention of Alex's name as his successor. The notion of being replaced by Alex, about whose abilities as a soldier Monty had often expressed doubt, was more than enough to persuade Monty to back down and ask Ike to tear up his earlier letter.

Unfortunately, it did not change Monty's mind about the right strategy for winning the war, as became apparent when, with Churchill's approval, he gave a rare press conference which was intended to smoothe the ruffled feathers of the Americans, but which had quite the opposite effect. Although Monty's speech had been carefully vetted by Churchill, the prime minister could hardly control Monty's delivery, which Monty's own intelligence chief described as "disastrous." "Oh God, why didn't you stop him?"[18] asked one distinguished British war correspondent who was an appalled witness to the press conference; but of course nobody could stop Monty, who had dressed for the occasion in a new red airborne beret, which, considering the feelings of the airborne troops toward him after Market Garden, was less than tactful. His praise for Ike and the American forces inevitably sounded "patronizing," and his description of British forces "fighting on both sides of American forces who have suffered a hard blow" was both untrue—British forces were hardly engaged at all—and infuriating to American commanders, since U.S. forces had suffered nearly 80,000 casualties while the British suffered less than 1,500. Perhaps worst of all was Monty's description of the fighting as "a most interesting little battle," one of those British understatements that would have been better left unsaid. Even nearly twenty years later, Ike, who was never one to exaggerate his feelings, was still able to write, "This incident caused me more distress and worry than did any similar one of the war."[19] The effect it had on Bradley and the other American commanders was even stronger, if anything, and is best summed up by Patton's description of Monty as "a tired little fart."[20]

It is typical of the deep division between British and American views of the battle even today that Monty's loyal biographer Nigel Hamilton quotes Ike as saying, "Praise God from whom all blessings flow!"[21] when Monty finally attacked—as if Ike had been deeply relieved and thankful. Americans will recognize Ike's comment as a combination of irony and incredulity—a sharp dig, by his standards, at Monty's slowness, proverbial caution, and unwillingness to attack in a battle that he thought was strategically pointless and badly managed.

Monty's press conference not only brought Anglo-American relationships to an all-time low but also served as fuel for Dr. Goebbels's Nazi propaganda machine, and threatened to undo more than two years of work on Ike's part to build a functioning alliance, in which British or American nationality was less important than victory. The British press, even without the help of Dr. Goebbels, worked itself up into a fury, much of it centering on whether Monty should be reappointed as Ike's ground forces commander. In retrospect much of this was inevitable—as the war in Europe approached its final stage, with hard fighting still ahead, the British were gradually waking up to the reality that they were no longer the senior partner, or even an equal partner of the Americans, and that British postwar interests were not even remotely similar to those of the United States, let alone to those of the Soviet Union. The principle of the Allies' unity had kept the two countries fighting together for over two years, in large part thanks to Ike's scrupulous and impartial attention to it; but the United States would soon have over sixty combat divisions in Europe, compared with fifteen British and Canadian divisions, while on the eastern front the Soviet divisions were numbered in the hundreds. In these circumstances, Monty's incautious and unfortunate remarks seemed to many people in Britain as if he were standing up for British pride and interests against "the Yanks," even though that had not been his intention. The alliance was far from unraveling, but it was no longer sacred.*

The fighting in the Ardennes continued until January 18, at which point the Germans had been driven back to their original positions, more or less. In the Ardennes, there was no equivalent of the German "pocket" at Falaise—Rundstedt's attack had been defeated and his losses in men, guns, and tanks were severe and could not be replaced; but there was no collapse or panic. The British and the Americans were still west of the Rhine, the Germans still held the Siegfried Line, and the question of how to deliver the final death blow to Germany still remained to be answered. In the east, the great Russian offensive began on January 12, and it would bring the Russians to within fifty miles of Berlin by the end of the month.

Ike was restless, and continually on the move. Kay Summersby reported that his knee was painful, he suffered from a cold and the flu, a cyst on his

* De Gaulle had never suffered from this illusion. He had always resisted Americans' interference in what he regarded as France's interests, and would shortly be in conflict with Ike again over the French occupation of the German city of Stuttgart, although it was not in the French zone. He was already looking ahead to an independent Europe freed from "Anglo-Saxon" hegemony, in which France would be the dominant power.

back required surgery, and "his temper was truly vile." Bradley was in low spirits and seemed to be bogged down. Monty was meticulously preparing his "thrust to the Rhine" while continuing to lobby for his appointment as the ground forces commander again, with all the resources that were going to Bradley and Patton redirected to him. Ike's friend and naval commander Admiral Sir Bertram Ramsay was killed when his airplane crashed. Ike's son John was about to be sent overseas, and Mamie was apparently complaining that Ike did not write often enough. "It always distresses me when I get a message from you indicating anxiety or impatience because I have failed to write,"[22] he replied to her, though considering what was going on around the supreme commander at the time, three letters home in two weeks doesn't seem that bad a record. He did note with wry amusement that although he had been promoted to five-star rank, and was entitled to an extra $255 a month, no decision had yet been reached officially as to what his insignia should look like—he had received three different sets of instruction on the matter from Washington before it was finally decided that the five stars should be arranged in a circle.

Cautiously, he wrote to Mamie that "we are on the road to victory," but he refrained from making any predictions about when victory might come. He reminded her that it was only three years since they had moved from San Antonio to Washington, and added, "I cannot remember the time when I was free of these continuing problems involving staggering expense, destruction of lives and wealth, and fates of whole peoples."[23]

The best advice of his engineers, who had been studying the river's flow and trying to decide where and how to build bridges across it, since the Germans would surely destroy the existing ones, was that he could not expect to get his armies across the Rhine before May 1, but Ike seems to have kept this piece of grim news to himself.

Or perhaps he just didn't believe them.

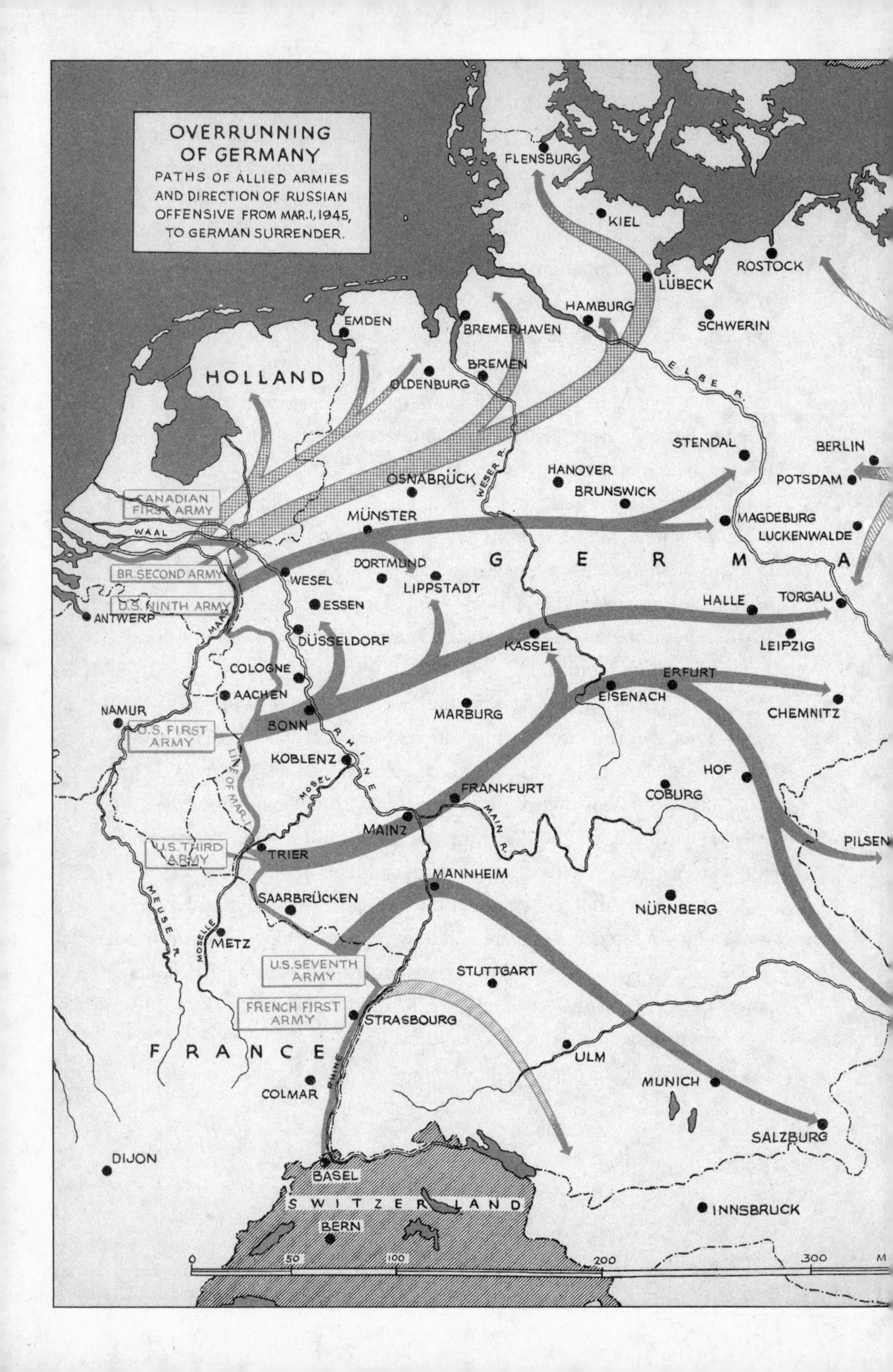
OVERRUNNING OF GERMANY
PATHS OF ALLIED ARMIES AND DIRECTION OF RUSSIAN OFFENSIVE FROM MAR. 1, 1945, TO GERMAN SURRENDER.
FLENSBURG
KIEL
ROSTOCK
LÜBECK
HAMBURG
SCHWERIN
EMDEN
BREMERHAVEN
BREMEN
OLDENBURG
HOLLAND
ELBE R.
STENDAL
BERLIN
POTSDAM
HANOVER
OSNABRÜCK
BRUNSWICK
WESER R.
CANADIAN FIRST ARMY
WAAL
MÜNSTER
MAGDEBURG
LUCKENWALDE
DORTMUND
G E R M A
BR. SECOND ARMY
WESEL
LIPPSTADT
U.S. NINTH ARMY
ESSEN
HALLE
TORGAU
ANTWERP
MAAS
DÜSSELDORF
KASSEL
LEIPZIG
COLOGNE
ERFURT
AACHEN
EISENACH
CHEMNITZ
NAMUR
MARBURG
U.S. FIRST ARMY
BONN
RHINE
LINE OF MAR. 1
KOBLENZ
HOF
MOSEL
FRANKFURT
COBURG
MAIN R.
MAINZ
PILSEN
U.S. THIRD ARMY
TRIER
MANNHEIM
SAARBRÜCKEN
NÜRNBERG
MEUSE R.
MOSELLE
METZ
U.S. SEVENTH ARMY
STUTTGART
FRENCH FIRST ARMY
STRASBOURG
F R A N C E
ULM
RHINE
MUNICH
COLMAR
SALZBURG
DIJON
BASEL
SWITZERLAND
INNSBRUCK
BERN
0
50
100
200
300
M

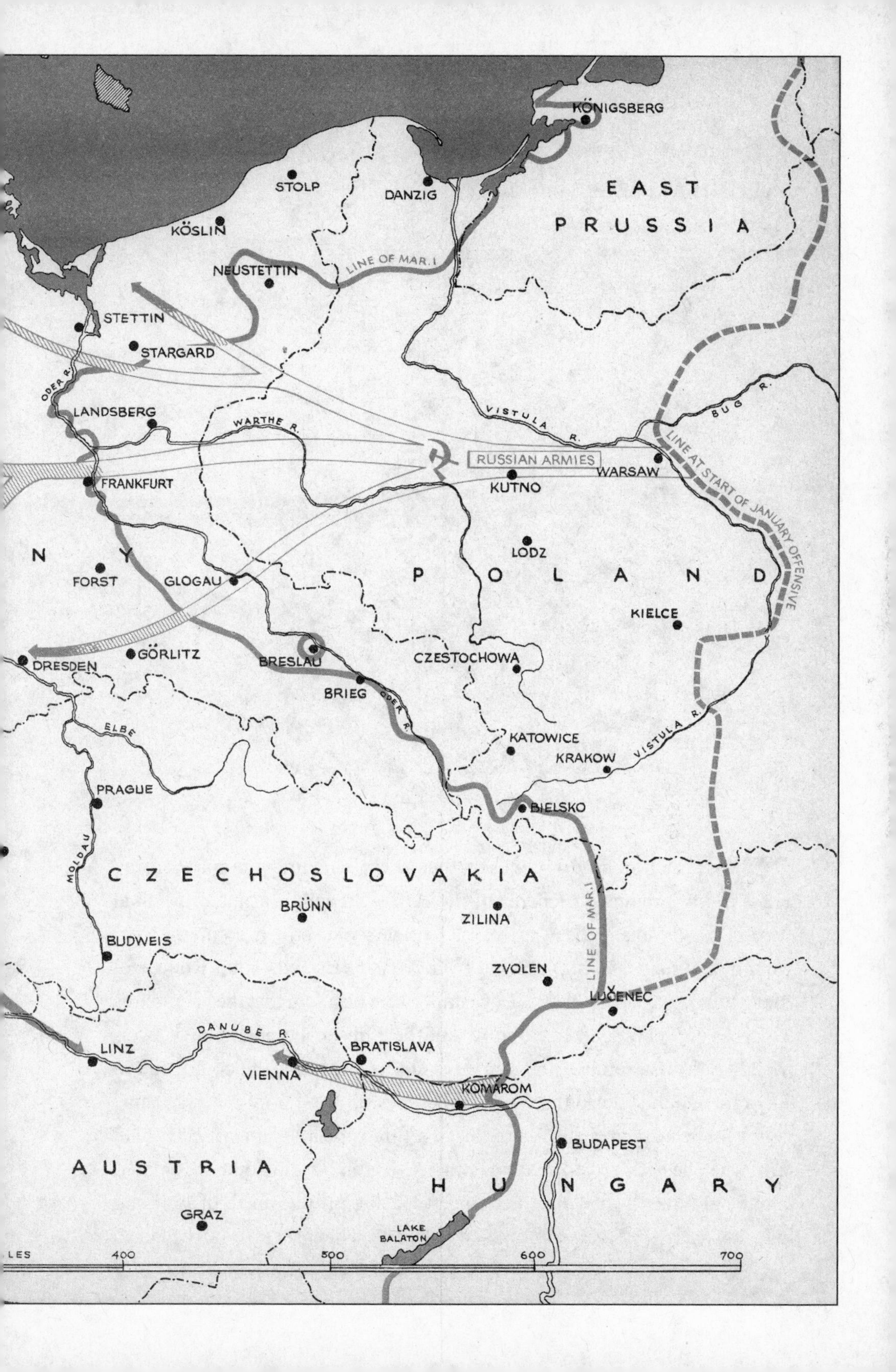
KÖNIGSBERG
EAST PRUSSIA
STOLP
DANZIG
KÖSLIN
NEUSTETTIN
LINE OF MAR. 1
STETTIN
STARGARD
ODER R.
LANDSBERG
VISTULA R.
BUG R.
WARTHE R.
RUSSIAN ARMIES
WARSAW
LINE AT START OF JANUARY OFFENSIVE
FRANKFURT
KUTNO
LODZ
N Y
POLAND
FORST
GLOGAU
KIELCE
GÖRLITZ
DRESDEN
BRESLAU
CZESTOCHOWA
BRIEG
ODER R.
ELBE
KATOWICE
KRAKOW
VISTULA R.
PRAGUE
BIELSKO
MOLDAU
CZECHOSLOVAKIA
BRÜNN
ZILINA
BUDWEIS
ZVOLEN
LINE OF MAR. 1
LUČENEC
DANUBE R.
LINZ
BRATISLAVA
VIENNA
KOMAROM
BUDAPEST
AUSTRIA
HUNGARY
GRAZ
LAKE BALATON
LES
400
500
600
700

CHAPTER 16

Armageddon, 1945

Bradley, Ike, Patton.

It is worth reminding oneself that as the supreme commander and newly promoted General of the Army prepared to make the final assault on Germany he was fifty-four years old, still a comparatively young man for a five-star general (MacArthur was sixty-four, Rundstedt sixty-nine); and although those who knew him best said that the six months since D-day had aged him badly, and he himself complained of overweight, lack of exercise, and numerous aches and pains, he could still, at any rate in public, flash the same exuberant smile and walk with the same loose-limbed urgency that were the first things people had noticed about him when he had arrived at West Point. Although Ike did not then, or later, aim for gravitas, he had managed since 1942 to grow into an air of authority which was impressive and unmistakable, and which seemed only natural in a man who had long since become used to making far-reaching military and political decisions. Even Monty at his most infuriating recognized that there were pressures on Ike far greater than those on any other general; indeed, everybody who came into contact with Ike was aware of this. The chief pressures, as the bitter winter fighting of 1944 carried on into January 1945, were how to bring the war to an end with as few casualties as possible, and how to keep the Anglo-American alliance together until it was over.

In neither of these problems was Ike much helped by Washington. Marshall was more Olympian than ever—after all, he was running a world war, of which Ike's European campaign, however important, was only a part—and one also senses, reading the messages between Marshall and Ike, a weakening of the pulse in the political direction of the war. Ike had not seen President Roosevelt since his brief trip home in January 1944, when the president had "kept him at his bedside for more than an hour," for a far-reaching conversation about strategy and the future occupation of Germany (over which they had disagreed). Roosevelt had been recovering from influenza, but his mind had been clear and even "the most obscure places in faraway countries were always accurately placed on his mental map."[1] Roosevelt told Ike that he had never felt better, and was being kept in bed only because his doctors were afraid he might have a relapse; but Ike made a point of this in his memoirs, as if he had felt a premonition that there was something more seriously wrong with the president. Those who were closest to the president—Harry Hopkins, General Marshall, his daughter Anna—closed their eyes to his physical decline, no doubt because they didn't want to see it. At Yalta, only just over a year after Ike had sat by the president's bedside, Sir Charles Moran, Winston Churchill's physician, with a doctor's cold, objective eye, was astonished at the change that had come over the president since he had last seen him. "The President," he wrote (in an absolutely correct diagnosis), "appears to be a very sick man. He has all the symptoms of hardening of the arteries of the brain in an advanced stage." He noted too how the president "sat looking straight ahead with his mouth open, as if he were not taking things in." Moran doubted "whether he is fit for his job here."[2]

The president had always managed, like a good stage magician, to control and direct events without appearing to do so, aided by his shrewdness, his amazing capacity for retaining information, his restless energy, his instinctive judgment about people, and his lively curiosity. Unlike Churchill, he had always been willing to delegate enormous authority, as he did to Marshall, but he still made sure that the policy of those who served him was his own policy, even down to the details. But by 1944 his energy had begun to fail him. Roosevelt, whose most noticeable and attractive quality had always been his good humor, became increasingly irritable; he was impatient with details; his attention span was dramatically curtailed; and his mind often seemed to wander. Harry Truman, his running mate in the election of 1944, noticed his decline and was shocked by it. But Truman had not been in the president's inner circle as a senator, nor would he be as vice president—so he saw Roosevelt's state of health more clearly than those around him did.

The effect of this carefully concealed and increasing vacuum at the heart of the American government was to place more responsibility for strategic and political decisions on the supreme commander, for it was in Europe that the most difficult decisions had to be made, and it was in Europe that the vital interests of the three principal allies differed most widely and were most likely to collide.* This tendency was increased by the simple fact of George Marshall's ascendancy over all but the most intimate of Roosevelt's advisers. Marshall did not radiate or reciprocate warmth, or indulge in gossip and small talk, all of which the president enjoyed. But on the other hand he did not try Roosevelt's patience, and he did not have his own agenda like the president's neighbor in Hyde Park, Henry Morgenthau, the secretary of the treasury; or his own constituency like Cordell Hull, the secretary of state. If Marshall had his own opinions,

* Britain's interest in the Pacific was limited—its attention was mainly focused farther east, on the restitution of its Asian colonies (i.e., Malaya, Burma, and Hong Kong) and the preservation of India's colonial status, to which the president was opposed, but in which he had reluctantly agreed not to interfere. He was more determined not to let France reimpose a colonial regime in Indochina and not to let Holland do so in the Dutch East Indies (soon to become Indonesia). De Gaulle had made it clear that he did not consider France's colonial empire any of the president's business. Since the Soviet Union was not at war with Japan, Stalin's views on the subject of the Pacific were for the moment irrelevant.

he kept them to himself, except on matters directly affecting Army administration and promotions. Marshall's integrity and common sense were undeniable—indeed the only people who criticized him at all (and then mostly in their diaries) were the British generals, who felt that he was neither a strategician nor a fighting general. ("Marshall clearly understood nothing of strategy,"[3] Brooke wrote dismissively about a meeting of the Combined Chiefs of Staff in Malta, in 1945.)

The president's distrust of the State Department, and his tendency to act as his own secretary of state when he felt like it, as at Tehran and Yalta, meant that responsibility for making important decisions about Europe was shifted to Marshall and to Ike in the absence of a strong secretary of state, and decisions were made on the basis of "military necessity." Thus Ike had effectively negated American policy on the subject of France in 1944 by agreeing to De Gaulle's objections before D-day about American plans for treating France as if it were being "occupied" rather than "liberated" by Allied forces. He allowed De Gaulle to land in France as soon as possible after D-day, and changed his strategy—at De Gaulle's demand—to liberate Paris in August 1944 (and to keep American troops out of the capital until Leclerc's French troops had taken the city and De Gaulle had made his entry there). The president, when he met with Ike for the last time, had expressed his firm determination that the United States would have as its zone of occupation the "northwest" of Germany so that American troops would be close to the big German ports and could be withdrawn no later than two years after the end of the war, which was the maximum length of time he wanted to keep an American military presence in Europe. But Ike said that he would prefer a "joint" occupation of Germany, rather than one divided into zones—a suggestion that would have prevented a divided Germany and the iron curtain (or at least put it more than 100 miles farther east). Ike lost that argument. Nevertheless, he later pleaded the "military necessity" of giving northwestern Germany to the British (with an American enclave around the port of Bremen), since their army was attacking in that direction, while the U.S. Army was advancing in the Rhineland and Bavaria; and he argued forcefully that it would be impossible for the armies to exchange zones of occupation without hopelessly snarling their lines of communication. Largely thanks to Ike, the British thus got the occupation zone that the president had wanted. On two major issues, therefore, Ike simply made American foreign policy based on his perception of military needs rather than on ideology, long-term American interests, or the president's wishes. In both these cases his decisions were, in fact, eminently sensible, but they were not necessarily those the president or the State Department would have made, and in the case of France his decision was exactly the opposite of what the president wanted.

When Monty had written his letter of apology to Ike, he remarked in it, "I am sure there are many factors which may have a bearing quite beyond anything I realize."[4] These words may have been, and almost certainly were, less than sincere, but they were truer than Monty supposed. Ike was beginning to act with imperial authority in matters far removed from the strictly military, and would indeed very shortly cause (and win) a serious row among the Allies by making direct contact with Stalin, as if he were one head of state negotiating with another. The British generals and, more significantly, the British government found this hard to understand, since even the most important of British commanders (Monty in northwest Europe, Alexander in Italy, "Jumbo" Wilson in the Middle East, Mountbatten in the Far East) were watchfully controlled by the British chiefs of staff, and above all by the prime minister—who was also the minister of defense and who interfered at will, and at length, in even the smallest of decisions, and reviewed every message and cable in detail (prompting Brooke to confide to his diary, "I don't feel that I can stand another day working with Winston").[5] This was, to begin with, not the American way. "American doctrine," Ike observed, "has always been to assign a theater commander a mission, to provide him with a definite amount of force, and then to interfere as little as possible in the execution of his plans."* Ike kept Marshall informed of his plans with daily "combined situation and intelligence

* It is worth noting that the increasing rapidity of communication has eroded this doctrine very considerably. Ike and MacArthur were perhaps the last American generals to enjoy that kind of trust and independence, and MacArthur was eventually fired by President Truman for exceeding his authority. In Vietnam both Johnson and Nixon tried to micromanage the war from the White House, with dire results, and in both wars in Iraq the commander in the field has been held on a very tight leash by the secretary of defense.

reports," called cosintreps, but these were brief and strictly factual, and his tactics and decisions were not dictated by Washington—it was up to him to allocate his resources and fight his battles as he saw fit. In the American way, if there was dissatisfaction with the results, you changed the commander; you did not tell him what to do or how to do it. This was a considerable gap between the Allies, and it caused endless misunderstandings.

In January 1945 these centered on Ike's final plan for conquering Germany. The plan was vigorously opposed by the CIGS, Brooke, whose view also represented that of Churchill and Monty. Brooke objected to Ike's "planned dispersion of forces"; this objection was a different way of once again urging on Ike a single "full-blooded" attack across the Rhine in the north, as well as the establishment of "an over-all 'ground commander,'" under Ike, to fight the battle. That was unacceptable to Ike, who "laboriously" explained to Brooke why he was wrong, without changing the CIGS's mind.

A further slight irritant was the fact that Ike had sent his deputy, Air Chief Marshal Tedder, to Moscow to exchange plans with Stalin and Stalin's military advisers. Tedder was received, perhaps not surprisingly, "with the utmost cordiality," and was given Stalin's assurance that the Soviet Union would launch a major attack on January 15 and—whatever the results—would "keep up a series of operations" to prevent the Germans from moving reinforcements from east to west. Although this approach to the Kremlin had been authorized by the Combined Chiefs of Staff, there seems to have been some apprehension that Tedder may have gone farther in his discussions with the Russians than the British expected him to—perhaps a natural consequence of Brooke's and Monty's distrust of Tedder. Since Tedder brought back from Moscow what seemed like good news, the Combined Chiefs of Staff further authorized Ike "to communicate directly with Moscow on matters that were exclusively military in character," a blank check which the British were soon to regret.[6]

"Rumblings" of British discontent reached Washington, and as a result General Marshall, who was to accompany the president to Yalta, broke his journey to meet with Ike at Marseille on January 25. Marshall was in full agreement with Ike, and urged him to send Bedell Smith to Malta to explain Ike's plans to President Roosevelt, Churchill, and their staffs at the military conference they were holding there before traveling on to Yalta. Smith did so, and he was backed by Marshall and the presence of the president, so the British eventually fell into line, although they were still not convinced. Indeed, Brooke raised so many objections to Ike's plan that Marshall was provoked into saying that he was "opposed to cramping Eisenhower's style" and expressing "his dislike and antipathy for Montgomery,"[7] as Brooke noted in his dairy.

In *Crusade in Europe* Ike later wrote that when the Allies finally reached the Rhine in March 1945, Brooke told him, "Thank God, Ike, you stuck by your plan. You were completely right."[8] This does not sound at all like Brooke, and it may be a rare example of Ike's painting a rosier picture than the truth. In fact, Brooke vigorously opposed Ike's plan at Malta, continued to do so, and much later, in his own book, claimed that Ike had "misquoted" him. "I never said to him, 'You were completely right,'" Brooke wrote, "as I am still convinced that he was 'completely wrong.'"[9] He went on to add that Ike's plan had retarded the defeat of Germany.

Thus, at the end of January 1945, the agreement between the two western allies was wafer-thin—they had really done no more than paper over their differences before the crucial attack. Ike's plan called for a four-pronged attack to destroy German resistance west of the Rhine, and to weaken the German army in the west so that once the Rhine was crossed the Ruhr could be cut off—though Ike attached rather less importance to this than the British did, since so much of Germany's war industry had already been dispersed. Partly in order to still British objections to his plan, the U.S. Ninth Army was attached to the British Second and Canadian First Armies (Twenty-First Army Group). This gave Monty a formidable force—though not nearly as formidable as he wanted—for the most northern attack, toward Wesel, Essen, and Düsseldorf, to be launched on February 8. That attack would be followed two weeks later by an attack on the German center, from Cologne to Koblenz, by the U.S. First Army and U.S. Third Army (U.S. Twelfth Army Group), and by two attacks in the south by the U.S. Sixth Army Group, with the U.S. Seventh Army attacking toward Kaiserslautern and Karlsruhe, and the French First Army toward Colmar, Freiburg, and Mulhouse. Brooke seems to have supposed that the attack on the center would be easier than the attack in the north, and that Ike's plan therefore favored the American armies—though there seems no obvious reason why that should be so—prompting Ike to tell him sharply, "I must tell you in my opinion there is no glory in battle worth the blood it costs."[10] These were harsh words, coming from Ike.

The plan had the infinite advantage of simplicity—like Grant's in 1864, it set into motion all the Allied armies to engage the enemy at every point, on the theory that superior numbers would either break the Germans or wear them down. It was also a plan that allowed Army commanders the freedom to improvise and to exploit opportunities as they arose. Monty was given plenty of manpower—Ike acknowledged the importance of the northernmost attack—but if Bradley or Devers hit the ground running and outpaced him, that was all right too. Monty and Brooke wanted the British to be assigned the role of crossing the Rhine first; but Ike, true to form as an old football coach, was willing to leave whoever was carrying the ball to score the point—the crossing of the Rhine would be made by whichever Army could do it first. Naturally competitive himself, Ike was a firm believer in the virtues of competition on the battlefield. In any case, just as Grant had aimed to destroy the Army of Northern Virginia rather than to take Richmond, Ike's intention was to destroy the German forces west of the Rhine rather than just to cross the river—Ike always resisted the idea of wasting the lives of his soldiers just to make headlines or capture places on the map. He had not been interested in taking Paris until De Gaulle made it clear to him that this was a military necessity because France might otherwise descend into political chaos or even civil war; and he was, if anything, even less interested in taking Berlin, or in whether Monty or Bradley would be first across the Rhine.

Ike took advantage of the pause before the big attack to erase the "Colmar pocket" in the south, which was taken on February 3 by two French corps and one American corps, inflicting 22,000 casualties on the enemy. Then, on February 8, the Canadians attacked in the north—Monty was held back from attacking with all the forces at his disposal by delays of his own making and others, like the flood tide of the Roer (Rur) River, which were not his fault. The Canadian forces soon ran into difficulty. In front of them was "a quagmire of flooded and muddy ground," and German resistance was stiff, and quickly became stiffer as the Germans began moving forces from in front of the U.S. Ninth Army to stop the Canadian advance.

With his usual infuriating insistence that what looked to lesser mortals like a setback was all part of his carefully worked-out plan, Monty claimed that his intention from the beginning had been to make the Germans weaken their forces in front of the American attack by shifting troops northward to resist the Canadians; but if this was the case, none of the American commanders seemed aware of it. In any event, on February 23 the U.S. Ninth Army, still attached on Ike's orders to Monty's Twenty-First Army Group, launched its attack and took München-Gladbach, the largest German city captured so far, six days later. Ike was there to enter the city in a Jeep, and as American antiaircraft guns opened up against a German

A rare photo of Ike wearing a helmet. *(Left to right)*: Alexander, Ike, Patton.

jet fighter showering him with red-hot shell fragments, he put on his steel helmet for one of the few times during the war.

Farther south, in the German center, Bradley launched a series of attacks that brought the U.S. First Army to the outskirts of Cologne by March 5, and into Remagen on March 7. At Remagen, in an astonishing and almost miraculous stroke of good luck, they found the Ludendorff Bridge over the Rhine damaged but still standing and usable. The German army engineers who were supposed to demolish the bridge had delayed blowing it up, to give retreating German army units time to cross, but U.S. First Army troops arrived so quickly that in the confusion the German engineers succeeded only in damaging the center span. Bradley, concerned that if he sent too big a force across the Rhine it would undo Ike's plan, called Ike at SHAEF headquarters in Reims, and asked what to do. Ike demonstrated once again that, for a commander, military genius lies in the ability to change his mind, forget about his plan, and seize an opportunity at once when it is offered to him. "Go ahead and shove five divisions over,"[11] he told Bradley, and by March 9 the Remagen bridgehead was three miles deep. The Germans made the mistake of attacking the bridgehead piecemeal, as units managed to arrive on the scene, instead of waiting to concentrate them for a big attack, and by March 9 there were enough American troops on the eastern bank of the Rhine to withstand any attack the Germans could make against them. Using long-range artillery and bombing attacks by the Luftwaffe the Germans managed to damage the Ludendorff Bridge, and eventually succeeded in dropping the center span into the Rhine (along with the unfortunate American combat engineers who were frantically trying to repair it under fire); but by that time Ike had put two Treadway floating bridges across the Rhine, each of them 330 yards long, and he was confident that the troops in the bridgehead could be supported.

On March 19, persuaded by General Marshall and by Bedell Smith that he was in desperate need of a rest, Ike finally gave in and agreed to take a few days leave at a villa called Sous le Vent, lent to him by a wealthy American. He commented to Mamie merely that he was taking "a five day trip," which was true enough; but he did not add that Kay Summersby was joining him, or that they were going to a luxurious seaside villa near Cannes. To be fair to Ike, he and Kay were accompanied by General Bradley and two American WACs, among others, so Sous le Vent was hardly a private lovers' nest. Still, Kay remarks that the two of them "would sit on the terrace all day long, looking out over the Mediterranean, chatting lazily, drinking white wine and sunbathing,"[12] while the others went to Monte Carlo, and that Ike's health improved rapidly. In one snapshot taken at the villa, Kay sits by the sea on a flagstone terrace carved out of the rock, next to Ike, seated in a deck chair, who is sipping an aperitif, with one of the two American WAC officers on his left, looking even more beautiful and glamorous than Kay. They are all in bathing suits and appear to be having a good time. Of course there is nothing wrong with any of this. Ike's health was poor; he was in constant pain from his knee injury; he was exhausted from nonstop stress, responsibility, and hard work; and nobody could begrudge him a few days' leave—and it is certainly possible

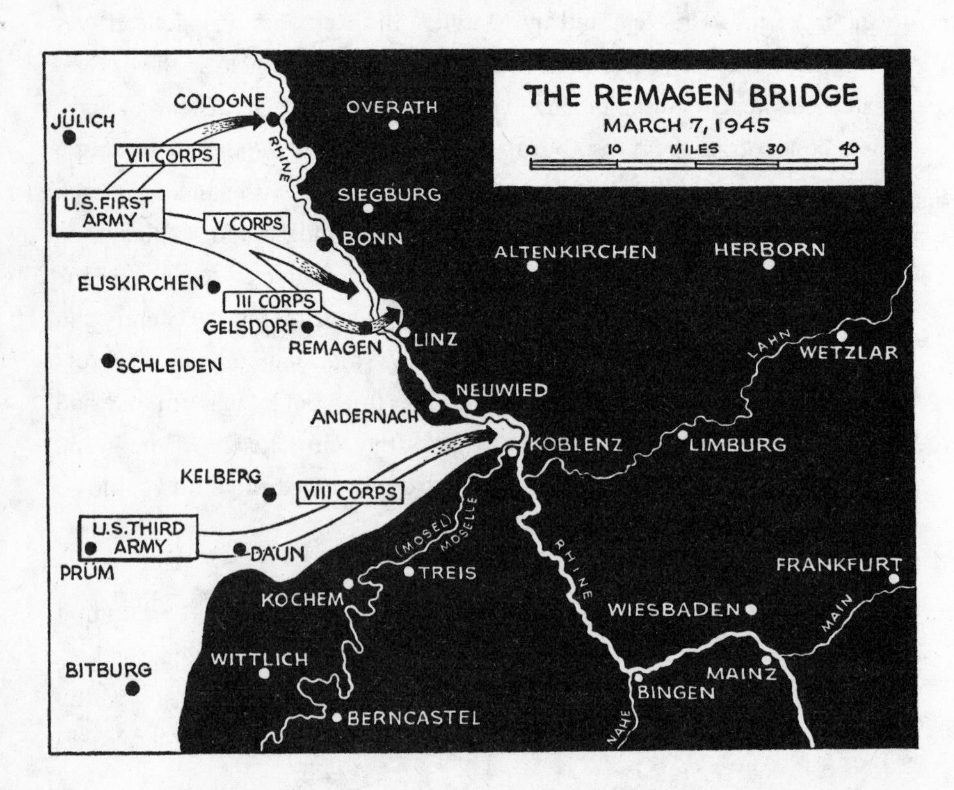

that Kay exaggerated the intimacy of this short holiday. But all the same it was the kind of thing that was bound to cause gossip and rumors, and that in the long run could not be kept secret, though many people objected to Bradley's presence rather than Kay's, believing that Ike was showing some favoritism toward one of his senior commanders. It was not Kay whispering sweet nothings into Ike's ear on the Riviera, if that had been the case, that the British generals feared, so much as Bradley using the occasion to persuade his old West Point classmate to let him, rather than Monty, make the final, major assault against Germany.

The success of Ike's plan so far did not, of course, stop Brooke from carping: "There is no doubt that he [Ike] is a most attractive personality and at the same time a very very limited brain from a strategic point of view,"[13] he wrote the day the First Army crossed the Rhine. But by now this was decidedly academic. Monty's Twenty-First Army Group crossed the Rhine on March 24—the prime minister, attended by Brooke, insisted on being there to see it happen in person and had a picnic luncheon on the west bank of the Rhine with Monty and Brooke—and by March 25 all German resistance west of the Rhine had effectively ceased. Ike watched from a church tower with General Simpson on the night of March 23 as more than 2,000 guns opened fire on the Germans across the river. At dawn he watched 1,572 transport aircraft towing 1,326 gliders drop two divisions of airborne troops on the east side of the Rhine while a fleet of naval landing craft, moved from the seaports by waterways and road, carried fifteen divisions across. Later the same day, Ike discovered to his horror that Churchill had managed to get an LCM to ferry him over to the east bank, despite German sniper fire and artillery. "Had I been present," Ike wrote, "he would never have been permitted to cross the Rhine that day."[14]

The Rhine, which only a few weeks earlier had seemed like such a formidable military obstacle, was now in the Allies' hands, and the Allied armies were capturing more than 10,000 German prisoners a day. Ike had three corps across the Rhine at Remagen; and Patton's Third Army was across the Rhine too, and about to move deep into southern Germany, taking full advantage of Hitler's autobahn network, while more than 9,000 aircraft struck at the German transportation system, inflicting almost unimaginable damage which Speer no longer had any hope of repairing.

The war, Speer told the Führer at last, was lost, thus setting in motion the first act of the tawdry melodrama of accusations, betrayal, and revenge in the *Führerbunker* that would end with Hitler's suicide. "The German people was not worthy of my great ideas," Speer quoted him as saying, "[it] was too weak to face the test of history, and was fit only for destruction." Faced with Hitler's order to destroy everything of value that remained standing in Germany—factories, bridges, railroads, mines, the entire infrastructure of the German nation—rather than to let it fall into the Allies' hands, Speer, for once, had delayed, prevaricated, and even lied to Hitler. He had even, according to his own account, toyed with the idea of assassinating his mentor and friend by dropping poison gas into the ventilator shaft of the bunker, but then he discovered that Hitler, alert as ever to any threat, had apparently foreseen this possibility and ordered the ventilation system modified to prevent just such an attempt on his life. So Speer, unnerved by what he had contemplated doing, drifted through Germany attempting to prevent Hitler's orders of destruction from being carried out, and trying to keep up the flow of equipment and ammunition to what remained of the German army.

In fact, the Germans were still fighting with undiminished skill and ferocity, but by now they were very close to collapse. The Führer had played a final, desperate card by replacing Field Marshal von Rundstedt (for the last time) with Field Marshal Kesselring as OB West, though even the magic of Kesselring's name and reputation could not halt or disguise the epic scale of Hitler's defeat in the west, or summon up reserves of fresh troops when there were none. In the meantime, in Kesselring's absence, the German army in Italy was already attempting to surrender independently to the western Allies. SS Obergruppenführer und General der Waffen SS Karl Wolff, Himmler's longtime confidant and trusted chief of staff, was negotiating directly with Allen Dulles of the OSS in Bern, Switzerland. When news of this leaked out, it set off alarm bells in the Kremlin and produced an angry and insulting stream of messages, first from Molotov and then from Stalin himself, to Churchill and Roosevelt, accusing the western Allies of trying to arrange a separate peace in the west while the Germans continued to fight in the east.* Churchill had shown some of these to Ike as they watched Monty's army crossing the Rhine, and they had deeply shocked and angered Ike. The Russians did not of course trust Wolff, who was one of the more sinister and politically flexible figures in Himmler's inner circle, and was deeply implicated in the "final solution." Worse still, they made it evident they did not trust their Allies either, and they said so in brutal terms. Ike told Churchill that he felt himself entitled to accept the unconditional surrender of any group of enemy troops from a company to an army "without asking anybody's opinion."[15] He would insist that all German troops under the officer surrendering must lay down their arms immediately and remain in place, and would not hesitate to advance through them toward the east. Churchill agreed, and remarked that he did not see "why we should break our hearts if, owing to a mass surrender in the West, we got to the Elbe, or even farther, before Stalin."[16]

Though Ike receives little credit for it, his strategy had in fact worked, despite the criticism of Brooke and Monty, or Patton's complaints that he was being deprived of fuel supplies which were going to Monty instead. Ike had not panicked when Rundstedt attacked in the Ardennes; he had massed his forces in the aftermath of the Germans' failure, then attacked across a broad front just as he had always intended to do, keeping up the pressure on the German army until it ceased to exist as a fighting force west of the Rhine. Now, less than six weeks after he had begun his offensive, he was across the Rhine in force nearly two months earlier than anybody had anticipated. What is more, he was about to give the Germans a final, knockout blow of his own devising, quite different from the strategy Monty wanted to pursue, but with the same end in mind.

Ike would use his bridgehead at Remagen not to attack to the east or the

* At the time there was speculation that Wolff was part of a Nazi plot to divide the western Allies from the Soviet Union, but it seems more likely that he was only behaving with common sense unusual in the Nazi Party in the face of defeat, and also trying to build up enough credit with the Allies so as not to be tried as a war criminal. In this he was successful. Both Dulles and Field Marshal Alexander stood up for him, and he was the only senior German officer allowed, as a special privilege, to wear his decorations during his brief imprisonment. He went on to land on his feet as a successful corporate public relations executive in postwar Germany.

north, but to pivot sharply to the right and advance to the southeast, taking both the Germans and the British by surprise, so that the First Army could join up with Patton's Third Army near Frankfurt-am-Main, on the east side of the Rhine. Then the two American armies would sweep toward the northeast in one powerful, concentrated blow to join up with Monty's Twenty-First Army Group in the north, cutting the Ruhr off completely, surrounding what remained of Model's Army Group B—the last major German fighting force in the west—and bringing the combined British and American armies within striking distance of the Elbe River. By now Ike had nothing to fear from attacks on his flanks—the Germans were running out of fuel, aircraft, and manpower, and many of the units Hitler was moving about on the map were ludicrously under strength: divisions with the strength of a regiment; regiments with the strength of a battalion; battalions reduced to one or two companies, and manned by drafts of old men from the Volksturm (the German equivalent of the British Home Guard) or boys fifteen or younger from the Hitler Youth, the former apathetic, the latter fueled by fanaticism, but both so poorly trained and armed as to constitute a negligible fighting force. Even so, Ike's plan was for an operation of astonishing daring, a huge curving left hook that would bring all meaningful German resistance to an end and destroy what was left of the Wehrmacht with one blow. Despite the chorus of complaints from Monty and from Brooke about his not being a strategist, Ike's strategy was simple, well thought out, and well suited both to the U.S. Army's strengths and to America's political goals—he had defended it stubbornly against attempts by the British to make him change it, and it was about to prove victorious.

By now Patton's Third Army was advancing more than thirty miles a day, and Hodges's First Army, to his north, scarcely less; the two armies, side by side, were sweeping east, then pivoting north in the "left hook" Ike had envisaged. On the afternoon of April 1, forward elements of the U.S. Ninth Army moving south made contact with U.S. First Army units moving north; and by April 4 the Ruhr was effectively surrounded, trapping Field Marshal Model and whatever remained of Army Group B—more than 300,000 men—in "a pocket of some 4,000 square miles," containing the bulk of Germany's heavy industry. Ordered by Hitler not to withdraw or retreat, and to destroy all the Ruhr's industrial installations, Model was for once unable or unwilling to obey the Führer. He could not break out of his encirclement and, like Speer, he refused to destroy the industrial heart of Germany. For almost three weeks he fought off the surrounding American forces in a steadily diminishing pocket, then calmly disbanded his army, burned his personal papers, and shot himself, aware that he would almost certainly be tried and executed for war crimes committed on the eastern front.

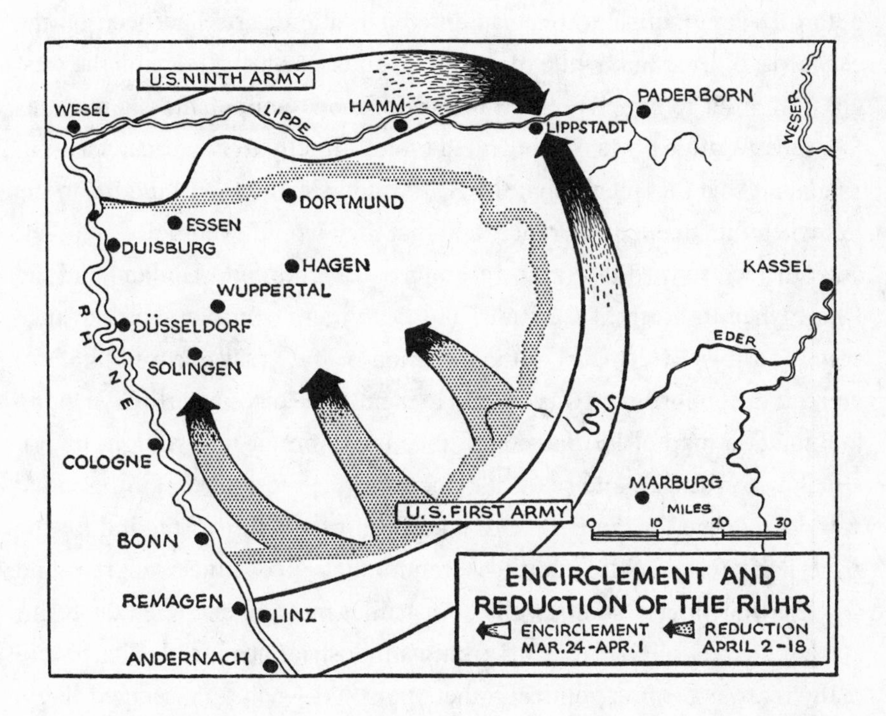

Thus by the beginning of April Ike's strategy was working just as he had intended. This is an important point to bear in mind, for once the cold war had begun in earnest and Ike entered politics as a potential candidate for the Republican presidential nomination, he came under increasing criticism from the right wing of his own party for not having taken Berlin, and thereby not having prevented the division of Germany between east and west or the isolation of Berlin deep within the Soviet zone of occupation.

Winston Churchill too criticized Ike for this, devoting no less than eight chapters of the last of the six volumes of *The Second World War* to this subject—indeed in the title of Volume 6, *Triumph and Tragedy,* the "tragedy" is above all Ike's failure—or refusal—to sanction an Anglo-American attempt to capture Berlin before the Red Army did. Ike had not been impressed by these arguments in 1945, and was still less impressed—and considerably more impatient—with them when he sat down to write his own war memoir, *Crusade in Europe*.

Berlin had been a source of friction between Ike and the British since before D-day. British and Canadian embarkation ports had been on the east coast of England, while those of the American army were on the west coast in order to simplify the transport of men, equipment, and supplies from the Atlantic ports of the United States directly to American bases in England. This fact meant that the British and Canadians would be on the left when they landed in France, and that they would eventually pivot left and advance toward the northwest, across Belgium and Holland and the Rhine, then to north of the Ruhr, and from there continue their advance across northern Germany in the direction of the great port of Hamburg and—if the opportunity was offered to them—the city of Berlin. Anybody looking at a map of Europe could trace this more or less straight line of attack with a finger, and see at a glance that the most natural thing in the world would be for the British to take Berlin from the north, and for the Americans to take Munich, then Prague, then pivot ninety degrees and take Berlin from the south, while both armies met and encircled the Ruhr, cutting the rest of Germany off from its industrial heartland. There were many reasons—some political, others personal—why this seemed like a good idea to the British. As the number of British troops declined in comparison with those of America and the Soviet Union, it seemed to many, especially the prime minister, that Britain's prestige and its place in the postwar world demanded that it take the German capital—this was elementary politics, and would have the additional advantage of keeping the Soviet Union as far to the east as possible, something which was already on Churchill's mind. As he said, he felt it was of the utmost importance that "we should shake hands with the Russians as far to the east as possible."

Then too, just as Monty wanted to cap his career with a victory parade in Berlin, Churchill wanted to appear on the reviewing stand as British troops marched by down Unter den Linden, celebrating the defeat of Nazi Germany after five years of fighting, two of them before the United States or the Soviet Union had entered the war. From the British point of view, there was thus every reason for the western allies to take Berlin.

From the American point of view, however, none of this held true. The occupation zones of Germany had already been agreed on between the Allies. Indeed, a copy of the map showing how Germany was to be split up fell into German hands during the Battle of the Bulge and was used by Hitler to show his generals just what was in store for them if they allowed themselves to be defeated, and Colonel General Jodl, Chief of the Operations Staff of the OKW, even showed it to his bride. So even if British or American troops advanced beyond these lines, they would only have to turn around and go back again once the war ended. Nobody seriously imagined that Stalin would fail to claim every square mile of territory allocated to him, whether it had been seized by his troops or not. As for the isolation of Berlin deep within the Soviet zone of occupation, it might be a mistake, but it had already been conceded.* The real misfortune was that the Russians had not been asked by their allies to give them an ironclad guarantee of free access to Berlin, but nobody at Yalta seems to have imagined that the Soviet Union might one day attempt to cut the city off from the west altogether. So far as Ike was concerned, no military purpose would be served by attempting to take Berlin, and he was not about to waste the life of a single American (or British) soldier to occupy territory that would only have to be handed over to the Russians the moment the war ended. He felt strongly about this in 1944 and 1945; he still felt strongly about it three years later when he wrote *Crusade in Europe;* and he continued to feel strongly about it as president.

* In 1944 President Roosevelt had expressed the strong desire to have Berlin occupied by the United States, but by the time he got to Yalta he either had changed his mind or was simply too exhausted and ill to bring the matter up again. In any case, the occupation zones as they were drawn placed the city in the Soviet zone, and it would have required the equivalent of the prewar Polish Corridor to link it with the American or the British zone—not a happy historical precedent, since the Polish Corridor had been a casus belli of World War II.

An unusual note of asperity creeps into Ike's narrative as he deals with this issue. "The future division of Germany did not influence our military plans for the final conquest of the country," he writes. "Military plans, I believed, should be devised with the single aim of speeding victory. . . . I decided, however that [Berlin] was not the logical or the most desirable objective for the forces of the Western Allies."[17]

Despite Churchill's admiration for Ike, on the subject of Berlin his tone becomes apocalyptic. Writing in 1953, five years after Ike's book was published, Churchill sounds as though he had already foreseen the cold war in 1944, but this was not, in fact, the case—at the time, the prime minister, though distressed by Stalin's obduracy, suspicion, and bullying, still held out hopes for cooperation between the two western powers and the Soviet Union. Churchill's postwar obsession with the capture of Berlin in 1945 was an expression of his desire to link the two periods by showing that he had been just as right about Stalin at the end of the war as he had been about

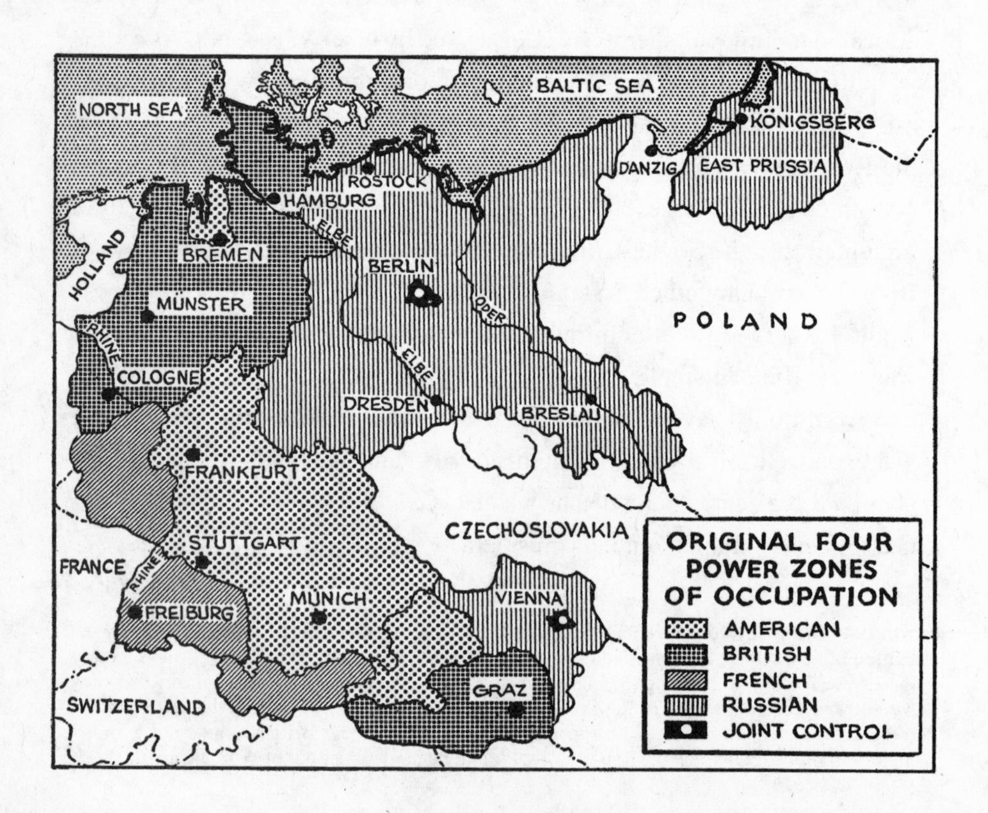

Hitler in the 1930s. This had the great literary advantage of giving his six-volume history of the war both a unifying theme and a satisfying ending that nicely reflected the beginning (*The Gathering Storm*), and it was also good politics at the height of the cold war; but for all that, it was an example of being wise after the event.

Churchill conceded that the United States had no strategic interest in Berlin, but he felt that it ought to have had some respect for British concerns and interests. He remarked, with his usual gift for the telling phrase, that "when wolves are about the shepherd must guard his flock, even if he does not himself care for mutton." This is true enough, but he made the leap from there to the claim that "Soviet Russia had become a mortal danger to the free world," which may have been true in 1953 but was hardly accepted as fact in 1944 and 1945. Neither British nor American public opinion was prepared to treat the Soviet Union as a "mortal danger" while the war against Germany was still being fought, nor was that the view of President Roosevelt or even Churchill himself at the time. Naive as it may now appear—and apparent as it was to everyone but starry-eyed leftists and fellow travelers in 1953—there was still a universal hope that victory over Germany would bring the western Allies and the Soviet Union closer together, not separate them into warring political blocs.

On March 30 the prime minister, who had inquired about the "strategy for the advance of the Anglo-American armies," received a detailed reply from Ike to say that his "main thrust" would be along the axis Kassel-Leipzig, and that he did not intend to cross the Elbe. The advance would be made by the U.S. Third, First, and Ninth Armies, under the command of Bradley, with an additional American army (the Fifteenth) at his command for "mopping up" behind the advance. Monty's task was to support Bradley's left flank, in the direction Hanover-Wittenberg; and the Sixth Army Group would stand by to advance toward Nuremberg and Regensburg, in south Germany, to deal with the possibility of a defiant last stand in a supposed "Alpine Redoubt" (also called the "National Redoubt") in the mountains of Bavaria. It was mistakenly anticipated by the Allies (and by many of Hitler's subordinates) that the Führer would choose to stage the last act of his own Götterdämmerung high in the mountains which had been the birthplace of Nazism, the home of its leader, and the source of many of the myths and legends sustaining the movement in good times and bad.

Ike's letter pleased nobody in London, where the notion of Monty and the Twenty-First Army Group being relegated to the role of spear-carriers as Bradley advanced into central Germany, rather than carrying out a full-blooded British-Canadian attack to reach Berlin, disappointed Churchill and infuriated the British generals. This reaction was compounded by the news that Ike had sent a telegram directly to Stalin to coordinate his attack with that of the Red Army. "He has no business to address Stalin direct," Brooke complained in his diary; "his communications should be through the Combined Chiefs of Staff, secondly he produced a telegram which is unintelligible, and finally what was implied in it appeared to be entirely adrift." From his TAC headquarters Monty fairly seethed at being deprived of the role he had anticipated, in favor of Bradley, and at the news that Ike was not only depriving him at last of the U.S. Ninth Army and giving it to Bradley, but apparently informing Stalin of his plans before communicating them to his senior commanders.

The British chiefs of staff were sufficiently upset by all this to draft and send a long message of complaint to Washington without the prime minister's having seen it first—a very unusual occurrence. It brought down on them an even longer message of reproof and corrections from 10 Downing Street, emphasizing Churchill's disappointment: "All prospect . . . of the British entering Berlin with the Americans is ruled out," he wrote—never mind that Ike had already made it clear that the Americans were not going there any more than the British.

The U.S. chiefs of staff replied quickly, making it clear that in their view Ike's communication with Stalin was well within his authority, and rejecting the British criticism of his strategy. Churchill responded to this with two very long messages on March 31: one to Ike, once again urging that Monty should retain the U.S. Ninth Army and advance to the Elbe and beyond it to Berlin; and the other, even longer, to the president, once again emphasizing that "Berlin remains of high strategic importance." Ike replied at length to the prime minister, politely but firmly sticking to his guns; and the prime minister finally capitulated on April 5 in a message to

Roosevelt, though he began it with the important qualification, "I still think it was a pity that Eisenhower's telegram was sent to Stalin."[18] By this time, the president's mind was perhaps not as firmly fixed on this dispute as the prime minister's—he was only five days away from his death of a cerebral hemorrhage at Warm Springs.

The war would be fought to its end as Ike had planned it. Neither he, Marshall, nor the president had any interest in attempting to reach Berlin. As he explained to Marshall, in a long message recapitulating the whole dispute (or, as Churchill described it, gracefully downplaying its importance now that he had lost, a "lovers' quarrel") between himself and the British, "Berlin is no longer a particularly important objective. . . . It's usefulness to the Germans has largely been destroyed and even his [Hitler's] government is preparing to move to another area."[19] Ike devoted an unusual amount of space to the subject in *Crusade in Europe*, demonstrating that he had not changed his mind about Berlin despite the cold war. His normally equable tone is disrupted only once, when, in his message to Marshall, Ike makes a brief but pointed response to three years of patiently—and by every account resentfully—listening to Field Marshal Alan Brooke lecture him tartly on the importance of concentrating his forces instead of dispersing them. "Merely following the principle that Field Marshal Brooke has always emphasized," he writes blandly, "I am determined to concentrate on one major thrust."[20] Marshall no doubt chuckled at that. This, together with Ike's lack of strategic skill, was Brooke's theme song, and now, when the war was almost won, what was Brooke demanding but the one thing he had always criticized Ike for—dispersion of the Allied forces by continuing to attack in the south and the center, while transferring the U.S. Ninth Army back to the Twenty-First Army Group and giving Monty the green light (and the lion's share of supplies) to make a full-scale assault on Berlin?

By April 12 the U.S. Third Army was near the prewar Czechoslovakian border and the U.S. First Army was close to the Elbe. Ike drove forward to meet with George Patton (to whom Ike had given specific and repeated orders *not* to take Prague before the Russians did), and saw sporadic but stubborn fighting going on throughout the day. He and Patton then visited a salt mine half a mile deep, discovered by American soldiers, in which the Germans had concealed a "hoard of gold" estimated to be worth about $250 millon,* much of it in the form of gold bars melted down from objects and personal belongings stolen by the Nazis throughout occupied Europe, mostly from Jews; and an amazing treasure trove of master paintings, mostly stolen from Jewish collectors. Later in the day, Ike visited his first concentration camp, Ohrdruf-Nord, a subcamp of Buchenwald, near the town of Gotha. He remarked afterward that he had "never at any other time experienced an equal sense of shock." Over 4,000 prisoners had just been murdered there by the SS guards before the guards abandoned the camp and fled at the approach of the Americans, and the bodies still lay scattered in grotesque heaps. The surviving prisoners were emaciated skeletons. Ike was shown the whipping block, the gallows, the ditches crammed with half-naked, decomposing bodies. There were no gas chambers or crematoriums—Ohrdruf-Nord was not a "death camp" as such, or even one of the major concentration camps. It was merely one of thousands of similar "forced labor" camps large and small in the vast SS concentration camp empire, but it was enough of a taste of the horrors of Nazi Germany to make Ike send to London and Washington and urge that "a random group of newspaper editors" should be sent to Germany instantly, so that evidence of Nazi atrocities could be "placed before the British and American publics in a fashion that would leave no room for cynical doubt." This was an eminently sensible step, which Ike seems to have been the first to think of, but which was perhaps natural to a man who in 1938 had been offered the job of resettling Jewish refugees from Nazi Germany in Asia as he was preparing to leave the Philippines and return home.

That evening he sat up late talking with Patton and Bradley, still stunned by what he had seen. As they were about to go to bed, Patton realized that his watch had stopped, turned the radio on to the BBC to get the time, and heard the news of President Roosevelt's death. "With some of

* This estimate is in terms of 1944 values. Since the price of gold has been multiplied by fifteen since then, a conservative estimate in present-day terms, factoring in inflation, would be at least $4 billion, and perhaps as much as twice that.

Mr. Roosevelt's political acts I could never possibly agree," Ike noted in *Crusade in Europe,* though unfortunately he did not go on to specify what these had been, but he added, "I knew him solely in his capacity as leader of a nation at war—and in that capacity he seemed to me to fulfill all that could possibly be expected of him."[21] This is something less than lavish praise for the late president; up to this point, the reader has been given to understand that Ike admired Roosevelt without reservations, and it is hard not to wonder if it was written with the Republican presidential nomination already in mind.

Less than 120 miles away to the northeast, the news of President Roosevelt's death had a more dramatic effect. Reichsminister of Propaganda Dr. Joseph Goebbels had returned to Berlin from a trip to visit General Busse's headquarters at Kuestrin. There, talking to Busse's officers (to give them what we would call a pep talk), he had reminded them that at the very ebb of Frederick the Great's life, when he was reeling from defeat after defeat in the Seven Years' War, and even contemplating suicide, he had received the news that his enemy the czarina had died—and by this miracle the House of Brandenburg was saved. This story was much on Goebbels's mind because he had been reading Carlyle's hero-worshipping biography of Frederick the Great aloud in the evenings to Hitler in the bunker. When he reached this episode in the king's life, the Führer had been moved to send for his "official" horoscope, which had been kept under lock and key by Himmler since 1938 as one of the "sacred" documents of the Third Reich. After the horoscope was produced, he discovered that it predicted many defeats "in the early months of 1945," followed by a miraculous and overwhelming victory in the second half of April, then Germany's rise to greatness again.* One of the staff officers asked Goebbels, with unusually frank skepticism, considering how risky express-

* This account of the event is based on that of Hugh Trevor-Roper (later Lord Dacre) in *The Last Days of Hitler.* Despite the heightened melodrama of Trevor-Roper's description of Hitler's last days, it remains by far the most accurate, and much of it is based on the Allies' interrogation of the survivors in 1945–1946, as opposed to later speculation or invention. Trevor-Roper ironically questions why Hitler did not consult his horoscope earlier, but most people do not as a rule go to faith healers, necromancers, quacks, or astrologers while things are going well for them, and the Führer was no exception.

ing any skepticism at all about a miraculous victory was in April 1945, "What czarina will die this time?"

Goebbels could not answer the question, but when he arrived home late that night and was told the news of Roosevelt's death, he immediately had a call put through to Busse at the front and told him, "The czarina is dead!" Then he ordered champagne and called Hitler on his private line in the *Führerbunker* to tell him the great news. "*Mein Führer,* I congratulate you! Roosevelt is dead. It is written in the stars that the second half of April will be the turning point for us. This is Friday, April 13. It is the turning point!"

Of course in the emotionally superheated atmosphere of the bunker, built fifty feet deep under the garden and the ruins of the Reichskanzlei, where the Führer had been living in cramped rooms not much larger than a walk-in closet since January 18 and where he was now trapped, any piece of good news was likely to provoke a brief, hysterical overreaction, and this was certainly true of Roosevelt's death. Euphoria soon gave way to renewed despair, however, as the news from the front came in over the next few days. The U.S. Ninth Army crossed the Elbe the next day, and the Red Army launched a powerful attack to the west, aimed at Berlin, before which division after division of the German army began to disintegrate. Nuremberg, the holy city of Nazism and site of the Nazi Party's famous rallies, fell to the Americans on April 16. Monty's Twenty-First Army Group reached the outskirts of the port of Bremen on April 19. And on April 20 Hitler's senior followers gathered grimly for the last time to celebrate his birthday in the bunker where he was to die.

Events were now beyond Hitler's control. "The *Führer*'s great life," Himmler told Count Folke Bernadotte, the Swedish diplomat who was trying to persuade him to begin negotiations for Germany's surrender, "is drawing to its close." On April 22, following a violent and terrible scene after Hitler learned that the attack he had ordered by SS Obergruppenführer Steiner which was intended to save Berlin and rescue the Führer had never taken place, he at last confirmed his decision to remain in Berlin to the end, and asked Goebbels to move into the bunker with his wife, Magda, and their six young children. By now, Bradley's rapid advance across Germany had cut off any further hope of moving away from Berlin, and the Russian encirclement of Berlin was almost complete. Germany was reduced to a narrow, broken corridor. On April 25 Soviet and American troops met for the first time at Torgau, on the Elbe, while in Berlin Hitler, calling to the last for the murder of prisoners of war and the *Prominenten* (famous people or the relatives of famous people, kept in the special care of the SS), made plans for his marriage to Eva Braun, for the writing of his personal and political testaments, and for the execution in the garden of the Reich Chancellery of his brother-in-law SS Gruppenführer Herman Fegelein, Himmler's representative in the bunker. Fegelein had married Eva Braun's sister Gretl, but prudently (though prematurely) decided to slip out of uniform and go home to his apartment, rather than sharing the family funeral pyre. On April 25 Patton's tanks were less than 100 miles from Berchtesgaden and already across the Czech border, to the alarm of the Soviets; and on April 30 American troops took Munich, the city where the Nazi Party was founded, and the scene of Hitler's early rise to fame and political power. Having eliminated from the succession both Göring and Himmler, because each of them was separately trying to negotiate a surrender, the Führer finally named Grand Admiral Dönitz as his successor. He then committed suicide, together with his bride, Eva, on the afternoon of April 30, shortly to be followed by Dr. Goebbels and his wife, who died after poisoning their six children.

The dark saga of the bunker was thus concluded just as Hitler had wished it to be, with a profusion of corpses appropriate to a Greek tragedy; and with his death, what remained of the German army's will to resist collapsed almost instantly. Despite the hopelessness of their situation and the almost complete destruction of their supplies and equipment, German army units fought with a courage and determination worthy of a better cause and leader so long as the Führer was alive. Even when he was invisible, silent, defeated, and trapped fifty feet underground, the mere fact of his being alive was sufficient to animate many of his soldiers; but by the same token the announcement of his death brought the entire sham colossus of Nazi Germany to an immediate and ignominious end. Neither Grand Admiral Dönitz nor the armed forces wished to continue the war, and one by one the major figures of the Third Reich (and countless minor ones) either surrendered or took their own lives. Berchtesgaden, Hitler's home, was taken on May 4, the same day on which troops of the U.S. Seventh Army met leading elements of U.S. Fifth Army advancing through the Brenner Pass from northern Italy. On May 5 German forces in Holland, Denmark, and northwest Germany surrendered to Monty on Lüneburg Heath, and a representative from Dönitz arrived at Ike's headquarters at Reims to inform him that all German U-boats at sea had been ordered to surface and return to port, and that the new German government wished to surrender.

Ike's views on Germany's surrender were simple—it was to be complete and unconditional. The Führer might be dead, and Dr. Goebbel's with him, but Goebbels's powerful capacity for mythmaking and outright lies survived beyond his funeral pyre. Almost every German official and senior officer wanted to surrender to the western Allies rather than to the Russians, and most of them believed, thanks to effective propaganda, that they would be allowed, and possibly even encouraged, to carry on fighting against the Red Army on the eastern front. Until the first week of May most of the significant offers of surrender had been made on this basis, and repeatedly rejected for that reason. Against all expectations, and despite four years of intransigence, lack of cooperation, suspicion, and hypersensitivity on the part of the Soviet Union toward its western allies, the war had been won without a major rift between them. Ike was determined not to disrupt whatever remained of that fragile alliance at the moment of victory. He therefore rejected Churchill's pleas to have American forces take Prague—it was the subject of one of the first messages to President Truman from the prime minister—which was in the Soviet sphere of influence, as agreed on between the Allies in Moscow and at Yalta. President Truman was no more anxious to liberate territory that had already been conceded to the Soviet Union's occupation than Ike was, so the matter was dropped.

So far as surrender was concerned Ike remained determined to keep the Russians informed and to give them no basis for complaints, though that did not of course prevent them from complaining or interfering. He informed the Soviet government that the Germans were offering to surrender, and asked it to send a senior officer to his headquarters as fast as possible to represent the Soviet Union. He had authorized his commanders to accept the surrender of German military formations of whatever size, but only on the condition that they surrendered unconditionally to all the Allied nations and were disarmed. Thus he had authorized Monty's acceptance of the German surrender at Lüneburg Heath, and made clear to Monty that the Germans were not to be allowed any illusions that they were surrendering to the British and Americans but not to the Russians. Ike had made that absolutely clear to General Admiral von Friedeburg when he and his small staff arrived in Reims—to Friedeburg's consternation, since he was authorized by Dönitz to discuss surrender only to the British and the Americans, not to the Russians. Friedeburg eventually managed to sort matters out with Dönitz's "government" in Flensburg, and was instructed that Colonel General Jodl, when he arrived, would be authorized to surrender "everything to everybody."

General Admiral von Friedeburg and an aide arrive at Ike's headquarters in Reims, May 5, 1945, to initiate the surrender of Germany's armed forces.

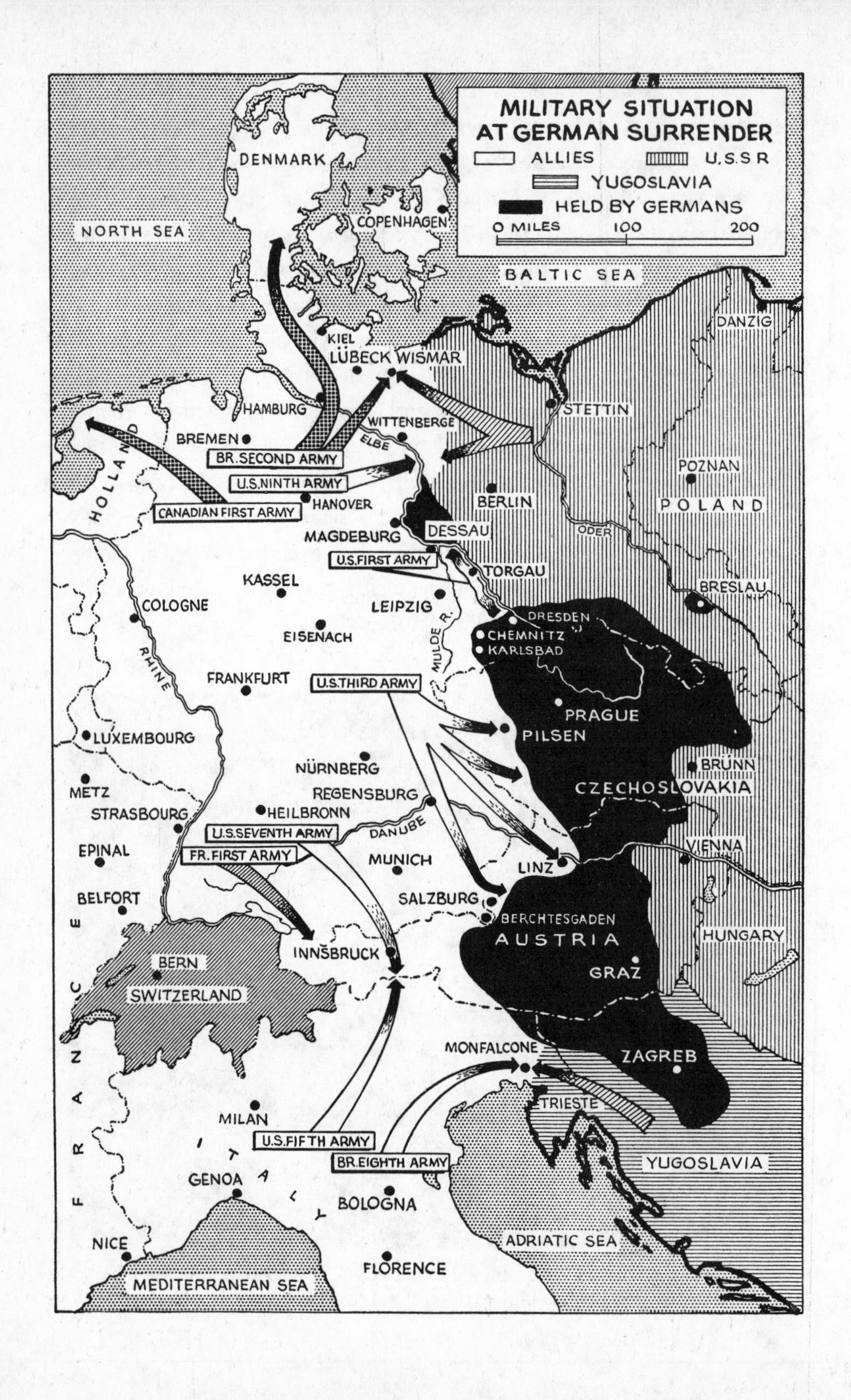

MILITARY SITUATION AT GERMAN SURRENDER
ALLIES
U.S.S.R
YUGOSLAVIA
HELD BY GERMANS
0 MILES 100 200
DENMARK
NORTH SEA
COPENHAGEN
BALTIC SEA
KIEL
LÜBECK
WISMAR
DANZIG
HAMBURG
STETTIN
BREMEN
WITTENBERGE
ELBE
HOLLAND
BR.SECOND ARMY
U.S.NINTH ARMY
HANOVER
CANADIAN FIRST ARMY
BERLIN
POZNAN
POLAND
MAGDEBURG
DESSAU
ODER
U.S.FIRST ARMY
TORGAU
KASSEL
BRESLAU
COLOGNE
LEIPZIG
DRESDEN
CHEMNITZ
KARLSBAD
EISENACH
MULDE R.
RHINE
FRANKFURT
U.S.THIRD ARMY
PRAGUE
PILSEN
LUXEMBOURG
NÜRNBERG
BRÜNN
METZ
REGENSBURG
CZECHOSLOVAKIA
STRASBOURG
HEILBRONN
U.S.SEVENTH ARMY
DANUBE
VIENNA
EPINAL
FR. FIRST ARMY
MUNICH
LINZ
BELFORT
SALZBURG
BERCHTESGADEN
AUSTRIA
HUNGARY
INNSBRUCK
BERN
SWITZERLAND
GRAZ
FRANCE
MONFALCONE
ZAGREB
TRIESTE
MILAN
ITALY
U.S.FIFTH ARMY
BR.EIGHTH ARMY
YUGOSLAVIA
GENOA
BOLOGNA
NICE
ADRIATIC SEA
FLORENCE
MEDITERRANEAN SEA

Though the British criticized Ike for not taking part in the surrender negotiations, it was, once again, Ike's long-held conviction that he would neither negotiate with nor talk to the enemy officers who surrendered. He left that task to Bedell Smith, who was eminently suited to take and keep a hard line; and to Major General Strong, SHAEF's chief intelligence officer, who spoke fluent German. In this, Ike was in part following, though the British could not be expected to know it, the example of Ulysses S. Grant at Appomattox Courthouse. Grant had made it absolutely clear to Lee from the beginning that no terms except unconditional surrender could be accepted; and Eisenhower, in part reacting to the sights he had seen at Ohrdruf-Nord concentration camp, was not willing to treat the German commanders as honorably vanquished foes, to whom were due the old-fashioned courtesies of war.

Friedeburg, Ike was convinced, was playing for time, in order to make possible the movement of as many troops as possible to the west and so avoid their falling into the hands of the Russians. Once Jodl had arrived, Ike ordered Smith to inform him that unless he surrendered at once the entire British and American front line would be "closed down," preventing any more German refugees, military or civilian, from crossing.

At two-forty-one on the morning of May 7 the instrument of surrender was signed; all hostilities were to end at midnight, May 8. General Jodl—who would be tried for war crimes, convicted, and hanged at Nuremberg—was brought to Ike's office just before three in the morning, and Ike asked him if he understood all the terms of the surrender document.

Standing rigidly at attention Jodl replied, *"Ja."*

Ike then said, "You will, officially and personally, be held responsible if the terms of this surrender are violated, including its provision for German commanders to appear in Berlin at the moment set by the Russian high command to accomplish formal surrender to that government. That is all."[22]

With that, Jodl saluted and left. World War II would be over in Europe just under forty-six hours later.

Kay Summersby reported that photographers rushed in—indeed it was there that the famous photograph of Ike holding the pens with which the surrender had been signed was taken. This is the photo in which, in the official version, Kay's presence behind Ike's right shoulder has been carefully airbrushed out. Ike then said he thought the moment called for a bottle of champagne, though, she recalled, with no triumph in his voice, and no elation.

Photos of Ike holding up the pens with which the surrender was signed, with and without Kay Summersby.

Grant's terse telegram to the secretary of war after Lee's surrender, announcing the end of the Civil War, was Ike's model: "General Lee surrendered the Army of Northern Virginia today on terms proposed by myself." Like Grant, Ike rejected all suggestions for a high-flown victory message, and simply cabled to the Combined Chiefs of Staff, "The mission of this Allied Force was fulfilled at 0241, local time, May 7th, 1945. Signed Eisenhower."[23]

CHAPTER 17

"I Had Never Realized That Ike Was as Big a Man"

Ike had a reputation for garbled syntax. Oliver Jensen, for example, would write a brilliant parody of "Eisenhowese"—Ike's version of the Gettysburg Address, which began: "I haven't checked these figures but 87 years ago, I think it was, a number of individuals organized a governmental set-up here in this country, I believe it covered certain eastern areas, with this idea they were following up based on a sort of national independence arrangement, and the program that every individual is as good as every other individual."[1] It is richly ironic, then, that his first step from victorious general to major political figure and statesman was marked by a speech.

The speech had an amazing impact at the time, and reading it today, one can see why. It was not an oration like Churchill's speeches; there was nothing high-flown, elegiac, pretentious, or polished about it. Nevertheless, it managed to convey, with evident sincerity, humility, simplicity, and moderation, the bonds that had linked the Americans and the British for four years of war and the meaning, for both peoples, of their shared victory. It was a message of hope, but it was not triumphalist—Ike simply expressed in words what everyone felt but nobody so far had said. Even the recently ennobled Field Marshal Lord Alanbrooke, Ike's confirmed critic, wrote in his diary that day: "Rushed off to the Guildhall for Eisenhower's presentation of the freedom of the City. Ike made a *wonderful* speech and impressed all hearers in the Guildhall including the Cabinet. . . . I had never realized that Ike was as big a man until his performance today."[2]

In fact, Ike made three speeches that day, one outside the Mansion House on being presented with the Freedom of the City of London* by the lord mayor; one at a luncheon in the Mansion House afterward; and one in the Guildhall, where he was presented with the sword worn by the duke of Wellington at Waterloo—quite an achievement for a man who didn't particularly enjoy public speaking. Of course, as Ike was fond of pointing out, he had written all of MacArthur's speeches during his years as the general's aide, but his own style was very different from that of his former boss. MacArthur spoke like a Shakespearean actor of the late nineteenth century, very slowly, his deep, sonorous, mellifluous voice giving his speeches the gravitas of an old-fashioned sermon—indeed, right-wing Republicans listened to him as if they were hearing the voice of God, and some people thought that MacArthur suffered from the same delusion himself. America's two most admired and successful generals, each now wearing five stars, could not have been more different on the podium, or elsewhere. People either loved or hated MacArthur—there was no middle ground—and many of his own soldiers were contemptuous of his poses, his vanity, his concern for his own comfort, and his majestic indifference to those of

* The City of London, "the square mile," is a different entity from "Greater London." Ruled over by a ceremonial lord mayor elected every year, and with its own police force, it contains most of London's financial district, many of the historical monuments, and almost none of the residential buildings. The custom of presenting the Freedom of the City goes back to 1237, and this is one of the greatest honors that can be bestowed on a foreigner. All "Freemen" receive a book called *Rules for the Conduct of Life*, written by a lord mayor in 1737.

lesser rank than himself. Ike, by contrast, was genuinely liked by the soldiers, American and British, who served under him, and he communicated a warmth and lack of ceremony that made soldiers ignore his high rank. He was one of them; and when he spoke, he spoke for them, in phrases which they could understand and with which they agreed. Civilians too admired Ike—he represented "the people," in the most traditional American meaning of that phrase. He was no faraway object of hero worship, as MacArthur liked to present himself. Like Grant and Lincoln, Ike was one of the people; and he had made good without ever losing sight of what he was and where he came from. When he spoke of those feelings simply and sincerely, his audience in London realized, as the rest of the world would soon realize, that they were listening to a victorious five-star general whose career was only beginning.

He began on a note which was not that of the conquering hero: "Humility must always be the portion of any man who receives acclaim earned in the blood of his followers and the sacrifices of his friends." Then, "This feeling of humility cannot erase, of course, my great pride in being tendered the Freedom of London. I am not a native of this land. I come from the heart of America. In the superficial aspects by which we ordinarily recognize family relationships, the town where I was born and the one where I was reared are far separated from this great city. Abilene, Kansas, and Denison, Texas, would together equal in size possibly one five-hundredth of a part of great London. By your standards those towns are young, without your aged traditions that carry the roots of London back into the uncertainties of unrecorded history. To those people I am proud to belong."[3]

To those people I am proud to belong. No words could better represent who Ike was and what he stood for. He went on to talk plainly about the values that bound the two countries, different as these countries were—freedom of worship, equality before the law, liberty—and declared that for these things "a Londoner will fight, so will a citizen of Abilene."

Toward the end of the speech, he put his own success as a victorious supreme commander in perspective: "Had I possessed the military skill of a Marlborough, the wisdom of Solomon, the understanding of Lincoln, I still would have been helpless without the loyalty, and vision, and generosity of thousands upon thousands of British and Americans."[4]

With this speech, made in the medieval Guildhall of London, before an audience that included every major political and military figure in Great Britain, Ike's march to the presidency began. He had rivals—for instance, MacArthur's desire for the presidency was well enough known to have given Roosevelt some concern, though MacArthur's quest for it was hopelessly compromised by his vanity and his arch-Republican ideas, as well as by the fact that he was an undisguised, unapologetic, and unreconstructed aristocrat who had lived most of his life outside the country. But in 1945 Ike was already widely recognized as having enormous political potential. Indeed there were only two reasons Ike could not have (as Grant had) the nomination of either party for the asking. First, Harry Truman was feisty and popular and knew his way around Democratic politics, including the big-city machines—after all, he had been placed in the United States Senate by "Boss" Pendergast's Kansas City machine—and around the labor unions, which still played a major role in those days. Second, on the Republican side there was some doubt about whether Ike, even though he hailed from Kansas or Texas, was really a Republican at all. This was not a question that the mostly British audience at the Guildhall asked, but his listeners there recognized "presidential timber" when they saw it.

So did everyone else. Ike's reception in London was tumultuous, perhaps the greatest ever given to an American since President Ulysses S. Grant's visit to London. Thirty years to the day since Ike had graduated from West Point as a second lieutenant, thousands lined the streets to cheer him as he was driven through the City of London, accompanied by his deputy Air Chief Marshal Tedder, in an open landau drawn by two matching bay horses and preceded by five mounted policemen on white horses riding abreast. The crowd of onlookers was enormous and enthusiastic; many people were waving American flags, and their cheering was so loud and continuous that it threatened to drown out the speeches. The next day, the *Times* of London, reporting the event, commented, "Many who already knew General Eisenhower as a great commander discovered yesterday for the first time that he is also an orator. His speech . . . had the moving eloquence which is native to the words of a sincere and modest man when he speaks from his heart of the ideals to which his life has been devoted."[5] Even the normally staid *Times* was moved to hyperbole by the man and the event: "If history now begins to take a course towards a happier order, in which the curse of war shall at last be exorcised from human relations, no man of our time will have made a nobler contribution to that issue than this great American."[6]

Churchill gave a formal dinner in Ike's honor at 10 Downing Street; and at her special request Ike called on Queen Mary, the mother of King George VI, at Marlborough House; he also had tea with the king and queen and Princess Elizabeth at Buckingham Palace. He was cheered by the audience when he went to the theater in London, and when he went out to dinner at a fashionable restaurant the band struck up "For He's a Jolly Good Fellow" as he came in and diners rose to their feet and sang it. On both these occasions he was accompanied by Kay in uniform—at the theater she was photographed by his side in his box, with his son John and John's date (a very pretty girl, found for him, of course, by Kay) beside them. After the dinner, Ike danced with Kay, though she makes it clear that dancing did not come easily or naturally to him. "There were a lot of adjustments to be made," Kay thought to herself after their nostalgic visit to Telegraph Cottage, where they sat on their favorite rustic bench and, if Kay is to be believed, held hands; and she was right, although they did not turn out to be the adjustments she had in mind.

It had already been made clear to Ike that he would command the American occupation forces in Germany—this entailed moving his headquarters from Reims to Frankfurt—and that when General Marshall left as Army Chief of Staff, after Japan was defeated, he would return to Washington and replace Marshall. All this was to be expected: it was the normal career path for an officer of Ike's seniority and distinction, and although he does not seem to have been enthusiastic about it, he accepted the inevitability of commanding the occupation forces and serving as Army Chief of Staff. What is harder to understand is the tangle of bad feelings that resulted from these moves. Ike had two personal concerns. One was to make a break with Kay Summersby in as decent and honorable a fashion as possible before his return to the United States. Whatever the nature of their relationship, it was clear to him, even if not to her, that it could not survive the war. The other concern—and no doubt the two were connected in his mind—was to bring Mamie over to join him in Germany. They had been separated for more than three years; her health worried him; and the time had come, as it was coming to millions of the men who had served under him, to resume his marriage.

He expressed all this in a letter to Marshall, asking for Marshall's reaction to the idea of Mamie's joining him in Europe. Ike did not exactly request Marshall's permission—after all, they were both five-star generals—but he made it clear that he wanted to feel he had Marshall's approval. It was symptomatic of Ike's enormous respect for Marshall that he approached him so openly on such a personal matter. "I would feel far more comfortable about her if she could be with me," he wrote, addressing the subject of Mamie's medical problems—this was about as personal as anybody ever got in correspondence with Marshall—and adding, in a revealing aside, that "the only other senior officer out here"[7] who wanted his wife to join him was Bradley. The aside speaks volumes about the way Ike's generals managed to console themselves with female company while serving overseas—Patton, for example, had been accompanied by his young "niece" throughout the war.

The most curious thing about Ike's letter was that he felt the need to write it at all. General MacArthur's wife, Jean, had long ago joined MacArthur at his headquarters without anybody's complaining about her presence—it is almost needless to add that MacArthur hadn't felt a need to ask for permission or to inquire what Marshall's feelings on the matter might be. If Ike had simply told Mamie to pack her bags and join him, it is hard to imagine that anybody would have been shocked or angered. Admittedly, millions of officers, NCOs, and enlisted men overseas were in the same boat, and there was something to be said in favor of their supreme commander's not breaking the rules for his own benefit. But it seems very doubtful that Ike's men would have objected to Mamie's presence now that the war was over, any more than they begrudged Ike his personal four-engine C-54, his Packard with the five stars front and back, or Kay Summersby in her WAC officer's uniform driving it. In all armies in all centuries the phrase "Rank hath its privileges" (or its equivalent in other languages) has always been well understood. Of course Ike was not MacArthur—he was always anxious not to be seen as taking advantage of his position, and would have found it hard to think of himself as a remote demigod in khaki, obeying nobody's rules but his own, as MacArthur did. Still, this was a battle which, had Ike cared to fight it, he would have won.

More curious still is that instead of simply telling Ike to use his own best judgment, or making the decision himself, Marshall took the matter to the new president. Perhaps Marshall had already detected in the former artillery captain from Missouri an instinctive distrust of generals (with the exception of Marshall himself), and a feeling that like big capitalists, "stuffed-shirt diplomats," self-important scientists, and blue-blooded snobs—big shots, in short—they should not be allowed to get away with anything merely because they thought they were better than the next fellow.

This element in Truman's character was unmistakable and deep-rooted, and even a less sensitive man than Marshall could hardly fail to have noticed it. Marshall may therefore have sensed that Truman wouldn't be comfortable with the notion that the Army Chief of Staff might be cutting a fellow five-star general a little more slack than a private or a lieutenant could expect, although the odds are that the president would have accepted any decision Marshall made about family matters among the generals of the U.S. Army, such was his respect for Marshall. Not surprisingly, the man who had once been a barefoot boy from Abilene shared the egalitarianism of the former haberdasher from Missouri, but Ike was less prickly, less censorious, and less self-satisfied about it—after all, he was a West Pointer and a general, and had grown accustomed to moving in the company of kings, presidents, prime ministers, the wealthy, and the privileged. Some of Abilene's raw, rough edges had been rubbed smooth by Ike's passage from West Point to general of the Army, whereas the president's sturdy Missouri skepticism about elites of any kind remained undiminished.

To Marshall's surprise, the president turned down Ike's request, thus depriving Mamie of the chance to rejoin her husband for another six months and giving Kay a reprieve, which can hardly have been what Marshall intended. Many years later, in 1973, when Merle Miller sat down with Truman to tape *Plain Speaking*, his "oral biography" of Truman, who was then fast approaching senility, the former president would claim that Ike had written to Marshall requesting permission to divorce Mamie and marry Kay, and that he and the Army Chief of Staff had discussed the matter, agreed it was out of the question, warned Ike that they would destroy his career if he went ahead, and then destroyed his letter.

This story is, to put it bluntly, both mischievous and unlikely.* Given the awe in which Ike held Marshall, it would have been improbable for him to request Marshall's permission or approval to divorce Mamie; in addition, Ike would have known what Marshall's reply would be, since Marshall was notoriously straitlaced about such matters. Then too, given Marshall's character and absolute sense of dignity, he would hardly have brought such a letter to the president's attention, or sat around with Truman in the White House gossiping about Ike and Kay Summersby. The fact is that by 1973—long before then, in fact as early as 1948—Truman had grown to dislike Ike, partly because Ike had made it plain that he was a Republican, partly because they had clashed over a good many issues, partly because as a general Ike had a natural distrust of "professional" politicians (this would surface again in his relationship with Nixon), and partly because the two men, despite certain superficial similarities, were simply oil and water.

It seems more likely that Truman, in his old age, was simply confounding the rumors he had heard about Ike and Kay with Ike's letter requesting that Mamie join him in Germany now that the war in Europe was over, and consciously or unconsciously putting a bad spin on something that was entirely innocent, and even praiseworthy. In any case, Ike fell into line; but

* There is also a possibility that it is a fake. In a trenchant article in *American Heritage* of May–June 1995, "Plain Faking?" by Robert H. Ferrell and Francis. H. Heller (which the presidential historian Michael Beschloss very kindly brought to my attention), the authors allege that Miller invented the story, along with several others in his book, and that no reference to it has been found in Marshall's files. Nor do Miller's own tapes of his conversations with Truman, now in the Truman Library, contain any mention by Truman of Kay Summersby, they allege.

he did not forget, and in the future the refusal to let Mamie join him would have echoes in his relationship with both Truman, whom he did not admire, and Marshall, whom he did.

In the aftermath of the war, every Allied country showered honors on the supreme commander. The Soviet Union gave him the Order of Victory, made of platinum and studded with diamonds and rubies; as well as the Order of Suvarov, First Class. France made him a Compagnon de la Libération and a Grand Officer of the Legion of Honor. Britain awarded him the coveted Order of Merit, the first time it had ever been bestowed upon a foreigner. Belgium gave him the Grand Cordon of the Order of Léopold. The Netherlands made him a Knight Grand Cross of the Order of the Lion. Poland made him a Knight of the Order of Polonia Restituta. Norway made him a Grand Commander of the Order of Saint Olaf. Denmark gave him the Order of the Elephant. Tiny Luxembourg awarded him the Grand Croix de l'Ordre Granducale de la Couronne de Chêne. Not since Wellington's victory over Napoleon at Waterloo in 1815 had any general been so widely honored—not to be outdone by the Europeans, Brazil, Chile, Ecuador, Haiti, Mexico, and Panama soon hastened to join in—or awarded so many titles and decorations.

All these paled into insignificance at the end of June, when Ike returned at last to the United States for the first time since his brief, secret visit home before D-day. He had the good fortune of being the first five-star general to return home victorious—MacArthur was still fighting the Japanese—but even allowing for that, his reception was extraordinary, in numbers, in affection, and in columns of praise in the newspapers. Nobody returning from Europe since Lindbergh had ever received such a welcome from such enormous crowds. Ike was greeted with victory parades in New York City, Washington, West Point, Kansas City, and Abilene. The outpouring of emotion was in part, of course, relief that the war in Europe was over, but was also a remarkable display of affection for Ike himself. In Abilene, a crowd more than four times the size of its population turned out to greet the general on his return to his hometown. In New York City the crowd was numbered in the millions. In Washington, D.C., a future chairman of the Federal Reserve Board, Dr. Arthur Burns, then "an economist at George Washington University, watched [Ike] drive past in his convertible, caught the friendliness he projected, turned to his wife and said, 'This man is absolutely a natural for the Presidency.'"[8]

Arthur Burns had hit the nail on the head—the same thought was occurring to many other people as Ike moved around the country imprinting himself on the minds of millions of Americans who watched him standing in the back of a convertible, tall, bronzed, grinning. The thought also occurred with exceptional force to politicians in both parties—for it was not yet clear what the general's politics were, or even if he had any. This situation would go on for some time. As late as 1947, if Ike's recollection is to be believed, President Truman offered to take the vice presidency if Ike would run for the presidency as a Democrat in the election of 1948—and perhaps it was so, since Truman's chances for election were thought to be slim. Ike wisely steered clear of politics in 1945—he was, in any case, still a serving officer—but his popularity was greater than that of any political figure in the United States, a fact which caused as much concern among Republican politicians as among Democrats, or in the White House.

From the very moment the door of his airplane opened at Washington airport and he descended to give Mamie a hug and kiss, they were dogged by the press, and given hardly a moment of privacy. He was called directly to the Pentagon, then flown to one victory parade after another for over a week before they got a chance to be in each other's company for any extended time. They went to White Sulphur Springs for a week, although even there they were accompanied by their son John, and by Mamie's parents, so the reunion was hardly private, secluded, or romantic, but then perhaps that was not what either of them really wanted. Inevitably, both of them found it disappointing, and also each of them now had a lot to learn about the other. Ike had become used to being surrounded by a large, devoted staff, while Mamie had been obliged to make up her own mind about things in her husband's absence, and did not take easily to being told what to do. She was occasionally resentful; he was impatient. The faint shadow of Kay Summersby still hung over them—even Ike's most devoted biographer, Stephen E. Ambrose, notes that Kay "was still much in the news and gossip, and still much on Mamie's mind."[9] This is putting it mildly, and certainly it cannot have been easy for Mamie to avoid thinking about the fact that at the end of their week together, Ike would be going back to the glamorous WAC lieutenant, whatever his relationship with her was.

Back in Germany, Ike faced an even more demanding schedule. He was responsible not only for the American forces in Europe, most of them yearning to get home as soon as possible, but increasingly for the starving and defeated Germans as well. The Germans now needed to be fed, housed, and given the wherewithal to restart their economy, because the Ruhr Valley, which the Royal Air Force had been at such pains to destroy for four years, was the industrial heart of western Europe as well as Germany; and as Ike eventually came to realize, without German industry, banking, mining, and above all prosperity there was no realistic possibility of rebuilding Europe. The first, faint stirrings of the postwar German *Wirtschaftswunder* (economic miracle) were taking place before Germany's former masters had even been tried for war crimes at Nuremberg.

Gifts and honors continued to pour in, including a platinum-and-gold cigarette case with five sapphire stars, engraved with General De Gaulle's signature—a personal gift from De Gaulle to the one American who had always taken his side, despite many disagreements between them. At his new headquarters in Frankfurt, Ike wrestled, rather glumly, with the endless problems of the occupation, the most intractable of which was the policy against "fraternization" between American troops and their former enemies. Ike's temper was quickly frayed both by criticism from newspaper reporters and by the number (and flagrancy) of transgressions against the nonfraternization rules, ranging from the fact that former Reichsmarschall Göring was being treated too leniently by his captors to inevitable fraternization by GIs with German women. This reached a peak of absurdity when GIs were strictly forbidden to hand out gum or candy to German children, at which point Ike sensibly decided that "small children" would be excepted from the rules. That was, of course, the proverbial thin end of the wedge—once soldiers were allowed to give gum and candy to small children, it would be hard to stop them from talking to the small children's mothers, and human nature (or commerce) would take care of the rest. Ike had been in favor of treating the Germans harshly, but he was also a realist, and soon came to realize that nonfraternization was impossible, and produced nothing except endless sensationalist stories in the press at home. Unusually for a general, he had always taken a liberal attitude toward the press and reporters, but now he no longer had any good news to give them—the only news coming out of occupied Germany that was likely to make the papers back home was critical of Ike. This, coupled with the fact that he was responsible for enforcing rules and regulations he did not necessarily approve of, and was subject to carping criticism by everybody from congressmen to the State Department, made his life a misery.

Ike was certainly in favor of denazification, as the policy of eliminating former members of the National Socialist Party from positions of responsibility was clumsily called, and this brought him into direct and painful conflict with his old friend George Patton, who was responsible for the occupation of Bavaria. Patton complained, not without reason, that it was impossible to find civil servants with any experience in Bavaria who had not at least paid lip service to Nazism, but then went on, in his usual bull-in-a-china-shop way, to tell reporters that while he knew nothing about politics, it seemed to him that the Nazi Party was just another political party, "just like a Democratic and Republican election fight."

This remark, not surprisingly, caused a sensation, particularly since Patton had also made no secret of his belief that the real enemy was now the communists, not the Nazis, and that the Russians ought to be pushed back out of Europe "to the Volga," if necessary with the help of the Germans. That was insubordination on a grand scale, and Ike could not ignore it even had he wanted to. Patton's opinions about Soviet Russia would have been unexceptional—indeed, they might even have made him immensely popular—in 1948, at the time of the Soviets' blockade of West Berlin and the Berlin airlift, when it became clear that the cold war was going to define East-West relationships for the foreseeable future; but in 1945 they were premature. Most people still hoped for good relations with the Soviet Union, and hardly anybody was as yet willing to forgive former Nazis merely because they were more efficient, or because they were anticommunist. Twice before, in Sicily and in England, Ike had called Patton in and given him a verbal hiding for his indiscretions, but this time there could be no forgiveness—Patton had gone too far, and there were no more battles to come in which he could redeem himself. Ike fired him from the command of his beloved Third Army and gave him a face-saving "paper command" of the Fifteenth Army, actually a headquarters without troops, where Patton's job was to study the military lessons of the war and draw up training manuals based on them. With that, their friendship, which had begun so many years ago in 1919 at Camp Meade, Maryland, when they had stripped a French Renault tank down to pieces and put it back together again, was over.

Ike's working relationship with his new commander in chief also got off to a rockier start than he expected, when he was summoned to the Potsdam Conference in Berlin, in mid-July. Informed for the first time by Secretary of War Stimson of the existence of the atomic bomb, Ike had surprised and angered Stimson by arguing against its use on Japan. Ike's strong feeling was, and remained, that the Japanese were already defeated, and that using the bomb against them was "completely unnecessary" and would only succeed in shocking world opinion. In addition, Ike argued that once the genie was out of the bottle, he could not be put back in again. The bomb could only increase world tension, he thought, at just the moment when there was a realistic possibility of ending it. Stimson was not impressed by this sensible line of reasoning, though perhaps he ought to have been, given his own reluctance to see Japan's cities destroyed—Stimson personally struck Kyoto off the list of targets proposed by the Air Force for the atomic bomb. Nor was Truman impressed three days later. Here was an issue in which civilians were more anxious than a five-star general to use a weapon. Actually, the reasons Ike put forward for not using the bomb against Japan more or less prefigured those raised by liberal revisionist historians of World War II about two decades later.* In

* Ike's apprehensions about using the bomb are startlingly similar to those raised in Gar Alperovitz's *The Decision to Use the Atomic Bomb*, published in 1966.

any case his advice was overruled, though it was a good example of the common sense, basic humanity, and good instincts that set Ike apart from so many military men, and that made him such a popular figure all over the world for so many years.

Invited to the Soviet Union by Stalin, Ike was accompanied to Moscow by Marshal of the Soviet Union Georgy Zhukov, with whom he had formed a respectful friendship. Zhukov was personable, a farm boy like Ike, and without contest one of the greatest commanders of the war—although he was shortly to be pushed out of the limelight, like all of Stalin's most successful generals, in order to further the postwar party line, which was that Stalin himself had been the great strategic genius, military leader, and "architect of victory." In the Soviet Union, unlike the United States, celebrated heroes could be (and were) shunted off into obscurity and written out of history to suit the needs of the party and Stalin's vanity, though they still fared better than the previous generation of Soviet military leaders, most of whom Stalin had had shot.

Like many another western visitor to the Soviet Union, Ike, who had brought his son John with him, came away with a rosier view of the country and of his host, Stalin, than either deserved. Ever since Prince Grigory Potemkin had ordered clean, happy, healthy villages full of well-fed, cheerful peasants hurriedly built along the route wherever the empress Catherine the Great was going to travel, a Russian specialty has been stage-managing things to impress visitors, and to hide every sign of misery and political repression. Stalin was, in fact, a good deal better at this than Potemkin had been, and Ike saw what every foreigner was shown: the famous Moscow subway with its glittering marble floors and crystal chandeliers, the GUM department store, the Stormovik aircraft factory, a collective farm, the Kremlin museum. He attended a soccer game, and at Stalin's invitation and standing beside Stalin himself—a rare privilege for a noncommunist—watched a vast and interminable "sports parade" from Lenin's tomb. "The groups of performers," he noted, "were dressed in the colorful costumes of their respective countries, and at times thousands of individuals performed in unison." Stalin watched for five hours, showing no sign of fatigue, while a 1,000-piece band played continuously. "This develops the war spirit," he told Ike, then added, after a moment, "Your country ought to do more of this."[10] Ike was given a magnificent banquet in the Great Hall of the Kremlin, and visited Leningrad, where more than 350,000 civilians had starved to death during the German's siege of the city in 1942. He was presented, at his own request, with a "generously inscribed" photograph of Stalin. He also had several discussions with Stalin, in one of which Stalin asked him to convey his apologies to General Marshall for a "sharp radio message" during the war over a "misunderstanding"—presumably a reference to, among other things, an Ultra intercept that Marshall had sent Stalin, which turned out to be wrong; as well as Stalin's angry outburst about the negotiations between General Wolff and Allen Dulles for the surrender of German forces in Italy. Whether Ike was fooled by all this is hard to say, but what impressed him most, in the wake of his disagreement with Stimson and the president about using the atomic bomb on Japan, was the incredible destruction that the war had brought on Russia. From the border of Poland to Moscow, there was hardly a house left standing, hardly a town or city that had not been fought over and destroyed, hardly a village that had not been burned to the ground by the Germans in their retreat. Ike recognized, as many westerners did not, the terrible price the Soviet Union had paid in World War II, and the unimaginable sacrifices its people had made. He was deeply struck when Zhukov, during a discussion about Soviet and Anglo-American tactics in the war, said that in the Red Army it was not the custom to have engineers clear minefields—the infantry advanced right through them, and those who were unlucky enough to step on a mine stepped on it.

Ike retained a fondness for Zhukov even after the Russians began to make difficulties about westerners' access to Berlin, and continued to describe Stalin as a "fatherly" figure long after it was no longer politic to do so. Though Ike was never, in the phrase of a decade later, "soft on communism," neither was he an ideological zealot. Although he would be the first Republican to win the White House since Herbert Hoover, his anticommunist credentials were never strong enough to satisfy those Republicans who would have preferred to see Senator Taft or General MacArthur in the Oval Office. Hence the choice in 1952 of Richard Nixon as Ike's running mate—Nixon was intended to reassure the conservative wing of the party that Ike's heart was in the right place.

As his correspondence with Mamie makes clear, her feelings on the subject of Kay Summersby had not moderated. She was apparently not pleased by the news that Ike had been invited to Northern Ireland, Kay's birthplace, and although Ike's letters to her do not mention that Kay would be accompanying him, she must have suspected it—wrongly, as it turns out, since Kay makes no mention of it in her memoir. In any event, it provoked a highly exasperated reply from Ike: "Saturday night I received your long letter. . . . What you could have meant by my being 'highly interested in Ireland' is beyond me. Jimmy, Sgt. Farr and I got there one evening, [and] carried out a schedule for the next 24 hours that would have killed a horse. . . . I certainly didn't stay in Ireland a second longer than I expected and only a third as long as the dignities desired."[11]

Outraged though he may have been by Mamie's letter, he wisely did not mention to her a quick trip he made back to the Villa Sous le Vent, near Cannes, with Kay. Once again, it was something less than a secluded romantic holiday à deux—Ike was accompanied not only by Kay, but by Ambassador Averell Harriman and his daughter Kathy—but Ike and Kay managed to get away in a motorboat for swimming expeditions, and even went out to nightclubs together. On their return to Frankfurt, he took her on a visit to see Mark Clark at Salzburg, and then up to Berchtesgaden to look at the ruins of Hitler's house. If Kay is to be believed, he even tried to give her the cigarette case from De Gaulle, which she refused on the grounds that if De Gaulle ever found out, he'd be furious (this was very sensible of her, and perfectly true). Nevertheless, as time went by, Ike was gradually being forced to make a decision about Kay, and it cannot have been a comfortable time for him. He had promised to help her become an American citizen—this was very much on her mind because she expected to be on his staff when he returned to the United States to replace Marshall. Ike told her that on his last day in Potsdam, he had spoken to President Truman about this, and that the president had responded sympathetically and had suggested Kay write to Secretary of State Byrnes about it. Both Ike and Kay were being naive if they thought that Truman was going to play Cupid on their behalf with the State Department, and in the event Kay's letter went into bureaucratic limbo, as no doubt the president intended.

By now it must have been apparent enough to Ike—nobody was more familiar with the unbreakable dos and don'ts of Army life than he was, unless it was Mamie—that there was no realistic possibility of his returning to Washington as Army Chief of Staff with Kay Summersby in tow, whatever her citizenship was. Nor was there any chance that Mamie would put up with Kay's presence for a moment—her letter about his trip to Ireland would have made it clear to him, if it had not already been, that this particular wound had not healed. He had gotten away with his relationship with Kay, innocent as it may have been, during the rough-and-tumble of war, partly because of the awe in which a supreme commander was held and the routine censorship of news stories; but by now there had been too much gossip and too many stories about them, and anyway the war was over. The attempt to give her De Gaulle's cigarette case should have warned Kay of what was coming—it has all the marks of a grand, guilt-inspired farewell gesture.

In October, Marshall finally prevailed on President Truman to accept his retirement as Army Chief of Staff, and Ike, with a certain reluctance, prepared to return to Washington. He had disliked being in Frankfurt, surrounded by people he still regarded as the enemy, and he had hated the problems of commanding an occupation force in which every soldier's one ambition was to get home and out of uniform as soon as possible. But he must also have been aware that whatever small measure of privacy he had managed to carve out for himself (and this included Kay) was going to be lost once he went home. As Army Chief of Staff, he would be a major figure in Washington, both socially and politically; his home would be at Quarters Number 1 in Fort Myer, where he would be surrounded by his Army staff and servants; and he would be constantly in the eye of the press. He would be living in a fishbowl, and the notion that Kay, freshly Americanized or not, would be sitting in an office outside his as his uniformed assistant, given the amount of speculation that had already gone on about the two of them, was a pipe dream.

He wrote—rather boldly, considering her feelings—to tell Mamie that Kay was "coming to the States along with [Lieutenant General Lucius] Clay and [Lieutenant Colonel James] Stack," to serve on his staff, but it was not to be. Shortly after he arrived home Mamie was stricken with bronchopneumonia; then Ike too was felled by the same thing, in his case exacerbated by exhaustion, and hospitalized for two weeks before taking up his post as Army Chief of Staff on December 3.

By then, he seems to have already reached his decision and done what needed to be done about Kay Summersby. Her name was abruptly dropped without explanation from the list of those members of his staff who were to join him in Washington, and the War Department telex was soon followed by a dictated, and in Kay's view "impersonal," letter from Ike explaining to her that according to Army regulations, she would have to be discharged from the WACs the moment she arrived in the United States, and it would therefore be impossible for her to work for him in the War Department, since she would be a civilian and a noncitizen.

At the bottom, he wrote in his own hand, "Take care of yourself and retain your optimism."[12] This was good advice, though perhaps not the most loving farewell on record.

This would not prove to be the end of the Kay Summersby story, however. She did not depart from his life (or from newspaper stories) easily, or for many years to come, but he had taken the first, small, necessary step toward the creation of something else—Ike the hero-president.

Ike had not expected to enjoy being Army Chief of Staff, nor did he. The problems facing the armed services were intractable—they included the rapid demobilization of many millions of men; the Army's desire for "universal military training" (UMT); the growing pressure for some "unification" among the armed services; the allocation between Army, Navy, and Air Force (which was then in the process of being spun off from the Army to become an independent service like the Royal Air Force) of rapidly—and drastically—reduced defense spending; and indeed the difficult question of just what kind of armed forces the United States needed now that the war had been won, and the first storm clouds of the cold war were appearing on the horizon. In addition, Ike was no longer commanding armies—he was merely one of three service chiefs competing for funding, attention, and political support. It says a lot about his new job that on his first day Ike got hopelessly lost in the Pentagon trying to find his way back to his own office after lunch, and that the story made headlines the next day. "No personal enthusiasm marked my promotion to Chief of Staff, the highest military post a professional soldier in the United States Army can reach," Ike commented with restraint. In private, he was more outspoken on the subject.

Ike was unusually forthcoming about his problems in correspondence with his old friend from Abilene "Swede" Hazlett, who had originally persuaded him to sit for the Annapolis examination, but whose own career in the Navy had been cut short by ill health. Ike's letters to Swede are long—and hand-typed by Ike—and unusually informative. Clearly, he liked and respected Hazlett, and perhaps also felt a little guilty that he himself had reached a lofty five-star rank whereas his friend had retired from the Navy with a serious heart condition at the rank of captain. They traded jokes about getting Ike's son John and Swede's daughter (whom Ike refers to as "Miss Swede") together. Swede reminded Ike that there was "a damned good-looking (and sensible) blonde out at Hood College in Frederick who would like to meet him"; and Ike expressed the hope that John would get to meet "that daughter of yours" before "she goes and gets herself married," though he sensibly qualified this by remarking that "the spectacle of two old-time Kansas farmer boys timidly sticking their noses into Cupid's business is, after all, a bit on the ludicrous side."[13]

More practically, while he was recuperating from pneumonia in the hospital ("flat on my back") in November 1945, Ike typed a letter of more than 2,000 words to Swede, responding to his old friend's misgivings about the idea of military unification, which the United States Navy was resisting with particular ferocity and a sustained attempt to capture the hearts and minds of members of the House of Representatives and the Senate. Ike rejected the notion that any one of the armed forces would be "swallowed up" in the process. "The American public," he wrote, "should understand that war has become a triphibious affair," and that henceforth "the closest possible kind of association between the individuals of these three forces throughout their service career is mandatory." He went on to point out in detail that each of the three services complements the others, and that there was no possibility of sensibly funding for defense "*unless the broad yearly program for all three services is presented . . . as a unit.*" He discussed combined training with the authority of a supreme commander who had given orders to admirals and air marshals as well as generals, pointing out that on this subject he was "almost a fanatic," and adding, from his own experience, "War is a matter of teamwork." Perhaps most important of all, he defined his philosophy of command—"*that in any war theater there should be one commander*, and that his authority should be so firmly established that there is no question as to his right to handle the three services as he sees fit."[14]

Ike dismissed what had become the burning issue of unification—that the three services might be obliged to wear the same uniform. He pointed out that while he didn't care what color uniform he wore, each service would no doubt keep its own uniform, traditions, and service academy, and that unification of the command structure of the services was not intended to end the Army-Navy football game.

Ike won Swede over to his point of view, but unification continued to be a troublesome issue (to the Navy "unification" was a code word for the Army's cherished desire to eliminate the Marine Corps), though not more so than such problems as what the balance should be between the three services—How much Navy did the United States need? How much Air Force? How big an Army?—and the thorny question of how to, and indeed whether to, go about integrating nuclear weapons into military strategy. Reading Ike's letters to Swede should be enough to dispel any notion that he was less than a fluent writer—they are crisply reasoned, well thought out, and phrased with unmistakable clarity. What is more, Ike was setting forth for his friend Swede (or perhaps trying out on him) a coherent and very definite philosophy of war. The heart of this philosophy was that each theater of war required a single commander with full authority to make decisions; unity of command so that air, land, and sea were seamlessly integrated in the commander's plans and operations; and the full resources required for victory. (Both in Korea and in Vietnam this formula would be ignored, with unfortunate consequences; and indeed the "Eisenhower doctrine" is still being ignored.)

Swede seems to have been one of the first people to bring up the subject of the presidency with Ike, commenting in a letter while Ike was still in Germany that "no matter what party you [are] affiliated with (and I have no idea if you're D. or R.), you could carry the country without even taking to the road."[15] Ike did not rise to the bait, but now that he was Army Chief of Staff, it was a subject that soon came up more and more often, and from people who were better informed and more influential than Swede Hazlett. As Army Chief of Staff Ike was deluged with invitations to speak, and much as he disliked making speeches, he felt obliged to accept a good many of them. He refused any payment for them—he considered speaking part of his duty—and in view of the reputation he later acquired for muddled speechifying, the speeches he made had a remarkable impact on people and were considered thought-provoking, challenging, and inspiring, even by the *New York Times*. He spoke on national defense, education, and foreign policy—it was only on the subject of politics, in which he had no interest, that his command of grammar and syntax apparently deserted him—and his speeches inevitably led him into the company of members of the East Coast political and financial establishment. These included heads of corporations and foundations, presidents of universities: the kind of solid, wealthy people who then still played a major role in both political parties, and above all in the Republican Party.

After nearly five years of separation, the Eisenhowers not only were reunited but had a house to live in. Quarters Number 1 at Fort Myer did not belong to them, but it was the best piece of housing the Army could offer, big enough to absorb all the furniture Mamie had been keeping in storage all these years while living in a small apartment, and elegant enough to let them entertain on a grand scale. This was old, familiar territory for Mamie, decorating government housing, however grand. Still, like the Grants before them, the Eisenhowers looked toward an uncertain future. In order to overcome Ike's reluctance to serve as Army Chief of Staff, President Truman had agreed to limit his term to two years—thus, in 1947, Ike would retire from the Army. Not only did the Eisenhowers not own a home of their own, but like the Grants they were by no means certain where their home should be when they bought one. Washington held no attraction for Ike; neither he nor Mamie contemplated going "home" to Abilene, or to Denver, where her parents lived; Ike loved the outdoors, gardening, golfing, and fishing; Mamie liked the glamour and excitement of a big city—and they had no roots anywhere to guide them in making a choice. Then too, though they lived on a fairly grand scale at Quarters Number 1, the Eisenhowers had very little money. As a five-star general Ike would have a pension of only $15,000 a year, no great fortune even then; and he had no savings, no capital, nothing to show financially for more than three decades of service as an officer in the U.S. Army. He would be fifty-seven in 1947, too young to retire to someplace where the weather was good, like Florida or southern California, and play golf, even if that had been what Mamie wanted—and it was not. They had discussed, rather vaguely and without much enthusiasm, the idea of living near San Antonio, perhaps with a summer fishing cottage on a lake in Wisconsin; but while this had a certain appeal for Ike—less, one supposes, for Mamie—he would have been hard pressed to afford it. Also, he and Mamie could not, or would not, accept any help, in this case very much unlike the Grants, who had accepted lavishly furnished houses offered to them by admirers in Philadelphia, New York City, and Galena, Illinois, in an age when there was more leniency about such gifts. Ike himself had ruled out the kind of "job" that retired generals were usually offered, like sitting on the board of major corporations as patriotic window dressing. Business did not interest him, and in any case he was determined not to trade on his reputation as supreme commander.

In the meantime, Ike did the best he could with what was, all things considered, a thankless task. His relationship with the president was mutually respectful, but distant—Truman did not rely on Ike for military advice the way Roosevelt had relied on Marshall, and in peacetime Ike's job was to cut the Army's size, not build it up. By one of those ironies that long service in the military occasionally produces, as Army Chief of Staff Ike now commanded, at least in theory, his old boss General Douglas MacArthur. While it was perhaps fortunate for Ike's blood pressure that there was some distance between the two of them, MacArthur's headquarters overlooking the Imperial Palace in Tokyo might as well have been Shangri-la, so remote and far removed was he from the War Department. Although he commanded the second largest contingent of American troops in the world, while also acting as America's proconsul in Japan—a kind of khaki-clad American equivalent of Great Britain's viceroy in India—MacArthur lived and worked in self-imposed, Olympian isolation, responding only reluctantly, when he responded at all, to orders and queries from home. His massive dignity and his reputation as a hero had given him an authority second only to that of the emperor himself in defeated Japan. Indeed, in 1945, for the first time in the history of Japan, the emperor left his palace to pay a call on another human being, General MacArthur; and by 1946 the general was so admired and respected by ordinary Japanese that a movement had sprung up to petition Congress for the admission of Japan to the union as the forty-ninth state. Given his view of himself as a godlike figure, which apparently coincided with the view the Japanese held of him, MacArthur was, not surprisingly, resistant to complying with messages from his former aide in Washington and the Philippines. In his infuriating way, he first argued that he needed at least 400,000 troops; then, when Ike had gone to bat for him against those in Congress who wanted to cut troop levels, and against the troops themselves, who wanted to come home, MacArthur calmly informed the newspapers that he could live with only 200,000, thus embarrassing Ike and making himself look good—which was perhaps MacArthur's purpose in the first place.

There were plenty of other incidents, and Ike felt that the only way to deal with them was to sit down and talk them over with MacArthur. Since there was clearly no question of the mountain's coming to Muhammad, Ike eventually flew to Tokyo to see the mountain. MacArthur, however, turned out to be more interested in the presidency than in troop levels, and pressed Ike hard to learn whether he had any intention of running in 1948, though Ike suspected that what MacArthur was really looking for was an expression of support from Ike for his own run for the presidency. The two generals fenced gently on this subject without reaching an agreement. Their conversation was formal and polite—it was probably enough of a victory for MacArthur that he had made the new Army Chief of Staff fly 10,000 miles to see him, something which he would never have tried to make George Marshall do—but MacArthur's opinion of Ike had not been changed by the latter's five stars. "He let his generals in the field fight the war for him," he told an aide contemptuously. "They were good and covered up for him. He drank tea with kings and queens. Just up Eisenhower's alley."[16] Whoever else might be about to get on the Eisenhower bandwagon, it would not be MacArthur. He remained silent and indifferent to Ike's attempts to win him over on many of the issues dear to Truman's heart; and when he was informed that a mountain in Canada had been named after Ike, called for an atlas, and remarked with satisfaction that considering the terrain, it was a pretty small one.

One consequence of the flight to see MacArthur was that Ike and Mamie began to travel together often—here at last was something which they could do together, and which made up, at least in part, for the years when Ike had hobnobbed with kings, queens, presidents, and prime ministers while Mamie stayed at home in Washington. The Army Chief of Staff traveled with impressive pomp and circumstance. In one year alone they visited ten countries, plus Hawaii and Guam, and during his two-year tour of duty as Chief of Staff they visited every state of the union at least once. They were invited to stay with the royal family at Balmoral Castle, where Mamie was placed on the king's right at dinner; and to dinner by Churchill, at Chartwell, his country home. Mamie was, at long last, getting some of the perks Kay Summersby had enjoyed for so long, and she would not have been human had she not felt a certain glow of triumph as she walked in the gardens of Balmoral with the king, the queen, and their corgis.

The ghost of Ike's relationship with Kay Summersby could not, however, be exorcised that quickly. Kay managed to get herself transferred to the United States as a WAC captain (her promotion had come through at last), and actually turned up in the Pentagon one day, with Telek, their Scottie, to say hello to Ike. He "got all red," Kay reported, and was politely evasive about her future; but when she finally received her orders she had been posted to a public relations job in California, about as far from Ike as the Army could send her and still keep her within the continental United States. She could read the writing on the wall, so she resigned from the WACs and moved to New York City, where she got a contract to write a book called *Eisenhower Was My Boss*—the first, and dramatically less sensational, of her two memoirs.

Kay's book deal cannot have been welcome news to Ike and Mamie—though in the event the book was admiring and this time around, Kay steered well clear of any claim that they were romantically linked. It was the first of many books by those who had once been part of Ike's "military family." First to come was one from his valet Mickey McKeogh, *Sergeant Mickey and General Ike*, which embarrassed Ike (who declined to write a preface for it). It demonstrated, however, that Madame Cornuel's famous remark "No man is a hero to his valet" is not always true, since the book was unashamedly hero-worshipping. Next, with a leaden thump of mortified embarrassment for Ike, came Harry Butcher's book, *My Three Years with Ike*. Butcher had been in need of money when he returned home from the war—he had divorced Mamie's best friend, Ruth, to marry the nurse with whom he had been having an affair overseas—and sold his diaries to Simon & Schuster. Since Ike had ordered Butcher to keep a diary, and many of the notes and entries in it were Ike's, not Butcher's, there was some question as to whether Butcher had the right to profit from it, but contesting this would have led to even more publicity than the book itself would produce. Ike forced Butcher to omit from the manuscript diary entries that were his own, and again refused to write a preface. Nevertheless, the book went on to became a best seller, and the serial rights were sold to the *Saturday Evening Post* for $175,000, a fortune at the time. References to Churchill's keeping Ike at the dinner table until the small hours of the morning, and to Ike's making fun of Churchill's atrocious table manners—Churchill was of that generation of the English high aristocracy which unashamedly slurped its soup, picked its teeth at the table, and held firmly to the old-fashioned view that a concern for good table manners was essentially middle class or frenchified—obliged Ike to write a letter of apology to Churchill. Churchill responded, "I must say I think you have been ill used by your confidential aide. . . . I am not vexed myself at anything he has said. . . . It is rather late at my age to reform, but I'll try my best."[17] Butcher's book also contained numerous unflattering remarks of Ike's about Montgomery, which prompted a pained letter from Monty to Ike. Further embarrassment came with a laudatory biography of Ike written by the well-meaning Kenneth S. Davis. Davis had interviewed Ike at length, and owing to some kind of slipup, printed several of Ike's unexpurgated remarks about Monty, obliging Ike to write another letter of apology.

It is difficult, perhaps impossible, to convey Ike's enormous popularity or the great curiosity the public had about him. There was no equivalent in 1945 of television's intrusive ability six decades later to overexpose famous people so that interest in them quickly peaks, and no equivalent of a news industry eager to catch the great and famous in unflattering poses and to reveal their human flaws. Ike was an American hero—not only in America but almost everywhere else—and the media treated him as one. It was not only his valet who worshipped him. He was not treated with awe, as MacArthur was, but he was accorded, perhaps more valuably, affectionate respect.

The problems Ike faced as Army Chief of Staff were insoluble, but he faced them with his usual combination of common sense and hard work. Demobilization was the hardest task. Even though Ike pledged himself to read every one of the special pleas sent in by soldiers who claimed to be hardship cases (there were several hundred of these a day), and even though he was kept personally informed of the number of soldiers discharged every day, the process still went too slowly to please the soldiers or their families. Ike discovered, to his dismay, that there was no way to know for sure how many men were left in the U.S. Army, and he was finally obliged to set an arbitrary date for an old-fashioned "hand count" census. Of course he was now protected—or perhaps insulated—by a large staff, whose primary purpose was to make him look good. He was astonished to find that somebody had even gone to the trouble of preparing a long, detailed, single-spaced memorandum to ensure that when he traveled he would be wearing the right decorations on his uniform:

> . . . hereafter when General Eisenhower's calendar indicates that he is to call at a foreign embassy or to participate in any event in connection with a foreign dignitary, action will be instituted immediately to insure that he wears the proper ribbons. . . .
>
> 1—When the engagement is made, Major Cannon will notify Major Schultz *immediately.*
>
> 2—Major Schultz will notify Sgt. Murray and is responsible that the ribbons worn by the General are proper and in order. *This will be done at once.*
>
> 3—Sgt. Murray will prepare the ribbons for General Eisenhower's blouse or jacket. *This will be done as rapidly as possible.*
>
> 4—Major Cannon will verify the propriety of ribbons.[18]

There is no denying that through no fault of his own Ike had long since grown used to being looked after on an imperial scale. Both as supreme commander and as Army Chief of Staff, he had a small army within the Army to look after his well-being and his wishes, which, even though they were less eccentric than those of General MacArthur (who had the cloth of his old uniforms carefully unstitched at the seams and preserved to be used again in tailoring new uniforms and service hats, so that he would always seem to be wearing well-used old ones), were nevertheless fiercely respected. Mamie chafed at the fact that she could no longer get up early and make Ike's breakfast—there was somebody standing by in the kitchen at Quarters Number 1 to do that—and that Sergeant Dry, Sergeant Murray, or Sergeant Moaney laid out Ike's clothes for him, looked after his belongings, and saw to his comfort. A large staff existed to draft Ike's speeches and correspondence—"*triple*-spaced, with wide margins at top and bottom, and both sides," since he was an inveterate reviser and tinkerer, and a fountain of second and third thoughts, who rewrote everything over and over again. Even the scraps of paper he threw away were carefully rescued from his wastepaper basket and filed.

Only four years had gone by since he had been a comparatively junior officer living a fairly ordinary life in married officers' quarters and dealing with the small problems of daily life like any other man. Now a chauffeur-driven car, a fleet of aircraft, and innumerable aides waited to take him wherever he wanted to go and do whatever he wanted done. The Army looks after its own senior commander (as do the Navy, the Air Force, and the Marines), and it is hardly surprising that Ike found it difficult to decide what he would do when he finally took off his uniform. The idea of running a small college, perhaps in Texas or Kansas, "in a rural setting," and living on a nearby ranch, had a certain appeal for him—Ike, after all, came from a place and a family where higher education was respected and taken seriously. Both his father and his mother had attended college, unusual for young people from farm families in those days (especially for girls); indeed, they had met at college. And Ike and his brothers had all benefited from an excellent school system and from a college education—in fact Ike had decided to try out for one of the two service academies because it was a way of getting a college education free, rather than from any interest in a military or naval career. The idea of becoming a college president was therefore not far-fetched, and he and Mamie had often discussed it.

As so often happened to Ike, his immediate future would be settled by two unexpected, or at least unsolicited, offers. The first came about when Douglas M. Black of Doubleday and William Robinson of the *New York Herald-Tribune* managed to persuade Ike to write a book about his experiences in command during World War II. Ike had never seriously considered *not* writing a book, and had received several offers, including one from Simon & Schuster, but he was bewildered and irritated by the complexity of serial rights, motion picture rights, foreign rights, etc. Black and Robinson, whom he instantly liked and respected, came up with a deal that solved this problem. They had discovered that a nonprofessional writer could sell his life story as a complete package for a lump sum, all rights included, and that the sale would be treated as a capital gain, rather than income, for tax purposes. After ascertaining from the secretary of the treasury that this was true, Ike decided to go ahead, and received $635,000 for his war memoir, which, after capital gains tax, left him nearly $500,000, which was an immense sum for the day, and which made him a rich man for the first time in his life. (Since *Crusade in Europe* went on to become a worldwide best seller, Ike would have made far more money with a normal royalty deal, but what he wanted was the security and independence of having a significant amount of money in hand.)

The second offer came when Thomas Watson, Sr., chairman of the board of IBM and also chairman of the board of trustees of Columbia University, offered him the job of president of Columbia. This was a happy outcome of Ike's speech making, which had brought him the acquaintanceship of many rich and powerful men over the past two years. Ike had, once again, not sought the job, but had positioned himself so that the job sought him. Of course Columbia was no small college "in a rural setting" such as he and Mamie had discussed together—it was and is one of the great American universities, situated in New York City. Accepting the offer would place Ike in a position of immensely high visibility, as well as automatically establishing him among the wealthy East Coast Republicans whose views on foreign affairs coincided with his own. If there was such a thing as a Republican elite—and there was—Ike would be joining it.

There were many attractions to the idea of writing his book and becoming president of Columbia, but the main one was something that Ike does not mention in his account of these events in *At Ease* or *Waging Peace*, two of his later books. Whatever else he was, Ike was no fool, even at presidential politics. He might win the Republican nomination in 1948, but it was not quite yet his for the asking. Other people—among them Senator Taft of Ohio and Governor Dewey of New York—had a prior claim, and both of them would put up a hard, bruising fight at the convention. Quite apart from that, Ike had a natural reluctance to resign from the Army and immediately run for the presidency against his own former commander in chief, as well as an instinct that Truman would fight hard, and perhaps even successfully, to win. Ike did not underrate Truman's abilities as a politician, and he shrewdly preferred to place himself in a well-paid, high-profile job, in New York City rather than Washington, and to write his book. If Ike had possessed the political cunning of an American Machiavelli, he could not have positioned himself better for the election of 1952, although, ironically, President Truman would shortly make it even more certain that Ike would run and win by putting him back in uniform—and in the news.

In the meantime, the presidency of Columbia also solved the Eisenhowers' housing problem, since it came with a house on Morningside Heights, a large and elaborate mansion of somewhat chilling grandeur, which had as an unexpected bonus a penthouse retreat on the roof. There Ike took up painting, as Winston Churchill had advised him to do during the war, finding, at long last, a pastime which soon became one of his favorite activities, and one in which he found, to quote his distinguished artistic mentor, that "work and pleasure are one." For the rest of his life, Ike sought peace and quiet in painting, and although he never took himself seriously as an artist, he was serious about his devotion to painting, "born," as he put it, "out of my love of color and in my pleasure in experimenting."

He worked equally hard—and with a productivity that many authors would envy—to complete his book. Ike had written all his life—orders, memos, correspondence—and often for discerning and demanding masters, like MacArthur or Marshall. His writing, though without flourishes, was always, crisp, unambiguous, and to the point; and his approach to *Crusade in Europe* was simple and pragmatic. He used his files, diaries, orders, and letters to guide him; worked sixteen hours a day, starting early in the morning and revising until his secretaries were exhausted; and finished the book in under three months—a remarkable accomplishment. He had prepared himself for the task by rereading Grant's memoirs, perhaps the greatest nonfiction book in American literature, and like Grant he aimed for simplicity and simple fact. As a result, *Crusade in Europe* was, and remains, one of the clearest and least opinionated books to come out of World War II, and by far the least self-exculpatory and least judgmental. If Ike had an ax to grind, he avoided grinding it in his book, and this is more than can be said of Churchill's six volumes, lofty as they are (they won Churchill the Nobel Prize for literature). It is also free of the small-minded bitterness and acid criticisms of their colleagues, unsweetened by victory, time, and second thoughts, which marked the books of Monty and Alanbrooke, and which set off a lasting transatlantic "war of the memoirs" that is still reflected in the difference between British and American histories of the war. Predictably, Monty took umbrage at the suggestion that things had not gone exactly as he had planned at Caen, but beyond that there were few complaints from those who had served with or under Ike in the war. Ike had the benefit of two gifted editors: Kenneth R. McCormick, editor in chief of Doubleday; and Joseph Barnes, then foreign editor of the *New York Herald-Tribune* and later editor in chief of Simon & Schuster (Barnes was the man who edited William L. Shirer's *The Rise and Fall of the Third Reich*). But for the most part Ike seems to have relied on his own passion for rewriting and correcting, on sheer hard work, and on his determination to get his facts right.

At Columbia Ike had the somewhat awesome task of succeeding Nicholas Murray Butler, who had presided there for almost half a century. Butler, a kingmaker in the Republican Party and a friend of Theodore Roosevelt and Elihu Root, had himself twice tried to win the nomination as the party's presidential candidate, and had been awarded, among countless other honors, the Nobel Prize for peace. His presence was so firmly fixed in everybody's mind at Columbia that when Ike attempted to enter his office on his first day as president, he was turned back by the guard because he was not Nicholas Murray Butler. Access to Butler's office on the second floor of Low Library had been by a small, carefully hidden private elevator, so that he could be reached only by means of an appointment made well in advance; but Ike changed this immediately, moved to a new office on the first floor, and kept his door open. To his dismay, faculty members and students did not drop in to talk with him, and his day was taken up by financial problems (despite its endowment, which included one of the largest real estate empires in New York City, the university was in desperate need of money), ceremonial functions, and dealing with the problems of a huge staff.

In later years Ike would reproach himself for not having found a way to break out of the routine of running Columbia to visit classrooms and get to know the faculty and students, and he was certainly hampered by not having the kind of staff he had surrounded himself with as supreme commander. Still, he managed to prevent Columbia's widely admired football coach, Lou Little, an old friend from Ike's own coaching days, from going to Yale; and to persuade the Nobel Prize–winning physicist Isidor Rabi to turn down an offer from Princeton's Institute for Advanced Study and stay at Columbia. Ike also oversaw a significant effort to put the university on a sounder financial footing; and he persuaded the very reluctant mayor of New York City to turn West 116th Street, where it passes through Columbia's campus, into the peaceful pedestrian mall it has remained to this day. His tenure at Columbia was made more difficult after the fall of 1948, when President Truman asked him "to go to Washington for two or three months to serve as a military consultant to the Secretary of Defense."[19] This was a tall order: commuting between New York and Washington exhausted Ike, as did the pressure of two important jobs, neither of which he felt he could focus on fully. He worked "immoderate hours"; as always, he fretted over details; he held his temper back at great cost to his blood pressure; and he continued to smoke four packs of cigarettes a day. Not surprisingly, by 1949 his health had deteriorated sharply, and after an alarming—and rather carefully concealed—cardiac crisis, followed by a four-week recuperation in Florida, he gave up cigarettes forever, and decided to stop commuting to Washington and to devote himself to Columbia. He took satisfaction in the belief that "Washington would never see me again except as an occasional visitor."[20]

This was optimistic, as it turned out. Ike was just beginning to find his feet as president of a great university—he negotiated successfully with the feared New York City labor leader Mike Quill to prevent a strike at Columbia, and was persuaded to give a lecture to students on "the major aspects of war," which, to his surprise and theirs, was a huge success—and at last he and Mamie bought a farm property for his retirement, which then seemed to them both to be fast approaching. The farm was about a mile from the Gettysburg battlefield, with an old house that would have to be totally rebuilt, or torn down to make room for another, but it finally settled for good the question of where they would live. Then, in 1950, on a Christmas trip to see Mamie's family in Denver, Ike received a call from President Truman asking him to return to Europe and become the first commander of the NATO forces.

It was not a request that Ike felt he could refuse—the NATO countries had unanimously requested Truman to offer the command of the fledgling military alliance to Ike; and Ike took the view, as he had done all his life, that for a military man the president's request is an order. The trustees of Columbia refused to accept his resignation, and insisted on giving him an "indefinite leave," in the hope that he would return to the university after his tour of duty at NATO. It pleased Ike to suppose that he would, but when he flew to Europe in January 1951, he must have already suspected that serving as Columbia's president until his retirement, then moving quietly to Gettysburg, was not in the cards. He had stepped back into uniform—and into the limelight.

NATO was, at the time, a wishful experiment, rather than an armed force, and it was by no means a popular one at first, on either side of the Atlantic. Whether twelve nations, ranging in size from "tiny Luxembourg" to the United States, could form a useful military alliance that would thwart the ambitions of the Soviet Union was by no means certain. In Europe, there was deep suspicion that the alliance would be dominated by the English-speaking countries (a new version of *l'Albion perfide*) to the detriment of France or the smaller European countries. To many Europeans on the left, the alliance seemed like deliberate saber-rattling directed toward Soviet Russia; a provocative way of dragging the European countries into America's quarrel with Russia; and a way of ensuring that the next war, possibly with nuclear weapons added to the horrors of the last one, would be fought on European, not American, soil. To Americans on the left it seemed like a way of cranking up the war machine again before it had even cooled after World War II. The Republicans, particularly those in the isolationist heartland (as opposed to Ike's new wealthy, internationalist East Coast friends), were deeply suspicious of maintaining American troops in Europe in peacetime: the troops might find themselves automatically fighting in a European war without congressional authorization—it seemed to Senator Robert A. Taft and many other Republicans like giving the president a blank check to get the United States into war.

All this was being fought out against a background of darkening hostility between the major communist countries and the rest of the world. In the Far East, North Korea, with the support of the People's Republic of China and the Soviet Union (the two were not yet open ideological foes), had invaded South Korea, dragging the United Nations and the United States into a war nobody wanted (and bringing Douglas MacArthur back into the field as supreme commander again). Meanwhile French Indochina was slipping from a guerrilla war between the communist followers of Ho Chi Minh and those who supported, or tolerated, government by the emperor Bao Dai and the French into a full-scale colonial war between the French army and the Vietminh. In Eastern Europe, one by one, the Soviet Union had replaced bourgeois or multiparty governments with one-party rule by communist governments backed by massive and apparently permanent Russian occupation forces. In Warsaw, Prague, Budapest, Bucharest, and Sofia the story was the same, as Stalin imposed on the Eastern European countries the familiar apparatus of show trials, purges, secret police control of every aspect of life, collectivized agriculture, and the nationalization of all industrial production. In Western Europe, poverty, deprivation, and a general resentment against the vast power and prosperity of the United States, combined with the alarming fact that western Germany was recovering from the war faster than its former European victims, moved intellectuals and organized labor to turn toward communism in increasing numbers, leading many to suppose that Italy and France might end up with communist (or fellow-traveling) governments in the not too distant future. The British were rapidly (though unwillingly) divesting themselves of much of their colonial empire, at any rate those parts of it which were neither profitable nor contained large amounts of oil; and Britain was reluctant to join with Europeans, who, on their side, were equally reluctant to admit Britain as part of Europe.

In September 1949 the Soviet Union had announced the successful test of its first atomic bomb, setting off a race to be first with the "super," the hydrogen bomb that would dwarf the power of atomic bombs. Thus the atomic monopoly of the United States had lasted only about four years; and with its end the world, seen from Washington, and indeed from the capitals of Western Europe, suddenly looked like a much more frightening place. The cold war made it necessary for Western Europeans to defend themselves—they could not hope to do so without the full participation of the United States (and to a lesser degree, of the United Kingdom), but they could not expect the Americans to do it for them. Nor could America, having fought for four years to liberate Europe from the Germans, abandon the Europeans to domination by the Soviet Union and withdraw once again back to the other side of the Atlantic. "The long-range plane," Ike astutely observed, "had moved America's frontier three-thousand miles eastward, to the heart of Europe." Under the circumstances, it is not surprising that the NATO powers chose Ike to command them—he had performed the miracle of getting the Americans, the British, and the French to fight under a single commander, and perhaps only he could get the twelve countries of NATO to form a fighting alliance that would deter the Russians. His prestige was such that he constituted, in 1951, NATO's most powerful weapon.

He visited all the NATO countries in a whirlwind eighteen-day trip; broadcast from Paris to all the members of the "Atlantic community," as he called it; and endured ceremony after ceremony in the bitter weather of one of the worst winters in anybody's memory—all this despite a bad reaction to a smallpox inoculation and a virus that kept him hospitalized in Germany for two days. Everywhere he went, he emphasized that the European countries could provide land forces sufficient, under a single command, to deter the Soviet Union; and everywhere his hosts asked him in turn how many divisions the United States would keep in Europe. President Truman had asked Ike to speak to an "informal joint session of Congress" on his return from his "fact-finding mission," and also to testify before the Senate Foreign Relations Committee and the House Foreign Relations and Armed Services committees, as well as make a national radio and television broadcast to the American people. This media blitz, as we might now call it, was an indication of just how controversial the whole subject of the NATO alliance was. NATO, of course, went on to become such a success story as the cornerstone of American and European security for many decades that it is hard to imagine the degree of opposition to it in the United States in 1951, or to overstate the importance of Ike's patient, determined explanations of its benefits. Truman had chosen the right man to sell NATO to the American public and, perhaps even more important, to the isolationist wing of the Republican Party—for one thing, it was virtually impossible for anyone not to take Ike seriously on the subject of European defense—and in the process he inadvertently established Ike as an ever more attractive and compelling contender for the Republican nomination.

Flying back from Europe, Ike landed at Stewart field, in Newburgh, New York, in an ice and sleet storm and suffering from a bad cold; met Mamie there; and proceeded to the Thayer Hotel in West Point, where he holed up for four days to write his speech. Few speeches, he noted later, had given him so much trouble—he had to persuade Americans that it was in their own best interest to help the Europeans defend themselves, and that the Europeans could do it, with American support and a substantial initial commitment of American troops. Any notion that Ike approached his speeches casually can be dismissed by the quantity of painstakingly revised drafts he went through, and the immense quantity of information he mastered and sifted down. Despite appearances, he was both a worrier and a perfectionist when it came to speech making, and if we read between the lines it is apparent that he was already aware that more was at stake than America's commitment to NATO. He was firing his first shot in the struggle for the heart and soul of the Republican Party, which would have to be torn from its isolationist roots, dragged kicking and screaming if necessary into the mid-twentieth century, and made to accept America's new international role as "leader of the free world." America's borders were no longer on the Atlantic or the Pacific; they were on the Elbe, or the Thirty-Eighth Parallel in Korea, or in far-away Taiwan—and when Ike talked of America's obligation to defend freedom, it is hard not to believe that the enemy he had in mind was as much Bob Taft as Stalin.

His return to Washington was an almost theatrical triumph, once again thanks in part to Truman, who had insisted on driving out to National Airport—despite another fierce ice storm, which froze the Potomac and made driving almost as dangerous as flying—to greet Ike and bring him and Mamie to lunch at Blair House, the president's home while the White House was being repaired and redecorated. Since the presidential limousine was delayed by ice on the roads, the supreme commander's aircraft had to circle around the airport in the storm until Truman finally arrived; despite his easygoing appearance Ike was a West Pointer and a general, and therefore a stickler for protocol and a respecter of rank. He was met as he descended from his new four-engine Constellation aircraft onto the icy runway by the president, members of the cabinet, an array of generals and ambassadors, and a small army of photographers and journalists, and driven to Blair House in the president's limousine.

The next morning, Ike spoke to both houses of Congress, assembled in the Library of Congress, and told them, as he had told the president, that for some time the United States would need to keep substantial forces in Europe—unwelcome news to many senators and members of the House in both parties. His speech was both eloquent and sincere, and was reported almost rapturously by the newspapers. The *New York Times* printed it in full—it filled a page of the paper—and commented, "It is this capacity to command respect and confidence; this combination of knowledge, experience and achievement; this quality of objectivity . . . this quality of directness, of toughness, of being the same inside and out that Washington has been seeking."[21] An editorial in the *Times* remarked, "It is further characteristic of General Eisenhower's point of view that he puts his reliance not in armies alone . . . but in things of the spirit, as manifested in national morale."[22] The *Washington Post* wrote, "Here was no hokum, no effort to mesmerize Congress, but a candid appraisal of the situation as he saw it. 'Ike' proved himself a master of psychology as well as a soldier-statesman."[23] *Time* magazine put Ike on the cover for the sixth time, reporting that he was the American public's choice for "man of the year," and calling his task "stupendous."

Ike cut through the doubts about Europe with precision and passion. The French had instituted compulsory military service "with almost no exemption for any cause whatsoever"; the Norwegians were determined never to be occupied again, and had flatly stated that "resistance to the point of destruction was preferable"; Luxembourg had expressed its determination to have universal military service with no exceptions. Europe was determined to defend itself, and America had an interest in helping it do so. In language that even the Taft Republicans could understand, Ike described what Lend-Lease had achieved in World War II. "It took a rifle and a man to go out and advance the cause of the Allies against the enemy we had. If the United States could merely provide the rifle and get someone else to carry it in order to do the work that was necessary, I was perfectly content."[24] This was practical, unsentimental soldiering, and the kind of blunt language that was the best answer to doubters, rather than high-flown sentiments.

All the same, and although the president had greeted Ike on his return as if he had been a conquering hero returning home from the wars rather than a successful negotiator and diplomat in uniform, many of the senators and congressmen were less respectful, or at least more wary. Many on the Republican right had returned to their prewar isolationism, out of which they had been briefly shaken by Pearl Harbor—they were certainly by instinct anticommunist, but it was native communists (and fellow travelers or communist sympathizers) they feared, not the ones in the Kremlin. Out of power for nearly twenty years, they were more interested in overturning Roosevelt's New Deal social reforms than in fighting the Red Army; and like the Bourbons returning to France after the revolution and the empire, they had, in the words of Talleyrand, "learned nothing and forgotten nothing." Apparently they had learned nothing from America's withdrawal from Europe after World War I, either—indeed many of them were inherently skeptical about Europe, and did not believe that tiny countries, or slightly larger ones whose vigor had been sapped by effete intellectuals, communistic labor unions, and socialistic politicians could possibly unite to defend themselves. Others were willing to provide money and arms for NATO but were reluctant to furnish troops; and even among those who were reconciled to keeping American forces in Europe as part of NATO, many thought that six divisions was too big and dangerous a commitment.

Of course, more than five decades later, when NATO and the Marshall Plan have been enshrined as America's most successful effort in foreign policy, and when Europe has long since risen from the ashes of World War II to become a major power block, while the Soviet Union and its "captive" or "satellite" nations, as the Eastern European communist governments came to be called, have altogether vanished from the scene in disgrace, these doubts about the wisdom of Truman's policy seem quaint and even farfetched; but that is not to deny the sincerity of people's doubts at the time.

The air of crisis that marked Ike's return and his appearance before the members of the House and Senate was caused not only by the overwhelming military strength of the Soviet Union and the corresponding weakness of Europe, but by the steadily more alarming situation in Korea. There, General MacArthur, whom Truman had once described as "Mr. Prima Donna, a brass hat," was urging the president to recognize a "state of war," and if necessary to drop thirty to fifty atomic bombs on mainland China's major cities.

The situation in Korea, which developed when North Korea invaded South Korea in June 1950, showed every sign of escalating out of control. By August 1950, the North Koreans had taken Seoul, the capital of South Korea, and pushed the United Nations forces all the way south to a small perimeter around the port of Pusan, where they were trapped with their backs to the sea. Despite the strong recommendation of two such very different people as General Omar Bradley and John Foster Dulles to the president "to save himself a lot of grief" and get rid of General MacArthur as the commander in Korea right at the beginning of the conflict, Truman declined to do so, fearing the political repercussions. Then, in September, MacArthur had persuaded the reluctant and doubtful chiefs of staff to let him make an "end run" by means of a brilliant, daring, and risky amphibious landing at Inchon. As he had promised, he defeated the North Koreans in a stunning victory that made him once again a national hero; recaptured Seoul; and drove the North Koreans back across the Thirty-Eighth parallel in disarray. In October MacArthur, confident that the Chinese would not send troops in any quantity to support their North Korean ally, crossed the Thirty-Eighth parallel; but in November the Chinese reacted by attacking with more than 260,000 men, forcing MacArthur's vastly outnumbered forces into one of the bitterest winter retreats in military history.

In October Truman had flown nearly 15,000 miles to meet MacArthur on Wake Island, only to find that MacArthur, having given him a two-hour briefing, would not even stay for lunch. His initial reaction on meeting

MacArthur had been intense irritation. As Truman descended from his aircraft onto the runway MacArthur did not salute him, but merely walked forward and shook his hand. In addition, the president, a stickler for dress, was offended by the battered old gold-braided peaked cap of a Field Marshal of the Philippine army, which had been MacArthur's trademark throughout the war, and by the fact that the general wore his shirt collar open, without a tie. Still, like so many people meeting MacArthur for the first time, Truman nevertheless came away impressed.

Once he was home again, that feeling faded fast when MacArthur demanded that the war be expanded and began to request the use of Chiang Kai-shek's "Chinese forces from Formosa" in Korea. This, along with MacArthur's suggestion about using atomic bombs, petrified the British and the Europeans, and alarmed Truman, who was determined not to let a limited "police action" in Korea escalate into World War III. On the day that Ike gave his speech to Congress, he shared the front page of the *Times* of London with a story about a combined Franco-American force in Korea surrounded by communist troops that had been obliged to defend its position all day long against savage enemy attacks, at times holding them off with bayonets in fierce hand-to-hand fighting. Fighting was escalating in Indochina too, though as yet this involved only the French; and in Malaya, where the British were fighting a communist insurgency. To many people it seemed clear that the real threat to the United States was in Asia, not in Europe, and that it was Douglas MacArthur the president should be listening to, not Ike.

To a startling degree, the arguments of those who were against an American commitment to NATO reflected other, older divisive issues in which Ike had played a significant role. Chief among these was the lost battle of those who had favored a "Pacific strategy" over an "Atlantic strategy" in World War II. Many of them still believed that Roosevelt's determination to give the defeat of Nazi Germany priority over the defeat of Japan had been a mistake, and had led him to "cave in" to Churchill, and later to Stalin. This division of opinion had pitted the Navy against the Army, MacArthur against Ike, and Republican isolationists against Democratic interventionists and wealthy easterners, for all of whom Europe loomed large as America's first concern. Although the war had been fought and won along the lines that Roosevelt wanted, the battle between those who looked eastward toward Europe and viewed America as an "Atlantic power" and those who looked westward to the Pacific was still going on in the early 1950s. It divided those who believed that Red China was the principal threat from those who thought the Soviet Union was; and it turned Ike and MacArthur into rival heroes—one the first and most prominent "Atlanticist" of them all, the other the revered champion of America's Pacific destiny. Nothing could illustrate this better than the fact that Ike's headquarters as NATO commander would be in Paris, among the difficult and fractious French, while MacArthur's was in Tokyo, among the hardworking and now compliant Japanese.

Ike's own commitment to NATO was unshakable; indeed, he was willing to put his political future on the line to ensure America's full participation. Since Republican opposition centered on the person of Senator Robert Taft, who was also the man who would benefit most if Ike decided not to seek the presidency in 1952, Ike decided to "kill two birds with one stone." He would meet privately with the senator and try to talk him into making a definite commitment to vote in support of American participation in NATO; at the same time, if Taft agreed, Ike would promise in return to remain in Europe and devote his time and energy over the next few years to establishing NATO on a firm footing, to the exclusion of politics. In short, if Taft agreed to deliver his support, and that of his followers, for NATO, and to make U.S. membership in NATO bipartisan policy, Ike would agree to repudiate any efforts to make him a candidate for the Republican nomination in the next presidential election. In Ike's account (in *At Ease*, which he wrote in 1966) of this unusual bargain he proposed to offer Taft, it is worth noting that he does *not* dismiss the presidency as something he didn't want—on the contrary, he makes it clear enough that he believed he could win the presidency in 1952, and that he was sorely tempted to try. Nevertheless, he was willing to sacrifice his presidential ambitions in return for Taft's agreement to vote for America's participation in NATO.

Taft strongly disliked, though not necessarily in that order, the military, war, and Europe, so of all the major political figures of the early 1950s he was the least likely to be swayed by Ike's rank and war record. He accepted Ike's suggestion of a meeting, but asked that it should be held at the Pentagon, "for a private and unannounced talk," rather than in his Senate office. Taft guessed correctly that the Army could find a way to slip him into the Pentagon unseen more easily than he could find a way of slipping Ike into his Senate office without anybody's leaking the news to the press.

Be that as it may, the meeting was not a success. Ike had prepared for it by writing on a slip of paper in pencil what he would say if he won Taft over to the support of NATO. He folded it up and placed it in his pocket, just as he had done with the note he wrote the night before D-day about what he would tell the press in the morning if the invasion failed. Then, he had taken full responsibility on himself for the failure of the landings; now, clearly echoing Sherman's famous refusal to run for the presidency—"If nominated I will not run, if elected I will not serve"—he wrote out a brief statement that would eliminate him entirely as Taft's rival: "Having been called back to military duty, I want to announce that my name may not be used by anyone as a candidate for President—and if they do I will repudiate such efforts."[25]

Taft was "whisked" up to Ike's office by a staff officer, and they had what Ike described as "a long talk," but for all Ike's charm, powers of persuasion, and mastery of the facts, Taft remained unmoved. Ike tried to put Taft at ease by saying that he didn't know how many American divisions were needed, and wouldn't until he had studied the matter more closely. His concern was simpler—he wanted to go back over there with the knowledge that both parties supported the idea of "collective security" as the best defense for Europe and the United States, and to be able to assure the Europeans that the formation of NATO would have bipartisan support in Washington. To this, Taft glumly kept replying, "I do not know whether I shall vote for four divisions or six divisions or two divisions."[26]

Again and again Ike tried to steer the senator away from the numbers game, and make him understand that it wasn't the number of American divisions that mattered; it was America's commitment to ensure the security of Western Europe, and America's willingness to take its place as an equal partner in an alliance of European countries. If there was one matter about which Ike could talk with absolute authority, it was making an alliance work, but he failed to budge Taft. Taft thought that the alliance was more likely to provoke the Soviet Union to fight than to deter it, and that membership in NATO was an "interventionist" policy that would involve the United States in the old quarrels of Europe, and force it to take sides in matters in which it had no business being involved. The result, Taft was deeply convinced, would be to turn America into "a garrison state." Also, pouring money into NATO would create inflation and higher taxes at home, undercutting the economic strength of the United States, which was its defense. So far as foreign affairs were concerned, Taft was much moved by former President Herbert Hoover's plea to turn America into a kind of giant Gibraltar, a fortress protected from the rest of the world by its prosperity and its arsenal of nuclear weapons. A later apologist for Taft would see in the senator's doubts about NATO the seeds of opposition to the Vietnam War, and praise him for seeking "a way to defend the country without destroying it, a way to be part of the world without running it." But even if that were so—and Taft's ideas were narrower and more rigidly isolationist than that—it overlooks the extraordinary success of NATO and of America's pledge to become "a good European." Few policies have ever paid a higher dividend, and few military alliances have survived for over half a century, in the end defeating the enemy without bloodshed—indeed Russia's ambition is now to join NATO, as most of its former captive nations have done.

There does not appear to be much evidence of friendship or mutual admiration between Ike and Taft; and this was unusual for Ike, who usually had a good word to say even for those who, like Monty and Alan Brooke, publicly disagreed with him. In any case, he did not read aloud to Taft what he had written on the piece of paper in his pocket. After the senator had left his office, Ike called a couple of his aides in, and with them as witnesses, carefully tore the note into shreds and discarded it. "In the absence of the assurance I had been seeking, it would be silly for me to throw away whatever political influence I might possess," he wrote later—in other words, while Ike was not going to throw his hat into the ring, he was not going to imitate Sherman either.

In Irving Berlin's smash hit musical of 1950, *Call Me Madam*—based (very loosely) on the life and times of Mamie's best friend and neighbor at the Wardman Tower, Perle Mesta—the number that got the most applause night after night was not sung by the star, Ethel Merman. The showstopper was sung by a chorus of senators and congressmen, and was called "They Like Ike." People went out of the theater singing it:

The presidential year will soon be drawing near;
The people soon will choose their fav'rite son.
I wonder what they'll do in nineteen fifty-two?
I wonder who they'll send to Washington . . ."
[Chorus of senators:]
"They like Ike
And Ike is good on a mike,
They like Ike—
—But Ike says he won't take it!
—Ah, that makes Ike
The kind o' feller they like,
And what's more,
They seem to think he'll make it.

Whether Taft had heard the song or not—he doesn't seem to have been the kind of man who spent a lot of time going to Broadway musicals—its sentiments were no doubt familiar to him. What is more, he is mentioned in it as one of the hopeless Republican pretenders:

They won't take Saltonstall and Stassen's chance is small,
The same would go for Vandenberg and Taft,
And Dewey's right in line with William Jennings Bryan.
There isn't anyone that they can draft. [27]

Irving Berlin had gotten the Republicans' situation right. Whoever the Democrats nominated—assuming Truman didn't run for a second full term, in which he showed no interest—Ike was the man to beat. Saltonstall was unelectable; Stassen was such a perennial candidate that his name was a national joke; Vandenberg and Taft* were senatorial heavyweights with not an ounce of presidential charisma between them; and Dewey had already been defeated in two presidential elections, once by Roosevelt and once by Truman.

Ike returned to Europe, with Mamie, to take up his duties and turn NATO into a working military alliance under his command. Mamie had to put to one side plans to rebuild their new house in Gettysburg, and throw herself once more into decorating a house that was not their own, although this time the French government had done the Eisenhowers proud, with a country house called Villa Saint-Pierre, a few miles outside Paris, which had once belonged to Emperor Napoleon III. It had a dining room that could seat forty, and the French contributed paintings, tapestries, and furnishings—including, as a kind of gesture of military homage to NATO's Supreme Allied Commander, the bed of Napoleon I from Fontainebleau, which Ike declined to sleep in. The main attraction of Villa Saint-Pierre for Mamie and Ike was its beautiful gardens. *Better Homes and Gardens* noted, with a political comment rare for the magazine, that Villa Saint-Pierre "looked curiously like the White House."[28]

Ike had taken up painting again in these idyllic surroundings. Nancy Mitford, the English novelist, quoted her friend Stephen Tennant, the outrageous aesthete whom she used as the model for the androgynously beautiful homosexual Cedric Hampton in *Love in a Cold Climate*, as remarking with some admiration, "General Eisenhower . . . paints rather well."

Actually, in photographs of Mamie taken for the magazine, the villa looks a good deal more glamorous than the White House. And although Ike carefully confined himself to military matters and the problems of turning NATO into a real army, and to paying ceremonial visits on heads of state with Mamie, the same thought that had occurred to the editors of *Better Homes and Gardens* was bringing him an unending stream of visitors from the United States, all of whom were asking the same question—Would Ike run? Even the president wrote to ask Ike that question, a

* Though this was not widely known at the time, Taft had cancer; he would die of it in 1953.

handwritten letter in which Truman asked, a trifle plaintively, "I wish you would let me know what you intend to do."[29]

He was not alone. Ike's brother Milton and his son John both wrote to ask the same question, and to give Ike their candid view about why he should run. As John put it, the American people deserved a better break than the choice between Truman and Taft; and Milton thought that if the nominees were going to be Truman and Taft it was Ike's duty to run.

As always with Ike, the word "duty" caught his attention, but still he hesitated. He disliked politics—he was outraged that the Democrats had tried to blame the initial disasters of the American Army in the Korean War on the Army budget of 1949 when he had been Chairman of the Joint Chiefs and adviser to the secretary of defense—and he felt strongly that his obligation was to turn NATO into a serious fighting force, which was no small or easy task. Just as he had dragged his feet when the call to promotion and higher command had come in 1941 ("As soon as you get a promotion they start talking about another one. Why can't they let a guy be happy with what he has?"), he dragged his feet now at the prospect of running for the presidency. Throughout his whole career he had always been fiercely ambitious, but he had just as fiercely never wanted to be seen as ambitious. It had never been Ike's way to reach for promotion; he let promotion reach out for him. He would not be hurried into making the decision about the presidency. As Mamie knew better than anyone else, Ike had to find his own way to his destiny—nobody could do it for him.

In April 1951, Truman, deciding at last that he could no longer tolerate MacArthur's insubordination, fired MacArthur—who without consulting anyone in Washington had issued a "communiqué" from his headquarters that read to most people like a direct challenge and threat to communist China. His replacement was General Matthew B. Ridgway. This incident set off a spectacular, if short-lived, political firestorm. A wonderful photograph exists of Ike as an aide informs him of the news. Ike is in uniform, cap and overcoat on, holding his reading glasses and a glove in one hand, and the expression on his face is one of doubt, shock, and dismay. Much as he and MacArthur had differed over the years, it is difficult not to see in his face an instinctive fellow feeling on the part of one five-star general for another at being fired by the former National Guard artillery captain in the White House—this, and perhaps a suspicion that Truman had finally burned his bridges behind him and made his reelection improbable if he chose to run again. Wisely, Ike refused to comment on the firing, though he and MacArthur exchanged gracious notes.

The immense problems of NATO absorbed Ike completely. The British were deeply opposed to the idea of a "united Europe," which they had fought for centuries to prevent; the French, in their proud, touchy way, were ambivalent about placing their Army under the command of an American; the idea that the Germans might take up arms again as part of the alliance scared everyone and outraged the Russians; and the Italians, though professing themselves eager to serve, were in fact prohibited from doing so by the terms of their peace treaty. American conservatives

Ike hears that Truman has fired MacArthur.

doubted that twelve European and American divisions could stop 175 Soviet divisions, and argued that it would be cheaper to use atomic bombs to defend Europe than to waste good money building a European army. But Ike was adamantly opposed to using atomic bombs—he believed that NATO could defend Europe with conventional weapons, and he was appalled at the idea of the atom bomb, which, he thought, would automatically transform even a modest border clash into World War III.

Tirelessly, Ike traveled from NATO capital to NATO capital, soothing bruised nationalistic feelings, making speeches, assuring his listeners that the United States was their partner, and had no aspiration to be their master. "We cannot be a modern Rome guarding the far frontiers with our legions," he told them and American visitors—the idea of an American "empire," attempting to hold in check to American policies and interests whole areas of the world by force of arms, was repugnant to him. The aim was to defend freedom, and that could be done only by free people, certainly with America's help and support, but never by large numbers of American troops alone.

In the meantime, however much Ike might have wanted to avoid politics, his name was still at the center of political speculation at home. In February 1952 an "Eisenhower for President" rally organized by John Hay Whitney and Tex McCrary filled Madison Square Garden—overfilled it, in fact, for nearly 33,000 people crowded into a space with only 16,000 seats—and went on boisterously until the early hours of the morning as the crowd chanted "We want Ike!" The famous American aviator Jacqueline Cochran brought a film of the rally to dinner at Villa Saint-Pierre and screened it for Ike, who "was flabbergasted by what he saw." He wrote a long letter to his old friend Swede Hazlett the next day, describing the experience: "In any event, the two hour film brought home to me for the first time something of the depth of the longing in America today for change. . . . I can't tell you what an emotional upset it is for one to realize suddenly that he himself may become the symbol of that longing and hope." Mamie felt that it was while watching the film of this rally that Ike made up his mind to run for the presidency, but if so, he continued to keep the decision to himself.

He occupied himself, in his few moments of spare time, with gardening and painting. He and Mamie flew to London for the funeral of King George VI; and to Turkey and then Greece, where he and Mamie dined with the king and queen. Much as he might repudiate the comparison of America to the Roman Empire, Ike's role was in every way that of a proconsul, representing America not to one country, but to twelve, while being at the same time the leader and the supreme commander of their military alliance. He dealt with kings, heads of state, and prime ministers as an equal, and was a better and more convincing spokesman for the defense policies of the Truman administration than the president himself, to the great annoyance of Republicans, who wanted Ike to attack Truman on that subject.

Political events were moving faster—it was beyond anybody's ability to slow them down in an election year. In January, Henry Cabot Lodge had announced the creation of an "Eisenhower for President" campaign, later to be named "Citizens for Eisenhower," for the purpose of entering Ike's name in the New Hampshire primaries. Ike had responded with a certain amount of ambiguity and resentment, angered that Lodge had put him on the spot, but affirming, at last, that he had indeed cast his vote as a Republican in previous elections, and agreeing that Americans certainly had a right to band together for a common cause, even if the cause was himself. He would not under any circumstances abandon his post of duty and return home to campaign in New Hampshire, however, despite the pleas of Republicans as different as Harold Stassen and Henry Cabot Lodge—there was great fear among moderate Republicans that Taft would win there.

In the end, it made no difference. On March 11, 1952, Ike won the New Hampshire Republican primary by an overwhelming majority (he received 50 percent of the vote to Taft's 38 percent), and a week later he received over 108,000 votes in the Minnesota primary even though his name was not on the ballot.*

* Senator Estes Kefauver beat Truman in the New Hampshire Democratic primary, leading Truman to declare at the annual Jefferson-Jackson Day dinner eighteen days later that he would not be a candidate for election—and of course leaving the race for the Democratic nomination wide open.

On March 20 Ike announced that he would reconsider his previous refusal to campaign for the GOP nomination. Mamie once again set about the task of packing up their furniture and belongings without knowing where she was going—for the house in Gettysburg was not finished and there was, of course, no guarantee that Ike would win the White House—while Ike prepared to hand over his command of NATO. At the end of May, France presented him with its highest military honor, the Médaille Militaire, at a grand ceremony at Les Invalides, where Napoleon is buried, Winston Churchill gave Ike and Mamie a farewell dinner at 10 Downing Street, and on June 1, 1952, Ike and Mamie returned home.

PART IV

The White House Years

CHAPTER 18

"I Like Ike!"

There is no "royal road" to the White House, not even for a national hero and victorious general, as George Washington was the first to discover. Ike and Mamie arrived home to be greeted by a storm of vicious stories about them released by the Taft forces. It was not the Democrats who were attacking Ike, but his fellow Republicans, thus giving him his first direct taste of politics now that he was a presidential contender and no longer a supreme commander and five-star general. The Kay Summersby story was rehashed, of course, as was the old story about Mamie's alleged drinking problem. In addition, the Taft people spread the rumor that Ike was Jewish, and circulated a photograph of Ike drinking a toast with Marshal Zhukov, "his Communist drinking buddy."[1] Ike did not mind being called Jewish or a communist, though he was neither, but he deeply resented the stories accusing him of having—and continuing—a "secret" affair with Kay, and about Mamie's drinking, no doubt because both of them would pain Mamie deeply. He even complained about them to President Truman, when paying a courtesy call on him at the White House, but that veteran politician merely remarked, from personal experience, that if those were the worst things the Republicans could find to say about Ike, he was a lucky man.

The fact was that Ike faced a rare and peculiar political problem—nobody in either party seriously doubted that he could win election to the presidency, but there was a good deal of doubt about whether he could win his party's nomination. The Republicans had been out of power for twenty years—and had been cut off from all those things that come with the presidency, from Supreme Court appointments to local postmasterships. As with any political party that has been out of power and patronage for a long time, ideological purity had become more important to many Republicans than winning elections. If Ike had run for the Republican nomination in 1948, when the laurels of victory were still fresh, he might have won it easily, but by 1952 the Republican Party seethed with the determination to do away with twenty years of Democratic rule, and to undo as much as they could of the legacy of Franklin D. Roosevelt. Domestically, the party was obsessed with the mission of overturning the New Deal and exposing left-wing "subversives" in government; in foreign and defense policy, it was anxious to undo the consequences of Yalta and to seek out the culprits who had "lost" China and Eastern Europe to the communists. General Marshall had gone to China in 1948 at President Truman's request to salvage whatever could be salvaged from the collapse of Chiang Kai-shek's army, so for Republican diehards, he played the unlikely role of the villain in the "loss" of China. And since Ike (however he may have voted) had served Roosevelt loyally and had "allowed" the Russians to take Berlin and Prague in 1945 instead of getting American troops there first, he too was suspect to those on the right who were eager to revise history, or at least to pin the blame for history on those who had made it.

Thus although Ike had handily defeated Taft in New Hampshire and Minnesota, his campaign to win the Republican nomination would be fiercely and bitterly contested. There were whole areas of Republican ideology about which Ike's opinion was unknown, even to himself; and from the point of view of party stalwarts, he had done little or nothing for the Republicans over the years. In this he resembled Winston Churchill in the late 1930s—the Conservative Party and Tory members of the House of Commons understood Churchill's ambition, and some even admired his brilliance, but over the years he had done very little to further their own (much more modest) objectives, and he therefore had no wide base of support among the party rank and file, whose most cherished ideas he was in any case suspected of not sharing.

Ike was in a similar fix. He had done nothing for rank-and-file Republicans, unless you counted winning the war in Europe; his opinions on a whole range of subjects dear to their hearts were unknown; and his views on foreign policy were deeply suspect. It was all very well for the British to have proclaimed Ike the "First Citizen of the Atlantic," in the words of the former Labour foreign secretary, Herbert Morrison, but that cut no ice with those who didn't trust Britain, who thought that British socialists like Morrison were pinkos, and dreamed of a Taft-MacArthur ticket for 1952.

Nor was Ike easily or instantly converted into a political candidate. As a general, he did not respond well to the backslapping physical familiarities of politics, and was observed to wince or flush with anger when they took place. Ike had always operated best out of consensus, even as a supreme commander, and his skill was in reaching that consensus, so he was not well suited for dealing with the arguments, angry dissent, name-calling, and posturing of politicians. This is not to say that he was politically naive, however. He understood from the beginning that he could win the Republican nomination only by reaching out to those who supported Taft, the conservative midwestern core of the party, in contrast to the wealthy, sophisticated easterners who supported Ike; and he did his best to convince the conservatives that he stood beside them on such matters as repealing labor legislation, lowering taxation, decentralizing the federal government (for the real old-guard Republican right, Washington, not Moscow or Peking, remained the enemy and the source of all evil, and the New Deal was synonymous with Marxism), and repudiating Yalta. Ike neither looked nor felt comfortable in this role, and when the subject of General Marshall was raised, his face turned dark red with tightly suppressed fury. Ike's views on social issues were moderate and sensible. He strongly opposed the demonization of such programs as Social Security and remarked, "Should any political party attempt to abolish Social Security . . . you would not hear of that party again in our political history. There is a tiny splinter group, of course, that believes you can do these things. Among them are Texas oil millionaires. Their number is negligible and they are stupid." In many areas the relationship between Ike and the Republican Party was, at best, a shotgun marriage, and one of those that gave him the most trouble was the rising popularity of Senator Joseph R. McCarthy, and of the ideas that had brought McCarthy to notoriety and prominence.

Not that Ike was, in the phrase of the day, soft on communism, despite his friendship with Marshal Zhukov and his cordial relationship with Stalin; but he was familiar enough with Washington during the Roosevelt and Truman years to realize that most of the senator's victims were small fry, and many of them were guilty of nothing worse than having a left-wing point of view at a time when the Russians were our allies. More difficult was the seemingly absurd notion that General of the Army George C. Marshall, of all people, was part of an infamous left-wing conspiracy so "immense" and dark as "to dwarf any in the history of man," in McCarthy's intemperate words. Not only had Marshall been Ike's mentor; he was a man of broad vision, impregnable integrity, and total, selfless honesty—a patriot who had served his country well as Army Chief of Staff, secretary of defense, and secretary of state; in many respects the architect of the Allies' victory in World War II, and of Europe's recovery after it. His loyalty to two very different presidents, as well as to his country, was beyond any rational doubt, and yet McCarthy had picked shrewdly in attacking him as a traitor, banking on the fact that Marshall was too remote, too dignified, and too sure of his own place in history to respond, or to attempt to justify himself. Marshall would not stoop to McCarthy's level by answering the attacks against him, and therefore, inevitably, some of the mud stuck, and stuck all the more easily because to many Republicans

Marshall stood for a whole list of things they hated—remote, patrician generals; Washington bureaucrats; the foreign policy of President Roosevelt; the "secret deals" made at Yalta; the inexplicable and catastrophic "loss" of China to the communists; the shameless subordination of America's interests to those of the wily Europeans and the scheming, imperialist Britishers. To those who believed in this particular conspiracy theory, Marshall's prominence, and indeed the Marshall Plan itself, by means of which billions of dollars had been poured into Europe to jump-start European industry and repair the ravages of the war, made him seem like the chief conspirator. Hard as it may be to believe today, given the enormous and enduring success of the Marshall Plan and NATO, both of which were central to the recovery of Europe and to America's sustained growth and prosperity in the second half of the twentieth century, Taft and his followers voted against both, with the fervent support of the bulk of the GOP. Far-fetched and hysterical as McCarthy's accusations against Marshall might be, they did not seem so to many people, perhaps most of those who would be voting for the Republican candidate for the presidency at the convention in Chicago in 1952.

Under these circumstances Ike had his work cut out for him to win over enough delegates to beat Taft, despite his victory in New Hampshire—after all, he himself had been close to Marshall and had carried out the policies of Roosevelt and Truman in Europe. He made two decisions that were to have a strong effect on his presidency, however. The first was to adopt as a principle "Thou shalt speak no ill of other Republicans,"[2] even if the other Republican was Joe McCarthy. The second was to pick Governor Sherman Adams of New Hampshire as his political equivalent of General Bedell Smith. Adams was flinty, granite-faced, abrasive, and, like Smith, a man who found no difficulty in saying "no" to people—in short, much as he got on other people's nerves, Adams was exactly what Ike needed in the campaign, and would need even more in the White House.

Ike and Mamie traveled from New York to Denver, via Detroit. Initially, his campaign for the nomination suffered badly from poor organization and his own diffidence. In Abilene television cameras caught rows of empty seats in a downpour so strong that Ike had to slog through ankle-deep mud to get to the podium, with his rain-soaked trousers rolled up to his knees, and once there he seemed unable to decide whether he had wanted to go to Berlin and been prevented from doing so by Roosevelt or had himself concluded that it wasn't a military objective. This prompted Nancy Mitford to write to Evelyn Waugh, "General Eisenhower has turned into pure knockabout, hasn't he. I am greatly enjoying it all—heavenly . . . when he lost his place in his speech." Eventually, Ike's own competitive spirit took hold, while his backers on the East Coast scurried to produce a competent campaign staff for him. Photographers were firmly discouraged from taking any pictures of Mamie drinking, even from a glass of ice water. Since the question of Berlin seemed to matter so much, Ike wisely decided to answer it sensibly and truthfully—it would, in his judgment, have cost the lives of at least 10,000 American soldiers to get there, and he did not think their families, or those in the audience who had had sons in the war, would have thought their deaths worthwhile in order to capture a "worthless objective." With that frank answer, the question went away, and ceased to trouble anyone but revisionist historians. It cannot be said that Ike enjoyed campaigning—though his grin was as big as ever—but he was getting better at it. All the same, he was learning on the job, and it was an uphill struggle. Observers, even as far away as England, felt that his election to the presidency was "an event thought to be decreasingly probable,"[3] as John Colville, the prime minister's private secretary, reported Churchill's opinion. Churchill retained a sentimental regard for the Democratic Party, although if the Republicans were going to win, he preferred Ike to Taft.

For nearly two weeks Ike and Mamie traveled around the Midwest, Taft country, where Ike was applauded warmly as a general but not nearly so warmly as a political candidate—curiously, the part of the country where he was born and raised was considered least likely to want him as the Republican nominee. The general view there of his chances of defeating Taft were about the same as the view at 10 Downing Street, and a poll by the Associated Press, generally regarded as accurate, predicted that Taft would have 530 delegates to 427 for Ike on the first ballot. In those days, before the national conventions had been redesigned to make for better television, and also before the number of states holding primaries and caucuses had increased to the point where the result of the conventions tended to be decided before they took place, the process of nominating a presidential candidate was still entwined with arcane parliamentary procedures and old-fashioned backroom politics—last-minute surprises were still possible. To paraphrase Bismarck, national conventions, like the making of sausages, were still a process that did not bear looking at too closely.

The Republican convention of 1952 was no exception; although it was the first convention to be televised, most of the significant decisions were made offstage. Since the keynote speaker was none other than General MacArthur, there was still a good chance, by clever stage management and astute politics, for a Taft-MacArthur ticket to emerge if the balloting went on long enough. Ike continued on his whistle-stop tour through the Midwest, declaring himself to be against inflation, dishonesty and corruption in government, world communism, socialized medicine, and sending a U.S. ambassador to the Vatican, and in favor of lowering the voting age to eighteen and not tinkering with price supports for farmers. (As one politician tartly observed, "It looks pretty much like he's for home, mother, and heaven," but Ike needed nobody to teach him that the road to the White House is paved with clichés.) Meanwhile the real action revolved around what Ike referred to delicately as "charges of irregularity in the naming of certain state delegations."

What this amounted to was that Senator Taft, in his capacity as "Mr. Republican" and as a lifelong professional politician whose own father had served as president, had taken good care, well in advance, to pack the delegations from certain southern states with his own loyalists. Ike was made aware of this early on, and while he might not yet be a full-fledged politician, he was still a general—he decided right away to forget about the Georgia and Louisiana delegations, since however corrupted they might be, they were small, and to concentrate his attack on the much larger Texas delegation. The nub of the problem there was whether any citizens could vote in a party's precinct convention, or only those who could show proof of being a member of that party. The law in Texas looked clear enough—party precinct conventions were open to "any qualified voter"—but the pro-Taft's Republican regulars insisted that voters must first sign a pledge to "participate in Republican politics in 1952." When pro-Ike voters, who turned up in large numbers, agreed to sign the pledge, Taft's supporters "walked out, convened their own rump conventions, and elected a rival slate of representatives."[4] All over Texas there were fistfights and angry protests, but the credentials committee for the state Republican convention committee, firmly controlled by Taft's supporters, eventually gave a "lopsided" majority of the convention seats to delegates pledged to Taft, arguing that they were avoiding "mob rule" and protecting the party from domination by "Republicans for a day."

Ike, who had hoped to be "above politics," immediately found himself plunged deep into the sordid heart of the political process. He decided to go to Dallas, where he charged that "rustlers stole the Texas birthright instead of Texas steers,"[5] and made it clear to Texas Republicans that he would fight. Perhaps more important, he decided to attend the convention in Chicago, which he had hoped to avoid—in those days the custom of having the front-runners attend the convention in person was relatively new, and Ike's preference would certainly have been to imitate Grant, who sat at home on his front porch and was nominated by acclamation. But Henry Cabot Lodge and Sherman Adams argued that it was Ike's duty to let his supporters see him, and that his presence was necessary to win the fight over the composition of the Texas delegation, since there were now two rival delegations from Texas in Chicago: one pledged to Taft and one to Ike.

The problem of the two competing Texas delegations had been dumped in the lap of the Republican National Committee, which was largely composed of men and women who owed their presence there to Senator Taft. Word was spread by Taft's people that the pro-Ike delegates were not really Republicans at all, but Democrats who wanted to see a Republican nominated who could and would be easily defeated in November, as opposed to the Taft-MacArthur ticket.

The existence of the two rival Texas delegations dominated the news in the weeks before the convention. Ike, now that he had agreed to attend it, shrewdly staged his own arrival in Chicago by going directly from the station to address a reunion of the Eighty-Second Airborne Division before even checking into his hotel, and was greeted with rapturous cheers by men who had served under him on D-day.[6] Here at least was something Taft could not copy. Once in his hotel, however, Ike was faced again with the knotty question of who should vote on the "contested" Texas seats, since by the arcane rules of the convention many of those whose seats were contested could vote on who was and who was not qualified. Ike's team had been fighting for this question to be resolved by proposing a change of rule, which, in a brilliant stroke of public relations, came to be known as the "Fair Play Amendment," and which divided most of the more liberal eastern delegations from those who were for Taft before the convention had even begun. All the same, the Associated Press poll had been uncannily accurate; on the eve of the convention Ike counted 530 delegates for Taft and only 427 for him—Taft needed fewer than 100 votes to reach the total of 604 that would make him the nominee.

Why Taft and his associates assumed that Ike was a political innocent is hard to see—after all, Ike had held his own against such heavyweight political figures as Churchill, De Gaulle, and Stalin, and had very often gotten his own way. That Ike didn't like politics is clear enough, but this doesn't mean he wasn't good at politics or didn't understand politics. He moved quickly too. On his first night at the Blackstone Hotel he had the leaders of the Pennsylvania and Michigan delegations to dinner, sought their support for his nomination, and received the leader of the Minnesota delegation after dinner. To all of them, he stressed the importance of the Fair Play Amendment, and all of them were cautiously receptive, fully aware, as Ike himself was, that a vote for the amendment would amount to a vote for him. Many of the states that were happy to support any change in the rules which seemed fair, honorable, and likely to take control of the party out of the hands of Taft were also "favorite son" states, including Minnesota, whose favorite son was Governor Harold Stassen; and California, whose favorite son was Governor Earl Warren. If these states voted for the amendment at the beginning of the convention, most of the delegates would take it as a sign that they were going to switch their votes to Ike later on, in which case Taft might soon lose his position as the front-runner.

The amendment passed by more than 100 votes, but Taft's political skill and the favors owed him were equal to the defeat, and the next day the credentials committee voted to seat the pro-Taft Georgia delegation. Taft's victory was short-lived, however. Ike had been right to concentrate on Texas, a place where, in the eyes of many outsiders, dirty politics was the rule rather than the exception, and the exclusion of the Eisenhower delegates seemed a good example of using the rules to thwart the man with the most support. So long as a vote for clean politics was a vote for Ike, he could hardly lose. "The next day convention reversed the recommendations of the credentials committee and voted (607 to 531) to seat the pro-Eisenhower Georgia delegates; and voted by acclamation to seat all the Eisenhower delegates from Texas."[7] After that, it was all over but the shouting.

Mamie was in bed with an infected tooth, while Ike, with his brothers and a few friends, sat in the living room of his suite trying not to listen to the interminable polling of the delegations on television (even in those days, the television reporting of a convention was an exercise in sustained boredom second only to Oscar night). At the end of the first ballot, Ike had 595 votes to 500 for Taft and 81 for Earl Warren (California was still holding out for some reward for dropping its favorite son in favor of Ike), nine votes short of victory. But then the head of the Minnesota delegation demanded the floor and switched the nineteen votes his state had cast for its favorite son, Stassen, to Ike. Eisenhower was the Republican candidate for President.

He went down the corridor to tell Mamie the news, and later reported that neither of them had any doubt he would win the election in November. It was the nomination that had been a close call. Typically, Ike next fought his way through packed crowds to cross the street and pay a courtesy call on Taft—the crowd in the lobby of Taft's hotel, he noted, was mostly hostile, since it consisted of Taft's brokenhearted supporters. Many of the women wearing Taft buttons were openly weeping.[8]

Ike pushed his way through them politely, remarking that he sympathized with their distress, and went upstairs to meet with Taft and his sons. He shook hands with Taft, and told his rival that he hoped they could be friends. Taft, who seemed slightly dazed by the fact that he had lost, asked if Ike would mind giving the press photographers a chance to record the moment, and Ike said he wouldn't mind at all, so they were photographed together, apparently standing in a hotel corridor, surrounded by cops, reporters, and Taft's supporters, Ike looking younger than his sixty-two years and bemused, Taft looking glassy-eyed, like a man who has just been hit over the head and still can't believe what has happened to him. Then Ike made his way back to his own hotel to face his first big decision as the candidate—his choice of a running mate.

There is, of course, no rule in American politics that a president has to know, let alone like, his vice president. Roosevelt barely knew Harry Truman at all when he picked Trumen as his running mate in 1944; and Ike's successor, John F. Kennedy, could barely disguise his contempt for his fellow senator Lyndon B. Johnson (whom Jackie Kennedy often referred to unforgettably as "Colonel Cornpone"). Nor is it necessary for a vice president to be a memorable or beloved figure—Truman's vice president, Alben Barkley, is scarcely remembered at all except by political scholars and players of Trivial Pursuit, and was dismissed by Averell Harriman with the acid comment that the only person in politics more garrulous than Barkley was Lyndon Johnson.

Ike had managed to work for more than two years with Monty as his deputy commander, a man who sometimes seemed to exist for the sole purpose of irritating him, so in making his choice of a running mate he was hardly likely to have been seeking a friend and companion. He had seen as little of Monty as he could during the war—indeed, one of Bedell Smith's most important jobs was keeping Monty away from Ike—but the distance between the two men had been nothing compared with that between Roosevelt and Truman, whom the president managed to ignore almost completely.

In *Mandate for Change*, Ike's personal account of his first term in office, he devotes about the same amount of space to choosing his running mate as he does to his meeting with Taft after winning the nomination. This would seem to imply at least that the choice he had made wasn't much on his mind in 1963, when he wrote the book, and perhaps also that it didn't take up much of his time in 1952. That too is not unusual—the choice of a vice presidential candidate has often been, in American history, hasty, casual, and done at the last minute. Indeed, Ike's account of the choice has a slightly defensive ring to it, as if he still needed to justify it. He makes it clear, to begin with, that he would have preferred Henry Cabot Lodge, Thomas Dewey, or Herbert Brownell, all of whom were unavailable for one reason or another; and names on the handwritten list he carried in his billfold in case his first choices declined to accept are not exactly inspiring: Senator Richard Nixon, Congressmen Charles Halleck and Walter Judd, and Governors Dan Thorton of Colorado and Arthur Langlie of Washington. Of these, Nixon was certainly the best known, because of his vigorous prosecution of the case against Alger Hiss and his electoral victory against Congresswoman Helen Gahagan Douglas, in which Nixon had accused Douglas of communist sympathies, describing her memorably, if ungallantly, as "pink right down to her underwear."

Ike had met Nixon once, when the young senator paid a call on him in Paris. Part of his appeal for Ike was his youth. Ike would be sixty-two when he reached the White House, and Nixon was only thirty-nine—this was, without question, an actuarially more sensible way of selecting a vice president than picking an old party warhorse whose cardiovascular system was even more stressed than Ike's. Still, there were other, and more potent, reasons for selecting Nixon. First of all, it was Nixon, always sensitive to which way the political wind was blowing, who had helped persuade the California delegation to vote for the Fair Play Amendment, thereby earning the gratitude of everybody in the Eisenhower camp. (The political rewards for California's shift toward Ike at the convention would be considerable—Nixon himself would become vice president, and Earl Warren, who had been the California delegation's favorite son candidate, after considerable Sturm und Drang, would be given a consolation prize in the form of appointment to the Supreme Court as Chief Justice, though as Ike would later complain, it "was the biggest damn fool mistake I've ever made in my life.")[9]

Perhaps more important to Ike was the fact that the choice of Nixon as his running mate partially shielded him from what would otherwise have been strong doubts about his candidacy among right-wing Republicans.

Nixon was no McCarthy—he was cleverer in his choice of targets and more flexible in his opinions than the junior senator from Wisconsin—but his credentials as a witch hunter were strong enough to win him the respect of those in the Republican Party who still worried that the federal bureaucracy was riddled from top to bottom with communists and fellow travelers. The choice of Nixon would go a long way toward reassuring right-wing Republicans and Taft's loyalists that they were not about to be swamped by East Coast Ivy League Republicans. Besides that, Nixon, unlike Ike, lived, slept, and breathed politics—he had no other interests, no hobbies, and no form of relaxation, and required none. Politics was his consuming passion, his raison d'être, and his obsession; and his knowledge of the Republican Party was encyclopedic. If he had not existed, Ike would have had to invent him.

When Brownell called Nixon and asked him to come over to the suite, the meeting of the two men was "cold" and "formal," very much that of a five-star general briefly welcoming a bright young junior officer to the "team"—something which would not change in eight years, despite manful efforts on Nixon's part to make the arrangement sound like a partnership, or a father-son relationship. Ike had found what he needed, for the moment, in Nixon—he did not intend to pay for it in the phony coin of political friendship.

The next item to confront him was the party platform, and with this Ike made his final bid to hold the party together and mollify the supporters of Taft. The platform was, as is so often the case with such documents, enormously long, in part drafted by extremists, and virtually unread. It began with a stirring preamble charging the Democrats with twenty years of "vicious acts," in which they had tried to achieve "national socialism,"* fostered class strife, "shielded traitors," concealed "corruption in high places," and "shamed the moral standards of the American people." This was merely a warm-up to charges that the Democrats had "lost the peace,"

* In this context, "national socialism" should not be mistaken for German "National Socialism." The drafters of the Republican Party Platform of 1952, perhaps needless to say, had in mind socialism in America, as represented by Social Security and other "liberal" New Deal programs, rather than Adolf Hitler's National Socialism.

"abandoned friendly nations," weakened the Nationalist Chinese, and allowed the Chinese communists to win power, and much more. Going from the sublime to the ridiculous, it pledged the candidate to "a more efficient and frequent mail delivery service," the support of the family farm (but with reduced federal farm subsidies), and the abolition of federal rent control. To Ike's distress, the platform contained what was virtually a shopping list of Taft's ideas on foreign policy and defense, many of them drawn up by John Foster Dulles, who would become Ike's secretary of state. Yalta was to be repudiated, the "captive nations" of Eastern Europe were to be freed, the iron curtain was to be rolled back (but not by American arms or overt interference), and concern for freedom in Asia was no longer to take second place to concern for Europe (a covert way of protesting against the assumptions of NATO, which Ike shared).

Wisely, Ike's acceptance speech skirted many of these issues—he made no specific promise to improve mail service, for example, or to abolish rent control, which would have lost him a considerable number of middle-class votes in New York City—but by accepting the platform as it was written and picking Nixon for the vice presidential nomination, he had gone as far as he could go in placating the Republican old guard. They never became uncritical enthusiasts for Ike, even though he would hold the White House for eight years, and he tried as best he could to keep his distance from them. Ike was an American from Abilene, but he was also a good European, perhaps even a great one; and his view of life was rooted in common sense, decency, and tolerance, not in ideology—in short, he remained all his life his mother's son. It would be twelve more years before the Republicans nominated a candidate who shared their core ideals, and twenty-eight years until they finally managed to actually elect a candidate who claimed to share them.

The earliest national election Ike could remember, he later remarked, was that of 1896, in which William McKinley beat William Jennings Bryan. Ike had marched in a Republican parade in Abilene, carrying a flaming torch, on a night he remembered clearly.[10] Now, fifty-six years later, he was running for the presidency himself, while his son John was on his way to Korea, where the war was still sputtering on interminably. With his father's blessing, John had insisted on serving in a frontline infantry unit, and he and Ike gently discussed the pressures that might be used against him, up to and including torture, if he was captured by the North Koreans. John quietly promised that he did not intend to be captured alive. This conversation set the tone for Ike's thoughts on the subject of Korea, which was to become a source of angry disagreement between himself and President Truman in the last weeks of the campaign.

No subject loomed larger than the Korean war, which had been going on since June 1950, and had so far cost the lives of over 20,000 Americans (in addition, 91,000 had been wounded and 13,000 were missing or captured), with no end in sight. These were substantial numbers for what was described as a "police action"; and while twenty-one countries had contributed military contingents to the United Nations forces in Korea, the army of the Republic of Korea (ROK) and the United States armed forces made up the bulk of the United Nations command, and of course accounted for most of the casualties. On the subject of Korea, Ike had been cautiously reticent, reluctant to make matters more difficult for Truman, or to predict what he would do differently if he won the election.

In the half century since the Korean war, it has become customary to contrast Ike's caution with MacArthur's swagger and willingness to extend the war, even if that meant the use of atomic bombs, but this is fair to neither general. Certainly, Ike always sought to avoid MacArthur's arrogant manner and the contempt in which the general held civilian politicians; but this is not to say that they were in fundamental disagreement in terms of strategy. Like MacArthur, Ike was opposed to fighting any war incrementally—that is, by increasing allied forces by dribs and drabs as the enemy forces increased. As an American general, Ike, like Grant, believed—it was his core belief, militarily speaking—in using America's overwhelming superiority in weapons and the full strength of its manpower to wear down the enemy. He did not believe in fighting small wars, in which American forces might be outnumbered, or guerrilla wars, in which no decisive victory was possible. He believed that if America had to fight, it must do so by bringing its full strength to bear on the enemy, as it had done on D-day, at the Battle of the Bulge, and in the invasion of Germany; and also that the mission must be clearly defined, and that having achieved it the bulk of the Army should be brought home as quickly as possible. He opposed involving the United States in "colonial wars" with their endless insurgencies and need for an embattled long-term occupation, like the one the French were fighting in Indochina.

Ike did not think that the Truman administration had given MacArthur the troops or the support he needed to win in Korea, certainly once the Chinese came in; nor did he think that MacArthur was necessarily wrong, once China was in the war, to suggest the use of atomic weapons, since they were the only practical way to match the enormous masses of troops the Chinese could send across the border. Neither Ike nor MacArthur viewed the atomic bomb as an "unthinkable" weapon. Ike was fond of pointing out that every major advance in weaponry had once been described as "unthinkable"—the repeating rifle, the machine gun, gas, the tank, the bomber—but "all weapons in due course became conventional weapons,"[11] as Ike would tell Winston Churchill. And the atomic bomb was, in any case, not an "unthinkable" weapon; it had been used—against Ike's advice, true—on the Japanese, and arguably had brought about their surrender. If America was placed in the position of having to fight the communist Chinese, then Ike, like MacArthur, could see no way to prevail except by using atomic weapons to destroy their ports and their lines of communications.

This is not to say that Ike wanted to use atomic bombs, any more than MacArthur did. It was his strong feeling that fighting a "land war" on the Asian continent was a mistake, and one the United States should not repeat—that, indeed, the politicians should not in the first place have allowed the Korean war to escalate to the point where the Chinese were brought into it. As a military man, he believed that it was for the politicians to define the mission, and then to give the commander the means and the freedom of action he required to carry it out. In Korea, the mission had never been clearly defined. Ike's old critic, Field Marshal Montgomery, who had been his Deputy Commander in NATO, and still filled that role, his personality as querulous and irritating as ever, had defined very well the problem in Korea, and there is no evidence that Ike disagreed with him.

Politicians, Monty argued—he was thinking of Truman—must say what they wanted, they must ask for a plan to end the war. "Do you want the integrity of South Korea? Do you want to take North Korea and South Korea and weld them into one nation? Or do you want to take the boundary of South Korea and shift it northward to a better defensive line? If you will say what you want, the fighting men could produce a plan to achieve it and tell you how much it would cost in blood and money. Until you say what you want, they can't."[12] As was so often the case with Monty's ruthless logic—part of his genius was that he was a great *simplificateur*—he hit the nail right on the head. The Truman administration had never defined its goal in Korea, beyond teaching the North Koreans a lesson, and had allowed MacArthur, once he tasted victory, to overrun all of North Korea, thus bringing the Chinese into the war, which was exactly what everybody had most wanted to avoid.

Ike ruled out accepting the continuation of the status quo, which involved accepting heavy casualties over the indefinite future for little or no gain, or renewing the offensive to the north by "conventional means," which could only be costly in lives. The choice was between using all the weapons available to United Nations forces, including atomic weapons, to deprive the enemy of its "sanctuaries" across the Yalu, and a negotiated settlement. Ike already guessed that it might be impossible to gain the latter without threatening the former—and making that threat believable enough that the North Koreans would be forced to the bargaining table by their Chinese and Russian allies—but he decided to hold off on making any decision until he had visited Korea himself and seen what was happening. Ike would not announce that he was going to Korea until two weeks before the election, when the announcement would almost guarantee his victory. This infuriated President Truman, who would complain that if Ike had a way of ending the war in the Korea, he should tell *him*, and save a lot of lives, instead of waiting. This was to underrate Ike, however. He did not have a "plan" as such, but he understood that the threat of a widened war without restraints on the type of weapons used would have more effect coming from a five-star general than from a civilian, and that his going to Korea was as much a warning signal for the enemy as the election ploy that Truman thought it was.

Although significantly less important than Ike's views on Korea, the two things about the campaign of 1952 that almost everybody still remembers were Nixon's famous "Checkers speech," and the deletion of a few phrases praising General Marshall from a speech he was about to give in Joe McCarthy's home state of Wisconsin. Since one of the major themes of the Republican campaign was putting an end to twenty years of Democratic cronyism, corruption, negligence, and "crookedness," it was a severe body blow to Ike's "crusade" for morality in Washington when the *New York Times* revealed that Nixon was the beneficiary of a slush fund provided by a group of Californian millionaires. Matters were not helped when Nixon instinctively claimed that the story was a communist "smear." Almost everybody in the party wanted Ike to drop Nixon from the ticket, and no doubt Ike would have been happy enough to do so, but he recognized very realistically that this would be tantamount to scuttling his own chances of winning. It was Ike who advised Nixon to make a nationwide television address (the actual idea came from Thomas Dewey), and when he did—it reached a record audience of 60 million people—it was also Ike who let his vice presidential candidate twist in the wind before he made up his mind what to do. Eisenhower's biographer Stephen E. Ambrose claimed that Nixon finally told Ike it was "time to shit or get off the pot,"[13] and that Ike never forgave Nixon for using that kind of language to him. That is possible—Ike himself could curse as fluently as any other soldier, but he had always been treated with respect as a general, and he would certainly not have appreciated a phrase like that spoken to him by a former temporary wartime lieutenant commander of the United States Navy. In any event, Nixon's speech, in which he brought tears to the eyes of millions of voters by mentioning the gift to the Nixon family of a little dog named Checkers and proudly declared that Pat Nixon owned a "respectable Republican cloth coat" instead of a mink, produced a flood of mail in his support—he had asked people to write or wire to the Republican National Committee—and convinced Ike that he should stay on the ticket. The incident did not create any warmth between the two men. Ike resented being dragged into the subject of Nixon's financial affairs, and was furious that as part of the fallout from the Checkers speech both candidates and their vice presidential running mates had to reveal their own financial statements and tax returns; and Nixon never forgave Ike for delaying the decision to keep him on the ticket. But then there had not been much warmth between them to begin with.

The incident regarding Marshall was a greater embarrassment for Ike than Nixon's Checkers speech, since it made him look like a captive of the right, as well as disloyal to General Marshall. The accepted version of the story is that Ike's speechwriter included some lines defending Marshall in a speech Ike was to make in Wisconsin, and that Senator McCarthy and some of the old-guard Republicans then persuaded Ike to remove the lines, which constituted a deliberate affront—or challenge—to McCarthy in his own state. Unfortunately, the text of the speech had already been released (or leaked) to the press, and Ike's decision to appease McCarthy made front-page news all over the country. Ike may have underestimated how sharply the media would react, and perhaps he hadn't yet faced up to the fact that leaks to the press were the common currency of politics, even among those he trusted; but much as Ike admired Marshall and disliked McCarthy, he knew he had to keep the two sides of the Republican Party together at any price. Without any claim to being a political philosopher, Ike would have agreed with Disraeli's advice, "Never complain and never explain," and was wise enough to know that the best thing for him to do was to hunker down, say nothing, and wait for the storm to blow over, which it did. If Marshall's feelings were hurt, he was too dignified to show it, or say so, and in the meantime the right-wing Republicans who hero-worshipped Joe McCarthy were mollified. There are those who still criticize Ike for this today, but it is hard to see why: he was, as he had been during the war and at NATO headquarters, the leader of a coalition, and his first duty was to hold the coalition together. He had made greater compromises before, and with people whose reputations were a good deal more checkered than Senator McCarthy's (Admiral Darlan, for one); and if nothing else, he had learned that these things tended to sort themselves out in time, if you were patient. Given enough rope, McCarthy would hang himself, and Ike would not need to pull the trap himself—and that is exactly what happened.

Despite these embarrassments Ike could probably still have beaten his opponent, Governor Adlai Stevenson of Illinois, merely by sitting at home on his front porch, if he had had one, but that was not in his nature. Leaving his campaign headquarters in Denver (he had chosen Denver because it was Mamie's hometown) on August 24, he set out to cover as much of the country as he could, putting in a remarkable 30,505 miles by air and 20,871 miles by rail, appearing in 232 town and cities in forty-five states. In the course of all this he also transferred his headquarters to New York City—the house at 60 Morningside Drive that was provided for the president of Columbia University was still as much of a home as the Eisenhowers had. "Each of those miles cost something in toil and sweat, and each day ended invariably in weariness,"[14] Ike later remarked. But in fact he was indefatigable, whether he was speaking at the National Plowing Contest in Kasson, Minnesota (where he gave the audience his views on agriculture); or being serenaded at two o'clock in the morning by a crowd of students in Austin, Texas, singing "The Eyes of Texas Are Upon You" to the candidate and his wife; or in Salisbury, North Carolina, where he and Mamie were photographed in their bathrobes at five-thirty in the morning on the rear platform of their railway car, oddly nicknamed "Look Ahead, Neighbor."[15]

Mamie's presence was an enormous asset for Ike, especially since Stevenson was divorced and had not remarried; and it is very apparent that, having been kept from his side during the war, Mamie was determined to make up for lost time by sticking close to him in her new role as the candidate's wife. It was one of the miracles of the campaign that Mamie, who had hitherto been thought shy, was transformed into a brilliant and enthusiastic political campaigner, perhaps as a way of demonstrating to Ike that she had nothing to learn from Kay Summersby. Two campaign theme songs were written for her; she was interviewed by Hedda Hopper, in the days when Hollywood newspaper columnists still counted for something; and she spoke at countless Republican women's clubs along the way. In her own way Mamie was as tireless as Ike, and even more approachable and friendly. This was probably the last old-fashioned, whistle-stop presidential campaign in American history, and surely the last in which one of the candidates was already a hero and a figure of international importance. When Ike visited Boston in mid-October, his motorcade was mobbed by students as it passed by Harvard University, not protesting, but so enthusiastic that they brought it to a complete halt, plucking off Ike's buttons and pieces of his coat and his car as souvenirs. Whatever chance Stevenson had of winning the election evaporated on October 24, when Ike announced, in Detroit, that if he was elected he would go to Korea.

To nobody's surprise, Ike won—by a margin of more than 6.5 million (roughly 55 percent of the popular vote), and 442 of the 531 electoral votes.

There were plenty of voters alive in whose memory there had never been a Republican president.

Ike had broken twenty years of defeat with a landslide victory.

Because Ike's main concern was peace—his memoir of the second term of his presidency would be called *Waging Peace*—and because we know that he succeeded in keeping the United States out of war for eight years, a kind of peaceful interregnum between the Korean war and the Vietnam war, it is hard for us to understand that placing a general in the White House, even so popular a general as Ike, caused a great many people anxiety. Even someone who had worked as closely with him as Churchill was alarmed at the prospect. Churchill wrote Ike a letter of congratulations, but it was considerably less warm in tone than his letter to Truman, and he told John Colville, "For your private ear, I am greatly disturbed. I think this makes war so much more probable."[16]

In fact Ike's first move, as he had promised, was to put an end to the war America was already in. Less than a month after the election, he left for Korea, accompanied by General Omar Bradley, Chairman of the Joint Chiefs of Staff, and a mixed bag of civilians and brass hats: Charles E. Wilson, the former CEO of General Motors and secretary of defense–designate;

Herbert Brownell, attorney general–designate; James Hagerty, Ike's new press secretary; and General Wilton B. Persons, an old Army buddy of Ike's. Along the way, they would pick up Admiral Arthur W. Radford, the Commander in Chief, Pacific, at Iwo Jima. Ike stayed in Korea only three days—the shortness of the visit was one reason President Truman, still smoldering over Ike's promise to go to Korea, condemned the trip as "a piece of demagoguery"—but it was enough to tell Ike what he needed to know, and to make him comment, "War is the saddest of all human activities."[17] He flew over the front, much of it along the Thirty-Eighth Parallel, in a small Army light plane used for artillery spotting, and took a good look at the terrain below. It was mountainous, rocky, snow-covered, and desolate, and the Chinese had gone to the enormous trouble of digging tunnels through the mountaintops so that they could hide their artillery pieces from air attack and then draw them out again unharmed to resume fire when the aircraft left. Everywhere there was evidence of their industry and enormous manpower. As a general Ike had no doubt about what any attack against the Chinese fortified positions would produce: "It was obvious that any frontal attack would present great difficulties." What Ike was looking at, squeezed in behind the pilot in what amounted to the military equivalent of a Piper Cub, made Tunisia look easy, and he applied to Korea the lesson of that campaign, and drew the logical conclusion: "Small attacks on small hills would not end this war." The second conclusion he drew, after meeting with President Dr. Syngman Rhee of South Korea, was that here was an Asian Charles De Gaulle—a stubborn, determined, courageous man who would not hesitate to give his American ally as much trouble as he gave the enemy, and to whom the word "compromise" was entirely unknown.

From these two conclusions Ike decided that the war would have to be ended. Pressure would have to be put on both sides to agree at the truce negotiations, for President Rhee was as unwilling to give up his dream of a renewed full-scale attack as the North Koreans and the Chinese were to accept the Thirty-Eighth Parallel as the limit of their ambitions in the Korean peninsula. "We could not tolerate the indefinite continuance of the Korean conflict," Ike decided. "The United States would have to prepare to break the stalemate."[18] He flew from Korea to Wake Island, and there boarded the cruiser USS *Helena*, and while aboard on his way to Pearl Harbor he completed his choice of cabinet appointments.

Korea, Guam, Wake Island, the *Helena*, Pearl Harbor—there was something crisply military about the president-elect's itinerary, but it was not reflected in his cabinet. The most prominent and most influential member was John Foster Dulles, Ike's choice for secretary of state, who was not everybody's favorite—Winston Churchill was soon saying that "he would have no more to do with Dulles, whose 'great slab of a face' he disliked and distrusted." Dulles was self-righteous, boring, opinionated, and loquacious; but even those who disliked him could not deny his qualifications. He was the grandson of one secretary of state and the nephew of another, had served as a diplomat since the age of nineteen, and was a distinguished international lawyer. The Europeans to a man would hate him—partly because he talked too much about "rolling back" the iron curtain, which they thought would lead to a nuclear war; and partly because he was humorless and preached to them, reminding them of President Woodrow Wilson at the Versailles peace conference. Ike recognized in him an unambiguous honesty and an unlimited capacity for hard work and details.

"Engine Charlie" Wilson's reputation has been considerably diminished by his famous remark, "What's good for General Motors is good for America," a statement that was politically unwise in the 1950s and rings hollow today, now that GM has become the beached whale of American industry. But Ike was not wrong in thinking that a secretary of defense ought to have some experience of running a big organization, and in those days GM was the biggest in the world.

The rest of the cabinet was not such as to raise eyebrows—the members were solid, white-bread Republicans. Indeed the only ones who were at all unusual were Ezra Taft Benson, the secretary of agriculture, who was not only a Taft loyalist but a member of the Council of Twelve, the governing body of the Mormon Church; and Martin Durkin, the labor secretary, who was the head of the AFL plumbers' union, and whose selection Senator Taft called "incredible," a judgment shared by the vast majority of Republicans. Still, Ike deserves credit for breaking with tradition—he tried to pick the best men he could find, and didn't worry much about the fact that one of them was a strong union man and another a Mormon elder.

The path to the inauguration was roughened, as is so often the case, by increasingly bad feelings between the outgoing and the incoming president. Truman's peppery nature did not help matters, of course—Truman was still angry about Ike's trip to Korea and was also offended by Ike's decision that he and the members-designate of his cabinet would wear homburg hats at the inauguration instead of the traditional silk top hat. Truman felt that such sartorial decisions were for the president to make, not the president-elect. Ike had dismissed the idea of wearing a top hat by arguing that tradition was not involved. "If we were going back to tradition, we would wear tricornered hats and knee britches,"[19] he said. When the Eisenhowers arrived at the White House, Ike turned down the president's invitation to come in and have a cup of coffee—Truman had already turned down Ike's invitation to go home in the *Independence*, then the presidential airplane, instead of by train. The two men rode in "frosty" silence together in the back of the presidential limousine, each clutching his own hat. Ike's son John had been ordered home from Korea to attend his father's inauguration, and John was concerned about whether he would be returned to his frontline unit, while Ike, though pleased to have him home, was worried that his presence in Washington might be read as favoritism. He asked Truman who had ordered his son home—perhaps putting the question a little more brusquely than he might ordinarily have done, given the atmosphere between the two men—and Truman snapped back, "*I* did!" Truman later remembered that Ike had explained his absence from Truman's inauguration in 1948 by saying that he had not wanted his presence to deflect attention away from the president, to which Truman replied sharply that he had not invited Ike, and that if he, the president, had asked him to be there, he'd have *been* there. That seems an unlikely exchange—it is hard to imagine Ike saying that his presence would have drawn attention away from the president—but there seems no doubt that it was a trip neither of them remembered afterward with pleasure.

Ike began his inauguration speech with a prayer that he had written specially for the occasion, and it was noted by both Republicans and Democrats that the speech itself was notably ecumenical—it made few, if any concessions to the idées fixes of the Republican right wing (including, but not limited to Yalta, treason in high places, rolling back the iron curtain instead of "containment," lowering taxes, or balancing the budget) and in fact suggested no sweeping changes in the social programs of the New Deal. The parade that followed was the longest in history—the Republicans were, after all, celebrating the end of twenty years in the political wilderness, as well as all the judgeships, ambassadorships, and postermasterships to come—and had the usual mixture of solemnity, as represented by the West Point Cadet Corps, and corn, which included a cowboy riding a trained horse who lassoed the president, a file of elephants, and the twenty-nine members of the Palomino Mounted Police Patrol of Colorado. (The West Point cadets reminded Ike that he himself had marched as one of them in Woodrow Wilson's inaugural parade forty years ago.) It was traditional for the president to leave the parade and return to the White House with the vice president, but it was perhaps indicative of Ike's feelings about Nixon that he decided to return with Mamie sitting beside him instead.

The next morning, punctually at seven-thirty (the new president kept a soldier's hours), Ike entered the Oval Office to learn that the Chinese had shot down an American B-29 reconnaissance aircraft over Manchuria, killing three of the crew and capturing eleven.* The week of the election, the United States had tested its first hydrogen bomb at Eniwetok atoll in the Pacific, an explosion many hundreds of times more powerful than that of the bomb used at Hiroshima. Nobody supposed that it would be long before the Soviet Union followed suit—the nuclear stalemate between the "free world" and the communist world was being ratcheted up a significant number of notches, with no end in sight.

Now that communism has virtually vanished (except in Cuba and, in a very different form, the People's Republic of China) and the Soviet Union

* Given the number of the crew—a normal B-29 crew would have had a crew of eight at most—the aircraft was engaged in radio and electronic surveillance. The author was doing exactly the same thing against the Soviets at the time in the RAF, flying in B-29s borrowed from the USAF for the purpose.

and its "satellite nations" have imploded, leaving a collection of weak and mutually hostile nations—a situation much like that which followed the dismemberment of the Austro-Hungarian Empire—it is hard to recall how the cold war dominated the political scene in the 1950s and 1960s, almost to the exclusion of everything else. The subjugation of the Eastern European nations one by one, the division of Germany, the blockade of Berlin, the Korean war, the insurgencies in French Indochina and British Malaya all seemed to have erased whatever hopes the victory of 1945 had engendered, brutally and jaggedly dividing the world as if it had been cut by a knife. Whatever else people expected of Ike, the most important thing was to navigate his way through the many danger spots of cold war strategy, so it is hardly surprising that his first few days in office were largely dominated by questions of defense and foreign policy. Indeed, the major domestic issue on his mind was the failure of the Senate to confirm Charlie Wilson as secretary of defense, a situation brought about largely by Wilson's habit of lecturing the members of the Senate Armed Services Committee as if they were recalcitrant GM salesmen and by his reluctance to sell his GM stock.

Apart from the downed B-29 and its unfortunate crew, Ike had to deal with the difficulties of the French in Vietnam—already an ominous issue. Successive French governments—France was then in a period of fast-revolving governments, rather like Italy more recently—had been grimly prosecuting the war in Vietnam. They were hampered by a natural reluctance to commit political suicide by sending conscripted young French soldiers to Indochina, so the war was being fought by French regulars, native "colonial" troops, and the famous French Foreign Legion, which at the time was dominated by former German soldiers. The French were in desperate need of modern aircraft and weapons, which could come only from the United States, but the difficulty was that while the French portrayed the war in Indochina as a front in the larger struggle of the West against communism, everybody else saw it merely as a French colonial war. American policy, insofar as anybody could define it, had been to make material support for the war conditional on France's promising the Indochinese independence when it was won, but since the entire purpose of the war was to retain France's hold on its Indochinese colonies, this had not been an easy sell in Paris. Ike had even more reservations about Indochina than he had about Korea, and although the French had sent his old comrade in arms General de Lattre de Tassigny, an impressive military figure indeed, to America on a kind of PR tour for the war, Ike was reluctant to sign a blank check, and would remain so during both his terms in office. In this, he showed a good deal more common sense than his successors.

The Eisenhowers were among the few presidential couples in modern times who did not find entering the White House a strange experience. They had nearly always lived in government quarters, and although the White House was larger than anything they had had before, it was not so very different from Quarters Number 1 at Fort Myer or from Villa Saint-Pierre, except that it also contained Ike's office. The Trumans had been obliged to have the White House rebuilt from the ground up, since decades of neglect had left it in ruinous shape, but the rebuilding had not included any serious attempt at restoring the interior, which was largely furnished by B. Altman.[20] The campaign to decorate the White House with museum-quality period pieces and art dates from Jacqueline Kennedy's time as first lady—in the Eisenhowers' years there, the rooms resembled those of a not particularly luxurious hotel. When the Roosevelts had lived there, the interior of the White House was unapologetically worn and shabby, rather like their own home in Hyde Park, New York—neither Eleanor nor Franklin Roosevelt had any interest in appearing in the pages of *Better Homes and Gardens*—and the Trumans had shown little interest in redecorating it once they moved in. When shown the family quarters, Mamie had been stoic rather than enthusiastic. The only point on which she dug in her heels was Mrs. Truman's small separate bedroom with a larger sitting room, which Mamie insisted on changing around so that she and Ike could sleep in the same big double bed. "Now I can reach over and pat Ike on his old bald head any time I want to," she told the White House deputy chief usher, a remark which swiftly made its way around Washington and from there to the rest of the country, and increased the already warm affection in which Mamie was held by most Americans. As an Army wife, she had no trouble running the White House smoothly—she understood all about hierarchies; she had spent her adult lifetime dealing with staff officers and NCOs; and she was used to the lack of privacy, for the White House was like a hotel placed within a headquarters, and there were always people around, day and night; household servants, the Secret Service, Ike's aides. Twenty years later, Ike's granddaughter-in-law Julie Nixon Eisenhower would remark that when she opened the door of her bedroom every morning in her nightgown to pick up her copy of the *Washington Post,* there were always half a dozen people in the hall to see her.

Ike himself was indifferent to his surroundings, provided they were comfortable and Mamie was happy with them, and the White House suited him perfectly. It was like his headquarters, only bigger, and since for the past ten years he had been surrounded by a large staff of people entirely devoted to his comfort and convenience, his was a seamless transition. President Truman, in an oft-quoted moment of Schadenfreude, had predicted that Ike would find the presidency difficult: "He'll sit here, and he'll say, 'Do this! Do that!' And nothing will happen. Poor Ike—it won't be a bit like the Army. He'll find it very frustrating." But this was to misread Ike's strengths. His skill lay precisely in getting stubborn, difficult, and bloody-minded people—Churchill, De Gaulle, Montgomery, and Patton among them—to do what he wanted despite their objections. Of course in small things he expected crisp obedience, but so did Truman, and so does every president. In the larger sense, contrary to Truman's prediction, Ike did not find the presidency much more difficult than being supreme commander in Europe—it was a continuation of his former job, on a larger scale. Nor was he a neophyte in Washington politics. As MacArthur's aide, as Marshall's protégé, as Supreme Commander of Allied Forces in Europe during the war and of NATO after it, and as a former Army Chief of Staff, Ike had worked closely with Congress and with two presidents for more than twenty years. Truman was wrong: Ike knew how things got done.

Of course Ike's view of the presidency was very different from that of Truman's. As a five-star general, Ike believed in delegating authority, and then letting his subordinates get on with their job. He did not, as the modern phrase goes, "sweat the details"—he saw it as his job to set the course, lead, and take responsibility. Then too, Ike preferred to lead by consensus, and had a good eye for that narrow margin in which most American voters' views are likely to agree or overlap. When Senator Taft expressed his outrage with Ike's first budget, Ike replied, with equanimity, that he "could not agree that the country should have, or wanted, a tax cut ahead of a balanced budget or a balanced budget ahead of national security."[21] Taft, despite his rigid doctrinal belief in tax cuts and a balanced budget, eventually calmed down and accepted the president's commonsense view of the situation, which as it happened coincided with what most Americans thought then, and probably think now. That was one of Ike's great strengths—he didn't approach things with a rigid set of political ideas. He listened, he thought about an issue, and then he tried to do what made sense. He didn't much care whether the solution to a problem came from a Republican or a Democrat, or assume that one party had a monopoly on good ideas, an attitude which was perceived with considerable apprehension by the Taft Republicans. He believed strongly that a president should "take personal responsibility for mistakes and give subordinates credit for success," a belief that not every president since his time has followed as scrupulously as he did.

It has become customary to dismiss the "Eisenhower years" as a time when nothing happened—a peaceful and self-satisfied period between the excitement and the exigencies of World War II and the social ferment of the 1960s—but this is hardly the case. Giving the impression that the Eisenhower years were stodgy was a part—an important part—of John F. Kennedy's presidential campaign, intended to take advantage of the fact that he was young. Youth had hitherto been thought of as a disadvantage for somebody seeking the presidency; but very shrewdly, Kennedy's campaign sought to portray him as energetic, active, quick to make up his mind, determined to make things happen, in contrast to the aging president and his cabinet of dull, gray men who played it safe. Everything from the Kennedys' games of touch football to Jackie Kennedy's passion for riding was in obvious contrast with Ike's leisurely afternoons on the golf course (and his even more leisurely work schedule) or the fact that Mamie liked to have her breakfast in bed and seldom rose before noon. Jack Kennedy was of course running against Nixon, not against Ike, but the campaign strategy was to have him run against the Eisenhower years, and to make a vote for him a choice for excitement, change, glamour, and even fun.

Of course, we now know that despite his youth, Jack Kennedy was as much of an invalid in the White House as Ike became, and that much of the excitement and glamour of the Kennedy years was an exercise in public relations and canny media seduction. Still, there is no question that the eight years during which Ike was in the White House have come to be regarded as a period of dull, self-satisfied excess, and have been portrayed this way by intellectuals, scores of novelists, and a generation or perhaps even two generations of historians. Skirts were long; cars were bloated, two-tone pastel behemoths loaded with chrome and fins; the place for women was still thought by many to be in the kitchen; social, racial, and gender inequality was ignored or brushed under the rug; sexual hypocrisy was rampant; it was the "age of the suburbs," of *The Man in the Gray Flannel Suit*, of vapid consumerist advertising, of conformity; it was the age before the pill, before the student revolution, before long hair and short skirts, before the civil rights movement and women's liberation, and above all before the Vietnam war and a series of assassinations moved America into a new and more violent phase of its history.

And yet the ferment of the 1960s was not just a revolt against the perceived conformity of the Eisenhower years, or even a revolt of the children of those years against their parents and their parents' assumptions and institutions. It was also the logical outcome of what was already beginning to change in American society in those years. The roots of the student risings, the civil rights movement, and much of the Sturm und Drang of the 1960s and the early 1970s can be found in the 1950s, when, in fact, there was a good deal more going on in American life than anybody is conscious of today. Ike did not preside over an inert world of G.E. refrigerator commercials featuring Ronald Reagan, *The Fred Waring Hour* on television (though this was a favorite of Ike's), and barbecue grills smoking in suburban gardens on summer evenings (though Ike himself liked to broil steaks and corn over a charcoal grill in the solarium on the roof of the White House). He presided over a country going through seismic change as it accepted at last what it had refused to accept in 1919: global power with all the associated dangers, temptations, and risks. The veterans of World War II came home to a country with an unprecedented and rising level of prosperity and employment, but also to a country committed to the undisputed leadership of the "free world." American isolationism, for all practical purposes, had ceased to exist—Ike himself had done as much as any man to kill it when he landed in North Africa in 1942, and in Normandy in 1944. Americans were now linked to vital strategic and business interests in every corner of the world, from sub-Saharan Africa to the Middle East, and from Latin America to Asia; and in an age when distance was being rapidly abolished by television, by commercial jet aircraft, and by new technologies, Abilene, Kansas, and even Senator Taft's beloved Ohio were indissolubly part of the larger world—indeed, they could even be destroyed by somebody pushing a button in Moscow. It was not exactly the "one world" Wendell Willkie had had in mind when he wrote his best-selling account of the journey he had taken at Roosevelt's request in 1942, but it was becoming one world; and Taft's dream of retreating into "fortress America" was as outdated as crossing the country in a Conestoga wagon.

The president had hardly had time to settle into the White House before events overtook him. The most startling was the death of Joseph Stalin on March 5, 1953. Stalin had seemed an unshakable, permanent world figure for nearly thirty years; and though his crimes as a leader equaled and in some respects exceeded those of Hitler, his longevity and total, personal control of Soviet politics and most of the world communist movement was regarded by most people as a guarantee of stability. That he was a ruthless and evil dictator few in the western world except members of the Communist Party and left-wing sympathizers seriously doubted, although evidence of Stalin's crimes would not begin to be revealed for another three years; but he was neither rash nor an adventurer, and his approach to foreign policy had always been marked by extreme, even paranoid caution. Perhaps the most alarming realization for the world at large was that the Soviet Union had not solved the oldest and most basic problem of government: an orderly, transparent succession. In monarchies, of course, succession had been guaranteed for thousands of years—indeed that was the whole purpose of monarchy—and the men who wrote the United States Constitution had labored long and hard on the subject, and set an example for a reliable and predictable succession. In the Soviet Union, succession to power remained at the level of a banana republic or a Mafia family, and in fact the first news to come out about who would replace Stalin was that the person most people had expected as Stalin's successor, Lavrenty Beria, the head of the secret police, had been arrested by his rivals and executed. This was not good news coming from the country with the world's largest army and a rapidly increasing stock of nuclear weapons. More alarming still was Ike's discovery that the State Department had absolutely no plans, intelligence, or strategy in the event of Stalin's death—even though his death should hardly have come as a surprise, given his age and ill health. Ike was hardly better informed than anyone else about the person or persons who would be running the Soviet Union.

The second difficult problem to reach Ike's desk was an appeal for clemency from Ethel and Julius Rosenberg. By 1953 the Rosenberg case was a matter of worldwide interest. The Rosenbergs—who were the parents of two young boys—had been convicted of conspiring, during the 1940s, with Harry Gold and David and Ruth Greenglass to provide details about the design of the atomic bomb to the Soviet Union. They were sentenced to death, and their execution was and remains enormously controversial. The case is difficult to unravel, all the more so now that the kind of information the Rosenbergs were alleged to have passed to Soviet agents can be found on the Internet and in countless books. The controversy also, inevitably, involved other volatile issues—the rapid transformation of the Soviet Union from an ally during World War II to the enemy during the cold war; anti-Semitism, for both Rosenbergs and their confederates were Jewish; and McCarthyism, for many of those involved in the prosecution of the Rosenbergs, like Roy Cohn, were allies of Senator McCarthy. No subject divided the left from the right as sharply as the guilt or innocence of the Rosenbergs, and, once they had been convicted (or successfully framed, from the point of view of the left), their execution.

Over fifty years later, the death penalty for the Rosenbergs still seems, to many people, extreme, a product of cold war hysteria; but it must of course be borne in mind that the death penalty itself was not controversial in the 1950s, and that the Rosenbergs were only two among many who would not be executed for the same crimes today. The Rosenbergs' appeal was not, as it happens, the first time that Ike had been obliged to make a decision about a controversial death penalty case. In January 1945 he had ordered the execution of Private Eddie Slovik, the only American soldier since the Civil War to have been executed for desertion. Slovik had refused to go into combat with his unit, Company G, 109th Infantry Regiment, and "deliberately absented himself on 8 October with the intent of deserting the military service so that he would be tried by court-martial and incarcerated and thus avoid the hazardous duty and shirk the important service of action against the enemy."[22] This desertion, Slovik's second, took place at the time of the bloody battle in Hürtgen Forest. The rate of desertion was rising along with the casualty rate, and it was felt necessary to make an example of Slovik, who had not only refused to fight and deserted his comrades, but had done so in the deliberate expectation that he would be court-martialed and imprisoned, and therefore escape the danger of combat for the rest of the war. A best-selling book by William Bradford Huie, a motion picture, and an award-winning television movie starring Martin Sheen, have tended to give Slovik a victim's status; but in the context of the winter battles of 1944–1945, Slovik's action was impossible to overlook. If a man could save his life by deserting—counting on being court-martialed for it and going to military prison—then many men might start to think twice about fighting, and it was Slovik's misfortune to have detected this catch-22 of military law. When the case reached Ike—as supreme commander he had the final decision—he reaffirmed the court-martial's verdict to execute Slovik by firing squad, rejected Slovik's subsequent appeal, and ordered his execution by a firing squad composed of soldiers of his own regiment. Slovik was duly executed on January 31.

The last American to give such an order had been President Lincoln; the last general to give such an order had been Winfield Scott. To put Slovik's case in context, however, he was buried in an unmarked grave next to ninety-five other American soldiers who had been hanged for the murder or rape of unarmed civilians. Slovik was not alone in being executed; he was unique simply in that he was shot for cowardice and desertion in the face of the enemy.

Ike took the Rosenbergs' case with equal seriousness. Just as he had carefully reviewed the verdict of the court-martial in the Slovik case, he reviewed the case against the Rosenbergs in detail, and even drew up a list of questions and answers that he read aloud to the cabinet. In the end, he told the cabinet, he had decided to deny the appeal, since there had been "neither new evidence nor have there been any mitigating circumstances which would justify altering this decision," and he determined that it was his "duty . . . not to set aside the verdict."[23] There was no disagreement. The Rosenbergs' execution at Sing Sing prison in New York was briefly delayed when Associate Justice William O. Douglas issued a stay, but the next day the Supreme Court vacated the order in a six-to-two decision, and the Rosenbergs were executed that same evening.

Ike's mail swelled with letters of protest from all over the world—there was even a letter from Pope Pius XII—and Ike felt obliged to release a statement after the Supreme Court had vacated the stay. He said that the Rosenbergs had "received the benefit of every safeguard which American justice can provide," and made it clear that "when in their most solemn judgment the tribunals of the United States have adjudged them guilty and the sentence just. I will not intervene in this matter."[24] Two huge and mutually antagonistic demonstrations—the shape of things to come—of the Rosenbergs' supporters and hostile demonstrators heckling them virtually barricaded the White House on the evening of their execution, but Ike remained as calm as he had been when he ordered the execution of Private Slovik. He did not think it was for him to overturn a judicial sentence that had made its way up through the federal courts. This was the worst part of his job, but he would not and could not shy away from what he took to be his duty. As ever, duty was the most important word in his vocabulary.

In many respects, his approach to the presidency was new and unique in the twentieth century. Ike was a general; his way of dealing with things was to work through a well organized staff. Political intrigue, back channels, and informal meetings were anathema to him—and this perhaps helps to explain his reservations about Vice President Richard Nixon, and his reluctance to allow Nixon into the inner circle at the White House. It is hard to imagine Ike with an éminence grise like Wilson's Colonel House;

or surrounding himself, like Roosevelt, with competing advisers, each with a separate agenda (one thinks of Baruch, Harriman, and above all Harry Hopkins); or having political cronies like Truman's. Ike's intention was that Sherman Adams should play Bedell Smith's old role as White House "Lord High Executioner," while John Foster Dulles put into practice (albeit with a few additions of his own) Ike's views on foreign policy, and Richard Nixon, at least in theory, looked after the political problems of the Republican Party. Notably, and unfortunately, lacking was anybody that Ike could rely on to handle domestic affairs. Ike's relationship with all these people was correct and formal, rather than intimate—they were not friends or cronies or trusted, powerful advisers and go-betweens as House had been for Wilson. Unlike Winston Churchill, who liked nothing better than to hear all the political gossip of the day over a glass of whiskey and soda and a cigar with intimate friends like Max Beaverbrook and Brendan Bracken, Ike, though he served two terms as president, never became a politician. He devoted much time and effort to establishing good relations with Congress, and even began—an important innovation—a series of regular luncheons at the White House with members of the House and Senate, which impressed many of them who had never seen the private quarters in the White House before. But this did not materially improve the president's relationship with his own party, which was moving more and more swiftly toward the right, while Ike moved toward the center, where he always felt more comfortable.

The long furor over the "Bricker amendment"—sponsored by Senator John Bricker of Ohio, and intended to limit a president's ability to negotiate treaties without congressional approval—was typical of the breach opening up between Ike and his own party. The target of Bricker's amendment was, of course, Yalta. Ike had not been present there, but he was still opposed to any attempt by Congress to tie the president's hands in his dealings with foreign powers. Besides, his own relationship with Stalin and the Soviet Union in 1944 had been friendly and cordial, and there was nothing in the Yalta agreement that he had criticized at the time. He was equally opposed to attempts on the part of Republicans in the House and the Senate to investigate communist subversion in the churches; to attempts by Republican senators to block the nomination of Charles E. Bohlen as ambassador to the Soviet Union (Bohlen's opinions about Yalta were about the same as Ike's); and to Senator McCarthy's campaign to root out treason in the Voice of America and USIA libraries overseas. The last item so incensed Ike that he angrily rebutted it during a speech at Dartmouth College, saying, "Don't join the book-burners! Don't think you are going to conceal faults by concealing evidence that they ever existed. Don't be afraid to go in your library and read every book."[25] These were strong words, coming from a president to members of his own party, even if they were coming via Hanover, New Hampshire. Ike's words got stronger still when J. B. Matthews, who was staff director of the Senate Permanent Investigations Subcommittee, a former research director for HUAC, and McCarthy's hit man and protégé, published a claim that at least 7,000 Protestant clergymen had served the "Kremlin conspiracy," and that "the largest single group supporting the Communist apparatus in the United States today is composed of Protestant clergymen."[26] Ike fired off a telegram of support to an ecumenical group protesting this attack on the clergy; the telegram created "a sensation," and for once McCarthy had to back down.

The gap between the president and the Republican Party was at its widest and most embarrassing with regard to McCarthyism, tax reduction, a balanced budget, and the Bricker amendment; and in all these areas Ike lacked a strong deputy who could act for him—domestic policies and politics were his weakest suit, and interested him least. What he did have, however, was a strong sense of what seemed sensible, humane, and necessary, a kind of middle-of-the-road, pragmatic approach to government, strongly reinforced by his brother Milton and by the experience, which they shared, of having grown up poor in Abilene. He had the quality, rare among politicians, of worrying away at a piece of legislation until he understood what its impact would be on the average person, not in theory, but in fact. He had a reputation—partly justified—for hands-off management, but he did his best to make sure that the policies of his administration represented his own point of view, however remote the subject. "I feel pretty good when I'm attacked from both sides," he said. "It makes me more certain that I'm on the right track." It was fortunate that Ike felt this way, because he was attacked by right-wing Republicans at least as much as by Democrats.

He has been strongly criticized for not having used federal power more aggressively in civil rights cases, but this criticism overlooks both Ike's sense of justice and his determination to avoid empty gestures. The Truman administration had ended official segregation in the United States armed forces in 1948, but in reality living conditions on military bases, particularly in the South, remained as firmly segregated as ever. Whatever Ike's personal feelings in the matter were—and it is safe to say that these were typical of most sixty-two-year-old white men—he was determined "to clear out segregation where Federal money is involved," and pledged to the principle that there "must be no second-class citizens in this country."[27] He moved swiftly to end continued segregation in the armed forces, including the navy yards, where there were still "COLORED and WHITE signs over drinking fountains."[28] It was typical of Ike's approach to desegregation that where possible a blitz scheme of redecorating mess halls and barracks was carried out: rather than being removed or obliterated, the signs were simply painted over and never relettered—their disappearance, in other words, was made part of a larger plan, so that there was no chance of creating an issue or making news. Aware that hardly any city in the South was more firmly segregated than Washington, D.C., Ike moved to end the Capital Transit Company's ban on black bus drivers and streetcar operators; to get the Chesapeake and Potomac Telephone Company to hire black telephone operators; and to end segregation in Washington's restaurants, lunch counters, hotels, and movie theaters, as well as in the fire department and the police department. He went on from there to ban segregation "on interstate trains and buses and in terminals," a sweeping move for which he neither sought nor received credit. In November 1953 Ike's attorney general, Herbert Brownell, "acting on behalf of the administration as a friend of the court, filed a strong brief contending that the Fourteenth Amendment prohibited racial segregation in the public schools and that the Supreme Court had the power to settle the issue." This led to the historic decision of May 17, 1954, in which the Supreme Court ruled unanimously that segregation in the public schools of the United States was unconstitutional. These were great and important decisions, and Ike never deviated from his duty to enforce the decisions of the courts on the subject of racial segregation. He was not any more anxious than his successor John Kennedy would be to confront white anger and violence in the South, and his personal view was that such momentous changes in the social structure would take time and patience to achieve—people had been living for a long time under a previous Supreme Court ruling that "separate but equal" facilities were constitutional, and it would take a good deal of time for them to accept the new ruling. But he never doubted that the Supreme Court ruling would have to be enforced, however much he later regretted having nominated Earl Warren as Chief Justice, or recoiled at the necessity of using the United States Army to escort black students into a high school.

Ike's attitude toward the presidency can best be summed up by a remark he made to Mamie as they left for Augusta, Georgia, in 1953 for a weekend of golf, in unpromising weather: "Nothing can get me mad today. Anything that will get me out of this place!"

The presidency understandably increased Ike's tendency to bursts of bad temper and high blood pressure—even the mention of the Bricker amendment or Senator McCarthy was enough to turn his face red and set the veins in his forehead throbbing. On the other hand, within limits, he and Mamie did not dislike life in the White House. It was, in any case (with the exception of Camp David, Franklin Roosevelt's rural retreat Shangri-la in the Catoctin Mountains* of Maryland, which Ike renamed after his grandson), their only home until the construction and decoration of their house in Gettysburg were completed. As befitted a general, Ike increased the formality of the White House to a modest degree, and Mamie showed more interest in entertaining than Bess Truman had. Ike reinstated the custom of the Easter egg roll on the White House lawn, which Roosevelt had discontinued; and he increased the number of formal White

* Though called mountains, these actually are good-size wooded rolling hills, like those in much of Vermont.

House dinners, which Mamie enjoyed. Though he himself preferred to eat upstairs in the private quarters quietly—just the two of them together—still, like many another American male, when his wife wanted to have company over, he sighed and gave in.

The president's love of golf—a sport he had come to late in life—was such that he practiced putting in the Oval Office and on a special putting green installed for him on the White House South Lawn, where, in those years before the assassination of John F. Kennedy, tourists could look through the iron railings and see the president swinging his club. His admirers built a cottage for the Eisenhowers overlooking the Augusta National golf course, and the president spent as much time there as he could. When he could not, he liked to leave his office early and play at Burning Tree, near Washington. Those who watched the president play remarked that he did so with more gusto than skill, but enjoyed every minute of it—indeed, he enjoyed anything that took him out of the White House for a few hours, and had no hesitation in showing it. Being in the White House was like being in a pressure cooker, he sometimes complained.

The pressure came mostly from the "big stuff" that the country had chosen Ike to deal with: the cold war, Korea, China, the nuclear arms race, and the French war in Indochina. These were also matters which are still familiar to us today as crises—the survival of Israel; the rise of Muslim extremism; threats to our oil supply; the murderous chaos that in so many places followed willing or unwilling "decolonization" by the European powers; and the questions of how big the armed forces of the United States needed to be, how they should be armed and for what kind of war, and at what point their sheer size would damage the country's economy and become as much of a threat to the survival of the United States as to a potential enemy.

Ike knew more than anybody else about the armed forces, and as much as anybody else about foreign policy. In no area was his judgment sounder, or more respected, and his decisions more critical. It was a great advantage that Ike knew closely so many of those in power—Churchill, Eden, and Macmillan in the United Kingdom; De Gaulle (after 1958, when he returned to power); and most of the French generals and politicians, as well as those of every western European country. Perhaps no president has ever entered office knowing so many other world figures, and with such broad sympathy for and lack of prejudice against foreigners, no matter what their politics. What is more, Ike understood the uses (as well as the limitations) of power, having worked closely with Churchill.

He did not share the right-wing Republicans, fear of communist China—he was already thinking about the possibility of a Sino-American rapprochement in 1952, twenty years before Nixon actually brought it about, and was confident that the United States had little to fear from China. He further incensed his own party by refusing to agree to a congressional attempt to ban communist China forever from taking over the Chinese seat on the UN Security Council, by suggesting that increased trade was the best way of improving relationships between the two countries, and by remarking that diplomatic recognition did not have to imply moral approval of a country's government or regime, regretting that Woodrow Wilson had set a precedent by refusing to recognize Huerta's regime in Mexico in 1913. He mused to his associates that opposing the communist Chinese would simply drive them closer to the Soviet Union. All this would be the basis for Nixon and Kissinger's negotiations with the Chinese—Ike was simply too early, and congressional opposition from within his own party was too strong.

His first important move was to force the Chinese back to serious negotiating over a Korean truce by showing them that the United States was not afraid to renew the war. The Air Force squadrons in Korea were reequipped with new Saber jets; another Marine division was sent to Korea; atomic weapons were dispatched to Okinawa, with considerable publicity; the United States Seventh Fleet, which had been interposed between mainland China and Taiwan in part to prevent Chiang's troops from raiding the mainland, was withdrawn, since it made no sense for American warships to be protecting communist China from attack; and the threat that the United States would bomb Manchuria if the stalemate continued was artfully leaked to India by Secretary of State Dulles, in the expectation that from there, it would reach the Chinese quickly. The result was to bring the Chinese and the North Koreans back to the bargaining table at Panmunjom; and although matters did not go swiftly, or without setbacks (chief among these being the obstinacy of Syngman Rhee and his deliberate release of Chinese POWs in South Korean hands), the truce was finally signed in 1953. Like some odd relic of the past, it remains in effect today. Right-wing Republicans were furious—Senator Knowland of California spoke for the majority of them when he denied that this was "peace with honor" and predicted that "we would inevitably lose the balance of Asia." But Ike summed up the feelings of most Americans when he said to the press, "The war is over, and I hope my son is going to come home soon."[29]

Despite this triumph, Ike was criticized constantly on the subject of defense. Those who are familiar with Stanley Kubrick's film *Dr. Strangelove* will remember the phrase the "missile gap," which had its origins in this period. Ike's critics, both Democrats and right-wing Republicans, alleged that he was allowing a dangerous gap to grow between Soviet and American strength in nuclear weapons, though this turned out much later not to have been the case. Ike preferred to spend money on building the Boeing B-52 heavy bomber—a wise investment, as it happens, since the B-52s are still flying, half a century later, after serving in Vietnam and in both Iraq wars. Ike also preferred more and larger hydrogen bombs, since the more B-52s the Air Force got, the more H-bombs were needed to arm them. The Soviet Union's advantage in missiles was more apparent than real. The Soviets had captured the German rocket research facility at Peenemünde, the birthplace of Hitler's V-1 and V-2 "revenge weapons." Although it had been constantly bombed by the RAF, with as many as 600 Lancasters in some raids, it was still the repository of advanced knowledge about ballistic missiles, as they were coming to be known, at the end of the war.* The Russians had removed everything and everyone they could find, poured immense amounts of money and effort into rocket

* The United States got, among other rocket scientists, the famous Wernher von Braun. The title of his memoirs, when they were published in America, was *I Aimed for the Stars*, which prompted the historian Cornelius Ryan to comment that they should have been called *I Aimed for the Stars—But Sometimes I Hit London*.

research, and even built a special city for the captured German scientists and engineers; and all this would pay off both in showy technological triumphs and in the threatening rows of missiles paraded past the reviewing stand at Lenin's Tomb every year on the anniversary of the October Revolution. Ike doubted that rockets would change the balance of power between the United States and the Soviet Union in the immediate future, was (correctly) skeptical about the number of missiles that the Russians were presumed to have by the CIA and congressional hard-liners, and was reluctant to pour unlimited money into technology that was still unproved. He took a calmer view than his critics of the danger of nuclear warfare—the Trumans had had a presidential fallout shelter constructed in the rebuilt White House, but Ike showed no interest in seeing it. To the fury of the admirals and congressional Republicans, he was just as reluctant to fund nuclear supercarriers for the Navy, reasoning that in a large-scale war the enemy would certainly make every effort to find and sink them, and in a minor conflict they would serve no purpose. On the whole, for a general, Ike was dubious about defense spending.

He was even more dubious about Senator McCarthy, who was rapidly becoming his bête noire. Ike's strategy of not getting "in the gutter with that guy," as he put it, took a long time to pay off, and did not do so until McCarthy, driven by a combination of recklessness and overconfidence, finally took on the United States Army, charging that an Army dentist at Fort Monmouth, New Jersey, had been promoted from captain to major despite having communist sympathies. McCarthy demanded that the dentist, Major Peress, should be promptly court-martialed, but the Army honorably discharged him at Camp Kilmer instead. At this point McCarthy—whose supporters demanded vociferously to know "Who promoted Peress?"—went after the Army general commanding Camp Kilmer, and accused him of being "not fit to wear that uniform," and not having "the brains of a five-year-old child." Zwicker, the general, was a respected officer with a good war record; and McCarthy, in attacking him, had finally gone too far. McCarthy proceeded to make matters worse for himself by carrying the fight upward to Army Secretary Robert Stevens. For thirty-six days, while all over the country people stopped work and stared at the television set, the "Army-McCarthy hearings," as they were known, were televised live. They entered American legend, with a cast of characters who are still remembered today. One was Roy Cohn, McCarthy's chief counsel, who had threatened to "wreck the Army," glaring and reptilian under the bright lights of the television cameras. Another was Cohn's intimate friend, the hapless David Schine, of the Schine hotel fortune. Together, Schine and Cohn had junketed across Europe, staying in the best hotels, in a highly publicized and deeply embarrassing search for the books of authors with communist sympathies in USIA libraries. Schine had been drafted into the United States Army, and on his behalf McCarthy and Cohn had tried to blackmail the Army into relieving him of KP and drill, freeing him to search for "pro-Communist" propaganda in textbooks at West Point. And of course there was Joe McCarthy himself, with his heavy "five o'clock shadow," his sweaty face, his wild accusations, his menacing eyebrows, his beady eyes and guilty expression, all too clearly an alcoholic on a rampage. Even more memorable was the Army's counsel, Joseph N. Welch, a dignified, elderly lawyer unmoved by McCarthy's blustering and threats, who at one point looked the senator in the eye and asked, "Sir, have you no shame?" It was a question heard by millions of Americans; and even among those who still thought there were "communists in high places" that needed to be routed out, it reverberated as few other moments in live television ever have. Cohn and McCarthy's demands for favoritism for Private Schine—"this *private*,"[30] as Ike referred to him with scathing contempt at a press conference—effectively brought McCarthy's reign of terror to an end. A few months later the Senate voted to censure McCarthy for "conduct unbecoming a senator."

More alarming to Ike than McCarthy was the remorseless collapse of the French position in Indochina. In an effort to reverse a decade of guerrilla warfare and a rapidly worsening military disaster, the French decided to try to lure the Vietminh (as the Vietcong were then called) into a large battle in which they could be overwhelmed by European firepower, airpower, and the elite of the French professional Army and the Foreign Legion. This would be a conventional battle, like those of the two world wars, not guerrilla warfare; and the Vietminh would be forced into the open to attack well-sited French positions, and then destroyed. This was a bold move—very similar to what the United States would attempt at Khe Sanh more than ten years later with similarly disappointing results, the American generals having apparently learned nothing from the French example—and for a time it looked as if it might succeed. But General Vo Nguyen Giap (who would later command the People's Army of Vietnam against the United States) had heavy artillery hauled through the jungle and dug into the hills around the French positions and the vital airstrip at Dien Bien Phu (the French commanders had assumed that this was impossible to do), and the French soon found themselves surrounded and besieged. The siege was protracted, heroic, gallant, bloody, tragic, photographed and reported on in detail in newspapers and magazines all over the world (another foretaste of the American war in Vietnam), and doomed. The French had placed themselves in the center of a bowl of hills from which they were shelled and attacked day and night, and once they were cut off there was no way to supply them except by air, on an airstrip that was in plain sight of hundreds of Vietminh gunners. Their position resembled that of the French army at Sedan in 1870, which General Ducrot had colorfully described to Emperor Napoleon III in a famous French military bon mot: *Nous sommes dans un pot de chambre, et nous y serons emmerdés* (We are trapped in a chamber pot, and we are going to get shat upon in it). Truer words were never spoken. Sedan cost Napoleon III his throne, and Dien Bien Phu would cost France Indochina. Though Ike had declined to send American troops to fight in Indochina, he had supplied France with "material assistance" that included Bearcat naval fighter-bombers, A-20 bombers, and helicopters; but the French wanted more. They begged for American troops and for air strikes to relieve the pressure on Dien Bien Phu, and even suggested the use of atomic bombs.

Ike was unmoved. "No one," he said, "could be more bitterly opposed to ever getting the United States involved in a hot war in that region than I am." He "could not conceive of a greater tragedy for America than to get heavily involved"[31] in Vietnam. The French even sent General Paul Ely, their chief of staff, to Washington, to plead with Ike soldier to soldier; but Ike could read a map as well as General Ely, and knew a military disaster when he saw one. The French got no American soldiers and no American air strikes, and Dien Bien Phu fell on May 7, 1954, leaving Vietnam divided by a temporary political solution. A whole new war would ensue, in which the United States would take over the role of France and play it to an equally tragic conclusion. Ike, it must be said, saw the writing on the wall on the subject of Vietnam more clearly than John F. Kennedy or Lyndon B. Johnson.

In 1955 Ike at last attended the eagerly awaited summit conference in Geneva with President Bulganin of the Soviet Union and prime ministers Edgar Faure of France and Anthony Eden of the United Kingdom. Eden had at last succeeded in replacing his aged mentor Winston Churchill in 1953, by which time Churchill's deafness, increasing senility, and ruthless determination to cling to power long after he had become incapable of exercising it had driven even those Tories loyal to him into desperate plots to get him out of 10 Downing Street. Meanwhile Churchill himself had complained to one and all that he wasn't "sure Anthony was up to it," or "ready for it yet." Eden, once the dashing young hope of the Conservative Party, was now himself looking old and ill, and had, in the view of many, not fought hard enough for the succession and left it until too late to make his move. Faure was one of a series of rather colorless French premiers, and in photographs taken at the conference he looks as if he is not altogether sure he belongs there, and fears he may be asked to leave at any moment. France was in any case rushing headlong into another colonial war, this time in Algeria, from which it would not be rescued until De Gaulle was recalled to power in 1958. As for Nikolay Bulganin, though he was an impressive-looking man—and, for a Soviet leader, suave—he was already overshadowed by Nikita Khrushchev, in whose hands such power as could be salvaged from Stalin's death would shortly be placed. Perhaps the composition of the conference doomed it to failure—certainly it produced no great result except to make people feel better. All over the world, people took heart from the fact that Ike and Bulganin were talking face-to-face, even if they were not saying much. These were the days when summit diplomacy, as it came to be called, was still in its infancy as a solution to the world's problems.

On his return, Ike faced a problem of a very different kind. While playing golf in Denver, he complained of heartburn, which he attributed to too many sliced Bermuda onions on his hamburger at lunch, but which by the early hours of the morning had been diagnosed as a heart attack—"a moderate anterior coronary thrombosis," as it was described to the press. Although Ike had a robust, erect figure, his medical history was not, in fact, good. In that innocent age before cardiologists started telling people how to live, Ike flouted what would today be regarded as every sensible rule for longevity. He had been a heavy smoker, his only exercise was golf, his temper was volcanic, his sense of duty was unremitting, and his favorite foods were steak, fried chicken, and his own fiery chili and beef stew.* While Ike rested in Denver, Nixon and John Foster Dulles minded the store. The president's doctors had ordered that he should not be "handed any controversial decisions"—not an easy order to follow when the Cypriots were rioting against the British Army, the Soviet Union began to arm Colonel Nasser's Egypt, and the Chinese balked at releasing American prisoners and civilians they had been holding, while in domestic politics the issue of farm subsidies—then, as now, a potent source of trouble in an election year—was heating up. Although soothing music was provided to relax the president (choices included *Clair de Lune*, "Stardust," Benny Goodman playing "Time on My Hands," and selections from *The Student Prince*), the mere news that he was "a little tired" still sent the stock market into a decline that erased $4 billion in one day's trading. After seven weeks of rest in the hospital, Ike was released to return to the White House in November, the health crisis apparently behind him.

What he could not put behind him was a problem his own heart attack had created, both politically in the United States and diplomatically in the rest of the world—whether or not he should run for a second term. For the next three months the United States, the rest of the world, the stock market,

* Ike was an enthusiastic amateur chef, especially at the grill. Those who are curious will find his own detailed recipe for "vegetable" soup (actually a meat soup with vegetables) in the notes of his book *At Ease*. It is long, and complicated, and it would not be likely to find favor with any panel of doctors today; nor would Ike's famous recipe for chili.

the media, and the members of the Eisenhower family waited breathlessly and nervously for the answer to this question. Nobody waited more anxiously than Vice President Nixon, whose rivals for the candidacy if Ike decided not to run again, as seemed likely, were thought to include Earl Warren, Harold Stassen, Thomas Dewey, William Knowland, Henry Cabot Lodge, and even Sherman Adams. For Democrats, of course, the suspense was even sharper. Being nominated to run against Ike was a poisoned chalice but no other possible Republican candidate had Ike's appeal, which cut across party lines.

Recuperating in Gettysburg—the Eisenhowers' house had at last been finished—Ike could hardly ignore the fact that the summit meeting at Geneva had produced no useful results. Nikita Khrushchev was brash, abrasive, full of confidence that the Soviet Union would prevail over the United States and communism over capitalism. (As he would later remark, "We will bury you!") From the gray figures who had carried Stalin's coffin (minus Beria, whom they had murdered), Khrushchev had managed to emerge as the strongest and the most vital. Nobody in the Soviet Union doubted his toughness, even among men who were notably tough. He had eliminated the kulaks (prosperous farmers) in the Ukraine for Stalin, forcibly collectivized Ukrainian agriculture, survived Stalin's purges, and helped to retain the party's iron control over the generals during World War II—hundreds of thousands of people had been killed at his orders. This was not a side of him that received much attention in the West, where his red-cheeked, chubby, smiling face and earthy good humor made him seem more human than Stalin. Khrushchev looked like a man you could do business with, but in fact he was both ruthless and determined, as well as quite sure he was on the winning side—a formidable opponent, and, as it turned out, a notably reckless one.

In fact, Khrushchev's influence was already being felt, in an increasingly aggressive and adventurous "forward" Soviet foreign policy, which was a serious concern to Ike. Stalin had been content to digest his gains in Eastern Europe and to support, to a limited degree, communist China's objectives in Asia. Khrushchev now reached out to Central America and Latin America, hitherto areas in which American influence was unchallenged;

to Africa, where the British, the French, and the Belgians had hitherto controlled things; and more dangerously to the Middle East, where through enormous arms sales and vigorous propaganda, he hoped to replace the role that had hitherto been played by Britain, France, and the United States. These moves, particularly the huge arms sales to Syria and Egypt, alarmed the Israelis, and threatened the reactionary desert monarchies on which America was already becoming dependent for oil.

Ike saw clearly that Khrushchev had opened a whole new series of fronts in the cold war, and this was something he was bound to take into consideration in deciding about his own future. He himself had often expressed his wish to be a one-term president; and after the heart attack in Denver, he at first took it for granted that this would be the case. Mamie was "dead set against" his running for a second term, and even Ike's brother Milton, who had emerged as his adviser and perhaps his closest political confidant, seemed doubtful that Ike would or could run again. While he was in Denver, Ike had escaped the deep depression that often follows a heart attack, but it hit him badly once he was home in Gettysburg. The weather was bad; he stamped around the house using "a golf club for a cane." "His morale slumped. His spirits were low."[32] He exhibited all the classical symptoms of a deeply depressed heart attack patient, at a time before the use of antidepressants in such cases was commonplace. His physical recovery from the heart attack was proceeding well, but his mental attitude was not.

After Christmas he went to Key West in search of warmer weather. There he could play golf, walk, and bask in the sun; and as he cheered up and regained strength, he gradually began to think about whether or not he should run for a second term. A part of his recovery was, of course, an increasing interest in work, and an increasing amount of work was being sent to him. As he picked up the burdens of the presidency again, however tentatively, his depression began to lift. Ike was not the most reflective of men, but he could hardly help noticing that the more he resumed work, the better he felt; and somewhere along the way, in the first week or so of January, he made the discovery, like many another man of his age, that he did not really want to retire at all. Being Ike, he held an intimate dinner meeting of thirteen of his closest friends and advisers (the group included Dulles, Lodge, Adams, Brownell, and Milton Eisenhower) as soon as he was back in the White House in mid-January, and asked them what they thought he should do. Each of them spoke in turn, and then Milton Eisenhower summed up the pros and cons—though as it happened, everybody present thought that Ike should run again if his health permitted it. Dulles praised Ike's "God-given ability" to reconcile men and nations, and Lodge attributed the country's prosperity to him. Ike listened and did not argue or disagree, but neither did he commit himself one way or the other; as at the meeting before D-day, he listened patiently and would eventually make up his own mind.

One strong argument for his running again was that the Republicans could not hope to keep the White House without him. None of the other possible candidates had his appeal to the voters. Earl Warren's decision on school segregation had made his name anathema throughout the South. Dewey had already lost two elections: once to Roosevelt and once to Truman. Stassen was a perennial loser. Knowland was too far to the right to be elected and was opposed to Ike's foreign policy. As for Nixon, Ike had often remarked in a rather patronizing way that his vice president was "maturing," but left in doubt that this process had been completed to his satisfaction yet.

Just as Churchill had fought for years against handing over his office to his heir apparent, Anthony Eden, Ike was ambivalent about Nixon as his successor. Eden, at least, was a glamorous figure: he had been a war hero (he won a "good" Military Cross in Flanders in World War I, serving in the trenches as an officer in the Rifle Brigade); resigned as Neville Chamberlain's foreign secretary at the risk of his political future to protest against appeasement; served at Churchill's right hand in the darkest days of appeasement and the war; and served with distinction in some of the highest offices of state, foreign secretary and minister of war among them. Nixon's curriculum vitae had no such distinctions—his war service had involved no combat; he had never run any office of state; his fame came mostly from having defeated Helen Gahagan Douglas in a singularly nasty campaign that was a byword for smear tactics; and, at any rate from the liberal point of view, he was notorious for hounding Alger Hiss into perjury. Ike, though he was not a professional politician, was by no means immune from Churchill's feelings on the subject of succession, and it is notable that Nixon was missing from the list of those invited to the White House for the intimate dinner to discuss Ike's future. Not only did Ike not want Nixon as a successor, he was not even sure he wanted Nixon as his running mate if he decided to go for a second term. As for Nixon, though in later years he liked to stress the "father-son" nature of his relationship to Ike, it was oedipal at best.

In these circumstances, like Churchill, Ike soon decided that it was his duty to run for a second term, and his family decided that it was the best thing for his health. During the gloomy weeks of Ike's depression in Gettysburg, Mamie had in any case come to the sensible decision that "he wasn't ready for retirement yet."[33] The doctors gave their qualified approval, announcing—with a balanced judgment that would have pleased Voltaire's Dr. Pangloss—that although on the one hand a person who has suffered one coronary is more likely to have another, on the other hand retirement might frustrate the president so badly that it would place a greater burden on his heart, so it was six of one or half a dozen of the other. A week after the dinner party at the White House, Ike wrote to the deputy secretary of state of New Hampshire saying that he had no objection to having his name entered in the state primary, but pointing out that he had not yet decided whether or not to seek a second term. Thus, while he did not exactly throw his hat into the ring, he allowed it to be placed at the side.

At the beginning of February Ike received a good report from the cardiologists at Walter Reed Hospital and, perhaps more important, spent a few days in Georgia strenuously pushing his way through "waist-high grass" to shoot quail, and playing golf. On February 29 he scheduled a press conference, which was awaited with enormous expectation, and in a burst of the garbled and puzzling syntax he employed when he was obliged to talk about himself, he finally let it be known, once his words had been parsed and sifted through for meaning, that if the Republicans wanted to nominate him, he would accept.

The news was enough to send the stock market soaring, depress the Democrats, and inspire people all over the world to hope for world peace; but among the subjects the president did not touch on was whether he wanted Nixon on the ticket again. It was a measure of the "father-son" relationship between them, in the worst meaning of that phrase, that Ike not only suggested to an appalled and dismayed Nixon that it might be better for his future to take a cabinet-level post, like secretary of defense, in which he could gain experience of running a great department of the government, but then kept him hanging on for weeks. In early March Ike announced at a press conference that he had "asked [Nixon] to chart his own course," when he knew perfectly well that his vice president was in fact determined to cling on to his present job by his fingernails. Not until April 27 did Ike finally allow Nixon himself to announce that he would be happy to accept the vice presidential nomination if the president and the party should offer it to him. Even then, it was left to Press Secretary Hagerty to say that Ike was "delighted" with Nixon's decision. He could hardly have kept Nixon on the ticket less graciously or more indirectly.

In all likelihood, Ike was simply testing the waters to see what the reaction in the Republican Party would be to dropping Nixon from the ticket, while Nixon beat the bushes to produce support for himself among the Republican state chairmen. Still, cruel as the relationship between president and vice president has often been in the political history of the United States, Ike's treatment of Nixon from January 1956 to April 27, 1956, was in a class by itself, and would be equaled only by his reluctance to come out in full support of Nixon when the latter ran for the presidency in 1960.

With or without Nixon, there was no realistic chance of beating Ike in 1956. Adlai Stevenson and Estes Kefauver fought a lackluster campaign without much conviction, while Ike, who had been warned that he should not campaign as hard as he did 1952, found that it really didn't matter this time whether he campaigned or not. The result was a foregone conclusion—Ike doubled the margin of his popular vote and carried all but ten states of the old South. (Even those ten voted less for Stevenson than for the fact that he at least had not been the one who made Earl Warren Chief Justice of the United States.)

Ike's first four years in office had been full of notable successes, but as if to prove that no president can ever truly count himself lucky until he walks out of the White House on Inauguration Day to watch his successor sworn in, they were to end with a simultaneous diplomatic triumph and a defeat in the cold war. And even the triumph would bring Ike no joy.

For some months in 1956 the situation in Egypt, which was under the control of Colonel Gamal Abdel Nasser, now president (and often referred to in the hostile Western press as "strongman") of Egypt, had been causing deepening concern in the West. The United States had agreed to take on the lion's share of financing Nasser's favorite and most important project, the Aswan High Dam on the Nile, which was central to his plans for modernizing Egypt. But as Nasser increased the closeness of his relationship with the Soviet Union—and voted at the United Nations for the inclusion of communist China—John Foster Dulles had canceled further American financing and assistance for the construction of the dam, to teach Nasser a lesson. Unfortunately, this high-minded move backfired. The Soviet Union agreed to take over America's role in the construction of the dam, and Nasser, infuriated at the West and anxious not to lose face in what was beginning to be called the third world, decided to nationalize the Suez Canal.

To understand the outrage in Britain and France at this decision, it is necessary to remember that both countries had a long and tangled relationship with Egypt, as well as a century and a half of rivalry in this regard. The young Napoleon Bonaparte had conquered Egypt in 1798 and, with the *coup d'oeil de génie* for which he was already famous, surveyed the desert between the Mediterranean and the Red seas and ordered his engineers to make plans for the construction of a canal there. The destruction of the French fleet by Nelson at the Battle of Aboukir doomed the French occupation of Egypt (and for the time being doomed the canal), and Bonaparte returned home to other conquests and ambitions. But France's interest and influence in Egypt never diminished; and just as the

"Egyptian style" came to dominate the fashions, furniture, and above all the imagination of the French, the French language, French styles, and French culture were adopted by the Egyptian upper class, which would always look toward Paris, not London, as its spiritual home.

The British, whose imperial ambition in Africa was to extend their rule from "Cairo to Capetown," managed to secure domination over Egypt in the mid-nineteenth century, although it remained in theory an independent country with its own king; and the idea of building the Suez Canal remained one of the great ambitions of the century. It was achieved at last by a French engineer, Ferdinand de Lesseps (who would later make the first attempt to build a canal in Panama), and financed by two whole generations of middle-class Frenchmen who placed their savings in the bonds of the grandiosely named Compagnie Universelle du Canal Maritime de Suez, which were thought to be as good as gold, or better. The financial problems of the company, however, allowed the British government (*l'Albion perfide* again) in the person of Prime Minister Benjamin Disraeli to buy the controlling shares of the Suez Company overnight with the help of the London Rothschild Bank in 1875. The canal then became a predominantly British concern, and in time the "lifeline" of the British Empire, permitting swifter communications between Britain and India, as well as the British dominions and colonies in Asia and the Pacific. (After the loss of India, the canal would become even more important than before, as the pathway for oil from the Gulf sheikhdoms to reach Europe.) In the two world wars the British had fought to protect the canal as a "vital British interest"—it was the reason for the exploits of Lawrence in Arabia and for Allenby's conquest of Jerusalem in World War I, and the prize that brought Rommel's Afrika Korps to the borders of Egypt in World War II. The notion that the Egyptians themselves would seize the Suez Canal was unacceptable in London, and equally unacceptable in Paris.

The French were, if anything, even more hostile than the British toward Nasser. The loss of Vietnam had been followed almost immediately by war in Algeria, which was regarded by the French not as a "colony," like Indochina or Morocco, but as an integral part of France itself

(*La France metropolitaine*). Algiers was in their eyes a French city, just like Lyon; and the war in Algeria, with its terrorist bombings by the rebels and its revelations of torture by the French army (a formula which will not sound unfamiliar over half a century later), rapidly became as nasty as a guerrilla war can be. It was all the more painful because in Algeria the French were at last using French conscripts who were doing their two years of *service militaire*, as opposed to using just the French regular Army and the Foreign Legion as they had done in Indochina. The fact that the Algerian rebels were thought to be armed and supported by Cairo infuriated the French against Nasser, and therefore brought them closer to the British in a part of the world where the two countries had hitherto been rivals.

The British and the French were determined to teach Nasser a lesson very different from the one Dulles had had in mind when he ended American support for the high dam. To that end they collaborated on a plan whose most vital element was that it must be kept secret from the Americans until it was too late for them to stop it. Together, they would encourage the Israelis—who were infuriated by terror attacks from fedayeen fighters (the 1950s equivalent of suicide bombers) crossing the Sinai from Egypt into Israel—to attack the Egyptian army in the Sinai and advance toward the canal. At this point, the British and French would intervene militarily to "protect" the canal; Nasser would fall; and the canal would be back in Anglo-French hands. The United States and the United Nations would then have to be mollified, of course, but it was felt that Anthony Eden should have no trouble in bringing his old comrade in arms Ike around once the seizure of the canal was a fait accompli.

This was a substantial miscalculation—one of many. Dulles should never have withdrawn American aid for the high dam in the first place; the French and the British should not have set out to use force without informing the United States of their intentions; and everybody involved, particularly Eden, should have had a better sense of how Ike was going to react—and of the dangers of intervention by the Soviet Union now that somebody as impulsive as Khrushchev was running things there. Eden was, of course, moved to take action precisely because he was afraid of being thought weak as compared with his predecessor Winston Churchill, and also because he equated the Suez Crisis with Munich. He had been among the few who predicted that appeasing Hitler would lead to war, and now, too old and too ill for the job of prime minister for which he had waited so long, he saw Nasser as another Hitler, and refused to "repeat Munich on the Nile."

This disastrous combination of mistaken assumptions led to the beginning of hostilities at the end of October. The Israelis swept to and beyond the canal, crushing the Egyptian army, while the British and French massed their troops, naval forces, and air forces in Cyprus, ready to strike. This clumsily stage-managed war against Egypt coincided, most unfortunately, with the uprising in Hungary that toppled the communist leadership; sent students, factory workers, and Hungarian soldiers into the streets of Budapest to fight the Soviet armed forces; and inadvertently diverted attention from what was about to happen at Suez.

Thus, just as Election Day was approaching for Ike, he was faced with a double crisis. He was, understandably, appalled by the duplicity of the French and British, and realized at once that despite the old ties of war there was no way in which the United States could condone an Anglo-French attack on a significant third-world country, however much he disliked Nasser. He realized too that despite Dulles's rhetoric about "rolling back the iron curtain," intervention in Hungary was out of the question. The Hungarians might expect arms, support, supplies, and even intervention from America, and feel that they had been promised all that and more by Radio Free Europe; but Ike needed no threatening messages from Bulganin and Khrushchev to know that all this was out of the question, and would very likely trigger a nuclear war. In any case, as Ike pointed out sharply, for all practical purposes Hungary was as inaccessible to American military power as Tibet. At the same time, the Anglo-French war against Egypt risked bringing about massive Soviet intervention in the Middle East, and possibly nuclear war as well.

Any doubts that may still have been felt about Ike's health were effectively put to rest by his activity from the end of October 1956 through the last days of the election, and on to Christmas, by which time both crises had finally ebbed. Ike was constantly under stress, and sick at heart at the risks to which the Israelis had exposed themselves, at his own inability to come to the aid of the Hungarians, and at the collapse of the British and the French under pressure. His health was put to as severe a test as that of any president has ever been, and he came through unscathed. What is more, he achieved most of what he wanted. The Israelis had triumphed—General Sharon, then a young tank commander, had not only reached the canal but advanced to within sight of the suburbs of Cairo. And though Nasser remained in power, his army was humiliated, his air force had been destroyed on the ground, and his prestige was badly damaged. The attacks by the French and the British had been carried out with less success—they had no equivalent of Sharon—and Nasser had retaliated by sinking ships to block the canal, exactly the move that the French and British had acted to prevent. By Ike's decision to cut off oil supplies to the United Kingdom and to sell the rapidly sinking pound sterling short, the British were plunged into a financial crisis that forced a reluctant and ailing Eden to accept a humiliating withdrawal of British and French troops, a United Nations cease-fire, and UN control of the canal. Eden shortly afterward resigned, and was replaced as prime minister by Harold Macmillan, Ike's old political adviser from their days in Algiers, who had been among the first to express pessimism about Eden's adventure. It was the end of more than Eden's career—it was the end of Britain's remaining pretensions to independent, imperial power; it was the end of the fiction, still persisting from World War II, that the United States, Great Britain, and France were equal world powers; and it spelled the end of both Britain and France as colonial powers. (Britain would shortly abandon Malaya, Kenya, Uganda, Tanganyika, and much else besides; France would shortly lose Morocco, Algeria, and most of its African colonies.)

Ike had acted swiftly, decisively, and undeniably for the good; and although he felt great sympathy for his old friends in Britain, and even greater sympathy for the gallant but ill-advised Hungarians, he carefully managed events to avoid a clash with the Soviet Union, and he preserved peace—not a perfect peace, to be sure, or one without victims and compromises, but still peace. The Soviet Union had threatened to use atomic weapons on London and Paris at the height of the Suez Crisis, and in order to discourage American intervention in Hungary, but Ike had taken all this blustering calmly in his stride and kept a firm control of events.

No American president had ever exercised power more surely or more deftly, or under greater pressure of time and events.

CHAPTER 19

"And So We Came to Gettysburg and to the Farm We Had Bought Eleven Years Earlier, Where We Expected to Spend the Remainder of Our Lives"

It is hard to think of any American president whose second term was not a disappointment to himself and others. If that is a rule, Ike, in any event, was to prove no exception to it—in part for reasons that were not of his own making. The Soviet Union's blustering but effectual use of power in the Suez Crisis and its success in putting down the Hungarian Revolution despite worldwide disapproval led, as is so often the case in world affairs, to hubris. Khrushchev had bullied and won—now he expected to go on winning if he bullied harder. But of course in world affairs, as in personal life, bullying tends to provoke resistance, particularly as the bully's strength becomes overextended. It a nutshell, Nikita

Khrushchev was a bully, and his bullying worked for some time; but when it stopped working he was unceremoniously removed from power by his colleagues (and was lucky not to be shot) and the Soviet Union started on the downward slide that would end in ignominious collapse. Far from burying the West (and capitalism), the Soviet Union (and communism) would be buried unlamented in a pauper's grave less than thirty years after Khrushchev's boast.

As he got to know Khrushchev better, Ike came to the conclusion that he was a bluffer, and a good deal more dangerous and less jovial than he looked. This inference was in itself unusual, since Ike almost always had a good word to say for his fellow world leaders, however reprehensible others might find them. He managed to see the best in both Stalin and Franco, and had kinder words to say about Nasser than anybody else in the non-Arab world—and even about Nehru, who gave him crisp, high-minded lectures on international morality that, no doubt, Ike could easily have lived without.

The year 1957 began with what was in the 1950s and 1960s a familiar problem: trying to get a civil rights bill passed in some meaningful form. Ike has received very little credit for his efforts on behalf of civil rights, but that is largely because he regarded himself as a "moderate" on the issue, and was not ashamed to say so, and also because he avoided rhetoric and dramatic gestures, and instead quietly insisted on enforcing the law. He had always believed in "the right to equality before the law of all citizens . . . whatever their race or color,"[1] and during World War II he had moved to desegregate Red Cross clubs in his theater of command, and taken the even more radical step of sending "Negro replacements" into "previous all-white [combat] units," four years before Truman's order to desegregate the United States armed forces. He was firm in his belief that black citizens' right to vote had to be enforced; he had no doubt that the Warren court's decision on school segregation was right, and that Supreme Court decisions must have "a binding effect . . . on all of us if our form of government is to survive and prosper";[2] and he was impatient with senators (including Lyndon B. Johnson and John F. Kennedy) who were slowing down and compromising the passage of his civil rights bill by "interminable speeches" and amendments intended to disembowel it. For all that, he wanted the desegregation of the schools to proceed surely, but slowly ("with all deliberate speed," as the Supreme Court itself had ruled), with due regard for the feelings of everybody concerned, and without causing a constitutional crisis. In this he was to be bitterly disappointed—he underestimated the strength and the anger of segregationists in the South, and perhaps also the determination of blacks to have a showdown on the subject of schools.

The school crisis in Little Rock, Arkansas, could easily have been avoided. The Little Rock school board, without enthusiasm, but in accordance with the law, had already proposed a plan for desegregation, which had been approved by the United States District Court, Eastern District of Arkansas, and which represented pretty much what the Supreme Court and Ike had in mind. "The senior high schools were to integrate by the fall of 1957, the junior high schools by 1959 or 1960, the elementary schools by 1962 and 1963."[3] Certainly this was not rapid—in fact a number of black parents sued because "the process would take too long"—but had there been even a modest degree of goodwill (and common sense) in the white community, the plan would have allowed time for an orderly transition from the old social order to the new, however painful and difficult it might be.

Unfortunately Governor Orval Faubus, in whom both goodwill and common sense were notably lacking, was determined to prevent this, and so he handed Ike an unwelcome hot potato—a civil rights crisis. Faubus, who had a good old boy's strong streak of shrewdness, after a series of doomed legal attempts to delay or block the process, ordered the Arkansas National Guard, assisted by the Arkansas State Police "mobilized" by the governor to act as a "state militia," to stop "Negro children from entering the high school." A series of legal moves of increasing urgency and complexity (many of them intended by Faubus primarily to slow events down to a crawl) failed to change the position of the federal judge. The federal court reaffirmed its ruling that integration should proceed at once as ordered, while Arkansas state troopers and National Guard troops turned the black students back, against scenes of increasing violence, hysteria, and fury on the part of white mobs outside the school.

This was exactly what Ike had hoped to avoid. Integration in the armed forces nearly a decade ago had gone smoothly precisely because everybody in them was used to the idea of obeying an order even if it was unpopular. Ike was still enough of a soldier to assume that responsible government officials would obey a court order, but the politicians of Arkansas were not soldiers, and resistance to the wishes of Washington, D.C., was part of their heritage. Faubus had everything to gain and almost nothing to lose by making the process as difficult and long-drawn-out as possible. Eventually, Ike agreed to see Faubus, in Newport, Rhode Island, where the president was vacationing, and Ike warned the governor bluntly, "In any area where the federal government has assumed jurisdiction and this is upheld by the Supreme Court, there can be only one outcome: the state will lose."[4] This forthright statement did not impress Faubus, who went home and launched a whole new series of pleas to the court and delays; he then withdrew the National Guard from the school, precipitating several days of mob violence, to the horror of most of the nation and a good part of the rest of the world, which, thanks to television, was now as much a witness as the citizens of Little Rock to what was happening at the high school. Eventually, the mayor of Little Rock begged the White House for federal assistance, and after rejecting suggestions that he should send FBI agents, Ike did what came naturally to him: he federalized the Arkansas National Guard, placing them under his control, and sent troops from the 101st Airborne into Little Rock. The 101st Airborne—whose troopers Ike had said good-bye to with tears in his eyes before they took off to land in Normandy on the evening of June 5, 1944—finally brought the violence to an end, although one rioter was clubbed and another struck by the point of a bayonet. Then, "nine Negro students entered the high school doors, and, under Army guard, sat through a full day of classes."[5] The southern governors begged Ike to remove the troops, but he reminded them that his "own responsibilities under the Constitution were not subject to negotiation,"[6] and remained adamant. The school crisis in Little Rock had been resolved.

It is curious that Ike gets so little credit for his action in this first, brutal demonstration of just how fiercely change was going to be resisted in the states of the old Confederacy (and not just in them alone). He acted with more energy—and in a more straightforward way—than either President Kennedy or President Johnson, and his choice of the 101st Airborne placed the crisis in the hands of troops whom he knew he could trust to enforce the law and maintain order. This was in accordance with his belief that if force was to be used at all, it must be on a scale that was irresistible—hence his preference for armed paratroopers rather than FBI agents. This was sound judgment, and it worked. Perhaps the chief reason why Ike does not get the credit he deserves in this area is that, unlike Lyndon Johnson, he had no interest in being portrayed as someone at the center of a national psychodrama; and unlike Kennedy, he did not make decisions while surrounded by friendly newsmen and photographers. He preferred to operate quietly, with the absolute minimum of public posturing. Ike had not much wanted to see Faubus—and, as things turned out, he was right, since Faubus went back on his word the moment he was home in Arkansas—but having been persuaded to do so, he met Faubus with as little attention from the press as could be managed. Ike wanted results, not news stories about himself, but he demonstrated very effectively to the southern governors what the results of disobeying a federal court order—and of crossing him, as Faubus had done—would be.

Many of the problems Ike faced are with us still, and therefore do not seem unfamiliar or remote. He was forced into difficult negotiations with the Israelis over Israel's occupation of the Gaza Strip, then as now a flashpoint and a potent source of trouble and terrorist attacks. He had to deal with a rising tide of problems with Lebanon, Syria, Jordan, and Iraq, as well as the sultanate of Muscat and Oman, and Saudi Arabia, where the king was inclined to see any troubles in the Middle East as arising from U.S. support of Israel. Toward the end of 1957 there occurred, however, the first of several disappointments that were to sour Ike's second term far more than anything that was happening in the Middle East.

On October 4, 1957, the Soviet Union launched into orbit the world's first man-made satellite, Sputnik ("fellow traveler"), a name that instantly entered the vocabulary of every language. Although Ike calmly declared that Sputnik did not "raise his apprehensions,"[7] he appeared to be the only person in the United States who felt that way. A wave of shock, anger, humiliation, and recrimination swept across the country, as Sputnik sped in its orbit, beeping continuous radio signals to earth, and thereby proclaiming the superiority of Soviet rocket technology (although, as one British observer commented, it proved merely that the Russians had grabbed cleverer Germans than the Americans did in 1945). There was also a strong element of fear—if the Soviets could lob a satellite into orbit, then they could probably also lob hydrogen bombs onto American targets with intercontinental ballistic missiles or (as they were beginning to become known even to the general public) ICBMs.

Ike moved swiftly to seek out the best information he could, but it was not comforting. Between 1945 and 1952 the Truman administration had been reluctant to spend large sums of money on rocket research. Wernher von Braun, the director of Hitler's V-1 and V-2 programs, bristling at criticism, replied that for six years after the end of the war the United States "had no ballistic missile program worth mentioning," and that "our present dilemma is not due to the fact that we are not working hard enough now, but that we did not work hard enough during the first six years after the war."[8] Ike did not feel that any good could come from simply laying the blame at Truman's feet, or from sheltering behind Wernher von Braun; but as a result he ended up taking much of the blame himself. It turned out too, that the United States Army had rockets that would have been perfectly capable of launching a satellite into orbit over a year before the Soviets did; but the program for launching an American satellite was linked to an entirely separate rocket system, which was behind schedule. Unlike that of the Soviet Union, the American satellite program was not intended to be secret—the intention was to share the technology and the scientific results with the whole world—and this of course had precluded the use of a military ICBM, the construction of which was one of America's principal military secrets. Since the Russians had operated—and would continue to operate—with their usual secrecy, the use of their most powerful military rockets in their space program had of course presented no comparable problems for them. All this was sensible—before long the United States would have its own satellite in orbit (although not before the Soviets trumped the United States once again by sending a dog, Laika, into orbit, to the fury of animal lovers the world over when it was realized that Laika would die in space). But it did not satisfy Democrats, who complained that Ike had neglected the space program and warned of a "missile gap"; or right-wing Republicans, who predicted a nuclear Armageddon, and believed Ike should have followed Taft's advice and tried to build a "fortress America," complete with nuclear bomb shelters and a defense based on H-bombs, instead of seeking "mutual security," as Ike had done, by wasting money on NATO or its Asian equivalent, SEATO.

It would be nice to think that Ike was indifferent to this criticism, but of course he was not—on the contrary, it made him very angry. Still, it did not change his mind. He was willing to spend more money on "space," but he did not believe that there was any single, simple answer to the problem of defending the United States. The cold war required a mix of weapons, not reliance on a single weapons system, however devastating and sophisticated. It would require allies, with viable armed forces of their own, in many cases paid for and equipped by the United States; a powerful, modern Navy; the continued expansion of the Air Force's B-52 bomber force; and even the expansion of the United States Marines, which, like every Army general, Ike had once hoped to see sharply reduced in size.

At any rate, the combination of the school crisis in Little Rock and Sputnik ended once and for all the notion that Ike was above or beyond criticism; and as always with popular presidents, the public's mood turned sour and querulous, and not just among Democrats who were waiting for 1960 and another crack at the White House. Worse was to come.

At the summit meeting in Geneva in 1955, the first "Big Four" conference since the end of the war, Ike had presented what seemed to him a revolutionary step toward peace. The "Open Skies" proposal, as it came to be known, was his brainchild, and he presented it with a combination of enthusiasm and candor that might, in other circumstances, have persuaded the Russians of its value and sincerity. It also had the merit of military simplicity. Each of the four leading nations (the former Allies of World War II) would make available to the others a complete list and map of every one of its significant military installations, its bases abroad, and the whereabouts of its naval forces, and each nation would be entitled to maintain aerial inspection teams on the other nations' soil, and to have regular or surprise inspection flights. In short, Ike proposed abolishing a substantial number of military secrets altogether, and came up with an ambitious quadripartite solution to a principle that Ronald Reagan would later define as "Trust, but verify." Short of immediate and total world disarmament, this was about as radical a proposal as any president of the United States had ever made, and it so dumbfounded President Bulganin of the Soviet Union that he appeared to accept it at first. Then Nikita Khrushchev, the real power in the Soviet Politburo, swiftly and scornfully rejected it (inadvertently exposing the fact that Bulganin was merely a front man), and Open Skies died.

The central idea of Ike's plan, however, did not die. If the Soviet Union would not join in Open Skies, then Open Skies might be brought to the Soviet Union. In November 1954 Ike had personally authorized the construction of an experimental high-altitude reconnaissance aircraft, the Lockheed U-2, a fragile jet-powered glider which could fly great distances at the then remarkable altitude of 70,000 feet, far above the reach of Soviet fighters, and was assumed to be almost undetectable by radar. In test flights above the United States, the photographs made from the plane were so astonishingly sharp and clear that it was possible to count the number of cars in a parking lot, and to distinguish without difficulty "even the [painted] lines marking the parking areas for the individual cars."

With the Soviet rejection of Open Skies the decision was made to use the U-2 to fly secret reconnaissance missions over the Soviet Union. Ike took full responsibility for the decision, and indeed insisted on personally approving and being briefed on "each series of missions."[9] He was convinced, as was everybody else involved, that it was the only way to redress the balance of information, for while huge amounts of information were available in the United States, from highway maps to the daily press, including the location of every important military site and factory, almost everything in the Soviet Union remained secret and undocumented. Everybody else assumed that the Soviets would never reveal that they had been helpless to prevent overflights of their own country, but Ike did not. He believed that in the event of an accident there would be a worldwide storm of protest, and serious diplomatic problems. He considered that the information produced by the flights was important enough to make that risk worth taking, but he never imagined that there was no risk. In this, as usual when it came to military matters, he was more realistic than his advisers. He had been assured by the CIA and the Chiefs of Staff that in the event of a "mishap" of any kind, the fragility of the aircraft would mean that it would break into pieces as it fell, so the Soviets would be unable to recover any meaningful parts of it—or a live pilot. "This was a cruel assumption," he wrote later, but the pilots, recruited by the CIA, went into the operation "with their eyes wide open," and were motivated by "a high degree of patriotism, a swashbuckling bravado, and certain material inducements."[10] Later on, when Ike's worst fears about the operation came true—and when the CIA's assumption about the pilots' willingness to use their poison needle proved unfounded—many people jumped to the conclusion that Ike had not been paying attention, that the U-2 program was something that had been slipped past him by the more ardent hawks in his administration like John Foster Dulles and Dulles's brother Allen, the director of the CIA. But Ike makes it abundantly clear in his account of his second term, *Waging Peace*, that this was not the case. Far from seeking to shield himself from the responsibility, he admitted being involved in every detail of the operation.

Late in 1957, Ike's health, always a concern after two heart attacks and surgery for ileitis, suffered another setback, from a small stroke. He had been signing papers when he suddenly found that he could not write his signature; he then dropped the pen and was unable to pick it up. When he buzzed for his secretary, he found that he was unable to speak coherently. Helped upstairs to his bedroom, he was unable to remember the name or the painter of his favorite picture (a Turner, presented to him by the Scottish Trust).* The doctors were relieved when Ike had recovered

* Although Ike gets little credit for either his ability as a painter or his taste in pictures, his admiration for the Turner in his bedroom is clearly genuine, and evidence of very good taste indeed.

to their satisfaction in a couple of days, but he himself was aware henceforth that his "memory for words was not what it had been."[11] From that time on, he spoke more slowly and, if tired, haltingly, and sometimes reversed syllables in a long or unfamiliar word. Combined with Ike's tendency toward garbled syntax, this had the effect of convincing those who were not his admirers that his mind was beginning to go, as Churchill's had gone in 1951 (even Mrs. Churchill believed he should not have been prime minister for a second time), though at an earlier age. Mamie herself underwent a hysterectomy, and this fact, along with the inner-ear imbalance that had plagued her for decades, and Ike's health problems, had the general effect of making the Eisenhowers seem older and in worse health than they actually were. All this also promoted the legend that Ike was a hands-off president, more interested in his golf score than in affairs of state.

This was never the case, however. Shortly after his stroke, Ike went to Paris for a NATO summit, stood up in an open car to acknowledge the cheers of thousands of Parisians most of the way from Orly Airport to the United States embassy, in freezing weather; and managed to make a short, impromptu speech in front of his old office at NATO headquarters. Ike had decided to go to Paris as a kind of test, to see if he was still capable of serving as president, and the results more than satisfied him that he could still do the job.

On his return, he dealt with a succession of domestic and foreign crises—the consequences of a second term since George Washington's time—without any diminution in the amount of paperwork or the number of meetings he was accustomed to. Indeed, Ike's own account in *Waging Peace* of a three-week period in his presidency is enough to disprove the notion of him as less energetic than Kennedy or Johnson, or less involved with detail. His account of his daily work is nearly as exhausting to read as it must have been to live through, and one cannot help being amazed by his energy and attention to detail, considering that he was a man in his late sixties with any number of health issues. Like Winston Churchill's, "his eye was on the sparrow." No detail was too small to catch his attention, no report too long for him to read or listen to patiently. From time to time, relating to these events, he makes a remark that rings true today, such as pointing out that the United States has no business transforming itself into "an occupying power in a seething Arab world,"[12] still good advice; and that if we should ever do so, "I am sure we would regret it." The length, detail, and meticulous drafting of Ike's memos, cables, letters, and instructions are amazing, as is his calm, balanced judgment in difficult circumstances.

Nevertheless, Ike's presidency was reaching its most difficult period, and he was beset with problems, which kept Mamie constantly on the lookout for his health. At this period of his life Ike's physician, General Howard Snyder, described the president as becoming increasingly short-tempered and rude. If so, it is easily understandable—though members of the Eisenhower family point out that Dr. Snyder was crotchety himself, and growing so old that his hand trembled when he gave injections, causing Ike considerable pain. Ike continued writing his long letters to his old friend Swede Hazlett, whose own health was fast fading. Ike commented on the worrying illness of Secretary of State John Foster Dulles (who was already suffering from the cancer that would kill him), and remarked, "Apparently with strangers his personality may not always be winning, but with his friends he is charming and delightful," which was putting it very mildly indeed. Ike also remarked that he himself was still making progress, although he often used the wrong word—"I may say 'desk' when I mean 'chair.'"

In November 1958 Hazlett died, and Ike attended his funeral at Arlington National Cemetery. In a photograph taken at the ceremony, he is seated in the small group of mourners next to his naval aide, looking utterly bereft. His correspondence with Swede Hazlett had been like a pressure valve—he could be utterly frank with Swede, and their friendship went back for nearly fifty years and flourished despite the vast difference in their rank and their service careers. The months ahead were to bring worse. Ike was obliged to land troops in Lebanon to put down an armed uprising—a move that proved neither popular nor successful. The Hashemite monarchy in Iraq was overthrown in a coup d'état, and the pro-Western king and his prime minister were dragged out of the royal palace and murdered in the streets of Baghdad by the mob. The Chinese communists began to bombard the offshore islands of Quemoy and Matsu, threatening to bring about a full-scale war in Asia. The United States economy slipped into a cyclical recession. And Sherman Adams, Ike's chief of staff, was forced to resign after a messy series of financial charges, including the acceptance of a vicuña overcoat, which was to become as famous in politics as Pat Nixon's "Republican cloth coat." Ike had no doubt about Adams's integrity and honesty, but his confidence was not shared by anyone else; and Adams, the toughest and least appealing member of his staff, resigned, to Ike's intense and lasting regret (and resentment at the politicians who had brought Adams down). He was replaced by the retired Major General Wilton B. Persons, a more easygoing personality and, as a military man, thought to be less vulnerable to the temptations of politics. Still, it was a blow to Ike, both personally and politically—doubly so, since it resembled a similar scandal involving Harry Vaughan in Truman's White House, with which the Republicans had enjoyed a field day.

This was followed by a full-scale crisis in Berlin, reignited by Nikita Khrushchev, who was determined to recognize the Russian zone of Germany as a separate East German Communist Republic and to repudiate the Allies' rights of transit by road and air to Berlin. This had very nearly triggered a war in 1948, and was not a less volatile subject ten years later; and it took a series of dangerous confrontations and a long and acrimonious diplomatic dispute with the Soviet Union until Khrushchev backed down enough to return matters to the status quo. During the crisis, Dulles's illness became increasingly obvious, and he was soon obliged to resign for reasons of health, to be replaced as secretary of state by Christian Herter. Dulles's death, though it was expected, put an enormous strain on Ike. He was not only fond of Dulles but had trusted him absolutely, and when they were both in Washington Dulles was in the habit of coming over to the White House in the late afternoon and sitting down with Ike to have an informal but detailed review of the world. Ike would miss Dulles even more than he would miss Sherman Adams, and inevitably this loss would place on his shoulders the additional burden of acting to some degree as his own secretary of state.

. . .

"When sorrows come, they come not single spies, but in battalions." The sorrows of 1958 concluded with the Republicans' loss of control of both houses; with fierce senatorial opposition to the appointment of Clare Boothe Luce as ambassador to Brazil, after she charged that the erratic and outspoken Senator Wayne Morse of Oregon "had been kicked in the head by a horse"; then with the Senate's rejection of Admiral Lewis Strauss as secretary of commerce after lengthy hearings and a floor debate that would have had seemed outrageously partisan and acrimonious even today.

In 1959 a foreign ministers' meeting in Geneva, intended to be the precursor for a full four-power summit meeting, collapsed ignominiously when the Soviet Union refused to discuss either Berlin or nuclear tests. It was shortly followed by the famous "kitchen debate" in Moscow, in which Khrushchev and Vice President Nixon argued angrily in front of a modern

Ike with British wartime leaders, London, 1959.

American kitchen at the American exhibit about the merits of their respective societies—and during the course of which, much to Ike's chagrin, an invitation was issued to Khrushchev to visit the United States, which the Soviet chairman accepted with what seemed to many unseemly haste. Ike regretted having to meet the man who had deliberately sabotaged the foreign ministers' meeting, and allow him to tour the country. To prepare himself for Khrushchev's unwelcome arrival, Ike flew to Europe to talk to Harold Macmillan, Konrad Adenauer, and Charles De Gaulle—who was now president of France and was in the process of bringing the Algerian war, and the mutinous uprising of right-wing generals who had wished to continue it, to an end. Mamie did not accompany Ike on this arduous journey, for which he used for the first time the new presidential jet aircraft, a Boeing 707 that had replaced the aging Lockheed Constellation *Columbine*. Ike visited the British royal family in Balmoral, the queen's Scottish home; was taught by her how to cook "dropped scones" over an open grill; and in ferocious weather went for a wet and freezing sail in Prince Philip's racing yacht, with the prince at the helm. The Allies' attitude toward Khrushchev's visit to the United States was skeptical, and best summed up by De Gaulle, who said that it was a "futile gesture" but that while he doubted it would serve any purpose, it was entirely a decision for the United States alone.[13] As usual, De Gaulle had a wide a range of subjects to discuss with Ike, above all France's decision that no American aircraft armed with nuclear weapons might be based or land on French soil unless France had a veto power over their use, which Ike could not accept.

Khrushchev's visit to the United States was a curious mixture of irritation and provocation, which might have been designed to test Ike's temper and blood pressure. First there was the problem of Khrushchev's demand that he should be received as a "head of state," when in fact he was not. Ike made the Solomon-like decision that he should be described as coming "in the capacity of a head a state," which solved the problem. Then there was the Soviet ambassador's demand that he be allowed to go aboard the airplane to greet Khrushchev, which Ike turned down flat. Much as Ike already disliked the chairman, it was his prerogative to stand at the foot of the steps leading down from the airplane and be the first to greet him on

American soil. Easygoing as Ike liked to appear, in matters like this his feelings about precedence were those of a five-star general, and he was no more flexible over protocol than General MacArthur had been. The tour itself was a mixture of propaganda, barnstorming, and blustering, much of which Ike was spared by having ordered Henry Cabot Lodge to accompany the chairman. It reached its low point in Los Angeles, where for security reasons Khrushchev was prevented from visiting Disneyland and lost his temper. (At a luncheon for the Khrushchev family, when informed that Mrs. Khrushchev still wanted to go to Disneyland even without her husband, Frank Sinatra said, "Tell the old broad I'll take her this afternoon," causing yet another ripple of bad feeling when this was translated.) When the chairman, by now in a vile mood, was taken to see Shirley MacLaine rehearsing one of the dance numbers for *Can-Can* at the Twentieth Century-Fox Studio instead of visiting Disneyland, he remarked, "In Russia we think a woman's face is more interesting than her behind."

In Washington, Khrushchev was the guest of honor at a formal White House dinner, at which Ike had arranged to have played "some robust music by Fred Waring and his Pennsylvanians,"[14] always a favorite of Ike's, but it did not soothe Khrushchev. Khrushchev also spent time with Ike in the calmer (and cooler) atmosphere of Camp David, but as De Gaulle had predicted—and Ike had not disagreed with him—nothing of substance came out of the meeting. Although it became fashionable to talk of the "spirit of Camp David" as a sign of optimism in East-West relationships, Ike never used the phrase himself, or liked it when others did.

This debacle was followed by a steel strike. Then came an extended world tour on Ike's part (Mamie's health did not permit her to accompany him), during which he visited Italy, Tunisia, Turkey, Pakistan, Afghanistan (where, with an old soldier's eye, he remarked grimly on the "rugged . . . mountainous . . . bleak and forbidding landscape"), India (where he was greeted by huge crowds), Iran, Greece, France, Spain, and Morocco. Hardly any American president has ever visited more countries, or worked so hard to put his point of view across both to heads of state and political figures and to the people. It was a remarkable achievement for a man of his age and health.

As if to prove that no president can ever escape crises, Fidel Castro at last came to power in Havana after years of guerrilla fighting. After an initial burst of favorable publicity in the American media, he quickly moved to radicalize the country, expel the middle-class and wealthy people, and kill his former opponents. Castro would soon become something of a cult figure to Americans on the left, but Ike recognized him very quickly as an enemy—say what you will about Ike, his view of progressive politics did not include a firing squad without a trial for those who disagreed with you.

Castro's coming to power was another of those events that seemed to symbolize America's loss of power—and Ike's loss of control over world events. This was true especially because, shortly before, Nixon's car had been attacked and stoned by a mob in Lima, Peru, and was then smashed and nearly overturned by a howling, hostile mob in Caracas, Venezuela. Something was in the air, and it had nothing to do with Nixon. Burmese mobs were rioting against the United States; the French (and the Algerians) were blaming the United States for the troubles in Algeria; mobs burned down two American libraries in Beirut; a mob tried to invade the Canal Zone in Panama; rioters burned the American flag in La Paz, Bolivia, and then stoned the USIA and the American embassy. Much of this had little to do with Ike—he had been the beneficiary of an enormous amount of goodwill when he became president. Still, seven years is a long time to be in office; the goodwill was wearing thin; and the world was moving rapidly toward a new wave of leftist militancy. Some of this militancy was promoted and encouraged by the Soviet Union, but much of it was based simply on increasing impatience with reactionary regimes and inadequate opportunities for those in poor and third-world countries who had completed their education. The postwar period was now over, and most of the world was decolonialized for better or worse, as Franklin Roosevelt had intended. But whether in Indochina, India, the Middle East, Africa, or Asia, the heroic struggles for independence were over and the postcolonial regimes had not as yet been able to deliver the goods in the form of jobs, growth, or prosperity. Populations were expanding; technologies were changing rapidly; distances were being abolished by jet airliners and cheaper fares; television was becoming universal, spreading the good and the bad worldwide instantaneously. In short, the world was becoming like a pot bubbling on the stove, building up a huge head of steam in the form of demand for change while the American government was widely seen as struggling to keep the lid down firmly on the pot and preserve the status quo.

Despite the lack of any useful agreement between Ike and Khrushchev at Camp David, much of the world looked forward with anticipation to the long-awaited four-power summit meeting in Paris, scheduled for May 15, 1960. The preparations for the meeting had been arduous and meticulous; and in an age when summit diplomacy still seemed like the best way of achieving peace, and when people still remembered the momentous meetings of Roosevelt, Stalin, and Churchill during the war, the expectation was still very high that leaders could achieve in private conversation what normal diplomacy could not.

Unfortunately, on May 1 Ike had been informed that a U-2 aircraft had disappeared while on a photoreconnaissance flight over the Soviet Union. For a few days, it was assumed that, as planned, the aircraft would have been "destroyed either in the air or on impact, so that proof of espionage would be lacking. . . . Self-destroying mechanisms were built in."[15] The pilot was assumed to have vanished along with the fragments of his airplane, leaving no traces for the Soviets to recover. Everybody involved—except the president himself—believed that the Russians would keep silent about the whole matter, having no proof in their hands, and being unwilling to admit that they were unable to prevent these incursions of their airspace. They had raised no protest, for example, about a previous successful overflight on April 9, which had been tracked by Soviet radar.* This bubble burst on May 6, when Khrushchev revealed, in an angry speech to the Supreme Soviet, that the remains of the aircraft, including its cameras and incriminating film, had been found on Soviet soil; and that the pilot, Fran-

* The most important information brought back by U-2 overflights was that the "missile gap" was phony. The actual number of Soviet missiles and bombers was far below that possessed by the United States, or claimed by the Soviet government. Knowledge of this alone, Ike felt, made the flights worthwhile despite the risk.

cis Gary Powers, was alive and well in Soviet custody, and far from having used his poison needle had confirmed that he was on an espionage mission.

Despite the CIA's cover story that the plane had merely been gathering "weather data," Ike rejected any suggestion that he should plead ignorance of the flights, or lay the blame on lower-ranking officials. He had known about the program since its inception, he had authorized it, and he regarded it as having been made necessary by the Soviet Union's secrecy and nuclear threats. He would not apologize for it—the most he would do was cancel further flights over the Soviet Union, which were in any case rendered impractical by the Russians' newfound ability to shoot them down. "We will now just have to endure the storm,"[16] he told his colleagues, with his usual combination of honesty and stoicism, and it was not long in coming.

On May 15 Ike arrived in Paris for the four-power summit to find that Khrushchev had already sent a long letter to the chairman of the conference, President De Gaulle, and to Prime Minister Macmillan, which effectively scuppered the meeting before it could begin. Khrushchev would not participate unless Ike apologized publicly for the overflight, agreed to halt all further U-2 flights, and punished those responsible. This was posturing on his part for the benefit of his own colleagues in Moscow, since the United States had already made it clear that none of these demands was acceptable; and when Khrushchev was finally persuaded to attend a session, he merely repeated them more angrily and at greater length. There was an ugly moment when Ike smiled at Khrushchev's vehemence, provoking a furious outburst; there was another when De Gaulle remarked icily that since the acoustics in the room were excellent it was not necessary for the chairman to raise his voice; and there was an uglier one still when De Gaulle pointed out that even as they spoke a Soviet satellite was flying over France in orbit without his permission, doubtless collecting information.

Still, few moments had been more embarrassing or disappointing in twentieth-century diplomatic history than Ike's presence in Paris for a four-power meeting that was clearly not going to take place, or the need for him to remain seated in front of the world's media while Khrushchev ranted at him, glowering and pointing his finger at the president of the

United States. Since it was clear that the United States was not prepared to apologize, the meeting broke up; Khrushchev withdrew his invitation to Ike to visit the Soviet Union; and Ike returned home, via Lisbon, on May 19.

Most people in the West were sympathetic toward Ike, however much hope they may have invested in the summit meeting; but still, there was an element of embarrassment which was inevitable, and which the Soviets exploited skillfully. It was followed by a further embarrassment: the cancellation by Japan of Ike's planned visit to that country, when angry crowds took over the streets and almost overturned the car carrying the American ambassador and Jim Hagerty, the president's secretary, who had flown to Tokyo to complete the arrangements for the visit. Since the Japanese government could not guarantee the president's safety in Tokyo, it had no choice but to cancel the invitation; but that too left a certain impression of weakness and chaos, which was not outweighed by Ike's successful visits to the Philippines, Taiwan, Okinawa, and South Korea. The original intention of his trip had been to visit the Soviet Union and Japan, the two main destinations on his planned itinerary—and instead he was left with the task of visiting the relatively peripheral countries on the way. This episode was followed by the shooting down of an American YB-47 electronic surveillance aircraft in the Baring Sea, fifty miles off the coast of the Soviet Union—an incident which, coming after the U-2 incident, caused an enormous furor. The aircraft had been based in the United Kingdom, which received a long, intemperate, threatening note from the Soviet Union, prompting Harold Macmillan to defend in the House of Commons the right of all the NATO allies, including the United States, to fly intelligence missions on the periphery of the Soviet Union.*

Macmillan's remark in the House of Commons was surely responsible for an embarrassing scene several months later at the opening of the United Nations General Assembly in New York. Khrushchev, who had chosen to come as the head of the Soviet delegation, heckled the prime minister

* The author was engaged in such flights from 1951 through 1953 in Germany, while serving in the Royal Air Force, and can vouch for the fact that these operations represented a central asset of both the American and the British intelligence agencies, and in view of the secrecy surrounding Soviet nuclear missile and bomber capabilities, were surely a justified and necessary precaution against a surprise attack.

loudly during the latter's speech, then took off his shoe and banged it on his desk repeatedly to express his displeasure. Macmillan—not for nothing referred to by the British press as "unflappable Mac"—calmly asked for a translation of the chairman's insults into English, and having received it, lifted an eyebrow and continued speaking. But this incident, along with the descent of the Congo—recently (and cynically) left to its own devices by the Belgians, who had made few preparations for its independence—into appalling chaos and bloodshed (which continue to this day), and the presence in New York of Fidel Castro embracing Khrushchev and asserting that he had always been a communist, created a situation that was far from being the kind of conclusion Ike would have wished to his eight years as president.

For years wittier and better-educated Democrats had complained that Ike "reigned rather than ruled," that his presidency was a sort of constitutional monarchy, in which he played the figurehead monarch. In fact, as the 3,000 pages of *The White House Years* make very clear, Ike was, if anything, too much involved in every detail of his presidency. Indeed, it is hard to think of any other political figure, with the exception of Winston Churchill, who was so immersed in every detail of foreign defense and domestic policies, or who went to such lengths to ensure that what was done in his name represented exactly what he wanted done. Again and again, Ike imposed firm and sensible requirements on his subordinates. To take an example, he was briefed on every detail of the CIA's buildup of a Cuban anti-Castro force, and fully approved of the intention; but he insisted that no landing should take place until the Cuban exiles had produced a viable and popular government in exile, and that while the landing or landings must be Cuban, the United States should be prepared to supply strong and effective air cover. These two conditions were dropped by President Kennedy, much to Ike's undisguised amazement when he was later briefed by Kennedy on the Bay of Pigs disaster. On the even larger issue of Vietnam, Ike had always felt that the United States did not belong there, and resisted attempts by France to draw America into the war; however, as he would later keep insisting to President Johnson, once we were in it, the only thing to do was to give General Westmoreland what he needed to win it, and leave it to him. The war should not and could not be micromanaged from the White House, Ike believed, and the idea of building and holding "strategic hamlets" and fortified American bases and thereby abandoning the countryside to the enemy was exactly what had defeated the French. Although he was always reluctant to use force, nobody understood better than Ike how to use it.

The Democrats' choice of John Kennedy and Lyndon Johnson in 1960 seemed to Ike even worse than that of Stevenson and Kefauver in 1956; but it cannot be said that this inspired him put his entire weight behind Richard Nixon as the Republican nominee. His attitude was in part unavoidable—the relationship between a president and a younger vice president is oedipal at best—and in part a reflection of the fact that Ike wanted Nixon to campaign on the eight-year record of his presidency: in short, to promise four more years of the same, rather than an energetic new beginning. That Nixon could not do—not if he wanted to win. He had to promise the voters that he could do better, and Ike took this as the same kind of criticism that he was already fed up with hearing from John Kennedy and, on the Republican side, from Nelson Rockefeller. Ike was angered by talk of a "missile gap"—all the more so because, although he could not say it, the U-2 overflights of the Soviet Union had proved that there was none—and he took any criticism of his record on defense and foreign policy as a personal attack. Thus Nixon's fairly predictable effort to make peace with Nelson Rockefeller (and the liberal, eastern wing of the Republican Party), and to promise a new and more vigorous leadership, incensed Ike, who was in any case much more sensitive to criticism than a hardened professional politician might have been.

Ike had been slow to come out for Nixon in the first place, and managed to sound lukewarm when he did. Part of this reluctance was normal. Ike was sufficiently monarchist in his feelings about the presidency to share the ambivalence that every British monarch has felt for his or her heir apparent: the "prince of Wales syndrome," which has affected monarchs as different (and as far apart in time) as Henry II, Henry IV, George I, Victoria, and Elizabeth II. A larger part had to do with Nixon's need, which he perceived to be critical to winning the election, to demonstrate that he not only had been involved in making the big decisions of the past eight years, but had actually made some of them, or at the very least that Ike had not made them without consulting him. This, as it happens, touched a sensitive nerve in Ike, who had always bitterly resented the implication on the part of the Democrats and the eastern, liberal media that he was a do-nothing president. As a result, Ike dug in his heels at any attempt to extract from him examples of Nixon's decision-making potential or his participation in any of the decisions Ike had made.

On this subject, the "father-son" relationship which was said to exist between them (and for which little proof exists) became that of an irascible, overbearing, impatient father and his importunate son and successor—Henry IV and Prince Harry, as it were, except that Nixon's private life was a good deal duller than Prince Harry's. In press conference after press conference, and in interviews as well, Ike firmly denied that anybody had shared—or could share—in the process of making a decision, or that Nixon had ever influenced any decision of his. He capped this by telling a correspondent for *Time* who requested an example of Nixon's playing an important role in the decision-making process, "If you give me a week I might think of one."[17] This was agonizing for Nixon—it not only undercut his main claim to the presidency—which was that he had more experience than Kennedy, that he had been, in effect, a kind of "deputy president"—but also called into doubt his relationship with Ike, whom he could not persuade to make even the slightest positive remark on the subject.

The discomfort between the two men was increased when Ike advised Nixon not to debate Kennedy on television. When Nixon had decided that he must do so, Ike suggested that he should try to avoid appearing "too slick." Finally, when Nixon telephoned Ike after the first debate, Ike revealed that he had been one of the few people of voting age in the country to have missed it, which must have cut like a knife. When it was at last clear that perhaps only Ike's participation could stave off a defeat for the Republicans, he joined in the campaign, but he confined himself to his own record over the past eight years, rather than to what Nixon might do in the next four.

Much later, when Ike's grandson David married Nixon's daughter Julie, a kind of benevolent haze was drawn over the election of 1960, no doubt in the interest of family harmony and good feeling, and Ike and Nixon developed (or affected) a somewhat closer relationship. But in private neither Nixon nor Mrs. Nixon ever forgot that when Nixon had most needed him, Ike failed to deliver his full support.

In everybody's memoirs, certainly in Ike's and Nixon's, the whole subject is evaded, though this is not unusual. In Britain, Anthony Eden, when it came time to write his memoirs, glided gracefully over his bitterness and anger at Churchill's refusal to resign when he had promised to, and at the endless delays while Churchill found in each new international crisis another reason to stay on, until he finally left the stage to Eden when it was already too late. The role of successor and heir apparent is never a happy one, and there is no reason why Nixon should have been an exception to this rule.

Even to the very end, things remained sticky between Ike and Nixon. Ike finally offered to do "more campaigning" for Nixon, though this offer was qualified by his plea of "diminished physical endurance" and of incurring "Mamie's wrath." Mamie herself called Mrs. Nixon and asked her to persuade Nixon to turn the offer down because she was afraid it would endanger Ike's health. Ike's feelings were hurt—he blamed the rejection of his offer on Nixon, not realizing that Mamie and his own doctor had been behind it. Nixon then suggested that Ike should make a "goodwill tour" to the communist-bloc countries after the election, an idea which Ike turned down. But Nixon then made the promise in a broadcast on election eve anyway, infuriating Ike, who had thought it was a cheap political trick in the first place.

Nixon's defeat was narrow, but of course in politics the margin of defeat doesn't matter. Ike is reported to have been depressed by Kennedy's victory, but like most presidents he soon became friendly enough with his successor—indeed there was between them none of the bad feeling and backbiting that took place between Ike and Truman and marked the last weeks of Truman's presidency. In the meantime, Ike got on with his job, and in preparing for his farewell address to the nation, came up with the idea and the phrase that have been most closely associated with him in the public mind ever since.

It was on the subject he knew best, better, indeed, than anyone else alive—the dangers of a vast military establishment and arms industry. Ike had first delved deeply into the arms-producing potential of American industry in 1930, when at General MacArthur's order he had begun to study how best to convert a peacetime economy into a wartime one. This detailed study would become crucial when Franklin D. Roosevelt set out to make America the "arsenal of democracy," and still more so after Pearl Harbor. As a general, Ike had a genius for logistics and a single-minded determination to concentrate the enormous productive capacity of the United States on the battlefield. Though he would doubtless have disliked Speer, as he disliked all Nazis, even brilliant and repentant ones, he and Speer had in common an intuitive, nuts-and-bolts view of the importance of mass-production techniques to modern warfare. Ike also had a deep conviction that America should fight only wars in which its industrial superiority guaranteed victory with the minimum loss in lives. It was not for nothing that Ike and Patton had stripped an early tank to pieces in 1920 and put it back together again—an unusual way for an infantry officer and a cavalry officer to spend their spare time in those days. Machinery, industrial capacity, and technology interested Ike, both as a general and as president. Though his role in it is hardly even remembered, Ike regarded his determination to build the interstate highway system as one of the great achievements of his presidency (along with the construction of the Saint Lawrence Seaway, another hugely ambitious construction project on a pharaohnic scale), and there is no doubt that few physical changes have marked America so deeply as the building of the interstate highways. It is hard for most Americans to even imagine what the country was like before the late 1950s, when road travel from state to state was still an adventure, and when small towns all across the nation still remained comparatively isolated from each other.*

* The author still remembers a car trip from New York City to Lake Placid, New York, in 1944 as an adventure. It was a time when one still traveled on what are now rural "back roads," and when motels, twenty-four-hour filling stations, fast-food restaurants, and much else that we now take for granted did not yet exist.

Ike, after all, had helped take an Army convoy of vehicles across the country from Washington to San Francisco in 1919, when much of the journey still had to be made on rutted tracks and dirt roads, and his imaginative leap to give America a modern highway system that would surpass in every way Hitler's autobahns was, for better or worse, the foundation of the country we live in today. It was a public works project so large in scale that it eclipsed everything before or since, and it came from a Republican president who believed firmly in a balanced budget.

Ike was our last president to have been born in the nineteenth century, but his outlook was that of a twentieth-century American, who took industrial miracles for granted but still understood that the country's industrial might, above any other single factor, had won World War II and would in the end win the cold war. He remained calm when confronted by the showy successes of the Russians in space, confident that while the Soviet Union had to choose, to borrow Bismarck's phrase, "between guns or butter," the United States could have both.

Thus, when Ike rather surprisingly capped his farewell speech with a warning to "guard against the acquisition of unwarranted influence, whether sought or unsought, by the military-industrial complex,"[18] he was talking not about a conviction of long standing, but about fears that were as new to him as to his listeners, many of whom did not at first know quite what to make of it. Ike himself had been part of the "military-industrial complex" for over thirty years; indeed, he was one of its creators, and he knew how great a role it had played in his own victories. Yet as early as 1945, when he had argued against using the atomic bomb on the Japanese, he was beginning to have doubts about the immense influence of defense contracting and new weapons systems over American politics and policies. He did not want to see America transformed into a "garrison state." The day after the speech he complained about the proliferation of advertisements in the pages of American magazines showing Atlas and Titan rockets, as if they were the only things Americans knew how to make.

It was, perhaps, fitting that the man who had been a five-star general and had commanded two of the greatest technical and industrial achievements in the history of warfare, the landings in North Africa and on D-day, should end his eight years in the White House by warning his country against allowing the armed forces and the arms industry to become the dominating power of American life. Ike had his failings, but among his strengths was the ability to use and apply simple common sense to large and complicated problems. Also, like Roosevelt, he had a genius for seeing the big picture, and no reluctance to make major decisions or to accept full responsibility for them. Above all, he knew the difference between right and wrong, and tried to apply that knowledge to politics and diplomacy without preaching or boasting of any inherent, superior morality. If supreme command had taught him nothing else, it made Ike a true citizen of the world.

On January 19 he gave John F. Kennedy a sober final briefing (in which he sounded more like a general talking to a lieutenant than one president talking to another) on the president's power to make "immediate, split-second decisions" as commander in chief, and explained all about the significance of the officer who carried "the satchel" full of nuclear commands, and would "shadow the President for all his days in office." Then, to illustrate what he meant, he pushed a button on his telephone and said, "Send a chopper." This call produced a helicopter that landed on the lawn outside the Oval Office in exactly six minutes, impressing even Jack Kennedy.

The next day, in the middle of a heavy snowstorm, he and Mamie left quietly and unobtrusively after Kennedy's inauguration, grateful that they were no longer the focus of attention, and drove the eighty miles to Gettysburg. All along the way people lined the road to wave farewell to Ike.

They were saying good-bye to a man who, they knew, for fifty years had always done his duty, and had neither promised nor asked for more.

CHAPTER 20

"Lower the Shades!"

Ike lived for nine more years, and apart from health problems of increasing intensity, they were happy ones. He was spared the long, lingering senility of Winston Churchill's last years; or the obsessive, acerbic, outspoken, and increasingly embarrassing old age of Field Marshal Montgomery, which undermined much of the respect in which Monty had been held by his countrymen.

Ike's mind remained sharp, apart from the occasional slips of memory inevitable past a certain age. From time to time he was consulted by his successors, particularly Lyndon Johnson, and gave them good advice, which was seldom followed. He was deeply troubled by the assassination of President Ngo Dinh Diem of Vietnam in 1963, which, it was rumored, had been ordered by President Kennedy and carried out by the CIA; and

Ike painting, at Gettysburg.

he correctly predicted that it would do no good. He was sharply contemptuous of President Johnson's interference in the tactics of the war in Vietnam, urging him "to go for victory."[1] Ike did not believe in assassination as a matter of state policy, and while he had been against getting involved in a war in Vietnam in the first place, he did not believe in half-measures once you found yourself in a war.

He played Scrabble with Mamie on her beloved glassed-in porch at Gettysburg, looked after his small herd of Aberdeen Angus cattle—the boy from Abilene was still a farmer at heart—played golf and shot quail on the estates of his wealthy friends, indulged his grandchildren, wrote prodigiously. Between the income from his books, his pension as a former president, his pension as a retired General of the Army (Congress had restored him to the five-star rank he had resigned in order to run for the presidency), and the financial advice of his brother Milton and his friends, Ike was well off, even wealthy.

In November 1965 he suffered another heart attack, but he recovered from it well enough to play golf again. In April 1968 he had yet another heart attack. Nevertheless, perhaps remembering that he had failed to give Nixon his full support in 1960, he gave a televised speech to the delegates at the Republican convention in Miami on August 5, from Walter Reed Hospital; and the next morning suffered a heart attack again. He watched from his sickbed as Nixon finally won the presidency, and from it he also watched his grandson David marry Julie Nixon.

On March 24, he began to sink into heart failure; and on March 28, as he was dying, he was able to give his final order—"Lower the shades!"

His last words were: "I want to go. God take me."[2]

He had lived for seventy-eight years, and exercised in his lifetime leadership at a level seldom experienced by any other human being.

At his own wishes he was buried in a standard-issue GI coffin, in Abilene, the town where he had grown up, and whose imprint—modesty, hard work, devotion to duty, honesty—had always marked his life.

He was, in every sense of the words, an American hero.

Acknowledgments

First of all I would like to express thanks to my patient and long-suffering editor Hugh Van Dusen; to his assistants Marie Estrada and Robert Crawford; to Amy Hill for her design of the book; and to the art and production teams at HarperCollins. I am also grateful to Carol Bowie, for her tireless preparation of the manuscript, and to Susan Gamer for her firm and sympathetic copyediting of the book.

For sage advice (not always followed), and for help, I must thank first of all the incomparable Gypsy Da Silva, and also Michael Beschloss, Sir Martin Gilbert, Sir Alistair Horn, Morton L. Janklow, and Lord Thomas.

I am particularly grateful to Michael Hill, for his tireless and brilliant job of research, without which I could never have hoped to deal with so long and multifaceted a life as that of Dwight D. Eisenhower; to Kevin Kwan, whose keen eye and help in finding the photographs were invaluable; to Dawn Van Ee, Historical Specialist, Manuscript Division, Library of Congress, who saved me from many errors; and to the Eisenhower Library, and particularly to Valoise Armstrong, for her generous assistance, patience, and scholarship.

For enthusiasm and support in undertaking this biography of a man about whom so many biographies have already been written, I am indebted to Jane Friedman of HarperCollins, to Susan Eisenhower, and to my dear friends Henry Kissinger and David McCullough.

Michael Korda

Notes

Chapter 1: "Ike"

Among the books I consulted in writing this chapter, the following were particularly helpful: Winston S. Churchill, *Painting as a Pastime*; Ferguson, *The Pity of War*; Lukacs, *Five Days in London, May 1940.*

In general I have based British casualty numbers on generally accepted figures. Ferguson's *The Pity of War* analyzes these very accurately and in great detail. Casualty numbers for the First Battle of the Somme are based on those in Prior and Wilson, *The Somme*, 300–302.

1. Alanbrooke, *War Diaries*, 355–505. This is a constant theme in Alanbrooke's diary entries, though he does vary it from time to time with even more exasperated remarks about Ike, such as, "He is a hopeless commander" (638).
2. Winston S. Churchill, *Into Battle: Winston Churchill's War Speeches*, speech in the House of Commons, May 19, 1940, 212.
3. Larrabee, *Commander in Chief*, 347.
4. Winston S. Churchill, *Into Battle,* speech in the House of Commons, June 4, 1940, 219.
5. Calder, *The People's War,* 113.
6. Winston S. Churchill, *Never Give In!* speech of November 10, 1942, 342.

Chapter 2: "I Hope to God I Know What I'm Doing"

Among the books that proved invaluable to me in writing this chapter were: Winston S. Churchill, *The Second World War*, Vol. 4, *The Hinge of Fate*; Trevor-Roper, ed., *The Last Days of Hitler*; Sir John Wheeler-Bennett, *Nemesis of Power*; Jenkins, *Churchill*; Horne, *Monty*; Hamilton, *Monty: Master of the Battlefield 1942–1944*; Summersby, *Past Forgetting: My Love Affair with Dwight David Eisenhower*; D'Este, *Eisenhower: A Soldier's Life*; Dwight D. Eisenhower, *Crusade in Europe* and *Letters to Mamie*; Susan Eisenhower, *Mrs. Ike*; Larrabee, *Commander in Chief*;

Trevor-Roper, ed., *Hitler's Table Talk*; Esposito, ed., *The West Point Atlas of American Wars*, Vol. 2; Messenger, *The D-Day Atlas*; Ambrose, *D-Day*; Ryan, *The Longest Day*; Thompson, *The Imperial War Museum Book of Victory in Europe*; De Gaulle, *Mémoires de Guerre, Le Salut*; Hastings, *Overlord*; Richards, *The Hardest Victory*; Saward, *Bomber Harris*; and Gilbert, ed., *The Churchill War Papers* and *D-Day*. To all these authors, and many others, I am deeply grateful.

Among those who participated in the events described in this chapter (directly and indirectly), and with whom I have had an opportunity to speak over the years, have been: Marshal of the Royal Air Force Lord Sholto Douglas, GCB, MC, DFC; Group-Captain Peter Townsend, CVO, DSO, DFC, RAF; Cornelius Ryan; Général Édouard Corniglion-Molinier; the Rt. Hon. the Viscount Bracken, PC; and Flying Officer Phillip Sandeman, RAF.

1. As Stalin himself once remarked, "One death is a tragedy; a million deaths is a statistic." It is hard to get absolutely reliable statistics about Soviet casualties in World War II even today—many deaths went uncounted, particularly in the chaos of 1941 and 1942—but this seems to be the accepted rough figure. Military deaths alone are estimated to have been between 8 million and 9 million. The total figure for Soviet military and civilian deaths includes, among others, civilians who were killed or who died of starvation, deprivation, and disease; Soviet prisoners of war who were murdered or deliberately starved to death by the Germans; Jews and Gypsies systematically targeted for death by the Germans; Soviet soldiers and citizens shot in large numbers for various reasons by the NKVD on Stalin's orders; and deaths among racial and national groups singled out by Stalin for preemptive mass deportation, like the Chechens. In the Soviet Union (unlike the United States, which shot only one soldier for desertion and cowardice in the face of the enemy in World War II), it was common for very large numbers of soldiers (up to and including the rank of general) to be executed by NKVD firing squads stationed right behind the front line to deal with such offenses as retreating, failure to advance successfully when ordered, or lack of enthusiasm and fighting spirit—and no doubt, in Voltaire's famous bon mot in *Candide* about the execution of Admiral John Byng, *pour encourager les autres.*
2. Trevor-Roper, *The Last Days of Hitler*, Ch. 1, 53–90.
3. Von Salomon, *Der Fragenbogen.*
4. Ernst von Weizsäcker, *Errinerungen* (Dortmund, 1964), quoted in Watt, *How War Came*, 511.
5. Ambrose, *D-Day*, 54; Gilbert, *D-Day*, 40–41.
6. Gilbert, *D-Day*, 56.
7. Kennedy, *The Business of War,* 327.
8. Gilbert, *D-Day*, 98.
9. Ibid., 125.
10. Summersby, *Past Forgetting*, 188.
11. Ryan, *The Longest Day*, 64.

12. D'Este, *Eisenhower*, 527.
13. Kay Summersby to author.
14. Summersby, *Past Forgetting*, 190.
15. Ibid., 189. There appear to be two variations of what Ike said in the car, the other being, "Well, it's on. No one can stop it now." The remark I quote in the text is more generally accepted, and anyway makes more sense, since there would have been no way to stop the invasion after Ike gave the order to go on the night of June 4.

Chapter 3: "What a Man . . . Did as a Boy"

The two books that I found must useful in writing this chapter were D'Este, *Eisenhower*, an amazingly thorough, scrupulously researched, and richly detailed biography; and Dwight D. Eisenhower's own *At Ease*, which contains a vivid and very detailed account of the family history and of his own childhood in Abilene. D'Este's book is such an impeccable work of scholarship that any biographer of Eisenhower can only write in its massive shadow. As for *At Ease*, it is by far the most personal and engaging book that Eisenhower wrote, and a real pleasure to read.

1. D'Este, *Eisenhower*, 15.
2. Ibid., 18.
3. Dwight D. Eisenhower, *At Ease*, 31.
4. Ibid., 67.
5. Ibid., 31–32.
6. Ibid., 29–30.
7. Ibid., 30.
8. Ibid., 83.
9. Ibid., 71–72.
10. Ibid., 72–73.
11. D'Este, *Eisenhower*, 82.
12. Dwight D. Eisenhower, *At Ease*, 40.
13. Ibid., 43.
14. Ibid., 101.
15. Ibid., 88.
16. Ibid., 89.

Chapter 4: "Where Else Could You Get a College Education Without Cost?"

Although Eisenhower's own memories of West Point are available in *At Ease*, no doubt filtered by time and his promotion to General of the Army, D'Este is excellent on the subject, and perhaps more objective and informative about the details of life as a cadet. D'Este explains many things that Eisenhower took for granted or assumed the reader would know.

1. Dwight D. Eisenhower, *At Ease*, 108.
2. Ibid., 5.

3. Ibid., 4.
4. Ibid.
5. Ibid.
6. D'Este, *Eisenhower*, 64.
7. Dwight D. Eisenhower, *At Ease*, 18.
8. Ibid., 20.
9. D'Este, *Eisenhower*, 166.
10. Ibid., 79.

Chapter 5: Second Lieutenant Eisenhower

Susan Eisenhower's portrait of her grandparents' marriage, *Mrs. Ike*, is franker, and sometimes shrewder, than Eisenhower's own account in his various memoirs. I have relied on her book for the early years of Ike and Mamie's relationship, reconciling it where necessary with Eisenhower's own memories and chronology in *At Ease*.

1. Dwight D. Eisenhower, *At Ease*, 111–112.
2. Ibid., 116.
3. Ibid., 113.
4. Susan Eisenhower, *Mrs. Ike*, 34.
5. D'Este, *Eisenhower,* 99.
6. Susan Eisenhower, *Mrs. Ike*, 21.
7. Dwight D. Eisenhower, *At Ease*, 113–114.
8. Susan Eisenhower, *Mrs. Ike*, 15.
9. Dwight D. Eisenhower, *At Ease*, 118.
10. Ibid., 119.
11. Prior and Wilson, *The Somme*, 301.
12. Dwight D. Eisenhower, *At Ease*, 122.
13. Susan Eisenhower, *Mrs. Ike*, 40.
14. Ibid., 45.
15. Dwight D. Eisenhower, *At Ease*, 126.
16. Ibid., 130.
17. Ibid., 131.
18. Prior and Wilson, *The Somme*, 78.
19. Wright, *Tank*, 28–51.
20. Ibid., 109; Prior and Wilson, *The Somme*, 216–224.
21. Macdonald, *To the Last Man*, 235–238; Winter, *Haig's Command*, 183–184.
22. Macdonald, *1914–1918*, 281.
23. Gilbert, *The First World War*, 427.
24. Dwight D. Eisenhower, *At Ease*, 147.

Chapter 6: "The Greatest Disappointment and Disaster in My Life"

1. Dwight D. Eisenhower, *At Ease*, 152.
2. Ibid., 153.
3. Ibid., 157.

4. Ibid., 158–159.
5. D'Este, *Patton*, 304.
6. Ibid., 296.
7. Susan Eisenhower, *Mrs. Ike*, 64.
8. Ibid., 68.
9. Dwight D. Eisenhower, *At Ease*, 181–182.
10. Susan Eisenhower, *Mrs. Ike*, 70.

Chapter 7: The Education of a Soldier

1. Susan Eisenhower, *Mrs. Ike*, 74.
2. Ibid., 76.
3. Dwight D. Eisenhower, *At Ease*, 184.
4. Susan Eisenhower, *Mrs. Ike*, 79.
5. Ibid., 83.
6. Dwight D. Eisenhower, *At Ease*, 187.
7. Ibid. 193.
8. Susan Eisenhower, *Mrs. Ike*, 83.
9. Ibid., 84.
10. Dwight D. Eisenhower, *At Ease*, 199.
11. Ibid., 201.
12. Susan Eisenhower, *Mrs. Ike*, 88.
13. Ibid., 92.
14. Dwight D. Eisenhower, *At Ease*, 205.
15. Ibid., 207.
16. Ibid., 207–209.

Chapter 8: The MacArthur Years

1. Manchester, *American Caesar*, 791.
2. Ibid., 19.
3. Dwight D. Eisenhower, *At Ease*, 213.
4. Ibid., 214.
5. Manchester, *American Caesar*, 158–159.
6. Ibid., 182.
7. This story is told brilliantly in Arthur Schlesinger Jr.'s multivolume biography of Franklin D. Roosevelt, and also in numerous other accounts of the convention.
8. Dwight D. Eisenhower, *At Ease*, 216.
9. Ibid., 216.
10. Manchester, *American Caesar*, 166.
11. Ibid., 32.
12. Ibid., 167.
13. Martin, *Winston Churchill*, 5:456.
14. Winston Churchill, *Never Give In!,* 128 (speech in the House of Commons, March 24, 1936).

15. The original "man on a white horse" was General Georges-Ernest-Jean-Marie Boulanger, who was accused of plotting a failed coup d'état in Paris in 1889, and ended by committing suicide at his mistress's grave in Brussels, prompting one of Clemenceau's most savage remarks: "He died as he had lived, like a second lieutenant."
16. Manchester, *American Caesar*, 174.
17. Dwight D. Eisenhower, *At Ease*, 223.
18. Ibid., 219.
19. Susan Eisenhower, *Mrs. Ike*, 134.
20. Manchester, *American Caesar*, 190–191.
21. Susan Eisenhower, *Mrs. Ike*, 138.
22. Ibid., 143.
23. Winston Churchill, *Never Give In!,* 134 (speech in the House of Commons, April 6 1936).
24. Udet to the author's uncle, Sir Alexander Korda, who had given Udet, a celebrated World War I fighter ace down on his luck and a friend of Leni Riefenstahl's, help in getting a job doing stunt flying in one of the popular "mountain pictures" of the 1920s. This was before Göring came to the rescue of his old comrade in arms from the Richthofen Squadron—they had both been awarded the coveted Pour le Mérite, known as the Blue Max, imperial Germany's highest decoration for bravery. Udet's score of kills as a fighter pilot in World War I was sixty-two; Göring's was twenty-two—a difference that Göring neither forgot nor, in the end, forgave. Udet was made the scapegoat for Göring's failure to destroy the RAF in the Battle of Britain, and committed suicide in 1941 because of accusations that he had failed to produce the right equipment for the Luftwaffe's bombing campaign. The fact that Udet was sexy, irreverent, and a spectacular bon vivant with an outrageous sense of humor (he was a gifted caricaturist and cartoonist and made many enemies with his drawings), and had a legendary way with women, needless to say, did nothing to help his case. His death became the subject of a hugely successful postwar play and film, written by Carl Zuckmayer, *Der Teufelsgeneral* (The Devil's General).
25. Susan Eisenhower, *Mrs. Ike*, 159.
26. Dwight D. Eisenhower, *At Ease*, 228.
27. Ibid., 220–230.
28. Ibid., 230.
29. Susan Eisenhower, *Mrs. Ike*, 158.
30. Ibid., 159.
31. Manchester, *American Caesar*, 182.

Chapter 9: "As Soon as You Get a Promotion They Start Talking About Another One. Why Can't They Let a Guy Be Happy with What He Has?"

1. Winston Churchill, *Into Battle*, 108 (House of Commons, May 13, 1960).

2. Ibid., 234 (House of Commons, June 18, 1940).
3. Gilbert, ed., *The Churchill War Papers*, 2:221.
4. Dwight D. Eisenhower, *At Ease*, 236.
5. Ibid..
6. Ibid., 237.
7. Ibid., 242–243.
8. Susan Eisenhower, *Mrs. Ike*, 169.
9. Dwight D. Eisenhower, *At Ease*, 243.
10. Ibid.
11. D'Este, *Eisenhower*, 279.
12. Ibid., 282.
13. Dwight D. Eisenhower, *At Ease*, 245.
14. Larrabee, *Commander in Chief*, 388.
15. D'Este, *Eisenhower*, 284.
16. Ibid., 285.
17. Ibid., 286.
18. Ibid., 297.
19. Ibid., 296.
20. *Papers of Dwight D. Eisenhower, The War Years*, 1:96.
21. Ibid., 96–97.
22. MacArthur Archives and Foundation, Newport, Va., Record Group 44a: Papers of Brigadier General Bonner F. Fellers, Box 1, F. 15 "Letter from Bonner Fellers to Mr. L. A. Davidson, 21 December 1967."
23. Kennedy, *The Business of War*, 169.
24. Barr, *Pendulum of War*, 17.
25. Alanbrooke, *War Diaries*, 101.
26. Winston S. Churchill, *The Second World War*, 2:162.
27. Ibid., 189.
28. Larrabee, *Commander in Chief*, passim.
29. Dwight D. Eisenhower, *Crusade in Europe*, 27–28.
30. Susan Eisenhower, *Mrs. Ike*, 177.
31. Dwight D. Eisenhower, *At Ease*, 249; and D'Este, *Eisenhower*, 288.
32. D'Este, *Eisenhower*, 298; and Dwight D. Eisenhower, *At Ease*, 249.
33. D'Este, *Eisenhower*, 421.
34. Larrabee, *Commander in Chief*, 421.
35. Dwight D. Eisenhower, *Crusade in Europe*, 25.
36. Ibid., 48.
37. *Papers of Dwight D. Eisenhower, The War Years*, 1:5 (and elsewhere).
38. D'Este, *Eisenhower*, 290.

Chapter 10: London, 1942

1. Summersby, *Past Forgetting*, 17–22.
2. Ibid., 24–25.

3. Ibid., 32.
4. Dwight D. Eisenhower, *Crusade in Europe*, 56.
5. D'Este, *Eisenhower*, 307.
6. Barr, *Pendulum of War*, 20.
7. Bierman and Smith, *The Battle of Alamein*, 207.
8. Dwight D. Eisenhower, *Letters to Mamie*, 23.
9. Summersby, *Past Forgetting*, 35–36.
10. Dwight D. Eisenhower, *Letters to Mamie*, 105.
11. Summersby, *Past Forgetting*, 39.
12. Ibid., 161 (both sides of the card are reproduced).
13. Reynolds, *In Command of History*, 387.
14. Summersby, *Past Forgetting*, 46.
15. Ibid., 248.
16. Dwight D. Eisenhower, *Crusade in Europe*, 57.
17. Ibid.
18. The best source for Fortitude is Hesketh, *Fortitude, The D-Day Deception Campaign*.
19. Trevor-Roper, ed., *Hitler's Table Talk*. This is a constant subject in Hitler's conversation and is repeated several times in slightly different forms.
20. Bland, ed., *Papers of George Catlett Marshall*, 2:283.
21. Dwight D. Eisenhower, *Crusade in Europe*, 61.
22. Ibid., 83.

Chapter 11: Algiers

The Tunisian campaign, involving great distances and four different Allied armies, requires a certain amount of background to understand. The books that I found most useful were Atkinson, *An Army at Dawn*, a superb and relatively fair-minded narrative history, given the sharp differences between the American and the British view of events; Eisenhower's own *Crusade in Europe*, which, if anything, strives to be even more fair-minded; and Hamilton, *Monty: Master of the Battlefield,* the second volume of a three-volume biography, which is unapologetically pro-Monty, but which, on the other hand, often shows him at his crisp best in summoning up the military realities of the campaign (even from a distance) and assessing the strengths and weaknesses of the commanders.

1. Dwight D. Eisenhower, *Crusade in Europe*, 85.
2. Ibid., 86.
3. Alanbrooke, *War Diaries*, 101.
4. Atkinson, *An Army at Dawn*, 45.
5. Dwight D. Eisenhower, *Crusade in Europe*, 95.
6. Dwight D. Eisenhower, *Letters to Mamie*, 59.
7. Dwight D. Eisenhower, *Crusade in Europe*, 99–100.
8. Atkinson, *An Army at Dawn*, 65.
9. Dwight D. Eisenhower, *Crusade in Europe*, 101.

10. Atkinson, *An Army at Dawn*, 116–129.
11. Ibid., 69–78.
12. Dwight D. Eisenhower, *Crusade in Europe*, 104.
13. Atkinson, *An Army at Dawn*, 148.
14. Dwight D. Eisenhower, *Crusade in Europe*, 104–114.
15. *Manchester Guardian*, November 18, 1942.
16. Dwight D. Eisenhower, *Crusade in Europe*, 110.
17. Ibid., 128.
18. Atkinson, *An Army at Dawn*, 171.
19. Ibid., 175.
20. Dwight D. Eisenhower, *Crusade in Europe*, 119.
21. Ibid., 120.
22. Esposito, ed., *West Point Atlas of American Wars*, text for maps 83 and 84.
23. Dwight D. Eisenhower, *Letters to Mamie*, 68.
24. Susan Eisenhower, *Mrs. Ike*, 195.
25. Ibid., 196.
26. Esposito, ed., *West Point Atlas of American Wars*, text for map 84.
27. Dwight D. Eisenhower, *Letters to Mamie*, 75.
28. Atkinson, *An Army at Dawn*, 258.
29. Hamilton, *Monty: Master of the Battlefield*, 145.
30. Ibid., 64.
31. Ibid., 133 (Montgomery referred to Anderson, in a withering phrase, as "a good plain cook").
32. Horne, *Monty: The Lonely Leader*, 9.
33. Alanbrooke, *War Diaries*, 351.
34. Dwight D. Eisenhower, *Crusade in Europe*, 123.
35. Atkinson, *An Army at Dawn*, 268.
36. Moran, *Churchill*, 79.
37. Bryant, *The Turn of the Tide, 1939–1943*, 282.
38. Atkinson, *An Army at Dawn*, 281.
39. Alanbrooke, *War Diaries*, 365.
40. Macmillan, *War Diaries*, 10.
41. Dwight D. Eisenhower, *Letters to Mamie*, 86.
42. Dwight D. Eisenhower, *Crusade in Europe*, 140.
43. D'Este, *Eisenhower*, 390.
44. Dwight D. Eisenhower, *Letters to Mamie*, 93.
45. Susan Eisenhower, *Mrs. Ike*, 193.
46. Dwight D. Eisenhower, *Letters to Mamie*, 93–96.
47. Dwight D. Eisenhower, *Crusade in Europe*, 142 and elsewhere.
48. Summersby, *Past Forgetting*, 112.
49. Dwight D. Eisenhower, *Crusade in Europe*, 148.
50. Bland, ed., *Papers of George Catlett Marshall*, 3:564.
51. Atkinson, *An Army at Dawn*, 410.
52. Hamilton, *Monty: Master of the Battlefield*, 213.

53. Ibid., 210 (Eisenhower's message to George C. Marshall, April 5, 1943).
54. Ibid., 225.
55. Dwight D. Eisenhower, *Crusade in Europe*, 157.
56. Dwight D. Eisenhower, *Letters to Mamie*, 125.

Chapter 12: Sicily

1. Dwight D. Eisenhower, *Crusade in Europe*, 159.
2. Hamilton, *Monty*: *Master of the Battlefield*, caption to photograph 44.
3. Ibid., 143.
4. Ibid., 249.
5. Horne, *Monty*: *The Lonely Leader*, 63.
6. Dwight D. Eisenhower, *Letters to Mamie*, 104–105.
7. Summersby, *Past Forgetting*, 116–117.
8. Ibid., 118.
9. Ibid., 123.
10. Hamilton, *Monty*: *The Lonely Leader*, 267.
11. Ibid., 269.
12. Dwight D. Eisenhower, *Letters to Mamie*, 126.
13. Summersby, *Past Forgetting*, 146.
14. Dwight D. Eisenhower, *Crusade in Europe*, 194.
15. Ibid., 167.
16. Alanbrooke, *War Diaries*, 406.
17. Ibid., 416.
18. Dwight D. Eisenhower, *Crusade in Europe*, 172.
19. Ibid., 174.
20. Ibid., 177.
21. Ibid., 176
22. Ibid., 184.
23. Ibid., 180.
24. Ibid., 183.
25. Ibid., 180.
26. D'Este, *Eisenhower,* 453.
27. Esposito, ed., *West Point Atlas of American Wars*, 2: text for maps 90 and 91.
28. Hamilton, *Monty*: *Master of the Battlefield*, 390.
29. Ibid., 447.
30. Ibid., 436.
31. Kimball, ed., *Churchill and Roosevelt: The Complete Correspondence*, 2:482.
32. Gilbert, *Winston Churchill*, 7:548.
33. Dwight D. Eisenhower, *Crusade in Europe*, 194.
34. Summersby, *Past Forgetting*, 150.
35. Ibid., 152.
36. Ibid., 153.

37. Hamilton, *Monty: Master of the Battlefield*, 459.
38. Montgomery, *War Memoirs*, 181.
39. Dwight D. Eisenhower, *Crusade in Europe*, 208.
40. Ibid., 207.

Chapter 13: "Supreme Commander, Allied Expeditionary Forces"—1944

1. Winston S. Churchill, *The Second World War*, 5:371.
2. Dwight D. Eisenhower, *Crusade in Europe*, 211.
3. Ibid.
4. Ibid., 271.
5. Hamilton, *Monty: Master of the Battlefield*, 475.
6. Bland, ed., *Papers of George Catlett Marshall*, 4:215.
7. Summersby, *Past Forgetting*, 166.
8. Winston S. Churchill, *The Second World War*, 5:380.
9. Ibid., 383.
10. Dwight D. Eisenhower, *Crusade in Europe*, 212, 217.
11. Ibid., 217.
12. Susan Eisenhower, *Mrs. Ike*, 215.
13. Dwight D. Eisenhower, *Letters to Mamie*, 162.
14. Hamilton, *Monty: Master of the Battlefield*, 485.
15. Ibid.
16. Ibid., 489.
17. D'Este, *Eisenhower*, 471.
18. Winston S. Churchill, *The Second World War*, 5:396–397.
19. Ibid., 397.
20. Saward, *Bomber Harris*, 170.
21. Richards, *The Hardest Victory*, 131.
22. Dwight D. Eisenhower, *Crusade in Europe*, 232.
23. Winston Churchill, *The Second World War*, 5:465–468.
24. Ibid.
25. Ibid.
26. Richards, *The Hardest Victory*, 222.
27. Ibid., 226–227; and Horne, *Monty*, 94.
28. Horne, *Monty*, 81.
29. Ryan, *The Longest Day*, 28.
30. Horne, *Monty*, 123–125.
31. Ibid., 83.
32. Dwight D. Eisenhower, *Crusade in Europe*, 228.
33. Summersby, *Past Forgetting*, 180.
34. Blumenson, ed., *The Patton Papers*, 2:472.
35. Dwight D. Eisenhower, *Letters to Mamie*, 168.
36. Ibid., 176.
37. Dwight D. Eisenhower, *Crusade in Europe*, 238.

38. Ibid.
39. Chandler et al., *Papers of Dwight David Eisenhower*, Vol. 3.
40. Dwight D. Eisenhower, *Letters to Mamie*, 179.

Chapter 14: Triumph—D-Day, June 6, 1944

1. Ryan, *The Longest Day*, 215.
2. Ibid., 135.
3. Ibid., 174.
4. Ibid., 270.
5. Ibid., 254.
6. Frank, *The Diary of a Young Girl*, 244–245.
7. Ryan, *The Longest Day*, 273.
8. Ibid., 269.
9. Esposito, ed., *West Point Atlas of American Wars*, 2: text for map 49.
10. Ryan, *The Longest Day*, 273.
11. Ibid., 269.
12. Horne, *Monty*, 142.
13. Ibid., 170.
14. Ibid., 119.
15. Ibid., 120.
16. Summersby, *Past Forgetting*, 201.
17. Winston S. Churchill, *The Second World War*, 6:43.
18. Dwight D. Eisenhower, *Letters to Mamie*, 190.
19. Summersby, *Past Forgetting*, 196–197.
20. Dwight D. Eisenhower, *Letters to Mamie*, 190.
21. Ibid., 193.
22. Summersby, *Past Forgetting*, 201.
23. Thompson, *Imperial War Museum*, *Book of Victory in Europe*, 132.
24. Horne, *Monty*, 139.
25. De Gaulle, *Mémoires de Guerre*, 564.
26. Ibid., 564–565.
27. Ibid., 565.
28. Ibid., 436.
29. Dwight D. Eisenhower, *Crusade in Europe*, 269.
30. Messenger, *The D-Day Atlas*, 111.
31. Horne, *Monty*, 160.
32. Dwight D. Eisenhower, *Crusade in Europe*, 265.
33. Ibid., 268.
34. De Gaulle, *Mémoires de Guerre*, 636.
35. Ibid., 637.
36. Ibid., 650.
37. Ibid.
38. Ibid., 653.

39. Summersby, *Past Forgetting*, 211.
40. Alanbrooke, *War Diaries*, 585.
41. Hamilton, *Monty: Master of the Battlefield*, 818.

Chapter 15: Stalemate

1. Dwight D. Eisenhower, *Crusade in Europe*, 243.
2. Summersby, *Past Forgetting*, 216.
3. Dwight D. Eisenhower, *Letters to Mamie*, 212.
4. Dwight D. Eisenhower, *Crusade in Europe*, 332.
5. Hamilton, *Monty: The Final Years of the Field-Marshal*, 162.
6. Ibid., 163.
7. Alanbrooke, *War Diaries*, 635; Hamilton, *Monty: The Final Years of the Field-Marshal*, 173.
8. Dwight D. Eisenhower, *Crusade in Europe*, 350.
9. Ibid.
10. Hamilton, *Monty: The Final Years of the Field-Marshal*, 241.
11. Horne, *Monty*, 303
12. Hamilton, *Monty: The Final Years of the Field-Marshal*, 218.
13. Ibid., 217.
14. Ibid.
15. Dwight D. Eisenhower, *Crusade in Europe*, 341.
16. Hamilton, *Monty: The Final Years of the Field-Marshal*, 265.
17. Bland, ed., *Papers of George Catlett Marshall*, 4:720.
18. Horne, *Monty*, 310.
19. Ibid.
20. Ibid., 284.
21. Hamilton, *Monty: The Battles of Field Marshal Bernard Montgomery*, 531.
22. *Letters to Mamie*, Dwight D. Eisenhower, 229.
23. Ibid.

Chapter 16: Armageddon, 1945

1. Dwight D. Eisenhower, *Crusade in Europe*, 218.
2. Moran, *Churchill*, 226.
3. Alanbrooke, *War Diaries*, 653.
4. Hamilton, *Monty: The Final Years of the Field-Marshal*, 279.
5. Alanbrooke, *War Diaries*, 649.
6. Dwight D. Eisenhower, *Crusade in Europe*, 366–367.
7. Alanbrooke, *War Diaries*, 653.
8. Dwight D. Eisenhower, *Crusade in Europe*, 327.
9. Alanbrooke, *War Diaries*, 677.
10. Dwight D. Eisenhower, *Crusade in Europe*, 371.
11. Ibid., 380.

12. Summersby, *Past Forgetting*, 218.
13. Alanbrooke, *War Diaries*, 669.
14. Dwight D. Eisenhower, *Crusade in Europe*, 390.
15. Winston S. Churchill, *The Second World War*, 6:388.
16. Ibid., 389.
17. Dwight D. Eisenhower, *Crusade in Europe*, 396.
18. Winston S. Churchill, *The Second World War*, 6:409.
19. Dwight D. Eisenhower, *Crusade in Europe*, 401.
20. Ibid., 400.
21. Ibid., 409–410.
22. Ibid., 426.
23. D'Este, *Eisenhower*, 704.

Chapter 17: "I Had Never Realized That Ike Was as Big a Man"

1. Quoted in Macdonald, *Parodies*, 447.
2. Alanbrooke, *War Diaries*, 697.
3. *London Times*, June 13, 1945.
4. Ibid.
5. Ibid.
6. Ibid., editorial.
7. Ambrose, *Eisenhower, 1890–1952*, 415.
8. Ambrose, *Eisenhower: Soldier, General of the Army, President-Elect,* Vol. I., 413.
9. Ibid., 416.
10. Dwight D. Eisenhower, *Crusade in Europe*, 459, 461–464; Ambrose, *Eisenhower, 1890–1952*, 429.
11. Dwight D. Eisenhower, *Letters to Mamie,* 270.
12. Summersby, *Past Forgetting,* 243.
13. Griffith, ed., *Ike's Letters to a Friend,* 33.
14. Ibid., 27–31; italics in source.
15. Ibid., 33.
16. Manchester, *American Caesar,* 560.
17. Dwight D. Eisenhower, *At Ease,* 322.
18. Ibid., 322–323.
19. Ambrose, *Eisenhower: Soldier and President,* 242.
20. Dwight D. Eisenhower, *At Ease,* 355.
21. *New York Times,* February 2, 1951, 5.
22. Dwight D. Eisenhower, *At Ease,* 372.
23. *Washington Post*, February 2, 1951, 22.
24. *New York Times,* February 2, 1951, 4.
25. Dwight D. Eisenhower, *At Ease,* 372.
26. Ibid., 252.
27. Irving Berlin, *Call Me Madam.*

28. Susan Eisenhower, *Mrs. Ike,* 258.
29. Ibid., 263.

Chapter 18: "I Like Ike!"

1. Susan Eisenhower, *Mrs. Ike*, 268–269.
2. This has been variously attributed to Presidents Dwight D. Eisenhower and Ronald Reagan.
3. Colville, *Fringes of Power*, 654.
4. Dwight D. Eisenhower, *Mandate for Change*, 38.
5. Ibid., 39.
6. Ibid., 40.
7. Ibid., 43.
8. Ibid., 45.
9. *American Heritage,* August–September, 2003.
10. Dwight D. Eisenhower, *At Ease*, 75.
11. Colville, *Fringes of Power*, 685.
12. Hamilton, *Monty: The Final Years of the Field-Marshal*, 831.
13. Ambrose, *Eisenhower: Soldier and President*, 279.
14. Dwight D. Eisenhower, *Mandate for Change*, 56.
15. Ibid., 63.
16. Colville, *Fringes of Power*, 654.
17. Dwight D. Eisenhower, *Mandate for Change*, 94.
18. Ibid., 96.
19. Donovan, *Eisenhower: The Inside Story*, 5.
20. Susan Eisenhower, *Mrs. Ike*, 283.
21. Dwight D. Eisenhower, *Mandate for Change*, 130.
22. Huie, *Execution of Private Slovick*, 144.
23. Dwight D. Eisenhower, *Mandate for Change*, 224.
24. Donovan, *Eisenhower: The Inside Story*, 46.
25. Dwight D. Eisenhower, *Mandate for Change*, 275.
26. Donovan, *Eisenhower: The Inside Story*, 94–95.
27. Dwight D. Eisenhower, *Mandate for Change*, 234.
28. Donovan, *Eisenhower: The Inside Story*, 159.
29. Ibid., 129.
30. Ibid., 256.
31. Ibid., 265.
32. Donovan, *Eisenhower: The Inside Story,* 397.
33. Susan Eisenhower, *Mrs. Ike*, 294; Ambrose, *Eisenhower: Soldier and President,* 395.

Chapter 19: "And So We Came to Gettysburg and to the Farm We Had Bought Eleven Years Earlier, Where We Expected to Spend the Remainer of Our Lives"

1. Dwight D. Eisenhower, *Waging Peace*, 149.
2. Ibid.
3. Ibid., 162.
4. Ibid., 166.
5. Ibid., 170.
6. Ibid., 172–175.
7. Ibid., 205.
8. Ibid., 206.
9. Ibid., 546.
10. Ibid.
11. Ibid., 227–229.
12. Ibid., 99.
13. Ibid., 426.
14. Ibid., 440.
15. Ibid., 547.
16. Ibid., 552.
17. Dwight D. Eisenhower, press conference, August 24, 1960.
18. Dwight D. Eisenhower, Farewell Address to the Nation, January 17, 1961.

Chapter 20: "Lower the Shades!"

1. Ambrose, *Eisenhower: Soldier and President,* 559.
2. Deathbed scenes and the last words of great men are often difficult to determine exactly. Witness the pious "official" version of George V's last words ("How goes the Empire?") and several people's memory of his expressing a coarse seaman's epithet to his doctor. I have sifted through the various versions of Eisenhower's death, and it does seem likely that he asked to be lifted up, then asked for the shades to be lowered, since the light was in his eyes. That he said he was "ready to go" seems agreed on, and hardly surprising given the weakening of his condition, and his long illness. Susan Eisenhower (whose father was in the room until close to the end, when Ike was given a sedative, and Mamie remained alone with him as he died) is probably the most reliable source.

Bibliography

Alanbrooke, Field Marshal Lord. *War Diaries 1939–1945*. Berkeley and Los Angeles: University of California Press, 2001.

Alperovitz, Gar. *The Decision to Use the Atom Bomb.*

Ambrose, Stephen E. *D-Day*. New York: Simon and Schuster, 1994.

———. *Eisenhower: Soldier and President*. New York: Simon and Schuster, 1990.

———. *Eisenhower: Soldier, General of the Army, President-Elect, 1890–1952.* New York: Simon and Schuster, 1983.

Arthur, Max. *Forgotten Voices of World War II*. Guildford, Conn.: Lyons, 2004.

Atkinson, Rick. *An Army at Dawn*. New York: Holt, 2002.

Ball, Simon. *The Guardsmen*. London: Harper Perennial, 2005.

Barr, Niall. *Pendulum of War: The Three Battles of El Alamein*. Woodstock: Overlook, 2004.

Beevor, Anthony. *The Fall of Berlin 1945*. New York: Viking, 2002.

Beschloss, Michael R. *The Conquerors*. New York: Simon & Schuster, 2002.

———. *Mayday: Eisenhower, Khrushchev, and the U-2 Affair*. New York: Harper and Row, 1986.

Best, Geoffrey. *Churchill—A Study in Greatness*. London: Hambledon, 2001.

Bierman, John, and Colin Smith. *The Battle of Alamein*. New York: Viking, 2002.

Bland, Larry, ed. *The Papers of George Catlett Marshall*, Vols. 2–5. Baltimore, Md.: Johns Hopkins University Press, 1986–2003.

Blumenson, Martin. *The Battle of the Generals*. New York: Morrow, 1993.

———. ed., *The Patton Papers*, Vols. 1 and 2. Boston, Mass.: Houghton Mifflin, 1972.

Bradley, Omar. *A General's Life*.

Bryant, Arthur. *Triumph in the West*. London: Collins, 1959.

———. *The Turn of the Tide, 1939–1943: A Study Based on the Diaries and Autobiographical Notes of Field Marshal Discount Alanbrooke.* London: Collins, 1957.

Caldes, A. *The People's War.* London: Cape, 1969.
Cannadine, David, *In Churchill's Shadow.* New York: Oxford, 2003.
Chamberlain, Peter, and Hilary Doyle. *Encyclopedia of German Tanks of World War Two.* London: Cassell, 1999.
Chandler, Alfred, Louis Galambos, and Dawn Van Ee, eds. *The Papers of Dwight David Eisenhower*, Vols. 1–17. Baltimore, Md.: Johns Hopkins University Press, 1970–1996.
Churchill, Winston S. *The Churchill War Papers.* See under Gilbert, Martin, ed.
———. *Into Battle*: *Winston Churchill's War Speeches.* London: Cassell, 1943.
———. *Never Give In! The Best of Winston Churchill's Speeches.* New York: Hyperion, 2003.
———. *Painting as a Pastime.* Delray Beach, Fla.: Levenger, 2002.
———. *The Second World War*, Vol. 1, *The Gathering Storm.* London: Cassell, 1948.
———. *The Second World War*, Vol. 2, *Their Finest Hour.* London: Cassell, 1949.
———. *The Second World War*, Vol. 3, *The Grand Alliance.* London: Cassell, 1950.
———. *The Second World War*, Vol. 4, *The Hinge of Fate.* London: Cassell, 1951.
———. *The Second World War*, Vol. 5, *Closing the Ring.* London: Cassell, 1952.
———. *The Second World War*, Vol. 6, *Triumph and Tragedy.* London: Cassell, 1954.
Cohen, Roger. *Soldiers and Slaves.* New York: Random House, 2005.
Colville, John. *The Fringes of Power.* New York: Norton, 1985.
Cowley, Robert, and Thomas Guinzberg, eds. *West Point: Two Centuries of Honor and Tradition.* New York: Warner, 2002.
Crisp, Major Robert. *Brazen Chariots.* New York: Norton, 1959.
Cross, Robert. *The Battle of Kursk.* New York: Penguin, 1993.
De Gaulle, Charles. *Mémoires de Guerre: L'Appel, l'Unité, le Salut.* Paris: Plon, 1994.
———. *Memoirs of Hope: Renewal and Endeavor.* New York: Simon and Schuster, 1972.
D'Este, Carlo. *Eisenhower: A Soldier's Life.* New York: Holt, 2002.
———. *Patton: A Genius for War.* New York: HarperCollins, 1995.
Dickson, Paul, and Thomas B. Allen. *The Bonus Army.* New York: Walker, 2004.
Donovan, Robert J. *Eisenhower, The Inside Story.* New York: Harper, 1956.
Eisenhower, David. *Eisenhower at War 1943–1945.* New York: Random House, 1986.
Eisenhower, Dwight D. *At Ease*: *Stories I Tell to Friends.* New York: Doubleday, 1967.
———. *Crusade in Europe.* New York, Doubleday, 1952.
———. *Letters to Mamie.* New York: Doubleday, 1977.
———. *Mandate for Change.* New York: Doubleday, 1963.
———. *Waging Peace.* New York: Doubleday, 1965.
Eisenhower, John S. D. *The Bitter Woods.* New York: Putnam, 1969.

———. *General Ike*. New York: Free Press, 2003.
———. *Yanks*. New York: Touchstone, 2001.
Eisenhower, Susan. *Mrs. Ike*. New York: Farrar, Straus, and Giroux, 1996.
Ellis, John. *The Sharp End: The Fighting Man in World War Two*. New York: Scribner, 1980.
Esposito, Colonel Vincent J., chief ed. *The West Point Atlas of American Wars*, Vol. 2, *1900–1953*. New York: Praeger, 1959.
Ferguson, Niall. *The Pity of War,* New York: Basic Books, 1999.
Ferrell, Robert H., and Francis H. Heller. "Plain Faking?" *American Heritage*, May–June 1995.
Frank, Anne. *The Diary of a Young Girl*. New York: Bantam, 1993.
Gallo, Max. *L'Album De Gaulle*. Paris: Laffont, 1998.
———. *De Gaulle: L'Appel du Destin*. Paris: Laffont, 1998.
———. *De Gaulle: La Solitude du Combattant*. Paris: Laffont, 1998.
Gilbert, Martin, *Churchill and America*. New York: Free Press, 2005.
———. ed. *The Churchill War Papers*, Vol. 1, *September 1939–May 1940*. New York: Norton, 1993.
———. ed. *The Churchill War Papers*, Vol. 2, *May 1940–December 1940*. New York: Norton, 1995.
———. ed. *The Churchill War Papers*, Vol. 3, *1941*. New York: Norton, 2001.
———. *D-Day*. Hoboken, New Jersey: John Wiley & Sons, 2004.
———. *The First World War*. New York: Holt, 1994.
———. *Winston Churchill*, Vol. 5. London: Heinemann, 1967. See also Vol. 7, *Road to Victory*. Boston, Mass.: Houghton Mifflin, 1986.
Goodwin, Doris Kearns. *No Ordinary Time*. New York: Simon and Schuster, 1994.
Griffith, Robert W., ed. *Ike's Letters to a Friend, 1941–1958*. Lawrence: University Press of Kansas, 1984.
Hamilton, Nigel. *Monty: The Battles of Field-Marshal Bernard Montgomery*. New York: Random House, 1994.
———. *Monty: The Final Years of the Field-Marshal 1944–1976*. London: Hamish Hamilton, 1986.
———. *Monty: The Making of a General*. New York: McGraw-Hill, 1981.
———. *Monty: Master of the Battlefield 1942–1945*. London: Hamish Hamilton, 1983.
Hastings, Max. *Armageddon: The Battle for Germany 1944–1945*. New York: Knopf, 2004.
———. *Overlord: D-Day and the Battle for Normandy*. New York: Simon and Schuster, 1984.
Hesketh, Roger. *Fortitude: The D-Day Deception Campaign*. Woodstock: Overlook, 2000.
Hobbs, Joseph P. *Dear General: Eisenhower's Wartime Letters to Marshall*. Baltimore, Md.: Johns Hopkins University Press, 1999.
Horne, Alistair. *Monty: The Lonely Leader, 1944–1945*. New York: HarperCollins, 1994.

Huie, William Bradford. *The Execution of Private Slovik.* Yardley, Pa.: Westholme, 2004.

Ismay, General Lord. *Memoirs*. New York: Viking, 1960.

Jenkins, Roy. *Churchill.* New York: Farrar, Straus, and Giroux, 2001.

Keegan, John. *Winston Churchill.* New York: Viking, 2002.

Kennedy, Major-General Sir John. *The Business of War.* London, Hutchinson, 1957.

Kimball, Warren F., ed. *Churchill and Roosevelt: The Complete Correspondence.* Boston, Mass.: Houghton Mifflin, 1984.

———. *Forged in War.* New York: Morrow, 1997.

Larrabee, Eric. *Commander in Chief.* Annapolis, Md.: Naval Institute Press, 1987.

Lukacs, John. *Churchill: Visionary, Statesman, Historian*. New Haven, Conn.: Yale University Press, 2002.

———. *Five Days in London—May 1940*. New Haven: Yale University Press, 1999.

Macdonald, Dwight. *Parodies*. New York: Random House, 1965.

Macdonald, Lyn. *To the Last Man*. New York: Caroll and Graf, 1999.

———. *1914–1918*. London: Michael Joseph, 1988.

Macmillan, Harold. *War Diaries: Politics and War in the Mediterranean*. New York: St. Martin's Press, 1984.

Manchester, William. *American Caesar*, New York: Dell, 1979.

May, Ernest R. *Strange Victory: Hitler's Conquest of France*. New York: Hill and Wang, 2000.

McCullough, David. *Truman*. New York: Simon and Schuster, 1992.

Messenger, Charles. *The D-Day Atlas*. New York: Thames and Hudson, 2004.

Miller, Donald E. *The Story of World War II*. New York: Simon and Schuster, 2001.

Miller, Merle. *Ike the Soldier*. New York: Putnam, 1987.

Montgomery, Field Marshal the Viscount. *Memoirs*. Cleveland, Ohio: World, 1958.

Moran, Lord. *Churchill: The Struggle for Survival 1940–1965*. London: Constable, 1966.

Morelock, J. D. *Generals of the Ardennes: American Leadership in the Battle of the Bulge*. Honolulu, Hawaii: University Press of the Pacific, 2002.

Morgan, Lieutenant-General Sir Frederick. *Overture to Overlord*. New York: Doubleday, 1950.

Morris, Donald R. *The Washing of the Spears*. New York: Simon and Schuster, 1965.

Mosier, John. *The Blitzkrieg Myth*. New York: HarperCollins, 2003.

Nelson, James. *General Eisenhower on the Military Churchill: A Conversation with Alistair Cooke*. New York: Norton, 1970.

Norwich, John Julius, ed. *The Duff Cooper Diaries*. London: Weidenfeld and Nicholson, 2005.

Papers of Dwight D. Eisenhower, The. Baltimore, Md.: Johns Hopkins University Press, 1970.

Persico, Joseph E. *Roosevelt's Secret War*. New York: Random House, 2001.

Pogue, Forrest C. *George C. Marshall: Organizer of Victory 1943–1945*. New York: Viking, 1973.

Prior, Robin, and Trevor Wilson *The Somme*. New Haven, Conn.: Yale University Press, 2005.

Ralf, Georg Reuth. *Rommel: The End of a Legend*. London: Haus, 2005.

Reynolds, David. *In Command of History: Churchill Fighting and Writing the Second World War*. London: Penguin, 2004.

Richards, Denis. *The Hardest Victory: RAF Bomber Command in the Second World War*. New York: Norton, 1995.

Roberts, Andrew. *Eminent Churchillians*. New York: Simon and Schuster, 1994.

Russell, Douglas S. *Winston Churchill, Soldier*. London: Brassey's, 2005.

Ryan, Cornelius. *A Bridge Too Far*. New York: Simon and Schuster, 1974.

———. *The Last Battle*. New York: Simon and Schuster, 1966.

———. *The Longest Day*. New York: Simon and Schuster, 1959.

Saward, Dudley. *Bomber Harris*. New York: Doubleday, 1985.

Summersby, Kay. *Past Forgetting: My Love Affair with Dwight D. Eisenhower*. New York: Simon and Schuster, 1975.

Taylor, Frederick. *Dresden*. New York: HarperCollins, 2004.

Tedder, Marshal of the Royal Air Force Lord. *With Prejudice*. London: Cassell, 1966.

Thompson, Julian. *The Imperial War Museum Book of Victory in Europe: The North-West European Campaign 1944–1945*. London: Motorbooks International, 1994.

Tompkins, Peter. *A Spy in Rome*. New York: Simon and Schuster, 1962.

Trevor-Roper, Hugh. *The Last Days of Hitler*. Chicago, Ill.: University of Chicago Press, 1992.

———, ed. *Hitler's Table Talk, 1941–1944*. London: Enigma, 2000.

U.S. War Department. *Handbook on German Military Forces*. Baton Rouge: Louisiana State University Press, 1990.

Von Salomon, Ernst. *Der Fragebogen*. Reinbeck bei Hamburg: Rowohlt, 1972.

Watt, Donald Cameron. *How War Came*. New York: Pantheon, 1989.

Wheeler-Bennett, Sir John. *The Nemesis of Power: The German Army in Politics, 1918–1945*. London: Macmillan, 1964.

Winter, Dennis. *Haig's Command*. New York: Viking, 1991.

Wright, Patrick. *Tank*. New York: Viking, 2000.

Ziegler, Philip. *Soldiers*. New York: Knopf, 2002.

Index

Grateful acknowledgment is given for permission to reproduce images on the following pages:

Robert Capa © 2001 By Cornell Capa / Magnum Photos: iv–v, 475

© Bettmann/Corbis: vii, 185, 229, 268, 519, 582 (top and bottom), 721

Courtesy of the Dwight D. Eisenhower Library: 1, 56, 72, 84, 105, 112, 143, 157, 213, 231, 376, 382, 425, 427, 579, 584, 629, 635, 637, 664, 695, 722

Imperial War Museum: 310 (Negative number NA6110)

Karl Hoeffkes / Polarfilm: 503

Mirrorpix: 251, 288, 393, 707

National Archives: 3 (photo no. 111-SC-190642), 24 (photo no. 1040), 429 (photo no. 111-SC-319091), 560 (photo no. 111-SC-179180)

Public Domain / Wikipedia: 501

United States Army Signal Corps photograph, courtesy of the USMA Library: 552

Wolfgang Willrich, 1940. Courtesy of Dr. Ralf Georg Reuth: 352